Advanced Materials for Agriculture, Food, and Environmental Safety

Scrivener Publishing
100 Cummings Center, Suite 541J
Beverly, MA 01915-6106

Advanced Materials Series

The Advanced Materials Series provides recent advancements of the fascinating field of advanced materials science and technology, particularly in the area of structure, synthesis and processing, characterization, advanced-state properties, and applications. The volumes will cover theoretical and experimental approaches of molecular device materials, biomimetic materials, hybrid-type composite materials, functionalized polymers, superamolecular systems, information- and energy-transfer materials, biobased and biodegradable or environmental friendly materials. Each volume will be devoted to one broad subject and the multi-disciplinary aspects will be drawn out in full.

Series Editor: Dr. Ashutosh Tiwari
Biosensors and Bioelectronics Centre
Linköping University
SE-581 83 Linköping
Sweden
E-mail: ashutosh.tiwari@liu.se

Managing Editors:
Swapneel Despande and Sudheesh K. Shukla

Publishers at Scrivener
Martin Scrivener(martin@scrivenerpublishing.com)
Phillip Carmical (pcarmical@scrivenerpublishing.com)

Advanced Materials for Agriculture, Food, and Environmental Safety

Edited by

Ashutosh Tiwari and Mikael Syväjärvi

Scrivener Publishing

WILEY

Co-published by John Wiley & Sons, Inc. Hoboken, New Jersey, and Scrivener Publishing LLC, Salem, Massachusetts.
Published simultaneously in Canada.

Cover design by Russell Richardson

Library of Congress Cataloging-in-Publication Data:

ISBN 978-1-118-77343-7

Printed in the United States of America

10 9 8 7 6 5 4 3 2 1

Contents

Preface

The levels of toxic and microbial contamination in food and the environment are influenced by harvesting and slaughtering agro technologies and by the processes applied during the manufacture of food. With current cultivation methods, it is impossible to guarantee the absence of pesticides and pathogenic microorganisms in raw foods of both plant and animal origin. The increasing incidence of widespread foodborne diseases and the resultant socioeconomic impact on the world population have brought food and environmental safety to the forefront of ecological public health concerns. The emerging field of advanced materials based on functional architectures offers potential solutions to some key performance challenges, along with improved sensitivity, longevity, stability, miniaturization and ruggedness, while reducing complexity and production cost.

The overall purpose of this book is to generate new solutions to the technical challenges for easy and rapid detection of food toxicants, microorganisms and environmental pollutants. Since the consumption of food and water is an essential part of the live detection of contaminating organisms, the book is especially focused on monitoring the presence of various toxic molecules within water, food and beverages. Moreover, the development of fundamental methodologies and inventive nanotechnologies is a scientific and technological area affecting many aspects of energy and the environment. These methodologies range from clay materials to aluminium alloys commonly used in various applications—for example, in the aeronautics industry. Nanotechnologies have expanded from semiconductors, photonics and healthcare processes to include environmental technology to reduce pollution. This raises the possibility of using nanotechnology for environmental applications through nanomaterials, processes and tools. In all cases, the fundamental aspects of materials and methods are prerequisites for further use.

In Part 1 of the book, Fundamental Methodologies, the first chapter, "Layered Double Hydroxides and the Environment: An Overview," examines layered double hydroxides that can be used for environmental decontamination. Clay materials open up production possibilities since they can be easily synthesized by various cheap and ecofriendly methods. Common decontamination processes are by anion exchange, adsorption and catalytic remediation. Layered double hydroxides not only act to prevent the dispersion of pollutants in effluents or wastewater—for example, by precipitating agents of heavy metal cations—but recently by also targeting molecules and inorganic substances. In the next chapter, "Improvement of the Corrosion Resistance of Aluminium Alloys Applying Different Types of Silanes," approaches for protecting aluminium alloys from attack by using different surface treatments are described. The advantages of aluminium alloys are their low density and good mechanical features compared with certain other alloys which experience a high corrosion rate. Unfortunately, corrosion treatments that exhibit good corrosion performance are still commonly used even though they are very toxic, carcinogenic and allergenic. The movement toward a more environmentally-friendly process includes "green treatments," as exemplified by a reduced use of chromium solutions in favor of silanes. Silane coatings exhibit advantageous barrier properties due to the dense Si-O-Si network which substantially reduces the penetration of aggressive species to the metallic alloy. The barrier properties then need to be combined with an understanding of defects in the barrier that causes localized corrosion processes. The quality of the protection layer from the treatment is dependent on parameters like treatment time, temperature, and chemistry of silane molecules. Further on, such treated surfaces would ideally also be matched with organic paint—for example, by adhesion properties.

With increasing industrial and agricultural activities, arsenic needs to be considered since it can be mobilized into surface water. In particular, water treatment is an increasingly urgent matter. "New Generation Material for the Removal of Arsenic from Water," considers the recovery and sorbent regeneration when a sorbent is exhausted and addresses removal techniques such as solvent extraction and chemical precipitation as synthetic coagulants. The process to restore the sorbent close to its initial state from the metal recovery is a critical step. Subsequently, the arsenic needs to be properly handled and disposed of after recovery. Many metals may be recovered and reused, but this is not as straight forward for arsenic since it has limited markets. The chapter, "Prediction and Optimization of Heavy Clay Products Quality," illustrates various process parameters used to design and control brick production such as chemical composition, firing temperature, weight loss, and water absorption. A new technique that enhances the physical and mechanical properties of natural fibers and their composites is described in the chapter, "Enhancement of Physical and Mechanical Properties of Sugar Palm Fiber via Vacuum Resin

Impregnation." In the chapter, "Environmentally-Friendly Acrylates-Based Polymer Lattices," the development of suitable coating by synthetic polymers is discussed. Such polymers are widely used in packaging and construction. Acrylic resins have great durability and weather resistance. The barrier properties are obtained through copolymerization with more hydrophobic monomers. The type of monomer, its sequence length distribution and polymer weight affect the transition temperature and viscoelastic modulus of the polymer which leads to a film with high clarity, good stability and high mechanical strength.

Part 2 of the book, Inventive Nanotechnology, begins with the chapter, "Nanoparticles for Trace Analysis of Toxins: Present and Future Scenario," which defines the advances and applications of nanotechnology for removal of water pollutants. These novel nanotechnological approaches make it possible to explore various nanometallic particles for extraction of pollutants and also for nanoremediation with reduced clean-up time without the need for eliminating treatment and disposal of contaminated soil. Highlighted in the next chapter, "Recent Developments in Gold Nanomaterial Catalysts for Oxidation Reaction through Green and Sustainable Routes," is the design of reaction specific catalysts in which nanomaterials are used as a key technology for green chemistry. In this oxidation reaction large quantities of agents can be provided—for example, gold nanoparticles as catalyst for oxidation reactions in the gas phase—which is the result of a combination of gold particle size, nature of support material and type of reaction.

"Nanosized Metal Oxide-Based Adsorbents for Heavy Metal Removal: A Review," presents an overview of nanosized metal oxides whose high surface area and specific affinity for heavy metal adsorption from the aqueous phase make them attractive for water purification. The methods for fabrication, physicochemical and adsorption properties, as well as application for heavy metal removal from the aqueous phase are described. Other issues requiring further attention are large-size particle aggregation and capacity loss. Highlighted in the chapter, "Future Prospects of Phytosynthesized Transition Metal Nanoparticles as Novel Functional Agents for Textiles," are the biological systems used for green synthesis of nanoparticles. The use of plants for making metal nanoparticles is an environmental approach due to their biocompatibility, low toxicity and environmental nature. Plants have shown the most promise as they seem suitable for large-scale biosynthesis compared to microprobes and enzymes. Also, new functionalities upon integration into textile materials may appear. Some of the metals produced are gold, silver and palladium, and other nanoparticles like copper. Zinc and cadmium oxide have gained interest as catalysis, sensors and photonic devices. Since details about plants that produce nanoparticles are not known, this field is of great scientific interest. In the chapter, "Functionalized Magnetic Nanoparticles for Heavy Metal Removal from Aqueous Solutions: Kinetics and Equilibrium Modeling," magnetic nanoparticles are placed in the forefront as platforms for detection and separation applications. These particles can be composed of both inorganic and organic components, which position such systems to potentially tune properties of hybrid materials for appropriate functions due to their small size, biocompatibility and superparamagnetic properties.

Pathogens are responsible for diseases which cause the disability and death of millions. The chapter, "Potential Application of Nanoparticles as Antipathogens," discusses the interface of microorganisms (pathogens) and nanostructures which can find wide application for use as antipathogens—substances which are used to kill, deactivate and control the pathogens. Nanoparticles such as silver, gold, titanium dioxide and zinc oxide are receiving considerable attention as antimicrobials. In the chapter, "Gas Barrier Properties of Biopolymer-Based Nanocomposites: Application in Food Packing," biopolymers and protein-based nanobiocomposites are described in respect to their gas barrier properties, which are promising for lowering oxygen permeability. In food packaging most materials are more or less nondegradable. New biomaterials could be more environmentally friendly and even be both edible and biodegradable. Still, there are issues when performance, processing and cost are combined. In performance, they are water sensitive and have limited mechanical properties with high brittleness. Nanocomposites as packaging materials are still in their infancy. Zero-valent iron, one of the most widely studied nanoparticles, is presented in the chapter, "Application of Zero-Valent Iron Nanoparticles for Environmental Clean Up." The attraction of these nanoparticles lies in their combination of nanosize and excellent reducing capability, which results in a powerful remediation tool for reducing toxic and hazardous wastes. The zero-valent iron can be injected into groundwater and aquifers to treat contaminated systems. At the same time it interacts with the contaminant and biotic component of the system. In the chapter, "Typical Synthesis and Environmental Application of Novel TiO2 Nanoparticles," titanium dioxide nanoparticles are introduced and reviewed as a player in environmental protection and in the implementation of techniques to remove inorganic or organic pollutants from wastewater. Finally, ZnO nanostructures offer the most promising platform for fabrication of various optoelectronic devices. In the last chapter, "Zinc Oxide Nanowire Films: Solution Growth, Defect States and Electrical Conductivity," the solution growth processes to fabricate ZnO nanowire films are reviewed. Then, theoretically proposed and experimentally observed defect states and their origin are discussed.

This book is written for readers from diverse backgrounds across various fields such as chemistry, physics, materials science and engineering, medical science, pharmacy, biotechnology, and biomedical engineering. It can be used not only as a textbook for both undergraduate and graduate students, but also as a review and reference book for researchers in the field. We hope that the chapters of this book will provide readers with valuable insight into the major area of nanosafety, green materials and their respective technologies.

Editors
Ashutosh Tiwari, PhD, DSc
Mikael Syväjärvi, PhD

Part 1

FUNDAMENTAL METHODOLOGIES

1

Layered Double Hydroxides and the Environment: An Overview

Amita Jaiswal*, Ravindra Kumar Gautam and Mahesh Chandra Chattopadhyaya*

Environmental Chemistry Research Laboratory, Department of Chemistry, University of Allahabad, Allahabad, India

Abstract

Due to their versatility, hundreds of millions of tons of clay minerals currently find applications not only in ceramics, building materials, paper coating and fillings, drilling muds, foundry molds, pharmaceuticals, etc., but also as adsorbents, catalysts or catalyst supports, ion exchangers, etc., depending on their specific properties. There are two broad classes of clays: Cationic clays (or clay minerals), widespread in nature, and Anionic clays (or layered double hydroxides), more rare in nature but relatively simple and inexpensive to synthesize on a laboratory or industrial scale. Cationic clays have negatively charged alumina-silicate layers with small cations in the interlayer space to balance the charge, while anionic clays have positively charged brucite type metal hydroxide layers with balancing anions and water molecules located interstitially. The layered double hydroxides (LDHs) belonging to the general class of anionic clay minerals can be of both synthetic and natural origin. Also known as hydrotalcite-like compounds (HTLCs), these materials are interesting because their layer cations can be changed among a wide selection, and the interlayer anion can also be freely chosen. Like cationic clays, they can be pillared and can exchange interlayer species, thus increasing applications and making new routes to synthesize the derivatives.

This chapter deals with the brief history of layered double hydroxides, their structure, properties, synthesis by different methods and characterization, along with their applications mainly in the environmental field.

Keywords: Layered double hydroxides, anionic clays, cationic clays, brucite, interlayer species, heavy metals, dyes, greenhouse gases, surfactants

---

**Corresponding author*: amita_ecsl@rediffmail.com, mcc46@rediffmail.com

Ashutosh Tiwari and Mikael Syväjärvi (eds.) Advanced Materials for Agriculture, Food, and Environmental Safety, (1–26) 2014 © Scrivener Publishing LLC

1.1 Introduction

Layered double hydroxides (LDHs) have been known for a very long time. Around 1842, naturally forming LDHs minerals were discovered in Sweden. Crushing these minerals leads to a white powder similar to talc. These materials were first synthesized by a German scientist, W. Feithnecht (1942), through reaction between dilute solutions of metals with bases, which he named "doppelschichtstrukturen" or double-sheet structure. The LDHs are also known as hydrotalcite-like compounds (HTLCs). Hydrotalcite (HT) is a hydroxycarbonate of magnesium and aluminium which occurs in nature in foliated and contorted plates or fibrous masses.

During the discovery of hydrotalcite another hydroxycarbonate of magnesium and iron was found, which was called pyroaurite. Pyroaurite was later recognized to be isostructural with hydrotalcite and other minerals containing different elements, all of which were recognized as having similar features. Hydrotalcites have been studied for their use as catalysts and precursors to various other catalysts as early as 1970 [1, 2].

Allman and Taylor studied single crystal X-ray diffraction on mineral samples which revealed the main structural entities of LDHs and disproved Feitknecht's theory. These studies showed that the two cations were in fact located in a single layer and the interlayers were composed of water and carbonate ions. Although the main entities of LDHs have been elucidated, Evans and Slade [3] have suggested that several intrinsic details still remain to be fully understood. These include the possible stoichiometric range and composition, and the position and arrangement of metals within each cationic layer. Prior to the study by Evans and Slade, Miyata and Okada [4–6] described many structural features of LDHs/HTLCs which have different guest anions.

Layered double hydroxide materials appear in nature and can be readily prepared in the laboratory. In nature they are formed from the weathering of basalts [7, 8] or precipitation [9] in saline water. All natural LDH minerals have a structure similar to hydrotalcite, which has the formula $[Mg_6Al_2(OH)_{16}]\ CO_{3.}\ 4H_2O$. Unlike clays, however, layered double hydroxides are not discovered in large, commercially exploitable deposits [9]. The LDHs have been prepared using many combinations of divalent and trivalent cations including magnesium, aluminium, zinc, nickel, chromium, iron, copper, indium, gallium and calcium [10–31].

1.2 Structure of Layered Double Hydroxides

Layered double hydroxides (LDHs) are also known as hydrotalcite-like compounds (due to their structural similarities to that mineral) or anionic clays and host-guest layered materials [1, 3, 32–35], which are quite rare in nature. Most LDHs are synthetic phases and their structure resembles the naturally occurring mineral hydrotalcite $[Mg_6Al_2(OH)_{16}]\ CO_3.\ 4H_2O$, having the general formula of $[M(II)_{1-x}\ M\ (III)_x\ (OH)_2]\ (Y^{n-})_{x/n}.\ YH_2O$, where, M(II), M(III) = divalent and trivalent metals respectively, $0.2 < x < 0.33$, and Y^{n-} = the exchangeable anions between the layers [10, 36, 37].

The basic layer structure of LDHs is based on brucite $[Mg\ (OH)_2]$, typically associated with small polarizing cations and polarizable anions. It consists of magnesium ions surrounded approximately octahedrally by hydroxide ions. These octahedral units form infinite layers by edge-sharing with the hydroxide ions sitting perpendicular to a plane of the layers. The layers then stack on top of one another to form a three-dimensional structure.

When Mg^{2+} is replaced by a trivalent cation similar in radius, an overall positive charge results in the hydroxyl sheets and counter balance is provided by carbonate ions which are positioned within the hydroxyl interlayer. In addition to carbonate ions, water molecules are found in the interlayer gallery. The nature of the interlayer anion and the extent of hydration often determine the layer spacing between each brucite-like sheet [38]. The brucite-like sheets may occur in two different symmetries, namely rhombohedral and hexagonal. In nature, the rhombohedral symmetry is widespread. However, in mineral samples, the hexagonal symmetry is seen to favor the interior of the crystallite samples, while the rhombohedral symmetry is found on the exterior. This is a result of cooling during crystallite transformation, in which the extrerior surface of the crystallite cools much quicker than the interior and hence the interior hexagonal form cannot transform due to a higher energy transformation barrier at lower temperature. From these observations, it has been deduced that the hexagonal symmetry is favored by high temperature [1, 4]. Naturally occurring minerals that exhibit a LDH structure include manasseite, pyroaurite, sjogrenite, barbetonite, takovite, reevesite, desautelsite and stichtite. They differ from one another in the stacking arrangement of the octahedral layers [1, 39].

Conventionally synthesized LDHs are strongly hydrophilic materials, either amorphous or microcrystalline with hexagonal habit, with the dominant faces developed parallel to the metal hydroxide layers. Adjacent layers are tightly bound to each other. Figure 1.1 shows the structure of layered double hydroxides.

One of the advantages of LDHs among layered materials is the great number of possible compositions and metal–anion combinations that can be synthesized. Layered double hydroxides (LDHs) have high charge density. The charge density is dependent on the metal ratio. Since it comprises a divalent and trivalent metal cation, their ratio affects charge density of the layers. A lower divalent/trivalent ratio results in a higher charge density.

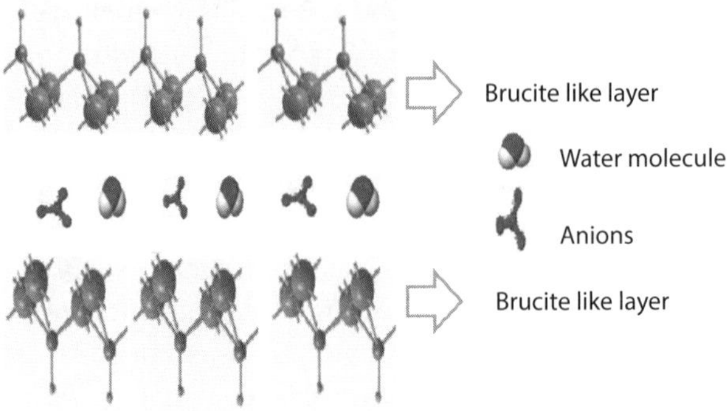

Figure 1.1 Structure of layered double hydroxide (LDHs).

1.3 Properties of Layered Double Hydroxides

Layered double hydroxides (LDHs) display unique physical and chemical properties close to those of clay minerals. Some interesting properties of these materials summarized by Del Hoyo [40] are:

- High specific surface area (100 ± 300 m^2/g)
- Memory effect
- Anion exchange capacities
- Synergistic effects

The LDHs exhibit anion mobility, surface basicity and anion exchangeability due to their positively charged layered structure. The anions and water, which fill the interlayer space, are labile. Therefore a variety of inorganic and organic anions can be intercalated in the interlayer of LDHs through anion exchange reactions [33]. The mixed metal oxides obtained on calcination of LDH usually exhibit properties such as high surface area, surface basicity and formation of homogeneous mixture with small crystallite size when heated to higher temperature [1]. The LDHs as well as the oxides obtained from them exhibit excellent catalytic activity. Structure reconstruction, or so called "memory effect," is another important property of LDHs which is unique to this class of layered solids. Structure reconstruction is usually achieved by first decomposing the LDH at suitable high temperature followed by treating the resultant mixed metal oxides with a solution containing a suitable anion [41]. These materials have a high capacity for adsorbing anions as well as cations [38, 42]. Magnetic properties of the LDHs depend on the space between the layers.

This space can be adjusted by insertion of organic anions with different chain lengths. This suggests that these hybrid materials would work as tunable magnets [43]. The LDHs intercalated with long-chain surfactant molecules such as dodecyl sulphate have the ability to swell in organic solvents. This property of delamination is exploited in the preparation of monolayers, which are used extensively in the synthesis of nanohybrids and nanocomposites [44].

The interlayer anions present in LDHs can be exchanged by other anions. The order of preference for some common inorganic anions is as follows:

$$NO_3^- < Br^- < Cl^- < F^- < OH^- < SO_4^{2-} < CO_3^{2-}$$

NO_3^- is an anion which can be easily replaced by a more strongly held one like CO_3^{2-}. Therefore when preparing precursor for interaction, nitrate salts are preferred and CO_3^{2-} is tried to be kept away. The interaction involves a guest molecule that is introduced into the host without affecting the structure of the host. Upon interaction of the guest molecule, the existing ion is replaced. The interlayer anions weekly bonded to hydroxide layers (e.g., NO_3^-, Cl^-) can be exchanged more easily [45–47].

1.4 Synthesis of Layered Double Hydroxides

Layered double hydroxides (LDHs) can be synthesized with a wide range of compositions and a variety of M(II)/M(III) cation combinations containing different anions in the interlayer spaces. A number of synthetic techniques have been successfully employed in the preparation of LDHs. There are number of methods used to synthesize LDHs which are presented below. [1].

1.4.1 Co-precipitation Method

Co-precipitation remains the most commonly used technique for the synthesis of LDHs. This technique involves the addition of aqueous solutions of M(II), M(III) and the anion that will be incorporated into the LDH structure. This method allows the direct preparation of LDHs for a variety of anions and cations. Therefore, this method is commonly used for the formulation of organic-anion LDHs, which are difficult to obtain by other methods [3]. Following are the three methods of co-precipitation which can be used for the synthesis of LDHs.

1.4.1.1 Co-precipitation by Titration

In this method, a metal salt solution is titrated against a basic solution during which not only solids of LDHs structure occur, but also other solid phases are formed, due to the precipitation of less soluble metal hydroxides prior to the precipitation of the LDHs.

1.4.1.2 Co-precipitation at Low Supersaturation

Co-precipitation at constant pH and at temperature between 60–80°C by the slow addition of mixed solution of M(II) and M(III) salts in the chosen ratio and the base solution. The rate of addition has to be controlled in order to obtain a more homogeneous phase. These conditions are best to obtain higher crystallinity of the material due to the fact that at constant pH the rate of crystal growth is generally lower than the rate of nucleation [3].

This method is most often used, as it is the most effective one. For example, Marchi *et al.* obtained an amorphous precipitate of Cu/Co/Al hydrotalcites catalysts by co-precipitation at high supersaturation while under the conditions of low supersaturation a crystalline HT was precipitated [48].

1.4.1.3 Co-precipitation at High Supersaturation

In this method, a solution containing a mixture of the metal ion is quickly added to the basic solution and mixed rapidly. This results in the formation of LDHs which are generally not very crystalline, but sometimes amorphous product is obtained due to the high number of nuclei formation. The precipitates obtained by this method are washed with distilled water in order to eliminate adsorbed ions and are dried at temperatures up to 120°C [1, 49].

1.4.2 Hydrothermal Synthesis

Hydrothermal preparation methods have been employed in order to control the particle size and distribution. There are two routes to hydrothermal treatment. In the first case, materials are treated at temperatures more than 373K in a pressured autoclave. Here, the LDHs may be synthesized from precursor such as MgO and Al_2O_3 or from mixtures obtained from the decomposition of the nitrate forms of these precursors [1, 50, 51]. More specifically, the precursors are placed in an autoclave set at temperature above 596K and pressure between 10–130 MPa for a fixed time period.

In the other case, LDHs are synthesized at low temperatures and undergo a process of aging. Aging involves refluxing the precipitate at a set temperature for 18 hours.

1.4.3 Urea Hydrolysis Method

Urea is a weak Bronsted base highly soluble in water which has been used for the precipitation of metal ions as hydroxides. The hydrolysis of urea gives a pH of about 9 depending on the temperature of the mixture, which allows its use as precipitating agent in the synthesis of LDHs. The crystallinity degree of LDHs has been observed to depend on the temperature of synthesis and the aging time. Larger particles are formed at low temperatures [52, 53] due to the lower nucleation rate which depends on the decomposition rate of the urea. However, this method is not indicated for the preparation of LDHs with low charge density, but allows the preparation of compounds with high charge density which are difficult to obtain with other procedures [52].

1.4.4 Sol-Gel Method

The sol-gel method involves the formation of a sol by hydrolysis and partial condensation of metallic precursor and followed by the gel formation. As metallic precursor, metallic alkoxides, acetates, acetylacetones and many inorganic salts are used. The properties of the LDH depends on the hydrolysis and condensation rates of the metallic precursors [54], which are modified by different parameters of the reaction like pH, nature and concentration of the precursor, solvent and temperature of the synthesis. The LDHs obtained by this method exhibit specific surface area larger than those obtained by co-precipitation method [55–57], but with controversial results regarding the basicity and the trivalent and divalent metal ion molar ratios [56–58].

1.4.5 Ion-Exchange Method

In this method, there is an exchange of the interlayer anions with anionic guest molecules, i.e., molecules one wishes to introduce into the LDH structure in order to produce the desired LDH–guest compound. The ion exchange in LDHs depends on several factors according to He *et al.* [59], affinity for the incoming guest anion, the medium in which the exchange occurs, the pH and the chemical composition of the LDH brucite-like layers.

In terms of thermodynamics, ion exchange in LDHs depends mainly on the electrostatic interactions between the positively-charged sheets and the interlamellar anion and, to a lesser extent, on the free energy involved in the changes of hydration [60]. Moreover, the equilibrium constants in the anionic-exchange process depend on the size of the anion. Hence, the anion exchange is favored for anions with high charge density (i.e., high charge and small ionic radius) [61].

The ionic exchange for simple inorganic divalent anions decreases in the order of $CO_3^{2-} > HPO_4^{2-} > SO_4^{2-}$ and for monovalent anions, $OH^- > F^- > Cl^- > Br^- > NO_3^- > I^-$. Synthesis of LDHs using ion exchange methods have been carried out by a number of researchers [5, 39, 62–65].

1.4.6 Rehydration Method

The rehydration method was reported by Miyata, who found that after calcination of the LDHs at 500–800°C, the mixed metal oxide rehydrated and reformed back to a LDH in the presence if anions. This unique "memory effect" of LDHs provides an effective synthesis route to obtain LDHs with desired inorganic and organic anions, and avoids the incorporation of competing inorganic counter anions [66–73].

1.4.7 Miscellaneous Methods

In addition to the above methods, several other methods which have been applied in the preparation of LDHs are presented below.

1.4.7.1 Salt-Oxide Method

In this method M(III) solution is slowly added into a suspension of M(II) metal oxide or hydroxide. The mixture is heated at room temperature for a few days in order to favor the reaction between the precursors. This method has been used to prepare LDHs having different combinations of divalent and trivalent cations and anions such as ZnCr-Cl, ZnCr-$NO_{3,}$ ZnAl-Cl and ZnAl-NO_3.

1.4.7.2 Template Synthesis

In recent years, this method has attracted more attention in the field of Materials Science. Using self-assembled aggregates as a template, inorganic materials can be directed to an ordered structure with specific morphology and size. There have only been a few reports of the synthesis of LDHs using this method [73].

1.4.7.3 Surface Synthesis

When a material is supported on the surface of a support, the resulting composite material is expected that presents improved properties such as mechanical performance, thermal stability and a high degree of dispersion. Some surface synthesis has been reported [74–76] using aqueous ammonia and precipitating agent urea or by impregnation of metallic salts following calcinations and hydrothermal treatment [77–79].

Various methods for the synthesis of LDHs can be used. The choice of the method will depend on the characteristics and application of the required material.

1.5 Characterization of Layered Double Hydroxides

Many physico-chemical techniques have been used to characterize LDHs. Some of the most commonly used techniques are XRD, FTIR and TG/DTA. Some other techniques are more specific for some LDHs such as SEM, DSC, NMR and Raman spectroscopy.

1.5.1 X-ray Diffraction

X-ray diffraction (XRD) is the main analytical technique for characterization of LDHs and crystalline phases. It is also used to determine the interlayer spacing and thickness of a single layer. The technique involves directing X-rays to the crystal. The radiation will be either reflected or diffracted at different angles. This technique is used to study the purity of the LDH intercalated material. The phase purity of the material is determined by the sharpness and/or the broadness of the diffraction peaks. The broader reflections correspond to the amorphous phase, while the sharper reflections correspond to the crystalline phase. From the diffraction data, the d-spacing of the intercalated LDH material can be determined. The reflections with the greatest d-spacing correspond to the d-spacing of the intercalated LDH [80].

1.5.2 Fourier Transform Infrared Spectroscopy

Fourier transform infrared spectroscopy (FTIR) is one of the molecular vibrational spectroscopic techniques used for both qualitative and quantitative analysis [81]. The FTIR analysis is not a diagnostic tool for LDHs, but can be useful to identify the presence of foreign anions in the interlayer between the brucite-like sheets. Besides this, information about the type of bonds formed by the anions and about their orientations can also be obtained. The infrared region is the region found in the wave number range 1.3×10^{4}- 3.3×10^{1} in the electromagnetic spectrum [82].

The FTIR is used to investigate the structural bonding and chemical properties of compounds [83]. On absorbing radiation by the molecule, the bond can stretch/vibrate or bend [84]. Each molecule absorbs a specific IR radiation depending on the type of functional group present. The resultant peaks in the IR spectra are called the "molecular finger print." Therefore, each functional group has its own frequency and this is useful for revealing the presence or absence of these groups from the spectra.

Some specific characteristic peaks of LDHs are: peaks at 3500–4000 cm^{-1} assigned as –OH absorption and Interlayer CO_3^{2-} (LDH-CO_3) [85, 86]; peaks at <1000 cm^{-1} assigned as the inorganic lattice vibrations of M-O and M-OH modes [86]; peaks at 1560–1400 cm^{-1} due to intercalated anionic species (e.g., Carboxylate-LDH) [80] and due to strong asymmetric and symmetric stretching bonds, and; peaks around 1360 cm^{-1} which can be assigned as carbonate impurity and nitrates [86].

1.5.3 Thermogravimetric Analysis and Differential Thermal Analysis

Thermogravimetric analysis (TGA) is used to measure changes in the mass of LDH and estimate the content of water and anion molecule, while differential thermal analysis (DTA) can be used to study thermal properties and phase changes that can occur upon heating the LDH/ LDH hybrids. Most thermal events correspond to water/anions removal from interlayer spaces.

1.5.4 Other Techniques

Other techniques include scanning electron microscopy (SEM), transmission electron microscopy (TEM), differential scanning calorimetry (DSC), X-ray fluorescence spectroscopy (XRF), Raman spectroscopy and nuclear magnetic resonance (NMR).

Scanning electron microscopy is used to study the surface topography of LDH intercalates by which particle morphology, porosity and phase composition within the material can be revealed. Both SEM and TEM provide general pictures of textural and crystal morphologies of LDHs. A hexagonal platelet morphology is usually shown by LDH [36, 87].

Both DSC and TGA are also used to check the chemical and physical properties of a material corresponding to temperature changes. It is used to detect the melting points (enthalpy of melting) and phase transitions of materials. This technique has not been explored very much in the study of LDH intercalates.

X-ray fluorescence spectroscopy is employed for trace element analysis and is also used for the determination of the divalent to trivalent cation ratios.

Raman spectroscopy has been mainly employed to study the structural accommodation of interlayer species and hydrogen bond network in LDH, particularly for oxoanions such as CO_3^{2-}, NO_3^{-}, SO_4^{2-} and CrO_4^{2-}, and organic carboxylate anions [22, 88, 89].

Nuclear magnetic resonance spectroscopy is an effective technique for studying the structural environments and dynamic behavior due to its unique ability to simultaneously probe element-specific local structure with high resolution, and to investigate atomic and molecular motion [90–92].

1.6 Applications of Layered Double Hydroxides

Layered double hydroxides (LDHs) represent one of the most technologically promising materials as a consequence of their low cost, easy preparation and the large number of composition variables that can be used. At present, a great deal of applications of LDHs in the field of heterogeneous catalysis, covering a wide range of areas, have made substantial progress, especially in the past decade. There are many possible applications of LDHs, not only as catalysts but also as scavengers of ions in water purification. Their applications in drug delivery systems are also well known. Some possible applications of LDHs are shown in Figure 1.2.

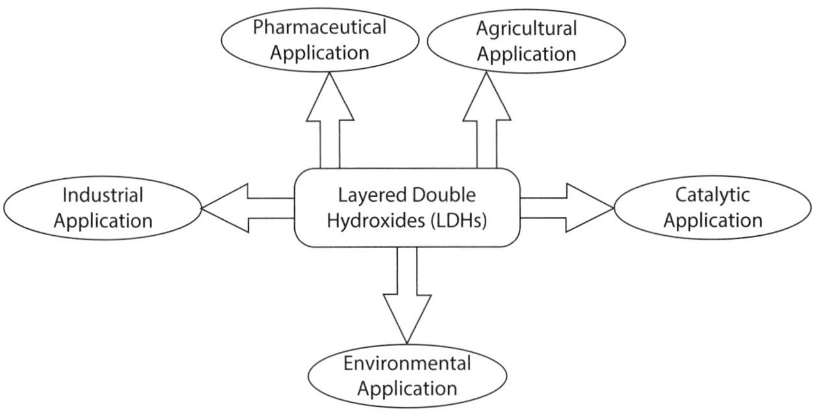

Figure 1.2 Applications of layered double hydroxides in various fields.

1.6.1 Catalytic Applications

Due to their flexible composition, the interest in layered double hydroxide materials has increased over the last decade. They are a relatively easy to synthesize, inexpensive, versatile and potentially recyclable source of a variety of catalyst supports, catalyst precursors or actual catalysts. The use of LDH materials in the field of catalysis is well known and has been reported for basic catalysis (e.g., polymerization of alkene oxide, aldol condensation, transesterification reactions, etc.), reforming of hydrocarbons with H_2O, hydrogenation (production of methane, alcohols, paraffins, olefins, etc.), oxidation reactions, support for Zieglar-Natta catalysts during polymerization of olefins, etc. [1, 93]. However, LDHs themselves have not found applications in the field of heterogeneous catalysis,, but when mixed metal oxides derived from LDHs precursors are used. They are found in numerable applications which are listed in Table 1.1.

Table 1.1 Compositions of catalyst precursors (LDHs) and their catalytic applications [1].

Compositions of catalyst precursors	Applications
$Mg_3Al_2\ CO_3\ (OH)_{16}.4H_2O$	Dehydration, carrier
$Ni_2A1_2\ CO_3\ (OH)16.H_2O$	Hydrogenation, dealkylation
$Ni_3Mg_3A1_2CO_3(OH)_{16}.4H_2O$	Hydrogenation, dealkylation, cracking
$Co_3\ Mg_3Al_2\ CO_3(OH)_{16}.4H_2O$	Hydrogenation
$CO_6Al_2CO_3(OH)_{16}.4H_2O$	Hydrogenation
$Ni_{0.9}\ Co_{0.75}Cu_{0.35}Mg_4Al_2CO_3\ (OH)_{16}.4H_2O$	Hydrogenation, dehydrogenation
$Cu_6Al_2CO_3\ (OH)_{16}.4H_2O$	Isomerization, hydrogenation of nitro groups, dehydrogenation of sec. alcohols to ketone
$Cu_3Zn_3Al_2CO_3(OH)_{16}.4H_2O$	Dehydrogenation of sec. alcohols to ketones, low temperature conversion
$Ni_3Mg_3Al_{1.8}Cr_{0.2}CO_3(OH)_{16.}\ 4H_2O$	Hydrogenation
$Cu_3Mg_3Al_2CO_3(OH)_{16}.4H_2O$	Dehydrogenation of sec. alcohols to ketones, hydrogenation of nitro groups

Since the catalytic behavior depends primarily on the chosen metals as the catalytic centers [1], metal cations present in the brucite-like layers of the LDHs play an important role for the catalytic properties of the final product. Moreover, the anions present between the layers of the LDHs can have an influence on the catalytic properties. The anions present in the LDHs can also have an effect on other properties such as the crystal size of the catalysts.

1.6.2 Agricultural Applications

Layered double hydroxides possess excellent potential as green carrier for plant nutrients, pesticides and growth regulators and as active principle in animal feeds. However, not much attention has gone into exploring the full potentiality of LDHs. They are one of the idealized inorganic materials for a wide range of agricultural fields not only because their framework can be composed of plant nutrients, but also because their structure offers charming features such as accommodation and controlled release of various active anionic agro-substances, high buffering capacity, high water retention ability and acid neutralizing potential.

Komarneni *et al.* [94] suggested nitrate-LDH as a potential slow-release fertilizer by synthesizing nitrate-LDH in ambient condition without any considerable contamination of carbonate-LDH. More attention has been given to pesticide formulation that consists mainly of various organic solvents. Lakraimi *et al.* [95] prepared pesticide-LDH hybrid with 2,4- dichlorophenoxyacetate, a broad leaf herbicide, by ion-exchange reaction with a chloride form of ZnAl-LDH for slow release formulation. The acid neutralizing potential and high anion adsorption capacity of LDHs could be explored for animal feeds as a complement to cationic clays, provided their nontoxicity is ensured.

1.6.3 Pharmaceutical Applications

Pharmaceutical applications of LDHs mainly depend on the acid buffering effect and anion exchange property. LDH-derived antacid and antipeptic are representative of their applications in pharmaceutics [96]. Powdered LDHs have been demonstrated to be one important type of green carrier or host for genes and drugs due to the excellent biocompatibility and nontoxicity or low toxicity. MgAl-LDH is used as an important component of drugs, or as nanocarrier for delivery of drugs and genes into cells [97, 98]. Wei *et al.* [99] demonstrated that LDH was able to be used as an effective nanocarrier by greatly enhancing the thermal- and photo-stabilities for L-Dopa and

L-Tyrosine [100]. Choi *et al.* have made significant investigations on the toxicity of LDH nanoparticles *in vitro* and *in vivo* in applications [97]. Xu *et al.* [101] recently reported an efficient LDH-based delivery for siRNA.

1.6.4 Industrial Applications

Layered double hydroxides have a large anion exchange capacity, making them useful for entrapping pesticides, surfactants, nitrates, chromates, etc., for the treatment of industrial and environmental wastewater.

As has already been pointed out, LDHs are used as catalysts. Their applications are increasing in the industrial production of organic materials and in conversion of natural gas. They also find applications in photochemical reactions. Many authors [102–108] also examined other applications of LDH materials. Some of these applications include the use of hydrotalcite-like compounds as additives in photochemistry, electrochemistry and in functional polymer materials. They have other applications in the pharmaceutical and biochemical industries [102].

1.6.5 Environmental Applications

In the field of environmental technology, much attention has been paid to the potential applications of synthetic materials as sorbents for organic and inorganic pollutants in water. Layered double hydroxides (LDHs) are used for the decontamination of the environment and prevent the dispersion of pollutants in nature. A wide range of contaminants can be removed from industrial effluents or wastewater by anion exchange, adsorption process and catalytic remediation, using LDHs. Recently, many other molecules such as pesticides, toxic organic chemicals, greenhouse gases, heavy metals and some other undesirable inorganic substances have increasingly been targeted for remediation by using LDHs for the protection of earth, health and the environment.

There has been a thorough reinvestigation of LDHs for anion exchange properties due to recent interest in developing the use of anionic clays for environmental remediation. The main characteristics have been studied to clearly characterize the adsorption properties of the materials under vigorous solid/liquid interface conditions. The effect of sorbent composition, surface and bulk adsorption, and concentration of adsorption site have been assessed. The adsorption capacity is deeply affected by the nature of the counteranion of the LDH layer.

In 1983, Miyata [109] first reported the ion exchange isotherm of a series of LDHs leading to the ion selectivity of layered double hydroxides for monovalent and divalent anions, $OH^- > F^- > Cl^- > Br^- > NO_3^- > I^-$ and

$CO_3^{2-} > SO_4^{2-}$. The reconstruction process of the LDHs has been investigated for the removal of environmentally undesirable anion. Parker *et al.* [110] have pointed out that the high selectivity for carbonate anion prevents LDHs to be used as an anion exchange material unless further treatment is made. They measured anion exchange capacities of hydrotalcite in single and mixed anion solutions by observing the amount of anion adsorbed by freshly prepared hydrotalcite. They have compared the relative preference of anions after 24 h as $SO_4^{2-} > F^- > HPO_4^{2-} > Cl^- > B(OH)_4^- > NO_3^-$. This application of layered double hydroxides can be explained by the points shown below.

1.6.5.1 *Removal of Heavy Metal Cations*

Layered double hydroxides can be used as precipitating agents of heavy metal cations for the decontamination of wastewater. The Mn^{2+}, Fe^{2+} and Cu^{2+} cations were removed by synthetic hydrotalcite-like compounds; zaccagnaite [31] and hydrotalcite thin films were used for the remediation of aqueous wastes containing hazardous metal ions [111].

1.6.5.2 *Removal of Nuclear Wastes*

Hydrocalumite or ettringite, the calcium form of hydrotalcite, have been evidenced as earlier alteration products of cements or basaltic glasses [112, 113]. They have been studied for their adsorption properties of actinides (Th, U, Np, Pu, Se and Am) [114–116]. The high ability of LDHs and their calcined products to adsorb/exchange anions from solution was also used for the removal of Tc, Re and Mo in their anionic form from radioactive wastewater [117, 118]. Synthetic hydrotalcite has been investigated for the sorption of iodine and iodine-containing anionic species [119–124].

1.6.5.3 *Removal of Greenhouse Gases*

Layered double hydroxides are efficient scavengers for acid gases recovery from hot gas streams due to their unique and strong basic properties. The recovery of CO_2 and SO_x from power plant flue gases is considered to be the first step in reducing total carbon and sulfur oxide emissions. Many papers and patents describe the use of calcined LDHs for the adsorption of CO_2 [125–128] and sulfur oxide [129–131] gases.

1.6.5.4 *Removal or Adsorption of Organic Compounds and Pesticides*

Recently, LDHs and their oxides have been investigated as scavengers to treat wastewater containing anionic organic contaminants. The materials are known to function as effective anion adsorbents in the uptake of phenols, terephthalates, anionic surfactants, ionizable pesticides and herbicides, humic and fulvic acids, anionic dyes, and colored organics from aqueous solutions [132–148].

Based on the ability of LDH to intercalate various types of anions in interlayers, organic–inorganic composite materials could be produced by intercalation of organic anions having functional groups in their structures into the interlayer of LDH. The LDHs modified with organic anions are expected to take up nonionic organic compounds from aqueous solution depending on the functional groups in the structure of the intercalated organic anions. For example, Mg–Al LDH intercalated with dodecylsulfate (DS) could adsorb hydrophobic pesticides such as atrazine, linuron and acephate, etc., from aqueous solution. This is due to the modification of the interlayer surface of the LDH from hydrophilic to hydrophobic.

You *et al.* [149] examined the surfactant-enhanced adsorption of organic compounds by Mg-Al LDHs. Organo-Mg-Al LDHs were prepared by incorporating anionic surfactants, octylsulfate, DS, 4-octylbenzenesulfonate, and dodecylbenzene sulfonate, into Mg-Al LDH via ion exchange. The anionic surfactants were intercalated into Mg-Al LDH with the surfactants oriented perpendicular in the interlayer. The octylsulfate formed bimolecular films, and other surfactants resulted in monolayer structure. Intercalation of surfactants into Mg-Al LDH decreased the surface area, whereas surfactants dramatically enhanced the LDH affinity for 1,2,4-trichlorobenzene and 1,1,1-trichloroethane in aqueous solutions. Adsorption potential depended on the type of surfactant as well as the configuration of surfactant molecules within Mg-Al LDH interlayers. The adsorption characteristics indicated that the retention of organic compounds by organo-Mg-Al LDHs was due to a partitioning mechanism.

1.6.5.5 Removal of Dye

The discharged effluents of industries contain a large number of dyes, increasing the total COD of wastewaters [150]. Although most of the dyes are nontoxic, many of them are highly toxic, like metals (Cr), with harmful consequences to aquatic life. Moreover, the persistence of color appearance (at concentration > 1mg/L) in treated wastewater prevents their reuse.

The LDHs have high adsorption capacities for dye molecule and can be very competitive with other sorbents. They have been demonstrated in environmental applications due to their high ability to removed color or dye by adsorption reaction [137–145].

1.6.5.6 *Removal of Surfactants*

Layered double hydroxides have strong surface and interlayer hydrophilic properties due to their high content of hydroxyl groups and water molecules. These properties limit the adsorption and the intercalation of non-ionic or hydrophobic contaminants. However, anionic surfactants such as alkyl carboxylate, alkyl sulphate or alkyl sulphonate are easily adsorbed and exchanged in LDH structure [151–153]. Dekany *et al.* [154, 155] compared the adsorption properties of both cationic surfactant/smectites and anionic surfactant/LDH antitype systems. The higher charge density of the anionic clays explains their higher adsorption capacities in terms of ion exchange capacity percentage. Pavan *et al.* [156–158] studied the adsorption of dodecylsulphate, octylsulphate, dodecylbenzenesulphonate and octylbenzenesulphonate by Mg-Al-LDH.

1.7 Conclusions

In this chapter we have established that LDHs are an important clay material. They can be easily synthesized in the laboratory by various cheap and eco-friendly methods. The LDHs can be synthesized by varying the composition of precursor according to their applications in various fields.

Layered double hydroxides have great potential for wide use in catalysis in a wide range of applications from refinery to fine chemicals and environment protection, and especially in the development of multifunctional catalysts with tailored properties and catalyst supporters. In this chapter we have examined some cases and discussed recent trends, limits and opportunities offered by this class of materials for application in the environmental and pharmaceutical industries. Attention was focused on the use of layered double hydroxides as an adsorbent for the treatment of water and wastewater from industrial effluents. Since several reviews have already been published, our objective was not to make a systematic review of the topic, but instead to offer a personal view of the more interesting recent direction of research and at the same time, in areas with an increasing number of publications which do not parallel the practical relevance of the topic.

Acknowledgements

The first author is extremely grateful to the University Grant Commission, New Delhi, for the award of Post Doctoral Fellowship for Women (Award Letter No. F. 15/53/12 (SA-II).

References

1. F. Cavani, F. Trifiro, and A. Vaccari, *Catalysis Today*, Vol. 11, p. 173, 1991.
2. F.J. Brocker, and L. Kainer, German Patent, 2024282, 1970, UK Patent 1342020, 1971.
3. D.G. Evans, and T.C.R. Slade, *Structural Aspects of Layered Double Hydroxides*, Berlin, Springer-Verlag, Vol. 119, 2006.
4. S. Miyata, *Clays and Clay Minerology*, Vol. 28, p. 50, 1980.
5. S. Miyata, *Clays and Clay Minerology*, Vol. 23, p. 369, 1975.
6. S. Miyata, and A. Okada, *Clays and Clay Minerology*, Vol. 25, p. 14, 1977.
7. I. Zamarreno, F. Plana, and A. Vazquez, *American Minerologist*, Vol. 74, p. 1054, 1989.
8. V.A. Drits, N.A. Cherkashin, and V.I. Doklady, *Akademii Nauk SSSR*, Vol. 284, p. 443, 1985.
9. F.M. Halcom-Yarberry, Layered double hydroxides: Morphology, interlayer anion and the origins of life, Ph.D. Dissertation, Denton, University of North Texas, 2002.
10. V. Rives, *Materials Chemistry and Physics*, Vol. 75, p. 19, 2002.
11. D.S. Tong, C.H. Zhou, M.Y. Li, J. Beltramini, C.X. Lin, and Z.P. Xu, *Applied Clay Science*, Vol. 48, p. 569, 2010.
12. J. Theo Kloprogge, and R.L. Frost, *Physical Chemistry Chemical Physics*, Vol. 1, p. 1641, 1999.
13. J. Theo Kloprogge, and R.L. Frost, *Applied Cataysis A*, Vol. 184, p. 61, 1991.
14. Q. Tao, B.J. Reddy, H. He, R.L. Frost, P. Yuan, and J. Zhu, *Material Chemistry and Physics*, Vol. 112, p. 869, 2008.
15. H.J. Spratt, S.J. Palmer, and R.L. Frost, *Thermochim Acta*, Vol. 479, p. 1, 2008.
16. V. Rives, *Journal of the Royal Society of Chemistry*, Vol. 10, p. 489, 1999.
17. Y.H. Lin, M.O. Adebajo, R.L. Frost, and J.T. Kloprogge, *Journal of Thermal Analysis and Calorimetry*, Vol. 81, p. 83, 2005.
18. J.T. Kloprogge, M. Weier, I. Crespo, M.A. Ulibarri, C. Barriga, V. Rives, W.M. Martens, and R.L. Frost, *Journal of Solid State Chemistry*, Vol. 177, p. 1382, 2004.
19. J.T. Kloprogge, L. Hickey, and R.L. Frost, *Material Chemistry and Physics*, Vol. 89, p. 99, 2005.
20. J.T. Kloprogge, L. Hickey, and R.L. Frost, *Journal of Raman Spectroscopy*, Vol. 35, p. 967, 2004.
21. J.T. Kloprogge, L. Hickey, and R.L. Frost, *Journal of Solid State Chemistry*, Vol. 177, p. 4047, 2004.
22. J.T. Kloprogge, and R.L. Frost, *Journal of Solid State Chemistry*, Vol. 146, p. 506, 1999.
23. T.E. Johnson, W. Martens, R.L. Frost, Z. Ding, and J.T. Kloprogge, *Journal of Raman Spectroscopy*, Vol. 33, p. 604, 2002.
24. L.M. Grand, S.J. Palmer, and R.L. Frost, *Journal of Thermal Analysis and Calorimetry*, Vol. 101, p. 195, 2010.

25. L.M. Grand, S.J. Palmer, and R.L. Frost, *Journal of Thermal Analysis and Calorimetry*, 2009, doi: 10.100/s100973.
26. L.M. Grand, S.J. Palmer, and R.L. Frost, *Journal of Thermal Analysis and Calorimetry*, Vol. 100, p. 981, 2010.
27. R.L. Frost, E. Horvath, E. Mako, J. Kristof, and A. Redey, *Thermochimica Acta*, Vol. 407, p. 213, 2003.
28. D.P. Das, J. Das, and K. Parida, *Journal of Colloid and Interface Science*, Vol. 261, p. 213, 2002.
29. K.H. Bakon, S.J. Palmer, and R.L. Frost, *Journal of Thermal Analysis and Calorimetery*, Vol. 100, p. 125, 2010.
30. A. Jaiswal, and M.C. Chattopadhyaya, *Desalination and Water Treatment*, Vol. 29, p. 252, 2011.
31. A. Jaiswal, and M.C. Chattopadhyaya, *Desalination and Water Treatment*, Vol. 12, p. 127, 2009.
32. V. Rives, ed., *Layered Double Hydroxides: Present and Future*, New York, Nova Science Publishers, 2001.
33. A.I. Khan, and D. O' Hare, *Journal of Material Chemistry*, Vol. 12, p. 3191, 2002.
34. K. Carrado, A. Kostapapas, and S.L. Suib, *Solid State Ionics*, Vol. 26, p. 77, 1988.
35. W.T. Reichle, *Solid State Ionics*, Vol. 22, p. 135, 1986.
36. R.L. Frost, J.M. Bouzaid, and W.N. Martens, *Journal of Thermal Analysis and Calorimetry*, Vol. 89, p. 511, 2007.
37. R.L. Frost, S.J. Palmer, and H.J. Spratt, *Journal of Thermal Analysis and Calorimetry*, Vol. 95, p. 123, 2009.
38. G.W. Brindley, and S. Kikkawa, *Clays and Clay Minerals*, Vol. 28, p. 87, 1980.
39. H.J. Wolk, and S. Muller, Patent application, Process for the preparation of nanocyrstalline hydrotalcite compounds, IPC8 class: AB32B516FI, USPC Class: 428–402, 2009.
40. C. Del Hoyo, *Applied Clay Science*, Vol. 36, p. 103, 2006.
41. A. de Roy, C. Forano, E. El Malki, J.P. Besse, M.L. Ocelli, and E.R. Robson, eds., *Anionic Clays: Trends in Pillaring Chemistry, Synthesis of Microporous Materials*, New York, Van Nostrand Rienhold, Vol. 2, p. 108, 1992.
42. S. Komarneni, N. Kozai, and R. Roy, *Journal of Material Chemistry*, Vol. 8, p. 1329, 1998.
43. M. Bujoli-Doeuff, L. Force, V. Gadet, M. Verdaguer, K. El Malki, A. de Roy, J.P. Besse, and J.P. Renard, *Material Research Bulletin*, Vol. 26, p. 577, 1991.
44. M. Adachi-Pagano, C. Forano, and J.P. Besse, *Chemical Communications*, Vol. 91, p. 303, 2000.
45. J.H. Choy, S.Y. Kwak, Y.J. Jeong, and J.S. Park, *Angewandte Chemie*, Vol. 22, p. 4041, 2000.
46. V. Ambrogi, G. Fardella, G. Grandolini, and L. Perioli, *International Journal of Pharmaceutics*, Vol. 220, p. 23, 2001.
47. K.M. Tyner, M.S. Roberson, K.A. Berghorn, L. Li, R.F. Gilmour, F. Carl, F. Batt, and E.P. Giannelis, *Journal of Controlled Release*, Vol.100, p. 399, 2004.

48. A.J. Marchi, and C.R. Apesteguia, *Applied Clay Science*, Vol. 13, p. 35, 1998.
49. F. Trifiro, and A. Vaccari, *Comprehensive Supramolecular Chemistry*, J.L. Atwood, J.E.D. Davies, D.D. Macniol, and F. Vogtle, eds., Oxford, Pergamon Press,1996.
50. S. Sharma, K. Kushwaha, V.K. Srivastava, S.D. Bhatt, and R.V. Jasra, *Industrial and Engineering Chemistry Research*, Vol. 46, p. 4856, 2007.
51. G. Renaudin, M. Francois, and O. Evrard, *Cement and Concrete Research*, Vol. 29, p. 63, 1999.
52. U. Constantino, F. Marmottini, M. Nocchetti, and R. Vivani, *Europian Journal of Inorganic Chemistry*, Vol. 10, p. 1439, 1998.
53. M. Ogawa, and H. Kaiho, *Langmuir*, Vol. 18, p. 4240, 2002.
54. J. Livage, M. Henry, and C. Sánchez, *Progress in Solid State Chemistry*, Vol. 18, p. 259, 1988.
55. F. Prinetto, G. Ghiotti, P. Graffin, and D. Tichit, *Microporous and Mesopoous Materials*, Vol. 39, p. 229, 2000.
56. M.A. Aramendia, V. Borau, C. Jiménez, J.M. Marinas, J.R. Ruíz, and F.J. Urbano, *Journal of Solid State Chemistry*, Vol. 168, p. 156, 2002.
57. M. Jitianu, M. Balaoiu, M. Zaharescu, A. Jitianu, and A. Ivanov, *Journal of Sol-Gel Science and Technology*, Vol. 19, p. 453, 2000.
58. T. Lopez, P. Bosch, E. Ramos, R. Gomez, O. Novarro, D. Acosta, and F. Figueras, *Langmuir*, Vol. 12, p. 189, 1996.
59. J. He, M. Wei, B. Li, Y. Kang, G.D. Evans, and X. Duan, *Structural Aspects of Layered Double Hydroxides*, Berlin Heidelberg, Springer-Verlag, Vol. 119, pp. 89–119, 2006.
60. N. Morel-Desrosiers, J. Pisson, Y. Israëli, C. Taviot-Ghèho, J.P. Besse, and J.P. Morel, *Journal of Material Chemistry*, Vol. 13, p. 2582, 2003.
61. V. Rives, ed., *Layered Double Hydroxides: Present and Future*, New York, Nova Sci. Publisher, 2001.
62. O. Saber, *Journal of Physics: Conference Series*, Vol. 61, p. 825, 2007.
63. E. Suzuki, S. Idemura, and Y. Ono, *Clays Clay Minerals*, Vol. 37, p. 173, 1989.
64. M. Drezdzon, *Inorganic Chemistry*, Vol. 27, p. 4628, 1988.
65. F.R. Costa, A. Leuteritz, U. Wagenknecht, D. Jehnichen, L. Hauoler, and G. Heinrich, *Applied. Clay Science*, Vol. 38, p. 153, 2008.
66. S. Carlino, and M.J. Hudson, *Solid State Ionics*, Vol. 110, p. 153, 1998.
67. L. Ukrainczyk, M. Chibwe, T.J. Pinnavaia, and S.J. Boyd, *Journal of Physical Chemistry*, Vol. 98, p. 2668, 1994.
68. F. Kooli, C. Depege, A. Ennaquadi, A. De Roy, and J.P. Besse, *Clays and Clay Minerals*, Vol. 45, p. 92, 1997.
69. H. Tagaya, S. Sato, H. Morioko, J. Kadokawa, M. Karasu, and K. Chiba, *Chemistry of Materials*, Vol. 5, p. 1431, 1993.
70. M. Chibwe, and T.J. Pinnavaia, *Chemical Communications*, Vol. 3, p. 278, 1993.
71. M.E. Perezbernal, R. Ruanocasero, and T.J. Pinnavaia, *Catalysis Letters*, Vol. 11, p. 55, 1991.

72. T. Kuwahara, O. Onitsuka, H. Tagaya, J. Kodokawa, and K. Chiba, *Journal of Inclusion Phenomenon and Molecular Recognition Chemistry*, Vol. 18, p. 59, 1994.
73. D.G. Evans, and X. Duan, eds., *Layered Double Hydroxides*, Berlin, Springer-Verlag, 2006.
74. E. Merlen, P. Gueroult, J.B. d'Espinose de la Caillerie, B. Rebours, C. Bobin, and O. Clause, *Applied Clay Science*, Vol. 10, p. 45, 1995.
75. J.B. d'Espinose de la Caillerie, M. Kermarec, and O. Clause, *Journal of American Chemical Society*, Vol. 117, p. 11471, 1995.
76. F. Zhang, J. Chen, P. Chen, Z. Sun, and S. Xu, *AIChE Journal*, Vol.58, p. 1853, 2012.
77. M.G. Alvarez, M. Pliskova, A.M. Segarra, F. Medina, and F. Figueras, *Applied Catalysis B- Environmental*, Vol. 113, p. 212, 2012.
78. F. Winter, A.J. van Dillen, and K.P. de Jong, *Chemical Communications*, Vol. 31, p. 3977, 2005.
79. F. Winter, V. Koot, A.J. van Dillen, J.W. Geus, and K.P. de Jong, *Journal of Catalysis*, Vol. 236, p. 91, 2005.
80. S. Carlino, *Solid State Ionics*, Vol. 98, p.73, 1997.
81. E. Pungor, *A Practical Guide to Instrumental Analysis*, Boca Raton, CRC Press, p. 384, 1995.
82. D.A. Skoog, D.M. West, and F.L. Holler, *Fundamentals of Analytical Chemistry*, Philadelphia, Saunders College Publishing, p. 502, 1996.
83. J. Madejova, *Vibrational Spectroscopy*, Vol. 31, p. 1, 2003.
84. G. Socrates, *Infrared Characteristic Group Frequencies*, New York, John Wiley and Sons, 1980.
85. F.M. Labajos, V. Rives, and M.A. Ulibarri, *Journal of Material Science*, Vol. 27, p. 1546, 1992.
86. G.R. Williams, and D. O'Hare, *Journal of Material Chemistry*, Vol. 16, p. 3065, 2006.
87. J.M. Fernandez, C. Barriga, M.A. Ulibarri, F.M. Labajos, and V. Rives, *Journal of Material Chemistry*, Vol. 4, p.1117, 1994.
88. H. Zang, K. Zou, S.H. Guo, and X. Duan, *Journal of Solid State Chemistry*, Vol. 179, p. 792, 2006.
89. J. Pérez-Ramírez, G. Mul, and J.A. Moulijin, *Vibrational Spectroscopy*, Vol. 27, p. 75, 2001.
90. Y. Kim, R.J. Kirkpatrick, and R.T. Cygan, *Geochemica and Cosmochiica Acta*, Vol. 60, p. 4059, 1996.
91. R.J. Kirkpatrick, Y. Ping, X.Q. Hou, and Y. Kim, *American Mineralogist*, Vol. 84, p. 1186, 1999.
92. X.Q. Hou, R.J. Kirkpatrick, Y. Ping, D. Moore, and Y. Kim, *American Mineralogist*, Vol. 85, p. 173, 2000.
93. D. Tichit, and B. Coq, *Cattech*, Vol. 7, p. 206, 2003.
94. S. Komarneni, and R. Roy, *Science*, Vol. 239, p. 1286, 1988.
95. M. Lakraimi, A. Legrouri, A. Barroug, A. De Roy, and J.P. Besse, *Journal of Material Chemistry*, Vol. 10, p. 1007, 2000.

96. A.I. Khan, L. Lei, A.J. Noirquist, and D. O'Hare, *Chemical Communications,* Vol. 22, p. 2342, 2342.
97. S.J. Choi, J.M. Oh, and J.H. Choy, *Journal of Material Chemistry*, Vol. 18, p. 615, 2008.
98. Z.P. Xu, H.Z. Qing, Q.L. Gao, and B.Y. Ai, *Chemical Engineering Science*, Vol. 61, p. 1027, 2006.
99. M. Wei, M. Pu, J. Guo, J. Han, F. Li, J. He, D.G. Evans, and X. Duan, *Chemistry of Materials,* Vol. 20, p. 5169, 2008.
100. M. Wei, Q. Yuan, D.G. Evans, Z. Wang, and X. Duan, *Journal of Material Chemistry*, Vol. 15, p. 1197, 2005.
101. K. Ladewig, M. Niebert, Z.P. Xu, P.P. Gray, and G.Q.M. Lu, *Biomaterials*, Vol. 31, p. 1821, 2010.
102. F. Li, and X. Duan, *Applications of Layered Double Hydroxides in Structure and Bonding*, Berlin, Springer-Verlag, Vol. 119, p. 193, 2006.
103. V. Prevot, C. Forano, and J.P. Besse, *Journal of Material Chemistry*, Vol. 9, p. 155, 1999.
104. R. Frost, A. Musemeci, J.T. Kloprogge, M. Adebajo, and W. Martens, *Journal of Raman Spectroscopy*, Vol. 37, p. 733, 2006.
105. M.J. Climent, A. Corma, S. Iborra, K. Epping, and A. Velty, *Journal of Catalysis*, Vol. 225, p. 316, 2004.
106. P.S. Braterman, P.X. Zhi, and F. Yarberry, "Layered double hydroxides (LDHs)," in: *Handbook of Layered Materials,* S.M. Auerbach, K.A. Carrado, and P.K. Dutta, eds., Marcel Dekker Inc.: Hoboken, 2001.
107. Y.X. Feng, N. Yin, Q.F. Li, J.W. Wang, M.Q. Kang, and X.K. Wang, *Industrial and Engineering Chemistry Research*, Vol. 47, p. 2140, 2008.
108. L. Ren, J. He, S. Zhang, D.G. Evans, and X. Duan, *Journal of Molecular Catalysis B: Enzymatic*, Vol. 18, p. 3, 2002.
109. S. Miyata, *Clays and Clay Mineral*, Vol. 31, p. 305, 1983.
110. L.M. Parker, N.B. Milestone, and R.H. Newman, *Industrial and Engineering Chemistry Research*, Vol. 34, p. 1196, 1995.
111. T.L. Grady, T. Hess, and H. Mcwhinney, The use of oxide thin film in the remediation of aqueous wastes containing hazardous metal ions, *Proceedings: NOBCChE*, Vol. 22 p. 115, 1995.
112. M. Zhang, and E.J. Reardon, *Environmental Science and Technology*, Vol. 37, p. 2947, 2003.
113. A. Abdelous, J.L. Crovisier, W. Lutz, B. Fritz, A. Mosser, and R. Muller, *Clays Clay Miner.*, Vol. 42, p. 526, 1994.
114. N. Van Suc, *E- Journal of Chemistry*, Vol. 9, p. 669–679, 2012.
115. X. Zhang, L. Ji, J. Wang, R. Li, Q. Liu, M. Zhang, and L. Liu, *Colloids and Surface A: Physicochemical and Engineering Aspects*, Vol. 414, p. 220, 2012.
116. M.-L. Chen, M.-I. An, *Talanta*, Vol. 95, p. 31, 2012.
117. M.J. Kang, S.W. Rhee, H. Moon, V. Neck, and T. Fanghanel, *Radiochimica Acta*, Vol. 75, p. 169, 1996.

118. M.J. Kang, K.S. Chun, S.W. Rhee, and Y. Do, *Radiochimica Acta*, Vol. 85, p. 57, 1999.
119. F.L. Theiss, M.J. Sear-Hall, S.J. Palmer, and R.L. Frost, *Desalination and Water Treatment*, Vol. 39, p. 166, 2012.
120. T. Toraishi, S. Nagasaki, and S. Tanaka, *Applied Clay Sci.*, Vol. 22, p. 17, 2002.
121. M.T. Olguin, P. Bosch, D. Acosta, and S. Bulbulian, *Journal of Nuclear Materials*, Vol. 359, p. 155, 2006.
122. H. Curtius, and Z. Kattilparampil, *Clay Minerals*, Vol. 40, p. 455, 2005.
123. G. Fetter, M.T. Olguin, P. Bosch, V.H. Lara, and S. Bulbulian, *Journal of Radioanalytical Nuclear Chem.*, Vol. 241, p. 595, 1999.
124. L. Kentjono, J.C. Liu, W.C. Chang, and C. Irawan, *Desalination*, Vol. 262, p. 280, 2010.
125. G. Mao, M. Tsuji, and Y. Tamaura, *Clays and Clay Minerals*, Vol. 41, p. 731, 1993.
126. X.P. Wang, J.J. Yu, J. Cheng, Z.P. Hao, and Z.P. Xu, *Environmental Science and Technology*, Vol. 42, p. 614, 2008.
127. M.K.R. Reddy, Z.P. Xu, G.Q. Lu, J.C.D. da Costa, *Industrial and Engineering Chemistry Research*, Vol. 45, p. 7504, 2006.
128. N.D. Hutson, *Chemistry of Materials*, Vol. 16, p. 4135, 2004.
129. T. Kameda, A. Kodama, Y. Fubasami, S. Kumagai, and T. Yoshioka, *Chemosphere*, Vol. 88, p. 250, 2012.
130. M Cantú, E López-Salinas, J.S. Valente, and R. Montiel, *Environmental Science and Technology*, Vol. 39, p. 9715, 2005.
131. F. Theiss, S.J. Palmer, A.G. Ayokond, and R.L. Frost, *Journal of Thermal Analysis and Calorimetry*, Vol. 107, p. 1123, 2011.
132. M.A. Ulibarri, I. Pavlovic, M.C. Hermosin, and J. Cornejo, *Applied Clay Science*, Vol. 10, p. 131, 1995.
133. M.C. Hermosin, I. Pavlovic, M.A. Ulibarri, and J. Cornejo, *Water Research*, Vol. 30, p. 171, 1996.
134. I. Pavlovic, M.A. Ulibarri, M.C. Hermosin, and J. Cornejo, *Fresenius Environmental Bulletin*, Vol. 6, p. 266, 1997.
135. C. Barriga, M. Gaitan, I. Pavlovic, M.A. Ulibarri, M.C. Hermosin, and J. Cornejo, *Journal of Material Chemistry*, Vol. 12, p. 1027, 2002.
136. E.L. Crepaldi, J. Tronto, L.P. Cardosoand, and J.B. Valim, *Colloids and Surfaces A: Physicochemical and Engineering Aspects*, Vol. 211, p. 103, 2002.
137. Y. You, H. Zhao, and G.F. Vance, *Applied Clay Science*, Vol. 21, p. 217, 2002.
138. J. Orthman, H.Y. Zhu, and G.Q. Lu, *Separation and Purification Technology*, Vol. 31, p. 53, 2003.
139. I. Pavlovic, C. Barriga, M.C. Hermosin, J. Cornejo, and M.A. Ulibarri, *Applied Clay Science*, Vol. 30, p. 125, 2005.
140. F. Li, Y. Wang, Q. Yang, D.G. Evans, C. Forano, and X. Duan, *Journal of Hazardous Materials*, Vol. 25, p. 89, 2005.
141. L.P. Cardoso, and J.B. Valim, *Journal of Physics and Chemistry of Solids*, Vol. 67, p. 987, 2006.

142. S. Vreysen, and A. Maes, *Applied Clay Science*, Vol. 38, p. 237, 2008.
143. Y.F. Chao, J.J. Lee, and S.L. Wang, *Journal of Hazardous Materials*, Vol. 165, p. 846, 2009.
144. L.E. Gaini, M. Lakraimi, E. Sebbar, A. Meghea, and M. Bakasse, *Journal of Hazardous Materials*, Vol. 161, p. 627, 2009.
145. J.S. Valente, F. Tzompantzi, J. Prince, J.G.H. Cortez, and R. Gomez, *Applied Catalysis B: Environmental*, Vol. 90, p. 330, 2009.
146. N.D. Setti, N. Jouini, and Z. Derriche, *Journal of Physics and Chemistry of Solids*, Vol. 71, p. 556, 2010.
147. Z. Sun, L. Jin, W. Shi, M. Wei, and X. Duan, *Chemical Engineering Journal*, Vol. 161, p. 293, 2010.
148. M.V. Villa, M.J. Sanchez-Martin, M. Sanchez-Camazano, *Journal of Environmental Science and Health*, Vol. 34, p. 509, 1999.
149. Y. You, H. Zhao, G.F. Vance, *Colloids and Surfaces A: Physicochemical and Engineering Aspects*, Vol. 205, p. 161, 2002.
150. S. Rajeshwari, S. Sivakumar, P. Senthilkumar, and V. Subburam, *Bioresource Technology*, Vol. 80, p. 233, 2001.
151. H. Kopka, K. Beneke, and G. Lagaly, *Journal of Colliod and Inerface Science*, Vol. 123, p. 427, 1988.
152. M. Meyn, K. Beneke, and G. Lagaly, *Inorganic Chemistry*, Vol. 29, p. 5201, 1990.
153. Z.P. Xu, P.S. Braterman, *Journal of Physical Chemistry*, Vol. 13, p. 268, 2007.
154. I. Dekany, and T. Haraszti, *Colloid Surface A*, Vol. 391, p. 123, 1996.
155. I. Dekany, F. Berger, K. Imrik, and G. Lagaly, *Colloid and Polymer Science*, Vol. 275, p. 681, 1997.
156. P.C. Pavan, E.L. Crepaldi, G.P. Gomes, and J.B. Valim, *Colloid Surface A*, Vol. 154, p. 399, 1999.
157. P.C. Pavan, L.C. Eduardo, and B.V. Joao, *Journal of Colloid and Interface Science*, Vol. 229, p. 346, 2000.
158. P.C. Pavan, G.A. Gomes, and J.B. Valim, *Microporous Mesoporous Materials*, Vol. 21, p. 659, 1998.

2

Improvement of the Corrosion Resistance of Aluminium Alloys Applying Different Types of Silanes

Anca-Iulia Stoica*, Norica Carmen Godja, Andje Stanković, Matthias Pölzler, Erich Kny and Christoph Kleber

Centre of Electrochemical Surface Technology, Wiener Neustadt, Austria

Abstract

Due to their low density and good mechanical characteristics, aluminium alloys are outstanding for use in aeronautics industry applications. Nevertheless, alloys commonly used in aerospace, such as AA 2024, show higher corrosion rates than pure aluminium and therefore need to be protected against corrosion attacks by different surface treatments. In the last decade, all novel studies concerning the corrosion protection of aluminium and aluminium alloys were based on a "green treatment," indicating that silanes are one of the most promising alternatives for chromates for the surface finishing of Aluminium and Aluminium alloys.

Presented here are the results of a silane surface treatment based on water soluble mixtures of silanes and alcohol soluble silanes which were used for the formation of corrosion protective coatings (conversion coatings) on the aluminium alloy AA 2024. Also, the quality and performance of the sealed anodic films before and after a salt spray test (SST) are elucidated by the application of different surface sensitive techniques, such as electrochemical impedance spectroscopy (EIS), scanning electron microscopy (SEM), focused ion beam (FIB), Fourier transformed infrared spectroscopy (FTIR) and contact angle (CA) measurements.

Keywords: Aluminium, AA2024, TSA, silane, corrosion protection, SEM, FIB, EIS, contact angle, SST

**Corresponding author*: anca.stoica@cest.at

Ashutosh Tiwari and Mikael Syväjärvi (eds.) Advanced Materials for Agriculture, Food, and Environmental Safety, (27–60) 2014 © Scrivener Publishing LLC

2.1 Introduction

With 8.1% by weight, aluminium is the third most abundant element in the Earth's crust, but due to its reactivity it is not found in an elemental state in nature. Aluminium and its alloys have been extensively studied due to their wide range of commercial applications in the aerospace, architecture, transportation, and manufacturing industries. One of the most prominent reasons for the applicability of aluminium alloys in aircraft construction are their low density and good mechanical characteristics. To improve the mechanical strength of aluminium, further alloying with different elements such as copper, zinc, magnesium and others is necessary. While this alloying increases the mechanical strength, it decreases the resistivity against localized corrosion attacks—especially in chloride-containing environments. The native oxide film formed when exposed to ambient atmospheres protects aluminum and its alloys to a certain degree, but it is heterogeneous and does not provide adequate corrosion resistance in many environments. Thus, in the aeronautic industry, the corrosion protection of structural aluminium alloys, such as the 2XXX and 7XXX series, requires an additional surface treatment which normally involves the use of Cr(VI) salts to obtain a high corrosion resistance. Anodic films formed under these conditions exhibit a good corrosion performance and Cr(VI) species are considered as the "standard" corrosion inhibitors, even if they are very toxic, carcinogenic and allergenic. An alternative route to improve the anticorrosive properties of aluminium and its alloys would be the sample anodization [1]. In general, the anodizing processes in acidic media are based on the use of electrolyte baths composed of mixtures of inorganic acids or mixtures of organic/inorganic acids [2]. The anodizing process and the properties of the thereby obtained anodic films are also influenced by the alloying elements of the aluminium itself. During the past two decades, a lot of work has been carried out to develop process alternatives to CAA (chromic acid anodization) based on sulphuric acid, sulphuric-boric acid, sulphuric-tartaric acid, malonic and oxalic acid formulations [3–5].

Addition of different weak acids such as boric acid (Boening Co., EP 0,405,624 (1991) or tartaric acid (US patent 2002/0157961) to sulphuric anodizing baths has been proposed as one possible alternative. Alenia Aeronautica S.P.A has proposed a new anodizing procedure involving the addition of tartaric acid in diluted sulphuric acid electrolyte, i.e., tartaric-sulphuric acid (TSA) anodizing [6]. In particular, anodizing of aluminium in a mixture of sulphuric and L(+)-tartaric acid (TSA) has proven to produce oxide layers comparable to those formed in chromic acid baths [7].

The use of tartaric and sulphuric acids for anodization is a well-known modification of the standard sulphuric acid anodization technique [8] and typically provides a passivation oxide layer with high corrosion resistance. The sealing quality and the corrosion resistance of AA2024 T3 anodized in sulphuric acid with or without addition of tartaric acid were intensively investigated by field emission scanning electron microscopy (FESEM) and electrochemical impedance spectroscopy (EIS) techniques [6].

The main results obtained were:

1. Anodizing of AA2024 T3 in the presence of tartaric acid led to the formation of anodic films with lower porosity than those obtained in diluted sulphuric acid (DSA).
2. The characterization of hydrothermal sealing by EIS measurements showed that the sealing quality was better for the anodic films formed in TSA than in DSA in agreement with the lower pore volume of the anodic films formed in TSA.
3. The corrosion behavior evaluated by EIS measurements revealed that the sealed films formed in TSA are significantly more resistant to corrosion than the sealed films formed in DSA. This enhancement of corrosion resistance is mainly associated with the higher density of the formed porous layer and a higher resistance of the barrier layer obtained after hydrothermal sealing of specimens anodized in TSA.

Anodizing by even more diluted sulphuric acid has been introduced [6] to obtain thin anodic films (1–5 μm) which should increase the fatigue resistance for specific structural materials, but the corrosion resistance obtained by this treatment is lower than in the case of chromic acid anodization.

Here, the anodizing process was performed in a 20 L TSA bath (0.53 M $C_4H_6O_6$ / 0.41 M H_2SO_4) at 37°C following Iglesias-Rubianes *et al.*, who already discussed in detail the growth of anodic layers on AA2024 T3 by this process even though in a different scale [9]. Due to the fact that AA2024 normally contains 4–4.5% of copper, the influence of Cu^{2+} ions in the anodic layer is substantial with respect to the morphology, phase composition and dielectric properties. It is known from literature that within barrier films, Al^{3+} and O^{2-} ions are mobile [10]. When copper is present in the bulk aluminium it will be incorporated into the anodic films as CuO islands. The Cu^{2+} ions are able to migrate outwards towards the bulk about three times faster than the Al^{3+} ions. The presence of this copper oxide—sometimes also referred to as "conductive oxide"—locally increases the conductivity of the alumina layer [10]. Due to a relatively rapid migration of Copper into the electrolyte, Cu^{2+} ions might therefore also be present in the pores of the surface film together with the acid residuals at the present pH. So in the anodic film formed, copper islands will mainly be dissolved in the electrolyte. These Cu^{2+} cations in the solution would re-precipitate at this pH (≈ 5) but are bound in chelate complexes by the tartaric acid. So, tartrate or $Htart^-$ ions might be able to remove Cu^{2+} not only directly from the pore surfaces but also through migration from the pores themselves. This reduces the heterogeneity of the formed anodic film and as a result leads to a more corrosive resistive layer [10].

Another alternative to Cr(VI) for the corrosion protection of aluminium and its alloys is the application of different silanes. Silane-based coatings usually exhibit good barrier properties due to the development of dense –Si-O-Si networks [11, 12], which inhibit the penetration of aggressive species towards the metallic substrate. Thus, the efficiency of the metal surface pretreatment based on silane coatings is strongly dependent on the barrier properties of the formed surface film [13, 14]. But even a small defect in the barrier layer may cause a localized corrosion process. Even for improved corrosion protection an additional sealing treatment of the formed porous oxide layer is required.

The purpose of sealing an anodic coating is to close the pores of the formed anodic layer, thereby at least causing the layer to become chemically inert. This is accomplished by a chemical change of the surface structure by hydration—experimentally realized by a cold seal and/or a hot seal procedure.

The process which takes place during hot water sealing (HTS) involves (1) the dissolution of anhydrous alumina from the pore walls, (2) the formation of a hydrated alumina gel inside the pores, (3) the saturation and subsequent precipitation of alumina hydrates—which eventually plug the entire length of the pores—and (4) the final process of crystallization and agglomeration, which might continue over months and years.

In detail the HTS process involves a complex reaction cascade leading to the formation of different (ad) layers in the surface film, namely:

- an outer layer of formed acicular boehmite crystals
- an intermediate layer of a high density hydrated oxide
- a near bulk and much thicker layer with a porous cell structure; and
- the innermost layer, also known as the barrier layer, being extremely thin (≈ 25 nm) and dense pore free.

However, in order to "close" very effectively (seal) the pores in the anodized layer to obtain an improved corrosion protection a process involving boiling water containing chromate is still commonly used. By means of the hot water sealing (HTS) treatment, the hydration of the aluminium oxide results in a volume expansion of the film resulting in an occludation of the surface pores. Additives in the sealing bath are sometimes used to increase the efficiency of treatment. The hazard of this treatment is due to the differential thermal dilatations between the substrate and the film, which might lead to the crazing of the film when thermally loaded [15].

Cold sealing techniques of anodic films via CCC (Chromate Conversion Coating) may offer advantages concerning the integrity of the anodized coating. Since the beginning of the 1990s, the high toxicity associated with Cr(VI) has imposed restrictions on its use in industrial applications and consequently has led to the development of Cr(VI)-free surface treatments. A number of promising "green" alternatives have been studied with the objective of replacing chromates, and one of the most promising alternatives to chromates in the metal-finishing industries are silanes [16–18]. Silicon compounds improve the properties of materials such as chemical resistance, mechanical strength and electrical performances.

Further improvement of the corrosion resistance is possible by applying HTS; different silane conversion coatings after TSA anodization with and without HTS were tested within this work.

2.2 Silanes for Surface Treatment

The goal of replacing the use of chromate for the surface treatment of metals with silanes has a lot of advantages from an environmental point of view, economical aspects, employment security and as far as the corrosion inhibitive effects on metals are concerned. Van Ooij and coworkers have studied the corrosion performances of a variety of silane films on different metals substrates [19]. Silane coatings can be used for various metals such as aluminium and its alloys [11, 20–22], copper [23], iron and steel [24], zinc [25] and magnesium [26] coating alloys and could efficiently protect metals against different forms of corrosion. Plueddemann [27] used the application of silicon compounds as coupling agents to promote adhesion in the surface treatment of metals. Silanes have the general chemical formula $(RO)_3Si(CH_2)_nR'$, where RO is a hydrolyzable alkoxy group (methoxy – OCH_3; ethoxy – OC_2H_5 or acetoxy $OCOCH_3$); R' is an organofunctional group (vinyl – C=C, amino – NH_2), and the value of n is around 3 [28].

2.2.1 Classification of Silanes

In general there are two possible ways to classify silanes:

a. **According to their structure:** mono-silane (Figure 2.1) (mono-silyl functional silanes) (see examples in Table 2.1) and bis-silane (Figure 2.2) (bis-silyl functional silanes) (see examples in Table 2.2). The *mono-silanes* have the general structure $R_3Si(CH_2)_nR'$, where R represents a silicon ester, R' is an organofunctional group such as chlorine, primary or secondary amines, or vinyl, and the value of n is around 3.

   Bis-silanes (dipodal silanes) (Figure 2.2) with the structure $R_3Si(CH_2)_nY(CH_2)_nSiR_3$ with Y- as a functional group, or without a functional group $R_3Si(CH_2)_mSiR_3$, generally assure better corrosion protection for different metals and alloys such as Al, Zn, Cu, Mg and steel [20–22,29]

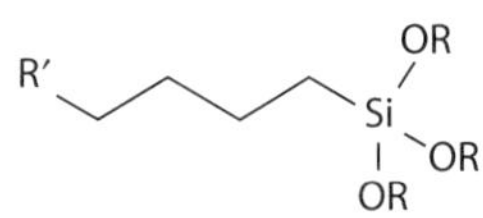

Figure 2.1 General chemical structure of mono-silane.

Table 2.1 Formula and chemical structure of typical monofunctional silanes.

Name of silane	Formula	Chemical structure
Vinyltriacetoxy silane (VTAS)	$CH_2{=}CH(CH)_3Si(COCOCH_3)_3$	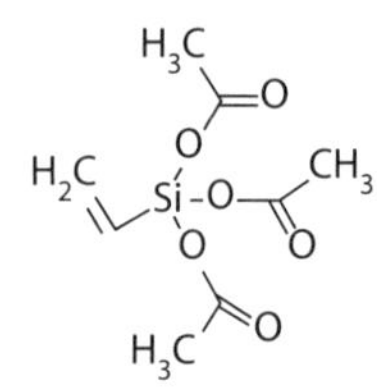
Vinyltriethoxy silane (VS)	$CH_2{=}CHSi(OC_2H_5)_3$	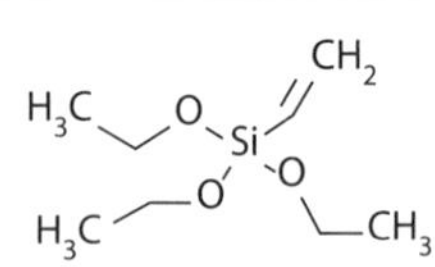
γ-aminopropyl triethoxysilane (γ-APS)	$(H_5C_2O)_3Si\text{-}(CH_2)_3\text{-}NH_2$	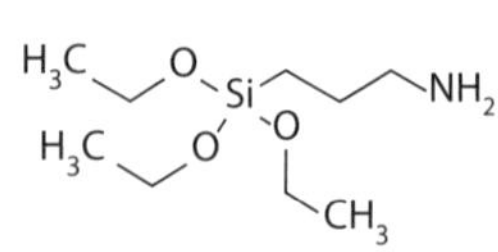

compared with mono-silanes due to the double number of hydrolyzable OR groups [22, 30].

Bis-silanes improve the anti-corrosion performance by depositing a more hydrophobic layer (SiOSi) on the substrate's surface during the condensation, and also through chemical bonding of the functional groups present in the bis-silane itself or with a mono-silane solution additive. It was observed that a mixture of silanes performed better corrosion protection than a single silane [22, 31]. Unfortunately, most of the bis-silanes are not soluble in water, limiting their industrial applications. Nevertheless, in optimal mixtures it could be proven that silanes are able to ensure equivalent

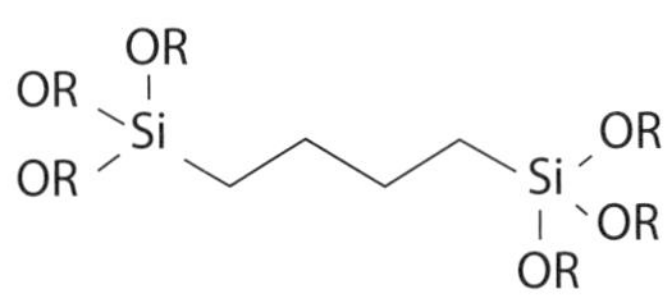

Figure 2.2 Chemical structure of bis-silane.

Table 2.2 Formula and chemical structure of typical bifunctional silanes.

Name of silane	Formula	Chemical structure
bis-1,2-[triethoxysilyl propyl] ethane (BTSE)	$(OC_2H_5)_3$-$Si(CH_2)_2Si(OC_2H_5)_3$	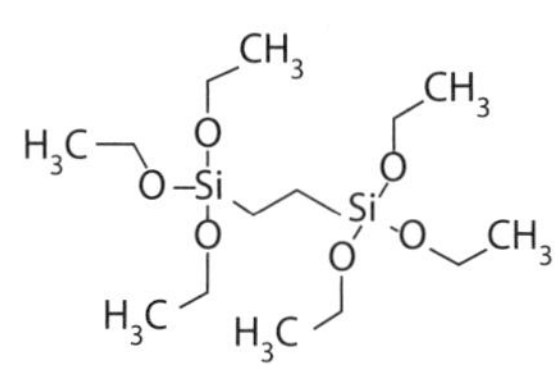
bis-1,2-[triethoxysilyl propyl] tetrasulfide (BTESPT)	$(OC_2H_5)_3Si(CH_2)_3S_4$ $(CH_2)_3Si(OC_2H_5)_3$	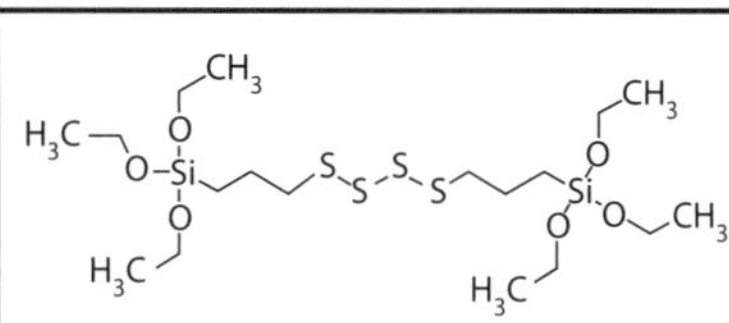
bis-[1,2-trimethoxysilylpropyl] amine, BTSPA	$(OCH_3)_3Si(CH_2)_3NH$ $(CH_2)_3Si(OCH_3)_3$	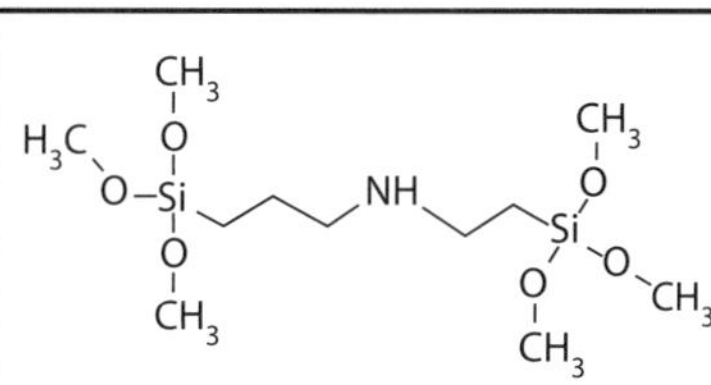

performances to Cr(VI) with respect to the obtained anti-corrosive properties.

Petrunin and his coworkers [20] studied the formation and properties of siloxane monolayer on metal surfaces using different silanes. The AlOSi covalent bonds occur in the presence of adsorbed water on the aluminium surface, and the silane monolayer on aluminium has, as a result, a decrease of the amount of adsorbed water on the surface leading to an inhibition of the hydration of the metal-oxide surface film. In the case of aminopropyltrimethoxysilane, the positively charged layer of the silane film activates local metal corrosion when exposed to chloride-containing media. Methacryloxypropyltrimethoxysilane (MAOS) [13] shows a good corrosion protection of Al alloys when immersed in NaCl solution due to the synergistic effect of Aluminum-oxides and of Al-siloxane compounds forming a polymeric passive film, together hindering ion diffusion to the surface.

b. **According to their hydrophobicity:** *water-based* (i.e., bis–[trimethoxysilylpropyl] amine (bis-amino silane, $(OCH_3)_3Si(CH_2)_3NH(CH_2)_3Si(OCH_3)_3$) and vinyl-triacetoxysilane (VTAS, CH_2=$CH(CH)_3Si(COCOCH_3)_3$) or *alcohol-based silanes* (i.e., bis-[3-(triethoxysilyl)] ethane, BTSE, $(OC_2H_5)_3Si(CH_2)_2Si(OC_2H_5)_3$ and bis [3 (triethoxysilyl)propyl] tetrasulfide - bis-sulfur silane, $(OC_2H_5)_3Si(CH_2)_3S_4(CH_2)_3$-$Si(OC_2H_5)_3$.

The most important advantages of these types of water-based silanes is that alcohol-free solutions can be prepared, that they hydrolyze instantaneously and completely, and that their corrosion protection performances are comparable with alcohol-based silanes and chromates, especially with topcoats. They have a broader compatibility with a lot of paints compared to monofunctional silanes, due to the fact that they have more organic functional groups which are responsible for paint adhesion [32, 33].

2.2.2 Surface Treatment and Silane Chemistry

Silane surface treatment is a simple process. This is accomplished by dipping metals into diluted silane solutions (e.g., 2–5 vol%) for a few seconds, followed by drying in air and curing by different temperatures. Silanes are only active in the hydrolyzed form (Si-OH). An access of SiOH in the solution ensures that there is no kinetic limitation of the reaction of the hydroxylized metallic surface (Me-OH) with the silanol groups (Si-OH) forming hydrogen bonds which can be dehydrated during the curing process to form the final Me-O-Si layer (Reaction 2.1) [34].

$$\text{Si-OH}_{(\text{solution})} + \text{Me-OH}_{(\text{metal surface})} \leftrightarrow \text{Si-O-Me}_{(\text{interface})} + H_2O \quad (2.1)$$

Excess SiOH groups which are adsorbed on the metals but not interacting with Me-OH groups condense among themselves and form a siloxane (Si-O-Si) surface film (Reaction 2.2).

$$\text{Si-OH}_{(\text{solution})} + \text{Si-OH}_{(\text{solution})} \leftrightarrow \text{Si-O-Si}_{(\text{silane film})} + H_2O \quad (2.2)$$

After the hydrolysis step, the silanol groups can be involved in condensation reactions, resulting in polymerization and/or precipitation. Silanol groups are important to form a covalent bond at the metal/film interface and also for crosslinking in the bulk of the silane layer. In the solution, the hydrolysed molecules can react to form dimers, trimers and bigger species through condensation [16]. The kinetics and equilibrium of hydrolysis and condensation of silanes in solution are influenced by the nature of the organofunctional groups, the pH of the solution, the concentration of the silane itself, the temperature, and the aging of the solution [29, 35,36].

In the case of a bis-sulfur silane (PSS) solution [30], the total hydrolysis reaction equilibrium is (Reaction 2.3):

$$(OR)_3Si(CH_2)_3S_4(CH_2)_3Si(OR)_3 + H_2O \leftrightarrow (OH)_3Si(CH_2)_3S_4(CH_2)_3Si(OH)_3 + 3ROH \quad (2.3)$$

where, S – sulfur atom (4 represents average number of the S atoms contained by each bis-sulfur silane molecule); OR – hydrolyzable alkoxy groups or OC_2H_5 in the case of the bis-sulfur silane; SiOH – silanol. During the silane surface treatment the curing/drying step is much more important than the immersion step compared to the conventional chromating process where the film thickness is proportional to the contact time. Due to the fact that hydrolysis and condensation reactions are catalyzed by acids or bases, the silane aqueous solution stability is strongly dependent on the pH [37, 38]. Thus, in acidic or alkaline medium, the rates of both hydrolysis and condensation reactions are high, while at or near neutral pH they are slow. If the reaction is catalyzed by HO^-, condensation is favored with rapid formation of a gel; in the case of a H^+ catalyzed reaction a high rate of hydrolysis is also favored but with slow gelation. These two processes can be controlled by the pH using different solutions, i.e., acetic acid. The bis-amino silane could be completely hydrolyzed in water in the pH = 4–9.5, the bis-sulfur silane requires at least 50% alcohol (ethanol) and a pH = 3.5–9.5. The hydrolysis of bis-sulfur silane could be performed without the addition of acid, but requires a minimum of 4 days for complete hydrolysis [12]. Different studies have shown [39, 40] that the hydrolysis of alkoxy groups proceed in steps. Thus, the hydrolysis of the first alkoxy group is slow, due to steric effects, and the others are rapid. Osterholtz and Pohl [37] studied the kinetics of hydrolysis and condensation of organofunctional alkoxysilanes and concluded that a change of pH in the alkaline or acidic part with one unit to the value of 7 results in an acceleration of hydrolysis rate; also, the condensation can be minimized by using silanes with a concentration below 1%. Arkes and coworkers reported that there is proportionality between the carbon chain of the alkoxy group and the hydrolysis rate. Thus, when the carbon chain of the alkoxy group is longer, the hydrolysis rate decreases [38].

Bis-sulfur silane (PSS) assured good corrosion protection of aluminium alloy, but it cannot be used for hot-dip galvanized steel (HDG). All former studies were focused on obtaining a universal silane mixture which is able to ensure proper corrosion protection for a range of metals and alloys [41]. The use of high percent alcohol content is not recommended for industrial processes due to economic, environmental and health reasons [23]. Therefore, correct hydrolysis can be achieved in solutions without or with little amounts of alcohol, leading to a good compromise between hydrolysis and condensation. Thus, in the case of a silane mixture, a synergetic effect is observed depending on the ratio between the two utilized silanes. Promising results for AA2024 with subsequent HDG were obtained from a mixture of bis-amino:bis-sulfur silane = 1:3.

As mentioned before, the nature and pretreatment of the metallic surface has a great influence on the interaction between the surface and the silanes. Therefore, the cleaning step before the treatment process has a strong effect on the characteristics of the deposited film. From different studies [42] it was concluded that alkaline cleaning represents the best pretreatment of samples prior to the application of silanes. According to the chemical bonding theory of Plueddemann [27], oxide surfaces which contain a high density of hydroxyl groups show a better ability to withstand corrosion. Another study [29] reported that acid or neutral cleaning solutions are less suitable in comparison with alkaline cleaning solutions because they lead to a lower concentration of hydroxyl groups on the surface. The effect of surface pretreatment on aluminium was also studied by Franquet [41], who concluded that the amount of hydroxyl groups on the surface, as a result of different pretreatments, influences the initiation and the formation of the silane films [28]. After alkaline degreasing of the metal samples, it is important that a water-free surface is present due to the fact that remaining contaminants may block the active sites by reacting with the Si-OH groups. Several studies have shown that the silane film thickness on metals increases with the concentration of the silane solution [31]. The relationship between the silane film thickness and the concentration of the solution is nearly linear [24]. It should also be noticed that in comparison to the conventional chromate layers (which are about 3 μm thick), silane films are typically much thinner. For example, a film from a 5% bis-amino/VTAS mixture formed silane film has a thickness of about 250 nm, while a film from a 2% silane solution is about 50 nm in thickness. Also, the thickness of the silane film and its chemical composition show a strong dependence on the nature of the used silane. In the case of bis-silanes the obtained surface film is thicker and more homogenous compared to one from a monosilane formed film, which is thinner and more heterogeneous.

The presence of sulphur in bis-sulfur silanes plays an important role in the improvement of the corrosion properties, which assures the formation of the silane film via the Cu-rich precipitates [43].

In the case of bis-sulfur silane (PSS) the possible Reactions 2.4–2.5 are:

$$xCu + R\text{-}S\text{-}S\text{-}S\text{-}S\text{-}R \rightarrow Cu_xS + R\text{-}S\text{-}S\text{-}S\text{-}R \qquad (2.4)$$

The breakdown of the sulfur bridges of the silane molecules on the sulphide surface can occur according to:

$$nCu_xS + R\text{-}S\text{-}S\text{-}S\text{-}S\text{-}R \rightarrow n_1(Cu_xS)\text{-}S\text{-}S\text{-}R + n_2(Cu_xS)\text{-}S\text{-}S\text{-}R \qquad (2.5)$$

where, $n = n_1 + n_2$ (n_1 and n_2 – corresponding stoichiometric coefficients).

2.2.3 Experimental Procedure

The silanization of metal surfaces can be accomplished by various experimental techniques, e.g., deposition from solution, from vapor, plasma and electrodeposition [44, 45]. Gandhi and van Ooij [46] demonstrated that by electrodepositing silanes on the metal substrate surface the obtained film is uniform and the interfacial layer is strong. Nevertheless, the most commonly applied method remains the deposition from solution because it is a very fast and cheap method and the film thickness does not vary significantly when the immersion time is in the range of 30 s to 30 min. Silane films inhibit corrosion attacks primarily because they act as a hydrophobic barrier, preventing the transport of water/ions to the metal/coating interface [25]. Also, the crosslinking of the siloxane film can be accelerated by increasing the curing temperature and time. Curing of the silane film also results in a decreased film thickness and an increase of the film density.

The selection of a suitable silane for the surface modification is influenced by one or a combination of the following listed factors [47]:

- desired concentration of surface hydroxyl groups;
- type of surface hydroxyl groups;
- hydrolytic stability of the bond formed; and
- physical dimensions of the substrate or substrate features.

The silane efficacy on inorganic substrates thereby varies from poor (carbon black, graphite) < slight (barytes, gypsum, marble) < good (lead, zinc, nickel, asbestos, steel, iron, inorganic oxides, talc, tin) < excellent (copper, silicon, aluminium-silicates, aluminium (AlO(OH)), glass, quartz, silica) [47].

Another major advantage promoting the usage of silane films in addition to the anticorrosive effects is that the mechanical properties such as resistivity against mechanical damages caused by impact, stretch and wear are improved. There are also some studies concerning the improvement of the mechanical properties of silane films by loading them with nanoparticles into the films. For example, to improve the corrosion protection of AA 5005 a mixture of bis-amino silane, VTAS 5% containing 50 ppm alumina, was used. The samples were exposed for 336 h in the SST and the results obtained were similar to the case of chromate corrosion protection [34].

There are also silane types, such as the bis-amino ones, which cannot be applied for corrosion protection of aluminium as a single agent because of the formation of protonated amine groups ($R\text{-}NH_3^+$) promoting the diffusion of Cl^- into the film, and therefore accelerating the occurring corrosive attack even though they act as a good primer for a subsequent painting step.

The application of bis-silanes (bis-1, 2-[triethoxysilyl] ethane (BTSE), bis-[triethoxysilylpropyl]) tetrasulfide (bis-sulphur or PSS) leads to thicker and more homogeneous films, whereas the usage of mono-silane (mercaptopropyltrimethoxysilane (MPS)) leads to thinner and more heterogeneous films. For example, BTSE is used in different commercial silane mixtures for thin-film deposition.

Table 2.3 Silane mixture solution; pH as a function of the content of bis-aminosilane in the bis-amino silane/VTAS mixtures (at a concentration of 5%) [32].

Bis-amino: VTAS	**pH**	**Solution stability**
VTAS	2.8	Hazy in a few minutes
1:5	3.0	Hazy in a week
1:2	3.3	Hazy in a week
1:1	3.5	Hazy in a week
2:1	4.0	Clear over one year
5:1	5.8	Clear over one year
6:1	7.5	Clear over one year
Bis-amino silane	10.8	Hazy in a few minutes

The quality of the obtained conversion coatings based on silanes also depends on some physical parameters such as curing time and curing temperature, as well as on the chemistry of the silane molecules themselves. Thus, the optimization of different processing variables is the foremost objective to obtain dense and functional protective silane films. Standard mixtures of silane applied for the aluminium alloy corrosion protection treatment are: bis-sulfur:bis amino = 3:1 and bis-amino:VTAS = 5:1 [31]. The pH value and the stability of the solution of the water-based silane mixtures strongly depends on the mixture ratio as shown in Table 2.3.

The treatment of aluminium and its alloys by silanes is less complicated compared to other metals, due to the fact that the applicable pH window is wider and aluminium oxide has a strong affinity to silicates. Furthermore, aluminium is a trivalent ion and can crosslink a Si-O-Si bond by inserting into it.

The most important properties of silane films are their hydrophobicity and barrier properties.

The factors which influence the ability of an organic silane to generate a hydrophobic surface are: organic substitutes, the extent of surface coverage, the residual groups (unreacted groups from silane and from the surface) and the distribution of the silane of the silane surface [47].

The silane groups are not electrochemically active in solution or in solid state, but some silanes contain functional groups with electrochemical activity. Due to the hydrophobicity of silane films, they reduce the rate at which water and electrolyte can reach the interface where these media would induce corrosion reactions.

Zhu and van Ooij have shown that the film deposited from the bis-sulfur silane solution on aluminium actually is made up of three different regions, a crosslinked outermost bis-sulfur silane film enriched Si–O–Si bond; an extensively crosslinked or dense bis-sulfur interfacial layer dominant with Si–O–Si and Al–O–Si bonds; and the inner Al oxide layer on the alloy substrate [28, 31,48]. Nowadays, for a robust coating and for good stability of the silane solutions some alternative methods were developed, such as the inclusion of additives in the silane film (inhibitors) [31]—rare earth metal salt, cerium nitrate [49], organic inhibitors, benzotriazole [50] and tolyltriazole [51]. Silanes precursors as starting materials are commonly used in sol-gel processes. The sol-gel derived coating provides good corrosion resistance of metals due to chemical inertness, reliable adhesion to metal substrates and also easiness of application at ambient conditions [52–54].

Our experimental research mainly dealt with a comparative study concerning the performance of BTSE, bis-sulfur silane (PSS), a mixture of bis-amino silane:VTAS, and a mixture of bis-amino silane:bis-sulfur silane as a post-treatment for the protection of an anodized AA 2024 T3 alloy against corrosive attacks induced by moisture and chlorine ions, and the detailed investigation of the obtained surfaces and interfaces by different surface-sensitive analytical methods, such as electrochemical impedance spectroscopy (EIS), contact angle, FTIR, ESEM/FIB and SST.

2.3 Materials, Methods and Experimentals

2.3.1 Materials

Commercially available AA 2024-T3 unclad aluminium alloy metal sheets (wt%: Cr 0.1; Ti 0.15; Zn 0.25; Fe 0.5; Si 0.5; Mn 0.3–0.9; Mg 1.2–1.8; Cu 3.8–4.9; others 0.15; Al balance) with the dimensions of 15 cm x 8 cm x 2 cm were used in this study. Before anodizing, the samples were degreased with acetone, dried in air, and afterwards an alkaline degreasing took place followed by alkaline etching and acid pickling processes followed by a final washing step with deionized (DI) water. The anodizing bath contained 40 g L^{-1} (0.41 M) H_2SO_4 solution, 80 g L^{-1} (0.53 M) of tartaric acid ($C_4H_6O_6$) and 4 g L^{-1} Al(III) as $Al_2(SO_4)_3$. The anodizing process was carried out at a constant cell voltage of 14 V for 20 min reached after a 5 min voltage ramp and the last step, 1 min ramp from 14 V to 0 V.

The H_2SO_4, tartaric acid ($C_4H_4O_6$) and $Al_2(SO_4)_3$ were purchased from Merck and Sigma. Bis 1, 2 [triethoxysilylpropyl] ethane (BTSE), bis-[3-(trimethoxysilyl) propyl] amine (bis-amino silane), bis-[3-(triethoxysilyl) propyl] tetrasulfide (or bis-sulfur silane) and vinyltriacethoxy silane (VTAS) were provided by ABCR (Germany) and were used as received without further purification (> 95%).

Hot water sealing (HTS) was used as sealing reference method at 95°C for a period of 30 minutes, whereby mainly the formation of Bayerite and Boehmite could be observed and the sealed films formed provided a good corrosion resistance proved for 192 h SST. But the energy efficiency with respect to the energy costs is too low in order to prolong the corrosion resistance up to 840 h in the SST. Therefore different procedures were applied such as TSA anodization and conversion with silane and TSA anodization followed by a HTS sealed by conversion coatings of silane.

2.3.2 Preparation of Silane Solutions

The water-based silane mixtures were prepared in a ratio of bis-amino silane:VTAS = 5:1 (v/v) (pH=4.9) and then diluted with deionized (DI) water to have a 2% solution, and stirred until the solution became clear. Bis-sulfur silane solutions (5% v/v) were prepared by adding silane to a mixture of DI and ethanol. The ratio of bis-silane/DI water/ethanol was 5/5/95 (v/v/v) (pH=5–6). The hydrolysis time is very important, for water-based silanes this is very short (15–60 min), and in the case of solvent-based silanes the hydrolysis time is up to 3–4 days. In the case of a silane mixture, bis-amino silane:bis-sulfur silane, the solution was prepared as follows: a water/alcohol solution of bis-sulfur silane 5% was prepared by adding five parts of bis-sulfur silane into a mixture of deionized (DI) water and ethanol. The ratio of bis-sulfur silane/DI water/ethanol was 5/5/90 (v/v/v). The solution pH was 6.5. The solution was stirred for 10 min, and then aged in ambient conditions for at least 2 days to ensure that the silane molecules were sufficiently hydrolyzed. A 0.5% bis-amino silane water/alcohol solution was made in a similar way. The ratio of bis-amino silane/DI water/methanol was 0.5/4.5/95 (v/v/v). The solution pH was modified from 10.5 to 7.5 by adding acetic acid in order to enhance the solution stability. The bis-amino silane solution was used after one day. The ratio between bis-amino silane and bis-sulfur silane was 3/1 and the pH=7.5.

2.3.3 Silane Treatment

After the anodizing step, the samples were immediately rinsed in deionized water; a part was HTS sealed (95°C, 30 minutes) as reference, the other part was treated with different silanes after HTS (mixtures bis-amino silane:VTAS, BTSE, bis-sulfur silane, bis-amino silane:bis-sulfur silane), and another part only treated with conversion coatings based on bis-sulphur silane without HTS sealing, in order to be sure to elucidate the best corrosion protection on AA 2024 by the application of silanes. The procedures applied are simple (which is an advantage too): the anodized samples (with and without HTS) were dipped into the silane solutions for 5 minutes at room temperatures. The silane-treated samples were dried at 100°C for 5 minutes. Both times were optimized.

2.4 Surface Analytics

Contact angles with deionized water (DI water), Diiodmethane and Ethylene glycol have been measured with a Krüss DSA10 MK2 drop shape analyzer.

The *SEM* investigations were performed using a Philips XL30 FEG ESEM environmental scanning electron microscope for characterization of nonconductive specimens, equipped with *EDX* for element analysis.

A FEI Quanta 200 3D, equipped with *FIB* (focused ion beam), was used for preparing (ion milling) and characterization of cross-sections through the oxide layer, determination of layer thickness, surface topography and structure, elementary composition and depth profiling (EDX), phase analysis, identification of segregations and inclusions.

The apparatus for the *Salt Spray Test* consists of a closed testing chamber, where a salted solution containing 50 g/L sodium chloride is atomized (converted to an aerosol) by forcing the solution through a fine nozzle. Thus a corrosive environment of sodium-chloride-containing fog is achieved so that parts exposed have to face a severe corrosive atmosphere. Tests have been previously performed in a chamber from Weiss Technik Type SC 450 according to DIN EN ISO 9227.

For *IR studies* an Equinox-55 (Bruker) equipped with a Golden Gate ATR-unit (ATR: attenuated total reflection where spectral information is gained by total reflection of the IR beam on the surface of the sample) was used. The typical penetration depth of the FTIR beam into the surface in this experimental set-up is 0.5 to 2 μm. The number of scans was 10 and baseline correction was also made.

Electrochemical *impedance spectroscopy measurements* (*EIS*) were realized to monitor the corrosion performance of the silane-treated AA 2024-T3 anodized with TSA as a function of immersion time in a 1% NaCl solution. The EIS measurements were carried out using a Zahner IM6 electrochemical workstation. The measured frequency range was from 100 kHz to 10 mHz, with an AC excitation amplitude of 10 mV. A standard calomel electrode (SCE) was used as the reference electrode and coupled with a glassy carbon counter electrode. The surface area exposed to the electrolyte was 1.00 cm^2. All the measurements were performed after 30 min of open circuit potential (OCP) measurements and the measurement potential was the OCP at that time.

2.5 Results and Discussion

2.5.1 Contact Angle

The contact angle measurements represent a simple method to define the relative degree of interaction of a liquid with a solid surface. Thus, if the contact angle of water is less than 30°, the surface is considered to be hydrophilic, and in the case where the value of contact angle is greater than 90° it is considered to be hydrophobic [42]. Thus, the contact angle of water on a substrate is a good indicator of the relative hydrophobicity. The contact angle value depends on the chemical composition (polar or non-polar surface) but also the roughness of the surface has a great influence. Thus, polar (hydrophilic) surfaces show a better wettability with polar liquids, such as water or water-based (or polar solvent-based) primers. The results of the contact angle measurements from the AA 2024 samples after being treated with different silanes are given in Table 2.4 and their wettability is presented in Figure 2.3.

The factors that have an important role in the ability of organosilanes to generate a hydrophobic surface are its organic functional groups, the extent of the surface coverage, possible residual unreacted groups (from the silane and/or the substrate surface), and the homogeneity distribution of the silane on the surface. The silane surface treatment depends on the number of hydroxyl groups on the treated metal surface and on their accessibility for bonding. A simple explanation is that the organosilanes react to form a monolayer and, therefore, in the case where all hydroxyl groups are covered by the silanes (which mean that the surface is protected) a hydrophobic surface is obtained.

In the case of the mixture of bis-amino silane:bis-sulfur silane, the wettability and the hydrophobicity could accurately be controlled, resulting in a film that wets very well and which is also sufficiently hydrophobic to provide excellent corrosion protection.

A comparison of the contact angle results obtained leads to the conclusion that the values increase as follows: HTS < BTSE < bis-amino

Table 2.4 Contact angle and surface energy values of DI-H_2O-deionized water, DIM-diiodomethane and EG-ethylene glycol.

Silane	Contact angle						Surface energy					
	DI-H_2O	±	DIM	±	EG	±	polar	±	Disper-sive	±	total	±
HTS	32.9	1.4	26.3	2.9	19.4	1.9	24.9	0.3	36.5	0.5	61.4	0.8
bis-amino silane: VTAS	71.0	2.2	44.9	1.2	45.1	1.2	7.6	0.1	34.0	0.1	41.7	0.2
BTSE	63.9	0.9	38.3	1.2	39.1	2.1	10.2	0.1	35.8	0.2	46.1	0.3
bis-sulfur silane	87.1	2.8	42.3	2.1	56.2	2.2	1.7	0.1	36.4	0.4	38.1	0.5
bis-amino silane: bis-sulfur silane	99.9	1.1	33.6	1.9	51.0	2.8	0	0	44.9	0.4	44.9	0.4

Figure 2.3 Anodized AA 2024 surface wetting for different silane treatment: (a) bis-amino silane:VTAS; (b) bis-sulfur silane; (c) bis-amino silane:bis-sulfur silane.

silane:VTAS < bis-sulfur silane < bis-amino silane:bis-sulfur silane. Therefore it can be stated that the chemical structure of the silane has a strong influence on the hydrophobicity of the layer, and for bis-sulfur silane and bis-amino silane:bis-sulfur silane, a good hydrophobic layer was obtained, increasing the corrosion protection properties (demonstrated also by SST); but from an economical and environmental point of view it is better to avoid the use of alcohol-based silanes or the preparation of the solution in a higher % of water. In addition, the influence of sealing time was studied (2, 5 and 10 minutes), and it could be observed that the sealing time has nearly no influence on the contact angle value. Comparing the results obtained by us for the contact angle in the case of water-based silane (bis-(trimethoxysilylpropyl) amine:vinyltriacetoxysilane = 5:1) (the lowest in the range 61.5 ± 5.8 and the highest 75.7 ± 2.8) with those obtained for contact angle in the case of alcohol-based silane (bis-(3-(triethoxysilyl) ethane (61.0±2.5 and 69.3±1.4), bis-[3- (triethoxysilyl) propyl] tetrasulfide) (87.1±2.9 and 103.8±1.6), one can conclude that chemical structure of silane has a strong influence on the hydrophobicity character of layers. In the case of the bis-sulfur silane, we observed a good hydrophobicity of the layers, increasing the corrosion protection properties (demonstrated also by SST). At the interface, a high density of Al-O-Si bonds are present, which are formed by the reaction of hydrophilic Al-OH groups, making the metal surface less favorable for water adsorption and eliminating the tendency for aqueous corrosion. The hydrophobicity of the film surface is destroyed in the case of exposure to a considerable amount of water, due to the fact that Al-O-Si hydrolizes back to re-form hydrophilic Al-OH and Si-OH groups. A surface film with a contact angle value in the order of magnitude of 90° remains at that order or will even decrease also in the case of immersion in water or exposure to salt-aerosols for a few weeks, indicating that Si-O-Si groups are not very stable and it is possible that they hydrolize back to much more hydrophilic silanol groups (2.6) according to:

$$\text{Si–O–Si} + H_2O \leftrightarrow \text{Si–OH} + \text{HO–Si} \tag{2.6}$$

There are different possibilities for improving the water resistance of the film, the most common being fully crosslinking the silane film and improving the intrinsic film hydrophobicity by using silanes with hydrophobic organic-inorganic groups, i.e., a sulphur chain -S4- (in bis-sulfur silane). Bis-[3-(triethoxysilyl)-propyl]-tetrasulfide used for conversion conferred hydrophobic properties on the surface, resulting in a protective action, which was proved also by the values of the contact angles and by SST (no pits for 192 h continuous salt spray exposure).

2.5.2 Characterization with SEM/EDX – FIB

The addition of tartaric acid to the anodization bath improved the layer structure. The layer structure consisted of an initial inner uniform barrier type oxide layer (not visible in Figure 2.5), a relatively thick middle layer (A) with a fine-structured morphology with enhanced corrosion resistance, and an outer oxide layer (B) with a coarse morphology increasing the adhesion of organic primers (Figure 2.5). The presence of Cu^{2+} in the anodic layer grown on aluminium copper alloys not only alters the ordered morphology of the anodic layer, but also changes its composition and dielectric properties. As stated in the introduction, Cu present in the solid solution oxidizes and leads to the generation of oxygen bubbles during film growth, including film disruption. A lateral porosity is caused by the changing direction of pore propagation. This contributes to the three-dimensional structure of the pore morphology.

The TSA anodic layers were measured by FIB and the layer thicknesses were between 2.5 μm (Figures 2.7 and 2.8), 2.6 μm (Figure 2.6) and 3.6 μm (Figure 2.9), appropriate for aerospace industry requirements. The TSA anodic layer could not protect the surface completely; the aluminium oxide layer was damaged by the action of Cl^-, assuring a gradual penetration of Cl^- through the barrier oxide layer. The silane conversion layers were very thin and the TSA pores' filling was dependent on the silane nature. For BTSE and bis-sulfur silane, the pores were not filled (Figures 2.4 and 2.5). We proposed a conversion coating base of silane mixture after HTS (the pores after HTS are filled) (Figures 2.10 and 2.11) for more improvement in corrosion resistance; e compared the results of the conversion coating with silane mixture with bis-amino silane:VTAS (Figure 2.8) and bis-amino silane with bis-sulfur silane (Figure 2.9).

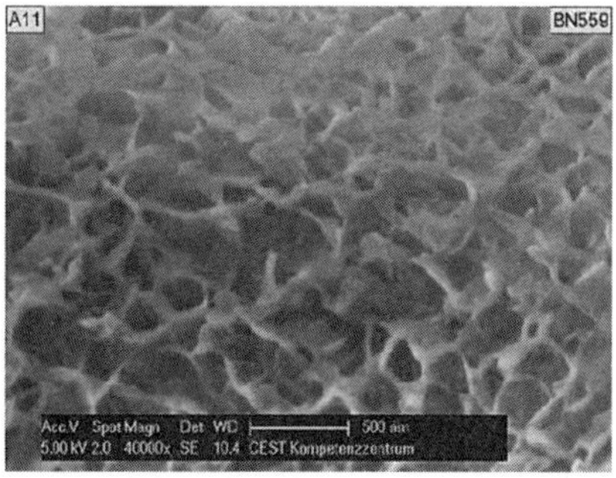

Figure 2.4 Surface with HTS sealing. Dense layer of acicular "petal"-shaped crystals. The pores are closed on the surface.

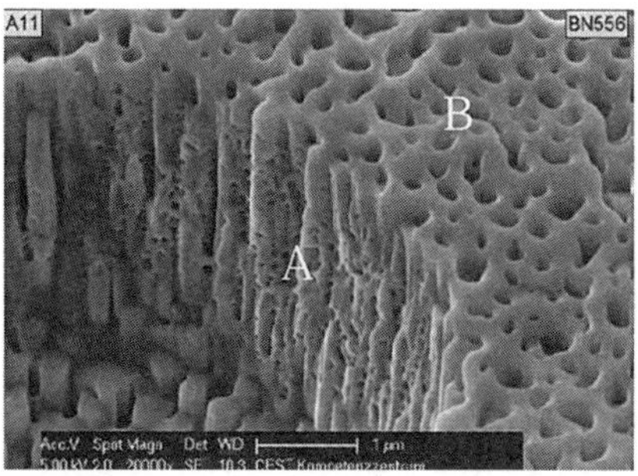

Figure 2.5 Surface etched with Ga ions. With the sealing removed, well distributed pores with high tortuosity became visible. Pore diameter ~170 nm (outer) and 15–44 nm (inner).

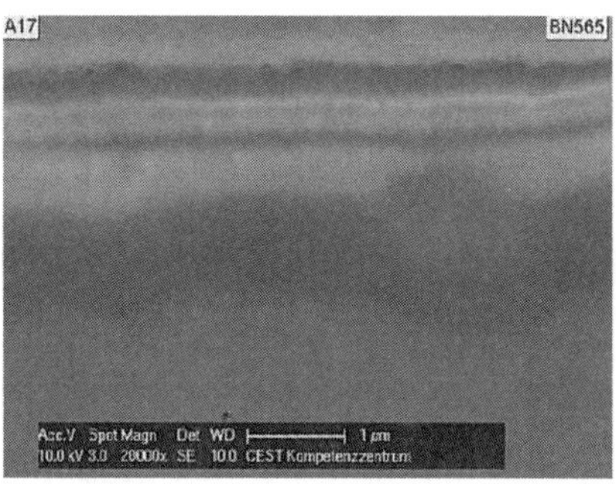

Figure 2.6 TSA with HTS. SEM imaging of the FIB cross-section confirmed the homogenous morphology (no pores) of the anodic layers. The white stripes within them are caused by the charging during SEM investigation. The intensity modulation of these stripes could be the result of different conductivity (composition) of the layers.

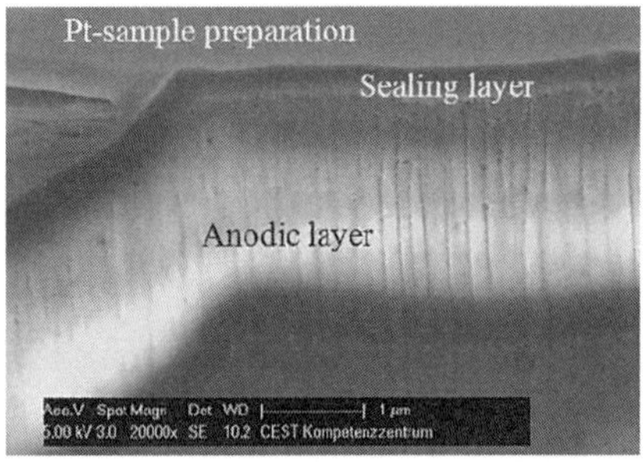

Figure 2.7 SEM of the FIB cross-section of anodized AA-2024 without HTS and treated with bis-sulfur silane 5% (20000x), anodic layer ≈ 2.6 μm. No filling of the pores. Conversion layer is very thin (≈ 360 nm).

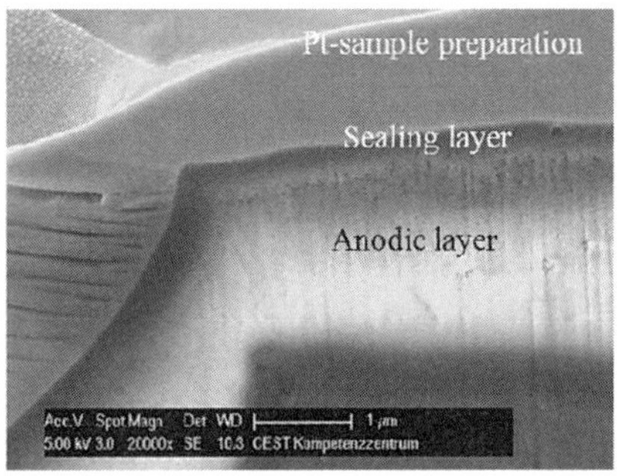

Figure 2.8 SEM of the FIB cross-section for anodized AA-2024 without HTS and treated with BTSE 5% (20000x), anodic layer ≈ 2.5μm, sealing layer is very thin. No filling of the pores. Conversion layer is very thin. (≈ 260 nm).

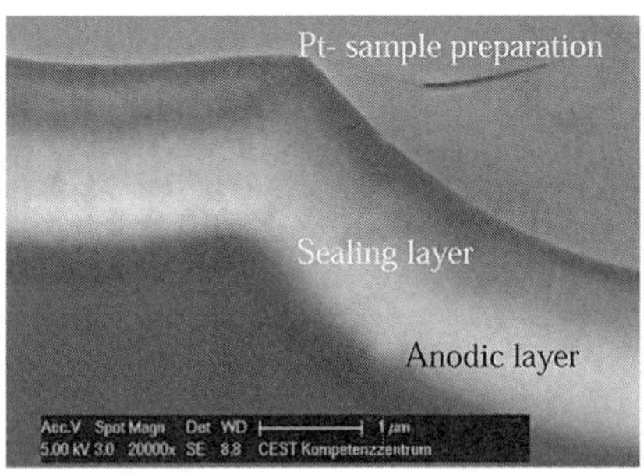

Figure 2.9 SEM of the FIB cross-section for anodized AA-2024 + HTS and treated with a mixture of bis-amino silane and VTAS (20000x), anodic layer ≈ 2.5 μm, sealing layer is very thin. No complete filling of the pores. Conversion layer is very thin.

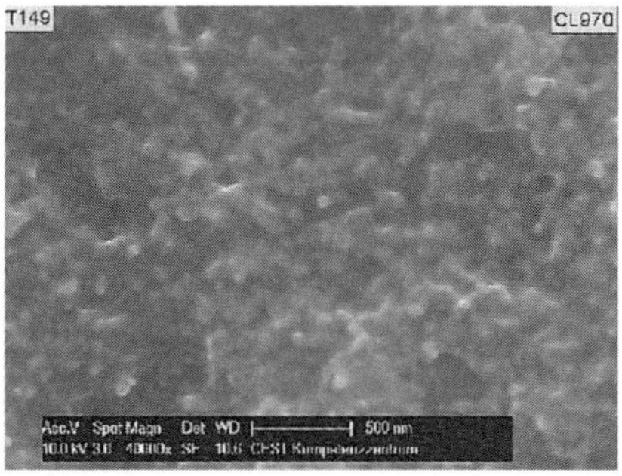

Figure 2.10 SEM of surface for anodized AA-2024 with HTS and treated with a mixture of bis-amino silane and bis-sulfur. The surface is completely covered.

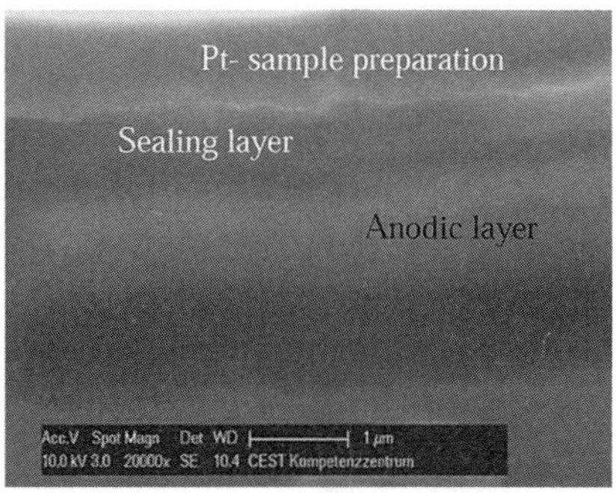

Figure 2.11 SEM of the FIB cross-section for anodized AA-2024 + HTS and treated with a mixture of bis-amino silane and bis-sulfur (20000x), anodic layer ≈ 3.6 μm, sealing layer is very thin. Complete filling of pores. Conversion layer is very thin.

The obtained EDX analysis results are shown in Table 2.5. The chemical composition (wt% and at%) of TSA-anodized AA2024 specimen presents higher oxygen contents due to alumina formation, while sulphur was incorporated from the bath into the anodic layer (which contained H_2SO_4 and $Al_2(SO_4)_3$). No copper could be detected here. The oxygen percentage content is higher for both sealed specimens due to hydration of the anodic layer during the sealing process.

The influence of the treatment time and the bis-sulfur silane concentration on surface film composition was systematically studied (Figure 2.12). It could be observed that the times of sealing as well as the silane concentration in the bath have a great influence on the results obtained.

Table 2.5 Composition of AA2024 sealed with silanes from EDX point analysis (wt% and at%).

Element	BTSE		Bis-amino:VTAS		Bis-sulfur silane		Bis-amino: bis-sulfur silane	
	wt%	at%	wt%	at%	Wt%	At%	Wt%	At%
O	47.1	60.5	40.7	53.9	40.3	54.2	38.8	52.5
Mg	0.8	0.7	0.6	0.5	0.7	0.6	0.5	0.4
Al	45.3	34.3	50.6	39.7	41.5	33.0	43.6	35.0
Si	4.2	2.8	5.8	4.4	5.0	3.8	5.4	4.2
S	2.6	1.7	2.3	1.5	12.5	8.4	11.7	7.9
Total	100.00	100.00	100.00	100.00	100.00	100.00	100.00	100.00

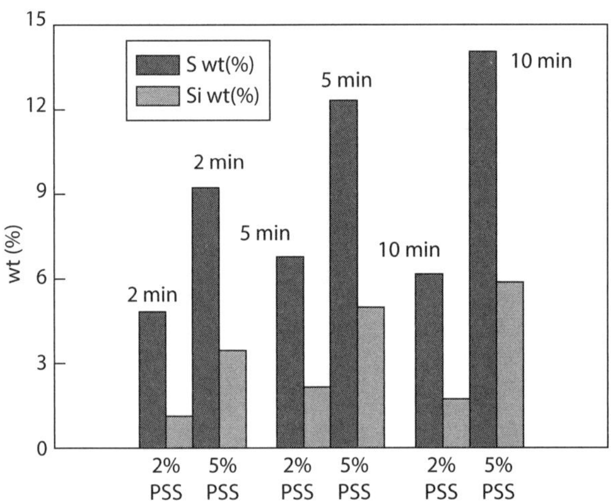

Figure 2.12 Variation of Si (wt%) and S (wt%) in the case of surface film obtained for conversion of AA 2024 anodized sample with bis-sulfur silane 2% and 5% at different times: 2 min, 5 min, 10 min.

2.5.3 Electrochemical Impedance Spectroscopy (EIS) Tests

The electrochemical behaviour of the aluminium oxide layer can be represented by the equivalent circuits shown in Figure 2.13. These circuits consist of a combination of resistance and constant phase elements (CPE): R_E: electrolyte resistance, R_P: porous layer resistance, Q_P: CPE parameters

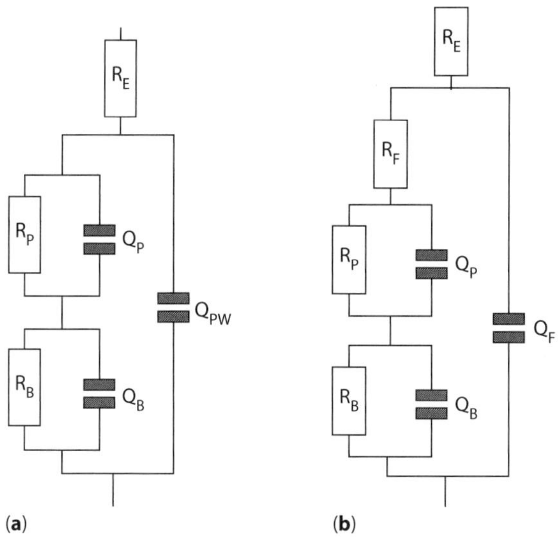

Figure 2.13 Equivalent circuit used to model the electrochemical behavior of (a) sealed anodic film with hot water and (b) sealed with organic additives (hydrophobic film). Abbrevations – indices capitals: B–barrier; P–porous layer; F–coating:conversion; E–electrolyte; R–ohmic resistance; Q–constant phase element.

associated with the porous layer, R_B: barrier layer resistance, Q_B: CPE parameters associated with the barrier layer, and CPW: capacitance associated with the pore wall; R_f: electrolyte resistance in the pores of the organic film/Q_f: CPE parameters associated with the organic film.

The clear differences observed between sealed and unsealed EIS spectra can be transformed into reliable information about the sealing quality and the stability of anodic films, according to previous studies which have proposed different equivalent circuits to simulate the properties of each layer (barrier and porous layers) in the anodic film.

Aluminium respective aluminium oxide layers are both high impedance systems. Application of an additional protecting layer increases the impedance further. Due to the organic nature of the coatings, high values of the resistance and the low values of the constant phase elements show good protective properties against corrosion (Table 2.6). The similar structures make it possible to use the same equivalent circuit for all coated samples, but the distribution of the values for the time constants is quite different. Therefore, although it seems that the Nyquist diagrams have miscellaneous shapes, that is not the case (see Figure 2.14). In general, it can be said that they are useful for comparing the corrosion resistance.

Table 2.6 Results of the impedance measurements for anodized AA2024 treated with different silanes after hot water sealing.

Treatment of sample	R_B[MΩcm²]	Q_B[F/cm²]	$Q_b$exp	R_p[kΩ/cm²]	Q_p[F/cm²]	$Q_p$exp	R_F[kΩ/cm²]	Q_F[F/cm²]	$Q_F$exp
TSA + HTS	42.23	1.91E-07	0.72	761	8.151E-09	0.01	–	–	–
TSA + HTS + bis-amino: VTAS	167.8	7.894E-07	0.97	3196	2.529E-09	0.68	1433	3.559E-09	1.00
TSA + bis sulfur silane	280.6	6.35E-08	0.8	37970	9.6E-10	0.9	17230	2.18E-09	0.90
TSA + HTS + bis amino: bis-sulfur	344.1	1.987E-07	0.85	7411	3.402E-09	0.48	42040	6.851E-09	0.88

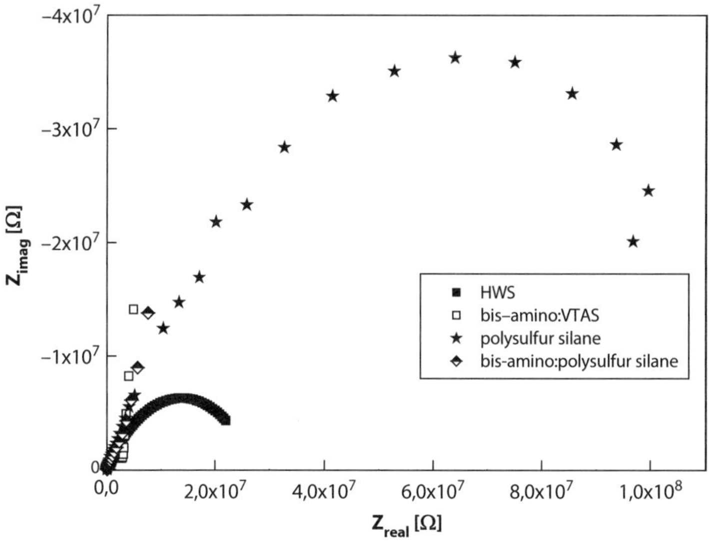

Figure 2.14 Representative Nyquist plots for surface films containing different silanes.

2.5.4 Salt Spray Test

The salt spray test (SST) is one of the most applied corrosion tests in the aircraft industry. It is used to compare the performances of different coated substrates, and the specimens are considered to fail if more than 2 pits per dm^2 are displayed.

Sample surfaces of 1 dm^2 were exposed to a continuous SST for 840 h and the obtained AA 2024 panels are presented in Figure 2.15.

The sample which was treated with bis-amino silane:VTAS after HTS at 95°C and 30 min showed some signs of corrosion after 72 h, while the sample treated with bis-sulfur silane after HTS at 95°C and 30 min had only one pit that appeared after 840 h of continuous exposure in SST. This is due to the fact that bis-amino:VTAS film is hydrofilic and contains amino groups (positively charged) which attract water and electrolytes to the siloxane/metal interface. In the case of sample treated with bis-amino silane:bis-sulfur silane, no signs of corrosion were observed after 840 h of exposure in SST. The results obtained after 840 h of continuous exposure from the sealed samples with different silane mixtures are in concordance with the results obtained from the same samples by EIS.

Alcohol-based silanes offer a higher corrosion resistance compared to the water-based silane system because the higher alcohol content removes

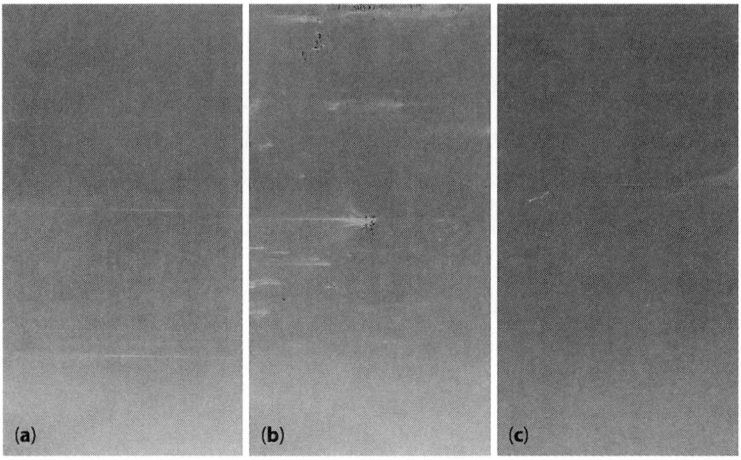

Figure 2.15 AA 2024 panels after 840 h of salt spray test: (a) TSA + HTS, 95°C, 30 min + bis-sulfur silane 5%; (b) TSA + HTS 95°C, 30 min + silane mixtures: bis-amino silane:VTAS; (c) TSA + HTS, 95°C, 30 min + amino silane:bis-sulfur silane.

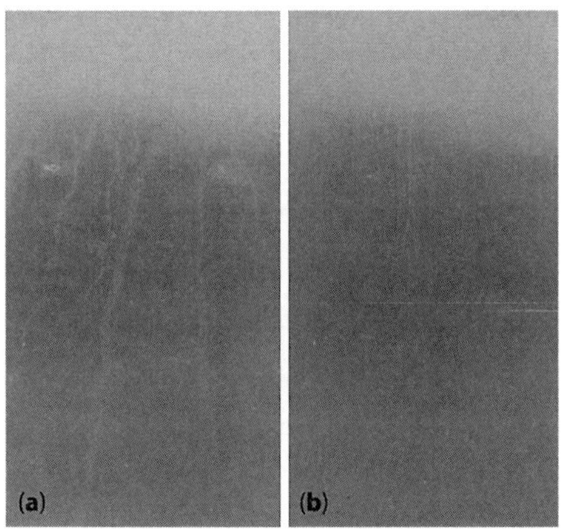

Figure 2.16 192 h SST: (a) AA 2024, TSA anodized + with bis-sulfur silane 5%, 5 min and drying 5 min at 100°C and (b) AA 2024, TSA anodized + HTS.

more water from the film upon drying and the silanol groups can more easily react to form a more crosslinked and denser film.

In the case of AA 2024 anodized samples treated 5 minutes with poly-sulfur silane 5% at RT followed by 5 minutes, curing at 100°C was observed. It was also noticed that up to 192 h in SST (Figure 2.16) no sign of corrosion appeared. The results at 192 h are comparable with TSA + HTS.

2.5.5 FTIR Spectroscopy

Fourier transformed infrared spectroscopy (FTIR) (Table 2.7) was used in order to observe the chemical structure and possibly occurring transformations in the silane films on the surface. In the case of samples treated with poly-sulfur silane and BTSE we registered the IR spectra (Figures 2.17 and 2.18). The bands obtained at 1100 cm^{-1} due to Si-O-Si became more intensive when the conversion time increased (especially for 5% poly-sulfur silane). The band for Si-O-Si has shifted slightly to higher frequency, increasing time.

One can observe that the sealing time is an important parameter; by increasing the time of contact the peak intensity corresponding to different substituents increases too. Furthermore, the concentration of the silane in the applied solution influences the film morphology (with respect

Table 2.7 FTIR absorption peaks corresponding to the silanes' film [31].

Wave number (cm^{-1})	Associated group
880	H- bonded Si-OH
1100	-Si-O-Si- assymmetric stretching vibration
1250	$-CH_2-$ wagging in $-CH_2-S$ (for PSS film)
1400	$-O-CH_2-CH_3$

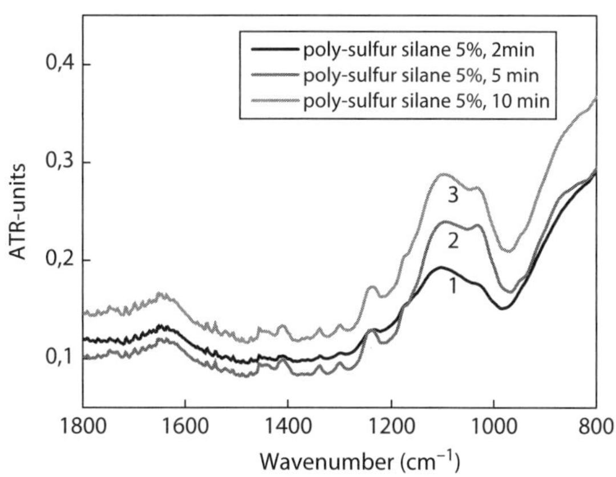

Figure 2.17 FTIR spectra of bis-[3-(triethoxysilyl) propyl] tetrasulfide film on TSA anodized AA 2024. Conversion at room temperature and drying at 100°C, 5 min. **1** – 2 min conversion; **2** – 5 min conversion; **3** – 10 min conversion.

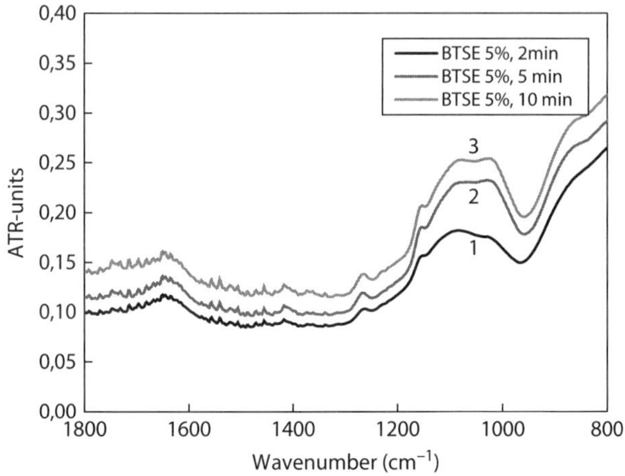

Figure 2.18 FTIR spectra of bis-3-(triethoxysilyl) ethane film on TSA anodized AA 2024. **1** – BTSE 5%, 2 min conversion, drying at 100°C, 5 min; **2** – BTSE 5%, 5 min conversion, drying at 100°C, 5 min; **3** – PSS 5%, 10 min conversion, drying at 100°C, 5 min.

to the layer density). In the case of conversion with BTSE and PSS 2%, it can be noticed that the treatment time has nearly no influence on the -Si-O-Si- peak intensity when compared with BTSE and PSS 5%, where from the spectra the peak intensity has an increase proportional with the conversion time.

2.6 Conclusions

It has been demonstrated that the silane mixture of bis-amino silane/bis-sulfur silane shows superior corrosion resistance properties when applied on the aluminium surface compared to a mixture of bis-amino silane/VTAS or only BTSE and bis-sulfur silane, respectively. A small volume of bis-amino silanes is sufficient to generate hydrophilic properties on the surface and to support the formation of a homogenous surface film on the anodized AA 2024 sample. On the other hand, a large portion of bis-sulfur silane enhances the hydrophobicity of the mixture film, being the basis for good protective performance properties of the mixture when applied on the AA2024 surface. The mixture of bis-sulfur silane with bis-amino silane in a ratio of 3:1 showed the highest corrosion resistance on AA 2024-T3. The improvement for this mixture is achieved by selectively overcoming the major drawbacks of the two individual silanes. The bis-sulfur/bis-amino mixture enhances the corrosion resistance of aluminium alloys due to the fact that a small portion of bis-amino silane makes the mixture solution hydrophilic enough to increase the wettability of the aluminium oxide, which facilitates the formation of a homogeneous film. A large portion of bis-sulfur silane enhances the hydrophobicity of the mixture film, which is the basis for the good protective performance of the mixture.

The bis-sulfur silane film is neutral and it is not expected to promote the penetration depth of Cl^-, but the mixture bis-sulfur:bis-amino (3:1) contains some $-NH_3^+$ group, but not too many, due to the fact that this mixture ensures very good corrosion resistance of AA2024. From the contact angle values observed it could be concluded that the surface film is hydrophobic and therefore somehow blocks the chlorine transport.

From the results obtained from the SST (840 h) tests it could be observed that the corrosion protection of AA 2024 TSA anodized, HTS sealed and conversion coatings with different silanes varied in this way:

Bis-sulfur silane:bis-amino silane > bis-sulfur silane > bis-amino silane:VTAS

Taking into account the obtained results, pretreatment using silanes represents a promising alternative to the Cr(VI)-based solution for AA 2024.

Acknowledgements

Financial support from the Seventh Framework Programme (FP7) (ECO-design) is gratefully acknowledged. The authors would like to acknowledge Dr. Jaroslaw Wosik (CEST GmbH) for SEM characterization.

References

1. G.E. Thompson, *Thin Solid Films*, Vol. 297 p. 192, 1997.
2. V.M. Lanker, *Metallurgy of Aluminum Alloys*, William Clowes and Sons, London, 1967.
3. G.E. Thompson, L. Zhang, C.J.E. Smith, P. Skeldon, *Corrosion*, Vol. 55, p. 1052, 1999.
4. L. Domingues, J.C.S. Fernandes, M. Da Cunha Belo, M.G.S. Ferreira, L. Guerra-Rosa, *Corros. Sci.*, Vol. 45, p.149, 2003.
5. G.W. Critchlow, K.A. Yendall, D. Bahrani, A. Quinn, F. Andrews, *Int. J. Adhes. Adhes.*, Vol. 26, p. 419, 2006.
6. G. Boisier, N. Pebere, C. Druez, M. Villatte, S. Suel, *J. Electrochem Soc.*, Vol. 155, p. C521, 2008

7. US patent 2002/0157961.
8. A. Dattilo, S. Tamiro, C. Romano, in: EP Patent 1233084, Alenia Aeronautica SPA, 2001
9. L. Iglesias-Rubianes, S.J. García-Vergara, P. Skeldon, G.E. Thompson, J. Ferguson, M. Beneke, *Electrochim. Acta*, Vol. 52, p. 7148, 2007.
10. M.A. Arenas, A. Conde, J.J. de Damborenea, *Electrochim. Acta*, Vol. 55, p. 8704, 2010.
11. A. Cabral, R.G. Duarte, M.F. Montemor, M.L. Zheludkevich, M.G.S. Ferreira, *Corros. Sci.* Vol. 47, p. 869, 2005.
12. W.J. van Ooij, D. Zhu, G.P. Sundararajan, S.K. Jayaseelan, Y. Fu, N. Teredesai, Surf Eng., Vol. 16, p. 386, 2000.
13. A. Franquet, H. Terryn, J. Vereecken, *Appl. Surf. Sci.* Vol. 211, p. 259, 2003.
14. H. Watson, P.J. Mikkola, J.G. Matisons, J.B. Rosenholm, *Colloids Surf. Physicochem. Eng. Aspects*, Vol. 161, p. 183, 2000.
15. Y. Goueffon, C. Mabru, M. Labarrere, L. Arurault, C. Tonon, P. Guigue, *Surf. Coat. Technol.*, Vol. 205, p. 2643, 2010.
16. I. De Graeve, E. Tourwe, M. Biesemans, R. Willem, H. Terryn, *Prog. Org. Coat.*, Vol. 63, p. 38, 2008.
17. R.W. Hinton, *Met. Finish.*,Vol. 89, p. 55, 1991.
18. A.N. Khramov, V.N. Balbyshev, N.N. Voevodin, M.S. Donley, *Prog. Org. Coat.*, Vol. 47, p. 207, 2003.
19. V. Palanivel, Y. Huang, W.J. van Ooij, *Prog. Org. Coat.*, Vol. 53, p. 153, 2005.
20. M.A. Petrunin, A.P. Nazarov, Y.N. Mikhailovski, *J. Electrochem. Soc.*, Vol. 143, p. 251, 1996.
21. A.M. Beccaria, L. Chiarttini, *Corros. Sci.*, Vol. 41, p. 885, 1999.
22. W.J. van Ooij, D. Zhu, *Corrosion*, Vol. 57, p. 413, 2001.
23. R. Tremont, H. De Jesus-Cardona, J. Garcia-Orozco, R.J. Castro, C.R. Cabrera, *J. Appl. Electrochem.*, Vol. 30, p. 737, 2000.
24. W. Trabelsi, L. Dhouibi, E. Triki, M.G.S. Ferreira, M.F. Montemor, *Surf. Coat. Technol.*, Vol. 192, p. 284, 2005.
25. W.J. van Ooij, A. Sabata, *Surf. Interface Anal.*, Vol. 20, p. 475, 1993.
26. F. Zucchi, G. Trabanelli, V. Grassi, A. Frignani, Proceedings of the Eurocorr 01, Associazione Italiana di Metallurgia, 2001.
27. E.P. Plueddemann, *Silane Coupling Agents*, New York, Plenum Press, 1991.
28. D. Zhu, W.J. van Ooij, *Electrochim. Acta*, Vol. 49, p. 1113, 2004.
29. W.J. van Ooij, T.F. Child, CHEMTECH, Vol. 28, p. 26, 1998.
30. D. Zhu, W.J. van Ooij, Corros. Sci., Vol. 45, p. 2177, 2003.
31. W.J. van Ooij, D. Zhu, M. Stacy, A. Seth, T. Mugada, J. Gandhi, P. Puomi, Tsinghua Science and Technology, Vol. 10, p. 639, 2005.
32. D. Zhu, W.J. van Ooij, *Prog. Org. Coat.*, Vol. 49, p. 42, 2004.
33. D. Zhu, Ph.D. dissertation, Department of Materials Science and Engineering, University of Cincinnati, 2002.
34. V. Palanivel, D. Zhu, W.J. van Ooij, Prog. Org. Coat., Vol. 47, p. 384, 2003.
35. G. Tesoro, Y. Wu, *J. Adhesion Sci. Technol.*, Vol. 5, p. 771, 1991.

36. F.D. Blum, W. Meesiri, H.J. Kang, J.E. Gambogi, *J. Adhesion Sci. Technol.*, Vol. 5, p. 479, 1991.
37. F.D. Osterholtz, D.R. Pohl, in: K.L. Mittal, ed., Silanes and Other Coupling Agents, VSP, Utrecht, p. 119, 1992.
38. B. Arkles, J.R. Steinmetz, J. Zazyczny, P. Mehta, J. Adhesion Sci. Technol., Vol. 6, p. 193, 1992.
39. K.J. McNeil, A.J. Dicaprio, O.A. Walsh, R.J. Pratt, *J. Amer. Chem. Soc.*, Vol. 102, p. 1859, 1980.
40. E.R. Pohl, in: *Proc. 38th Ann. Tech. Conf. Reinforced Plastics/Composites Inst.*, Section 4-B, 1983.
41. A. Franquet, H. Terryn, J. Vereecken, *Surf. Interf. Anal.*, Vol. 36, p. 681, 2004.
42. A. Sabata, B.A. Knueppel, W.J. van Ooij, *J. Test Eval.*, Vol. 23, p. 120, 1995.
43. A.M. Cabral, R.G. Duarte, M.F. Montemor, M.G.S. Ferreira, *Prog. Org. Coat.*, Vol. 54, p. 322, 2005.
44. H. Woo, P.J. Reucroft, R.J.J. Jacob, *Adhesion Sci. Technol.*, Vol. 7, p. 681, 1993.
45. B.C. Zhang, Adsorption of silane films on iron surfaces, Ph.D. dissertation, University of Cincinnati, Cincinnati, USA, 1997.
46. J.S. Gandhi, W.J. van Ooij, Pulsed electrodeposition of organofunctional silanes for improved corrosion protection of aluminium alloys, in: *ASM 2004 Intl. Surface Eng. Congress*, Orlando, FL, Aug. 2–4, 2004.
47. B. Arkles, Gelest Inc., Morrisville, PA, Reprinted with permission from: Paint & Coatings Industry, Oct. 2006.
48. D. Zhu, W J. van Ooij, J. *Adhesion Sci. Technol.*, Vol. 16, p. 1235, 2002.
49. K. Aramaki, *Corros. Sci.*, Vol. 43, p. 2201, 2001.
50. R. Rungta, *Metal Finishing*, Vol. 95, p. 114, 1997.
51. A.N. Onal, A.A. Aksut, *Anti-Corr. Met. Mat.*, Vol. 47, p. 339, 2000.
52. T.L. Metroke, J.S. Gandhi, A. Apblett, *Prog. Org. Coat.*, Vol. 50, p. 231, 2004.
53. J.D. Mackenzie, *J. Sol–Gel Sci. Technol.*, Vol. 2, p. 8, 1994.
54. G. Gupta, S.S. Pathak, A.S. Khanna, *Prog. Org. Coat.*, Vol. 74, p. 106, 2012.

3

New Generation Material for the Removal of Arsenic from Water

Dinesh Kumar* **and Vaishali Tomar**

Department of Chemistry, Banasthali Vidyapith, Rajasthan, India

Abstract

Arsenic (As) is a noxious constituent whose role in water pollution has received extensive attention in the scientific community. It is mobilized into surface water during industrial waste discharges, agricultural use of arsenical pesticides, herbicides, fertilizers, discharges from coal-fired thermal power plants, petroleum refining and ceramic industries, etc. In the last 20 years, naturally occurring arsenic has been found in the natural waters of the USA, Argentina, Taiwan, China, Hungary, Vietnam, Chile, Bangladesh and India. This is of concern because a high level of arsenic (>10 μg/L) may increase the risk for many health problems, some of which are discussed herein. However, numerous studies have established that arsenic removal can be achieved by a range of techniques, specifically solvent extraction, chemical precipitation as synthetic coagulants, iron co-precipitation, ion exchange and reverse osmosis, etc. In this chapter many of the adsorbents used for arsenic removal are reviewed along with several inexpensive adsorbents for advanced techniques including treated slags, carbons developed from agricultural waste, biosorbents, zeolites and some commercial adsorbents. Also included are the resins, gels, silica, and treated silica tested for arsenic removal, which have been found to be superior. Immobilized biomass adsorbents, which offer wonderful performances, are also discussed. By surveying the extensive list of adsorbents compiled from literature presented here it is evident that low-cost sorbents have demonstrated outstanding arsenic removal capabilities. In particular, nanomaterial-based adsorbents might be promising adsorbents for environmental and purification purposes.

Keywords: Removal, arsenic, adsorbents

**Corresponding author*: dschoudhary2002@yahoo.com

Ashutosh Tiwari and Mikael Syväjärvi (eds.) Advanced Materials for Agriculture, Food, and Environmental Safety, (61–86) 2014 © Scrivener Publishing LLC

3.1 Introduction

Arsenic is omnipresent and its contamination of groundwater is a worldwide issue. Groundwater contamination due to arsenic in the Bengal delta region of Bangladesh and West Bengal, India, is considered one of the biggest natural calamities of the 21st century. Indication of arsenic poisoning was observed in the mid-1980s in Bangladesh and West Bengal [1, 2]. Similarly, in different areas of Pakistan arsenic concentration was found to be high in surface and groundwater. For instance, in the Sindh province of Pakistan, 16–36% of the population have been exposed to arsenic contaminated water with over 10–50 μg/L. This high level of arsenic (>10 μg/L) may cause skin lesions and rhagades, and also damage mucous membranes, and the digestive, respiratory, circulatory and nervous systems. Furthermore, it is associated with skin, liver and lung cancers.

3.1.1 Properties of Arsenic [3–6]

Before going through a detailed study on the removal of arsenic from water, we will elucidate some properties about this toxic atom. Arsenic is a chemical element with symbol As and atomic number 33, and is ranked 20th in natural abundance, comprising about 0.00005% of the earth's crust, 14th in the seawater, and 12th in the human body. This metalloid is a silver-grey brittle crystalline solid with atomic weight 74.9 g/mol, specific gravity 5.73 g/L, melting point 817°C (at 28 atm), boiling point 613°C, and vapor pressure 1 mm Hg at 372°C; the arsenic chemistry in an aquatic system is more complicated because the element can be stable in four oxidation states (−3, 0, +3 and +5) under different redox conditions. Soil corrosion and leakage contribute to 612×108 and 2380×108 g/year of arsenic, respectively, in dissolved and suspended forms in the oceans. Arsenic(III) is a hard acid and preferentially complexes with oxides and nitrogen. Conversely, arsenic(V) behaves like a soft acid, forming complexes with sulphides [7]. Inorganic forms of arsenic most often exist in water supplies. Arsenic is uniquely sensitive to mobilization (pH 6.5–8.5) and under both oxidizing and reducing conditions among heavy metalloids [8]. Two forms are common in natural waters: arsenite and arsenate, referred to as arsenic(III) and arsenic(V). Pentavalent (+5) or arsenate species are, and H_2, while trivalent (+3) arsenites include $As(OH)_3$, AsO_2O and . Pentavalent species predominate and are stable in oxygen rich aerobic environments. Trivalent arsenites predominate in moderately reducing anaerobic environments such as groundwater [9].

The toxicity of arsenic varies widely according to its oxidation state; As(III) is far more toxic than As(V).

3.1.2 World Health Organization Guidelines

The World Health Organization (WHO) has predicted that 200,000–270,000 people will die of cancer after drinking arsenic-contaminated water in Bangladesh [10]. The maximum permissible limit for arsenic in the water of various countries according to WHO guidlines is given in Table 3.1.

3.1.3 Toxicity

Arsenic occurrence in the environment, its toxicity, health hazards, and the techniques used for speciation analysis are well known and have been reviewed. Long-term [24, 25] drinking water exposure causes skin, lung, bladder, and kidney cancer as well as pigmentation changes, skin thickening (hyperkeratosis) neurological disorders, muscular weakness, loss of appetite, and nausea [26–28]. This differs from acute poisoning, which

Table 3.1 Maximum permissible limit for arsenic in the water of different countries.

Country	Maximum Permissible Limit (μg/L)	Ref.
Argentina	50	[11]
Bangladesh	50	[12, 13]
China	50	[14]
Chile	50	[15]
India	10	[16]
Japan	-	[17]
Mexico	-	[18]
Nepal	50	[19]
New Zealand	10	[20]
Taiwan	10	[21]
USA	10	[22]
Vietnam	10	[23]

typically causes vomiting, esophageal and abdominal pain, and bloody "rice water" diarrhea [29]. Environmental toxins enter the food chain through irrigation water.

3.1.4 Technologies

A variety of methods have been used for the fast removal of arsenic in water and wastewater, some of which are discussed below.

3.1.4.1 Oxidation/Precipitation

Oxidation/precipitation is a simple, slow process using low-cost materials. It also oxidizes other inorganic and organic constituents in water. It kills microbes and oxidizes other impurities [30–36], while mainly by oxidation alone, As(V) can be removed and the, oxidation process can be accelerated. Oxidation steps and efficient control of pH is needed for this method.

3.1.4.2 Coagulation/Electrocoagulation/Co-precipitation [37–39]

In the alum coagulation process durable powdery chemicals are available. It is a low-cost technology that is very simple to operate, and is very effective over a wide pH range. In iron coagulation and lime softening, very common chemicals are used which are easily available and more efficient than alum coagulation on a weight basis. In contrast, it produces toxic sludge, low removal of arsenic occurs and pre-oxidation may be required. Also, readjustment of pH is required for this process.

3.1.4.3 Sorption and Ion Exchange Technique [40–43]

Activated alumina, iron-coated sand and ion exchange resin are very cheap and do not require regeneration. This technique removes both As(III) and As(V) and is pH independent. Also, it is relatively well known and commercially available. Beside all this, in this technique replacement of material after four to five regenerations is essential. It produces toxic solid waste. It is a high cost medium and high tech operation.

3.1.4.4 Membrane Technologies [44–51]

These techniques are well-defined and have high-removal efficiency and produce no toxic solid waste. They are also capable of removing other contaminants. But similar to other techniques, they also have some disadvantages like very high capital and running costs, pre-conditioning, and high water rejection.

3.1.5 Adsorption Process

Adsorption is evolving as a frontline of defense. It takes stuff (arsenic) out of water and puts it onto a solid adsorbent with micro and macro pores. Adsorption removes contaminants from water by attaching them onto the surface of a porous solid adsorbent. In physical adsorption, the process is low energy consuming and reversible, while the chemical adsorption process requires high energy and is irreversible.

3.1.5.1 Commercial Activated Carbons

In 1998, Allen and Whitten reviewed the production and characterization of activated carbon from many carbonaceous sources. Recently, the science and technology of charcoal production was reviewed [52]. Commercial activated carbons have been extensively used for As(III) and As(V) adsorption from water [53–56].

3.1.5.2 Synthetic Activated Carbons

Activated carbons are produced by carbonization employing slow substrate heating in the absence of air below 600°C. This removes volatiles. Then chemical or physical activation follows. Treatment with oxidizing agents (steam, carbon dioxide, or oxygen) at elevated temperature or with chemical activates ($ZnCl_2$, H_2PO_4, H_2SO_4, KOH, K2S, KCNS, etc.) completes the activation [57, 58]. Huang and Fu [59] examined the As(V) capacity of 15 brands of activated carbon over a wide pH range. The carbon type, total As(V) concentration and pH were major factors controlling As(V) removal. Manju *et al.* [60] prepared a coconut husk carbon (CHC) by carbonizing one part coconut husk with 1.8 parts by weight of sulphuric acid (18 M) at 150°C for 24 h. Recently, arsenic(III) was removed from aqueous solution (concentration range of 5–20 mg/L) by activated carbons developed from olive stones and solvent-extracted olive pulp [61]. Oat concentrations from 5 to 20 mg/L were also tested as adsorbent. Different parameters of activated carbons are given in Table 3.2.

3.1.5.3 Chars and Coals

A mixture of synthetic hydroxyapatite and barite, or bone char, has been used for arsenic(v) removal from water. Although the hydroxyapatite and barite mixture had a small influence on arsenic concentrations, bone char was found to be a very effective adsorbing agent for As(V) in the pH range of 2–5 [62]. Biochar byproducts from fast wood/bark pyrolysis, were investigated as adsorbents for the removal of As(III), Cd(II), and Pb(II) from water [63]. Oak bark, pine bark, oak wood, and pine wood

Table 3.2 Activated carbons with different parameters.

Adsorbent	pH	Concentration range	Capacity (mg/g) As (III)	As(V)	Ref.
Activated carbon	–	300 mg/L	2860		[53]
Activated carbon	–	–	25		[54]
Coconut shell carbon with 3 % ash	5	0–200 mg/L	2.4		[55]
Activated carbon			3.75		[59]
Coconut husk carbon	12	50–600 mg/L	146.30	–	[60]
Activated carbons from olive pulp and olive stone, carbon A	7	5–20 mg/L	1.393	–	[61]
Activated carbons from olive pulp and olive stone, carbon B	–	–	0.855	–	[61]
Activated carbons from olive pulp and olive stone, carbon C	–	–	0.738	–	[61]
Activated carbons from olive pulp and olive stone, carbon D	–	–	0.210	–	[61]

chars were obtained from fast pyrolysis at 400 and 450°C in an auger-fed reactor and characterized. Sorption studies were performed at different temperatures, pHs and solid to liquid ratios in the batch mode. Maximum adsorption occurred over a pH range of 3–4 for arsenic and 4–5 for lead and cadmium. The equilibrium data were modeled with the help of Langmuir and Freundlich equations. Overall, the data were well fitted with both the models, with a slight advantage for Langmuir model. The As(III) removal followed the order: pine wood char (1.20 μg/g) < oak wood char (5.85 μg/g) < oak bark char (12.1 μg/g) < pine bark char (12.15 μg/g). This study shows that byproduct chars from bio-oil production might be used as plentiful inexpensive adsorbents for water treatment (arsenic remediation) at a value above their pure fuel value. Further studies of such chars, both untreated and after activation, seem warranted. Part of the efforts to generate byproduct value from biorefineries are shown in Table 3.3.

Table 3.3 Chars and coals with different parameters.

Adsorbent	pH	Concentration range	Capacity (mg/g) As(III)	As(V)	Ref.
Hydroxyapatite and barite or bone-char	2–5	4–100 mg/L	–		[62]
Pine wood char	3.5	10–100 μg/L	0.0012	–	[63]
Oak wood char	3.5	10–100 μg/L	0.006	–	
Oak bark char	3.5	10–100 μg/L	0.0074	–	
Pine bark char	3.5	10–100 μg/L	12	–	

3.1.5.4 *Red Mud*

Red mud has been used for arsenic removal from water. Heat and acid treatments on red mud increased its adsorptive capacity. Arsenic adsorption on acid- and heat-treated red mud is also pH dependent, with an optimum range of 5.8–7.5 for As(III) and 1.8–3.5 for As(V) [64]. Adsorption followed a first-order rate expression and fit the Langmuir isotherm well. Isotherms were used to obtain the thermodynamic parameters. The As(III) adsorption was exothermic, whereas As(V) adsorption was endothermic [65]. Also reported was As(V) removal by using liquid phase of red mud (LPRM) [66]. Seawater-neutralized red muds (Bauxol) [67], Bauxsol activated by acid treatment, and by combined acid and heat treatment, and Bauxsol with added ferric sulphate or aluminium sulphate [68], activated Bauxsol (AB), and chemically modified and activated Bauxsol (AB)-coated sand [69–70] were all applied to arsenic removal. Modified calcined bauxite was also used for As(III) and As(V) remediation from groundwater [71–74] in batch and column modes. The optimum pH was 7.0 for both As(III) and As(V). Adsorption was unaffected by temperature variations. No appreciable ionic effects except from and EDTA were observed from the background ions including Ca(II), Fe(III), Cl^-, , and F^-. Sorption capacities were calculated using the Langmuir model are given in Table 3.4.

3.1.5.5 *Blast Furnace Slag*

Steel plants generate a large volume of granular blast furnace slag. It is being used as filler or in the production of slag cement. Recently, it was converted into an effective and economical scavenger and utilized for the remediation of aqueous arsenic [75, 76]. Blast furnace slag (BFS) for aqueous As(III) remediation has been used. The maximum As(III) adsorption

Table 3.4 Red mud with different parameters.

Adsorbent	pH	Concentration range	Capacity (mg/g) As(III) As(V)	Ref.
Red mud (RRM) Red mud (ARM)	7.25 for As(III); 3.50 for As(V) 7.25 for As(III); 3.50 for As(V)	33.37– 400.4 µm/L 33.37– 400.4 µm/L	0.663 0.514 0.884 0.941	[64]
Activated Bauxsol Bauxsol	4.5 4.5	7.03–220.9 mM for As(V); 2.04–156.7 mM for As(III)	0.541 7.642 1.081 -	[68, 69]
Bauxsol	4.5	0.54–20.34 mg/L	- 3.32	[70]
Modified Calcined Bauxite	7.0	0.5–8.0 mg/L	- 1.57	[71–74]

capacity by BFS was 1.40 mg As(III)/g of BFS at 1 mg/L As(III) initial concentration. The oxidation of As(III) to As(V) and its adsorption/precipitation onto BFS was the dominating mechanism [77] (see Table 3.5).

3.1.5.6 *Fe(III)/Cr(III) Hydroxide Waste*

Chromium(VI) compounds are used as corrosion inhibitors in cooling water systems in industries. Electrolytically generated Fe(II) reduces chromium(VI) in the wastewater to Cr(III) under acidic conditions. The Fe(III)/Cr(III) ions produced in solution are precipitated as Fe(III)/Cr(III) hydroxide by the use of lime. The resultant sludge is discarded as waste. Namasivayam and Senthilkumar [78] adsorbed As(V) from water onto a Fe(III)/Cr(III) hydroxide waste generated electrolytically in the treatment of Cr(VI)-containing wastewaters from fertilizer production. Chrome sludge, a waste material from electroplating, was tested to adsorb As(V) from aqueous solutions [79]. The maximum sorption capacity of chrome sludge for As(V) was 21 mg/g and the results are mentioned in Table 3.5.

Table 3.5 Blast furnace slag, Fe(III)/Cr(III) hydroxide waste and fly ash with different parameters.

Adsorbent	pH	Concentration range	Capacity (mg/g) As (III)	As(V)	Ref.
Iron(III) oxide-loaded melted slag	2.5	20–300 mg/L	2.9–30.1	18.8–78	[75]
Waste Fe(III)/ Cr(III)	4.0	20–100 mg/L	–	11.02	[78]
Iron oxide coated sand	7.5	100–800 μg/L	0.006	–	[80, 81]
Iron oxide coated sand	7.5	100–800 μg/L	0.028	–	

3.1.5.7 Fly Ash

Coal combustion produces a huge amount of fly ash byproduct, whose disposal requires large quantities of land and water. Currently, its applications are limited to civil engineering uses including cement and brick production and roadbeds. Bottom ash can also serve as an adsorbent [80, 81]. Resource recovery from coal fly ash is one of the most important issues in waste management worldwide. Since the major chemical compounds contained in fly ash are aluminosilicate, intensive efforts have been recently made to utilize this material as an adsorbent (see Table 3.5).

3.1.5.8 Soils and Constituents

3.1.5.8.1 Soils and Sand

Arsenate [As(V)] adsorption-desorption kinetics were reported on Olivier loam, Sharkey clay, and Windsor sand, followed by arsenic release using successive dilutions [82]. The As(V) retention rate was initially rapid followed by gradual or slower retention behavior with increased reaction time. Manganese greensand was effective for removing arsenic to < 25 μg/L [83]. The addition of iron was necessary to achieve an effluent arsenic level of 25 μg/L in the manganese greensand filtration system. Iron oxide-coated sand (IOCS) exhibited a high adsorption capacity (136 μg/L). Another study [84] achieved 285 μg/g of arsenic removal on iron oxide-coated sand. Sulphate-modified iron oxide-coated sand (SMIOCS) was also used for As(III) and As(V) removal [85]. The SMIOCS was prepared by coating $BaSO_4$ and Fe on quartz sand. The maximum As(V) removal was obtained in acidic pHs [86], while maximum As(III) removal was obtained at pH 7–9 [85], as shown in Table 3.6.

Table 3.6 Soils and constituents with different parameters.

Adsorbent	pH	Concentration range	Capacity (mg/g) As(III)	As(V)	Ref.
Olivier soil	5–6	5–100 mg/L	–	0.42	[82]
Iron oxide coated sand	–	100 μg/L	0.136	–	[83]
Iron oxide coated sand	–	325 μg/L	–	0.018	[84]
Modified iron oxide-coated sand	7.2	0.5–3.5 mg/L	0.14	–	[85]
(SMIOCS)	4–10	0.5–3.5 mg/L	0.13 (pH 4), 0.12 (pH 7), 0.08 (pH 10)	–	[86]
Goethite	–	5–25mg/L	–	5	[87]
ZMA (Sonora)	4	0.1–4 mg/L	0.0048	0.1	[90, 91]
Shirasu-zeolite (SZP_1)	3–10	1.3 mM	–	65.93	[92]

3.1.5.8.2 Clay Minerals and Zeolites

Clay minerals and oxides are widespread and abundant in aquatic and terrestrial environments. Finely divided clay minerals and oxides exhibit large surface areas. Clay minerals and oxides adsorb the cationic, anionic, and neutral metal species. They can also take part in the cation and anion-exchange processes. Their sorption capacities, cation and anion-exchange properties and binding energies vary widely. The goethite (surface area 103 m^2/g) prepared from the oxidation of ferrous carbonate from double decomposition of ferrous sulphate doped with sodium lauryl sulphate and sodium carbonate was also used for arsenate removal [87]. Adsorption capacity of 5 mg/g (pH 5.0) was achieved.

Zeolites have received increasing attention for pollution control as standard components in wastewater treatment [88]. Both ion exchange and adsorption properties of zeolites have been used for the selective separation of cations from aqueous solution. Zeolites are crystalline, hydrated alluminosilicates of alkali and alkaline earth cations, having infinite, three-dimensional structures [89]. Aqueous arsenic sorption by natural zeolites, volcanic stone, cretaceous powder CACMM and clinoptilolite-containing rocks was studied. The content of the zeolitic phases was: 55% clinoptilolite + 35% erionite in ZMA (Maxican, Sonora), 40% clinoptilolite + 30% mordenite in ZME (Maxican, Oaxaca) and 55% clinoptilolite + 30% mordenite in ZH [90, 91]. The adsorption of As(V) from drinking water by an aluminum-loaded Shirasu-zeolite (Al-SZP1) was slightly dependent on the initial pH over a wide range (3–10) [92]. The Al-SZP1's ability to adsorb As(V) was equivalent to that of activated alumina.

Competition arsenite, chloride, nitrate, sulphate, chromate, and acetate ions had little effect, but phosphate greatly interfered with the adsorption. A ligand-exchange mechanism between As(V) ions and surface hydroxide groups on Al-SZP1 was presumed. The adsorbed As(V) ions were desorbed by 40 mM aqueous NaOH. An iron-conditioned zeolite was prepared and used for arsenic removal from groundwater at pH 7.8 and temperature 145°C [93]; other parameters are shown in Table 3.6.

3.1.5.9 Oxides

3.1.5.9.1 Manganese Dioxide

Manganese oxide minerals have important environmental chemistry uses. Arsenic removal from drinking water by monocomponent fixed-bed adsorption of phosphate and arsenate using two natural manganese oxides was investigated [94, 95].

3.1.5.9.2 Activated Alumina

Activated alumina (AA) prepared by thermal dehydration of aluminium hydroxide has a high surface area and a distribution of both macro- and microspores. The United Nations Environment Programme (UNEP) agency classified AA adsorption among the best available technologies for As removal from water. Arsenic(V) sorption occurs best mostly between pH 6.0 and 8.0, where AA surfaces are positively charged. The adsorption of As(III) is strongly pH dependent and it exhibits a high affinity towards AA at pH 7.6 [96]. Arsenites from water were removed with AA and iron oxide-impregnated AA [97]. The effect of adsorbent dose, pH, and contact time were investigated. The removal of As(III) was strongly pH dependent. The adsorption capacity of iron oxide-impregnated AA was 12 mg/g. Precipitated Fe(OH)3 on the surface of activated Al_2O_3 supports was used. The Fe content of the AA was 0.31% m/m (56.1 mmol/g) having pHzpc = 6.9. The total capacity was 0.12 mmol/g. The adsorbent can be used for binding both anions and cations by varying the pH. If pHeq < pHzpc, anions are sorbed on the $Fe(OH)_3/Al_2O_3$ surface through surface

Table 3.7 Oxides with different parameters.

Adsorbent	pH	Concentration range	Capacity (mg/g) As III) As(V)	Ref.
AA	7.6	1 mg/L	0.180 –	[96, 97]
Iron hydroxide coated alumina	6.62–6.74 As(III) 7.15–7.2 AS(V)	0.1–1.8 mmol/L	7.64 36.64	[98]

OH group. The pH of the isoelectric points for these adsorbents (pHiep) were 6.1 for As(III) and 8.0 for As(V) [98] and the remaining parameters are given in Table 3.7.

3.1.5.9.3 Titanium Dioxide

The ability of nanocrystalline titanium dioxide (TiO_2) to remove arsenate and arsenite and to photocatalytically oxidize As(III) was evaluated [99]. Adsorption of As(V) and As(III) on commercially available TiO_2 suspensions (Hombikat UV100 and Degussa P25) was investigated versus pH and initial adsorbate concentration [100]. Bang *et al.* [101] studied a novel granular titanium dioxide for groundwater arsenic removal. More arsenate was adsorbed than arsenite on TiO2 at pH 7.0. The adsorption capacities for As(V) and As(III) were 41.4 and 32.4 mg/g TiO_2, respectively.

3.1.5.9.4 Lanthanum Hydroxide

Lanthanum hydroxide (LH), lanthanum carbonate (LC), and basic lanthanum carbonate (BLC) remove As(V) from aqueous solutions [102]. These lanthanum compounds were effective at a concentration of < 0.001 mM. Dissolution was appreciable at initial pH values < 4.3, < 4.3, and < 4.0 for LH, LC and BLC, respectively.

Arsenic removal followed first-order kinetics in the neutral pH range, and the order of the rate constants was LH > LC > BLC. The optimum pH range was 3–8 for LH, 4–7 for LC, and 2–4 for BLC. Two arsenic uptake mechanisms were proposed: (i) adsorption by the exchange of and/or OH groups with arsenic ions in neutral to alkaline pH where La does not dissolve, and (ii) precipitation of insoluble lanthanum arsenate, LaAsO4, in acidic pHs.

3.1.5.9.5 Ferrihydrite/Iron Hydroxide/Iron Oxides

Amorphous Fe(O)OH has the highest adsorption capability since it has the highest surface area. Surface area is not the only criterion for high removal capacities of metal ions and other mechanisms (ion exchange, precipitation) play an important role. Most iron oxides are fine powders that are difficult to separate from solution afterwards. Therefore, the EPA has proposed iron oxide-coated sand filtration as an emerging technology for arsenic removal at small water facilities [103, 104]. Granular ferric hydroxide (GFH) was investigated for arsenic removal from natural water [105]. The application of GFH in test adsorbers demonstrated high treatment capacity of 30,000–40,000 bed volumes before an arsenic concentration of 10 g/L was exceeded in the adsorber effluent. The sorption capacity was 8.5 g/kg. Badruzzaman *et al.* [106] evaluated porous granular ferric hydroxide for arsenic removal in potable water systems. Granular ferric hydroxide (GFH) is a highly porous (micropore volume ~0.0394 cm^3/g, mesopore volume ~0.10 cm^3/g) adsorbent with a BET surface area of ~235 m^2/g. The pseudo-equilibrium (18 days of contact) arsenate adsorption capacity at pH 7 was 8 gAs/mg dry GFH at a liquid phase arsenate concentration of 10 gAs/L. Adsorption and desorption of methyl arsenic acid [$CH_3AsO(OH)_2$], methyl arsonous acid [$CH_3As(OH)_2$], dimethylarsinic acid [$(CH_3)_2AsO(OH)$], dimethylarsinous acid [$(CH_3)_2AsOH$], arsenate [$AsO(OH)_3$], and arsenite [$As(OH)_3$] on iron oxide minerals (goethite and 2-line ferrihydrite) were studied by Lafferty and Loeppert [107]. Akaganeite [Fe(III)7.6 Ni (0.4) O (6.4) OH (9.7) C(11.3)] in either fine powder (nanocrystals) or granular forms can also be used to remove As(V) from water [108, 109]. Akaganeite powder was prepared by $FeCl_3$ hydrolysis in aqueous solutions and precipitation using ammonium carbonate. Removal of As(V) by akaganeite β-FeO(OH) nanocrystals was also reported [110]. Arsenic removal increased with increasing temperature and the Langmuir adsorption capacities were compared and the results are tabulated in Table 3.7.

3.1.5.10 *Biosorbents*

Biosorption is capable of removing traces of heavy metals and other elements from dilute aqueous solutions. Algae, fungi and bacteria are examples of biomass-derived sorbents for several metals. Such sorbents have produced encouraging results. There has been a review of how bacteria, fungi and algae take up toxic metal ions [111, 112]. It is important to differentiate biosorption or sorption from bioaccumulation. Biosorption (or bioadsorption) is a passive immobilization of metals by biomass. Mechanisms of cell surface sorption are independent of cell metabolism; they are based upon physicochemical interactions between metal and functional groups of the cell wall.

3.1.5.11 *Chitin and Chitosan*

Braconnot first described chitin in 1811 upon isolating a substance he called “fungine,” from fungi. The first scientific reference to chitin was taken from the Greek word “Chiton,” meaning an “oat of mail,” for the material obtained from the elytra of May beetles [113]. Chitin is first prepared from shells of Crustacea at low-cost by removing other components, such as calcium and proteins, by treatment with acids and alkalines. Chitin and chitosan are excellent natural adsorbents [114–116] with high selectivities due to the following reasons:

- Large numbers of hydroxyl and amino groups give chitosan high hydrophilicity.
- Primary amino groups provide high reactivity.
- The polymer chains of chitosan provide suitable configurations for efficient completion with metal ions

All of the results obtained from these adsorbents are given in Table 3.8.

3.1.5.12 *Biomass*

Removal of trivalent arsenic from solution was assessed by Kamala *et al.* [117]. The As(III) removal capability of fresh and immobilized G. cambogia biomass was estimated. The As(III) uptake was not greatly affected by pH, with optimal biosorption occurring at around pH 6–8. Common ions such as Ca(II) and Mg(II) did not inhibit As(III) removal at concentrations up to 100 mg/L, but 100 mg/L of Fe(II) caused a noticeable drop in the extent of As(III) removal. Immobilized biomass columns were recycled five times. Water lettuce (*Pistia stratiotes L.*) is an aquatic plant also used for the removal of arsenate [118] in water. Young plants were harvested from a pollution-free pond and hydroponically cultured, effectively absorbing arsenic in a range from 0.25 to 5.0 mg/L. From 22.8% to 82.0% of the As was removed for a biomass loading of 20 g/L at pH 7.0 after 144 h. Ridvan *et al.* [119] examined the fungus, Penicillium purpurogenum, for cadmium, lead, mercury, and arsenic ion removal from water. Heavy metal loading capacity increased with increasing pH under acidic conditions, presumably as a function of heavy metal speciation versus the H+ competition at the same binding sites. The adsorption of heavy metal ions reached a plateau at ~ pH 5.0. The fungus adsorption capacity for As(III) was 35.6 mg/g (Table 3.5). Metal ion elution was achieved using 0.5M HCl. This fungus was recycled through 10 adsorption cycles. The biosorption

Table 3.8 Biosorbents, biomass, water hyacinth (*Eichornia crassipes*) and human hair with different parameters.

Adsorbent	pH	Concentration range	Capacity (mg/g) As(III)	As(V)	Ref.
Chitosan dry water hyacinth plant leaf	4	g/L	58		[114]
Molebdate-impregnated chitosan gel beads	-	-	–	200	[116]
Fresh biomass	6	50–200 mg/L	128.1	-	[117]
Penicillium purpurogenum	5	10–750 mg/L	35.6	-	[119]
Biomass	2	1–10 mg/L	13.17	-	[121]
Human hair	-	90–360 μg/L	–	1.80	[125]

of cadmium, lead, mercury and arsenic ions by the Penicillium purpurogenum fungus has also been reported [120]. Selective As(III) adsorption on waste chicken feathers with a high fibrous protein content was studied [121]. The disulfide bridges present were reduced to thiols by thioglycolate. The As(III) adsorption was favored at low pH. Arsenic uptake was 270 μmol As(III)/g of biomass. The XANES analyses demonstrated that arsenic is adsorbed in its trivalent state. This is a major advantage over conventional As uptake, which usually requires a previous oxidation to As(V). Each adsorbed As atom was directly bound to three S atoms with estimated As–S distances of 226 pm based on EXAFS analyses. The results of this biomass are mentioned in Table 3.8.

3.1.5.13 Water Hyacinth (Eichornia crassipes)

The water hyacinth (*E. crassipes*) is a member of the pickerelweed family (*Pontederiaceae*). The plants vary in size from a few centimeters to over a meter in height [122]. Water hyacinths (*E. crassipes*) were used as a pollution monitor for the simultaneous accumulation of arsenic, cadmium, lead and mercury [123]. Arsenic removal by water hyacinths (*E.crassipes*) was also reported [124] and tabulated in Table 3.8.

3.1.5.14 Human Hair

Wasiuddin *et al.* [125] examined the ability of human hair to adsorb arsenic from contaminated drinking water. Both static and dynamic tests along with the numerical modeling have been carried out to test human hair as an adsorbent. The maximum adsorption capacity of 12.4 μg/g was reported at an arsenic concentration of 360 μg/L and other parameter effects are shown in Table 3.8.

3.1.6 New Generation Materials

For the removal of arsenic from water, hydrous zirconium oxide was loaded successfully onto polymeric adsorbent (D401) to obtain a new adsorbing material (D401-Zr) [126]. TiO_2 nanofibers (NFs) [127] have been used for arsenic removal, and the potential of chitosan-coated sand (CCS) and iron-chitosan-coated sand (ICCS) [128] toward the removal of both As(V) and As(III) from aqueous systems. A novel adsorbent, iron(III) chelate of an amino-functionalized polyacrylamide-grafted coconut coir pith (Fe(III)-A-PGCP) [129], was prepared and used for the removal of arsenic(V) from aqueous solutions. A high efficiency of contaminant removal by nanoscale zerovalent iron (nZVI) [130] and the first example of arsenite and arsenate removal from water by incorporation of arsenic into the structure of nano-crystalline iron(III) oxide [131] have often been reported. Composites of cellulose@iron oxide nanoparticles [132] have been used for arsenic removal and the efficacy results are given in Table 3.9.

3.2 Arsenic Desorption/Sorbent Regeneration

Once the sorbent becomes exhausted, the metals must be recovered and the sorbent regenerated. Desorption and sorbent regeneration is a critical

Table 3.9 New generation materials with different parameters.

Adsorbent	pH	Concentration range	Capacity (mg/g)		Ref.
			As(III)	As(V)	
Zr-Loaded Resin (D401-Zr)	6.3–9.2	10 mg/kg	– –	11.84	[126]
TiO_2 nanofibers (NFs)	7	0.0133 mmol/kg	–	–	[127]
chitosan-coated sand (CCS) iron–chitosan-coated sand	7	100 to 1000 μg/L	17 26	23 56	[128]
(Fe(III)-A-PGCP)	7	10 - 300 mg/L	–	107.8	[129]
cellulose@iron oxide NPs	7	50–100 mg/L	23.16	32.11	[132]

consideration and contributor to process costs and metal(s) recovery in a concentrated form. A successful desorption process must restore the sorbent close to its initial properties for effective reuse. Desorption can be improved by gaining insight into the metal sorption mechanism. In most of the arsenic sorption studies discussed earlier, desorption/regeneration was not discussed. Very few desorption studies are detailed in literature. Furthermore, once arsenic is recovered in the concentrated form, the problem of how to dispose of this concentrated arsenic product must be addressed. This is a difficult task. Few attempts have been made in the literature to address the handling of concentrated arsenic wastes. Various disposal options and their advantages and disadvantages were reviewed [133]. The methods frequently used for other metals and organics include combustion or recovery and purification for resale. These options are not feasible for arsenic due to the following reasons:

- Incineration is not practically feasible because arsenic oxides are volatile and can easily escape.
- Recovery and purification of arsenic is not cost effective because arsenic has limited markets.

3.2.1 Cost Evaluation

The cost of arsenic removal adsorbents developed from waste materials seldom appears in the literature. The cost of individual adsorbents depends on local availability, processing required and treatment conditions. These are not broadly and thoroughly discussed in any paper anywhere in the literature. Costs will vary when the adsorbents are made in (and for) developed countries, developing countries or underdeveloped countries. Numerous commercially available activated carbons have been used for arsenic adsorption, both as-received and after chemical modifications. However, chemical modification costs are seldom mentioned in the research reports. Furthermore, no consistency exists in the data presented. Most papers describe only batch experiments but not fixed-bed studies. Batch equilibrium adsorption isotherms cannot simulate or predict dynamic performances directly due to the following limitations:

- Isotherms are equilibrium tests so the time restrictions are not considered.
- Isotherms are based on carbon exhaustion-granular systems.
- Long-term chemical and biological effects are not evident.

3.3 Conclusions

The heavy metals such as lead have been serious polluters of water since Roman times and perhaps earlier. They have been major water pollutants during the 20th century and continue to create serious problems in the 21st century. Mercury is a serious source of danger to top-of-the-food-chain ocean fish. As we have documented here, arsenic in drinking water is having a major human impact in several locations. Many treatment technologies are available for arsenic remediation but none of them is found to be completely applicable. Successful separation/removal processes should have:

- Low-volume stream containing the concentrated contaminant(s).
- A high volume exit stream containing the decontaminated liquid, solid or gas.

Adsorption is a useful tool for controlling the extent of aqueous arsenic pollution. Activated carbon was studied extensively for arsenic removal. However, carbon only removes a few milligrams of metal ions per gram of activated carbon. Regeneration problems exist. Thus, activated carbon use is expensive. Activated carbon use in developing countries is even more problematic due to cost. Therefore, a definite need exists for low-cost adsorbents which exhibit superior adsorption capacities and local availability. Adsorption is but one tool in the effort to remove arsenic from drinking water. Currently, about 100 million people are consuming water with arsenic concentration up to 100 times the 10 g/L guideline of the World Health Organization. Recent articles in scientific publications [134–136] focusing on the drinking water problems in Bangladesh demonstrated that two different approaches have had maximum impact:

- Testing tube wells followed by switching away from contaminated wells to alternate uncontaminated water sources.
- Installation of deep wells that supply water from older aquifers that do not contain elevated arsenic levels.

Furthermore, three major recommendations were made:

- Stimulate the periodic monitoring of water quality no matter what mitigation options exist.
- Encourage the wise use of deep aquifers low in arsenic.
- Widely publicize the known effects of arsenic on the mental development of children.

This common-sense approach illustrates that a variety of low-cost approaches must be employed in many underdeveloped locations throughout the world. Only when these approaches are exhausted will adsorption be likely to contribute to further mitigation efforts.

Acknowledgement

The authors are thankful to the University Grant Commission (UGC), New Delhi, for its financial assistance, Grant No. F. No. 39–734/2010(SR).

References

1. M.A. Khan, and Y.S. Ho, *Asian J. Chem.*, Vol. 23, pp. 1889–1901, 2011.
2. A.M. Goswami, B. Nath, J. Jana, S.J. Sahu, M.J. Sarkar, G. Jacks, P. Bhattacharya, A. Mukherjee, D.A. Polya, J.S. Jean, and D. Chatterjee, *J. Contam. Hydrol.*, Vol. 99, pp. 22–30, 2008.
3. B.K. Mandal, and K.T. Suzuki, *Talanta*, Vol. 58, pp. 201–235, 2002.
4. S. Deng, Z. Li, J. Huanga, and G. Yua, *J. Hazard. Mater.*, Vol. 179, pp. 1014–1021, 2010.
5. P.L. Smedley, H.B. Nicolli, D.M.J. Macdonald, A.J. Barros, and J.O. Tullio, *Appl. Geochem.*, Vol.17, Issue 3, pp. 259–284, 2002.
6. E.T. Mackenzie, R.J. Lamtzy, and V. Petorson, *J. Int. Assoc. Math. Geol.*, Vol. 6, pp. 99–142, 1979.
7. S. Sarkar, L.M. Blaney, A. Gupta, D. Ghosh, and A.K. Sengupta, *Environ. Sci. Technol.*, Vol. 42, pp. 4268–4273, 2008.
8. J.A.G. Santos, M.I.S. Gonzaga, and L.Q. Ma, *J. Hazard. Mater.*, Vol. 180, pp. 662–667, 2010.
9. M. Kim, H.J. Um, S. Bang, S.H. Lee, S.J. Oh, J.H. Han, K.W. Kim, J. Min, and Y.H. Kim, *Environ. Sci. Technol.*, Vol. 43, pp. 9335–9340, 2009.
10. M.M. Karim, *Water Res.*, Vol. 34, Issue 1, pp. 304–310, 2000.
11. J. Ahmad, B. Goldar, and S. Misra, *J. Environ. Manage.*, Vol. 74, Issue 2, pp. 173–185, 2005.
12. C.F. Harvey, C.H. Swartz, A.B.M. Badruzzaman, N. Keon-Blute, W. Yu, M.A. Ali, J. Jay, R. Beckie, V. Niedan, D. Brabander, P.M. Oates, K.N. Ashfaque, S. Islam, H.F. Hemond, and M.F. Ahmed, *Science*, Vol. 298, No. 5598, pp. 1602–1606, 2002.
13. T. Agusa, T. Kunito, R. Kubota, I. Monirith, S. Tanabe, and T.S. Tana, *Biomed. Res. Trace Elements*, Vol. 13, pp. 254–255, 2002.
14. X. Guo, Y. Fujino, S. Kaneko, K. Wu, Y. Xia, and T. Yoshimura, *Chin. Mol. Cell. Biochem.*, Vol. 222, No. 1–2, pp. 137–140, 2001.
15. A.M. Sancha, *Water Supply*, Vol.18, No. 1, pp. 621–625, 2000.

16. S. Sarkar, A. Gupta, R.K. Biswas, A.K. Deb, J.E. Greenleaf, and A.K. Sen-Gupta, *Water Res.*, Vol. 39, Issue 10, pp. 2196–2206, 2005.
17. H. Kondo, Y. Ishiguro, K. Ohno, M. Nagase, M. Toba, and M. Takagi, *Water Res.*, Vol. 33, Issue 8, pp. 1967–1972, 1999.
18. I.J. Sobala, D. Ocinski, and E.K. Balawejder, *Ind. Eng. Chem. Res.*, dx.doi.org/10.1021/ie400478x, 2013.
19. R.R. Shrestha, M.P. Shrestha, N.P. Upadhyay, R. Pradhan, R. Khadka, A. Maskey, M. Maharjan, S. Tuladhar, B.M. Dahal, and K. Shrestha, *J. Environ. Sci. Health A: Toxic/Hazard. Substances Environ. Eng.*, Vol. 38, No. 1, pp. 185–200, 2003.
20. S.J. McLaren, and N.D. Kim, *Environ. Poll.*, Vol. 90, Issue 1, pp. 67–73. 1995.
21. W.P. Tseng, H.M. Chu, S.W. How, J.M. Fong, C.S. Lin, and S. Yeh, *J. Natl. Cancer Inst.*, Vol. 40, No. 3, pp. 453–463, 1968.
22. J.D. Ayotte, D.L. Montgomery, S.M. Flanagan, and K.W. Robinson, *Environ. Sci. Technol.*, Vol. 37, No. 10, pp. 2075–2083, 2003.
23. M. Berg, H.C. Tran, T.C. Nguyen, H.V. Pham, R. Schertenleib, and W. Giger, *Environ. Sci. Technol.*, Vol. 35, No. 13, pp. 2621–2626, 2001.
24. N.E. Korte, and Q. Fernando, *Crit. Rev. Environ. Control*, Vol. 2, No. 1, pp.1–39, 1991.
25. W. Tanga, Q. Li, S. Gaoa, and J.K. Shanga, *J. Hazard. Mater.*, Vol. 192, pp. 131- 138, 2011.
26. C.K. Jain, and I. Ali, *Water Res.*, Vol. 34, pp. 4304–4312, 2000.
27. A.M. Donia, A.A. Atia, and D.H. Mabrouk, *J. Hazard. Mater.*, Vol. 191, pp.1–7, 2011.
28. P. Mondal, C.B. Majumder, and B. Mohanty, *J. Hazard. Mater.*, Vol. 150, pp. 695–702, 2008.
29. J.C. Ng, J. Wang, and A. Shraim, *Chemosphere*, Vol. 52, Issue 9, pp. 1353–1359, 2003.
30. M. Borho, and P. Wilderer, *Water Sci. Technol.*, Vol. 34, No. 9, pp. 25–31, 1996.
31. O.X. Leupin, and S.J. Hug, *Water Res.*, Vol. 39, Issue 9, pp. 1729–1740, 2005.
32. M. Bissen, and F.H. Frimmel, *Acta Hydrochim. Hydrobiol.*, Vol. 31, Issue 2, pp. 97–107, 2003.
33. T.M. Gihring, G.K. Druschel, R.B. McCleskey, R.J. Hamers, and J.F. Banfield, *Environ. Sci. Technol.*, Vol. 35, No. 19, pp. 3857–3862, 2001.
34. P.K. Dutta, S.O. Pehkonen, V.K. Sharma, and A.K. Ray, *Environ. Sci. Technol.*, Vol. 39, No. 6, pp. 1827–1834, 2005.
35. H. Lee, W. Choi, *Environ. Sci. Technol.*, Vol. 36, No. 17, pp. 3872–3878, 2002.
36. M. Zaw, and M.T. Emett, *Toxicol. Lett.*, Vol. 133, No. 1, pp. 113–118, 2002.
37. H.K. Hansen, P. Nunez, and R. Grandon, *Miner. Eng.*, Vol. 19, Issue 5, pp. 521–524, 2006.
38. S.P. Panda, L.S. Desphands, P.M. Patni, and S.L. Lutada, *J. Environ. Sci. Health A*, Vol. 32, No. 7, pp. 1981–1987, 1997.
39. L.D. Benefield, and J.S. Morgan, "Chemical precipitation," in: *Water Quality and Treatment, 4th ed.*, R.D. Letterman, ed., AWWA, New York: McGraw-Hill.

40. D.A. Clifford, "Ion-exchange and inorganic adsorption," in: *Water Quality and Treatment: A Handbook of Community Water Supplies, 5th ed.*, American Water Works Association, McGraw-Hill, New York, 1999.
41. J. Kim, and M.M. Benjamin, *Water Res.*, Vol. 38, Issue 8, pp. 2053–2062, 2004.
42. D.V. Nenov, and D. Kamenski, *Environ. Prot. Eng.*, Vol. 12, No. 3, pp. 79–87, 1986.
43. R. Baciocchi, A. Chiavola, and R. Gavasci, *Water Sci. Technol.: Water Supply*, Vol. 5, No. 5, pp. 67–74, 2005.
44. B. Han, T. Runnells, J. Zimbron, and R. Wickramasinghe, *Desalination*, Vol. 145, No. 1–3, pp. 293–298, 2002.
45. M.L. Ballinas, E. Rodriguez de San Miguel, M.T.J. Rodriguez, O. Silva, M. Munoz, and J. de Gyves, *Environ. Sci. Technol.*, Vol. 38, No. 3, pp. 886–89, 2004.
46. S. Velizarov, J.G. Crespo, and M.A. Reis, *Rev. Environ. Sci. Biotechnol.*, Vol. 3, No. 4, pp. 361–380, 2004.
47. F.F. Peng, P. Di, *Ind. Eng. Chem. Res.*, Vol. 33, pp. 922–928, 1994.
48. S.D. Huang, J.J. Tzuoo, H.S. Gau, and C.F. Fann, *Sep. Sci. Technol.*, Vol. 19 No. 13015, pp. 1061–1072, 1984/1985.
49. Y.C. Zhao, A.L. Zouboulis, and K.A. Matis, *Sep. Sci. Technol.*, Vol. 31, No. 6, pp. 769–785, 1996.
50. L. Iberhan, and M. Wisniewski, *J. Chem. Technol. Biotechnol.*, Vol. 78, pp. 659–665, 2003.
51. I. Katsoyiannis, A. Zouboulis, H. Althoff, and H. Bartel, *Chemosphere*, Vol. 47, p. 325–332, 2002.
52. M.O. Corapcioglu, and C.P. Huang, *Carbon*, Vol. 25, No. 4, pp. 569–578, 1987.
53. P. Navarro, and F.J. Alguacil, *Hydrometallurgy*, Vol. 66, No. 1–3, pp. 101–105, 2002.
54. H.E. Eguez, and E.H. Cho, *J. Met.*, Vol. 39, pp. 38–41, 1987.
55. L. Lorenzen, J.S.J. Deventer, and W.M. Landi, *Miner. Eng.*, Vol. 8, No. 4–5, p. 557–569, 1995.
56. L. Jubinka, and V. Rajakovic, *Sep. Sci. Technol.*, Vol. 27, No. 11, pp. 1423–1433, 1992.
57. S.J.T. Pollard, G.D. Fowler, C.J. Sollars, and R. Perry, *Sci. Total Environ.*, Vol. 116, pp. 31–52, 1992.
58. D. Mohan, and C.U. Pittman Jr., *J. Hazard. Mater.*, Vol. 137, No. 2, pp. 762–811, 2006.
59. C.P. Huang, and P. Fu, *J. Water Poll. Contr. Fed.*, Vol. 56, p. 232, 1984.
60. G.N. Manju, C. Raji, and T.S. Anirudhan, *Water Res.*, Vol. 32, Issue 10, pp. 3062–3070, 1998.
61. T. Budinova, N. Petrov, M. Razvigorova, J. Parra, and P. Galiatsatou, *Ind. Eng. Chem. Res.*, Vol. 45, pp. 1896–1901, 2006.
62. R. Sneddon, H. Garelick, and E. Valsami-Jones, *Mineral. Mag.*, Vol. 69, No. 5, pp. 769–780, 2005.
63. D. Mohan, C.U. Pittman Jr., M. Bricka, F. Smith, B. Yancey, J. Mohammad, P.H. Steele, M.F. Alexandre-Franco, V.G. Serrano, and H. Gong, *J. Colloid Interf. Sci.*, Vol. 310, Issue 1, pp. 57–73, 2007.

64. H.S. Altundogan, S. Altundogan, F. Tumen, and M. Bildik, *Waste Manage.*, Vol. 22, pp. 357–363, 2002.
65. H.S. Altundogan, S. Altundogan, F. Tumen, and M. Bildik, *Waste Manage.*, Vol. 20, Issue 8, pp. 761–767, 2000.
66. H.S. Altundogan, and F. T¨umen, *J. Environ. Sci. Health, Part A: Toxic/Hazard. Substances Environ. Eng.*, Vol. 38, No. 7, pp. 1247–1258, 2003.
67. H. Genç-Fuhrman, J.C. Tjell, D. McConchie, and O. Schuiling, *J. Colloid Interf. Sci.*, Vol. 264, pp. 327–334, 2003.
68. H. Genç-Fuhrman, J.C. Tjell, and D. McConchie, *J. Colloid Interf. Sci.*, Vol. 271, Issue 2, pp. 313–320, 2003.
69. H. Genç-Fuhrman, J.C. Tjell, and D. McConchie, *Environ. Sci. Technol.*, Vol. 38, No. 8, pp. 2428–2434, 2004.
70. H. Genç-Fuhrman, H. Bregnhøj, and D. McConchie, *Water Res.*, Vol. 39, Issue 13, pp. 2944–2954, 2005.
71. P.B. Bhakat, A.K. Gupta, S. Ayoob, and S. Kundu, *Colloid Surf. A: Physicochem. Eng. Aspects*, Vol. 281, Issue 1–3, pp. 237–245, 2006.
72. P.B. Bhakat, A.K. Gupta, and S.Ayoob, *J. Hazard. Mater.*, Vol. 139, pp. 286–292, 2007.
73. S. Ayoob, A.K. Gupta, and P.B. Bhakat, *Sep. Purif. Technol.*, Vol. 52, pp. 430–438, 2007.
74. S. Ayoob, A.K. Gupta, and P.B. Bhakat, *Colloid Surf. A: Physicochem. Eng. Aspects*, Vol. 293, Issue 1–3, pp. 247–254, 2007.
75. F.-S. Zhang, and H. Itoh, *Chemosphere*, Vol. 60, Issue 3, pp. 319–325, 2005.
76. J.S. Ahn, C.-M. Chon, H.-S. Moon, and K.-W. Kim, *Water Res.*, Vol. 37, Issue 10, pp. 2478–2488, 2003.
77. F.-S. Zhang, and H. Itoh, *Chemosphere*, Vol. 65, pp.125–131, 2006.
78. C. Namasivayam, and S. Senthilkumar, *Ind. Eng. Chem. Res.*, Vol. 37, No. 12, pp. 4816–4822, 1998.
79. K.S. Lee, and C.K. Lee, *Environ. Technol.*, Vol. 16, No. 1, pp. 65–71, 1995.
80. V.K. Gupta, V.K. Saini, and N. Jain, *J. Colloid Interf. Sci.*, Vol. 288, Issue 1, pp. 55–60, 2005.
81. V.K. Gupta, A. Mittal, and L. Krishnan, *J. Hazard. Mater.*, Vol. 117, No. 2–3, pp. 171–178, 2005.
82. H. Zhang, and H.M. Selim, *Environ. Sci. Technol.*, Vol. 39, No. 16, pp. 6101–6108, 2005.
83. O.S. Thirunavukkarasu, T. Viraraghavan, K.S. Subramanian, O. Chaalal, and M.R. Islam, *Energy Sources*, Vol. 27, pp. 209–219, 2005.
84. T. Viraghavan, O.S. Thirunavukkarasu, and K.S. Suramanian, *Water Qual. Res. J. Can.*, Vol. 36, No. 1, pp. 55–70, 2001.
85. R.C. Vaishya, and S.K. Gupta, *J. Chem. Technol. Biotechnol.*, Vol. 78, pp. 73–80, 2002.
86. R.C. Vaishya, and S.K. Gupta, *Sep. Sci. Technol.*, Vol. 39, No. 3, pp. 645–666, 2004.
87. P. Lakshmipathiraj, B.R.V. Narasimhan, S. Prabhakar, and G.B. Raju, *J. Hazard. Mater.*, Vol. 136, No. 2, pp. 281–287, 2006.

88. S. Kesraoui-Ouki, C.R. Cheeseman, and R. Perry, *J. Chem. Technol. Biotechnol.*, Vol. 59, pp. 121–126, 1994.
89. Z. Li, J.S. Jean, W.T. Jiang, P.H. Chang, C.J. Chen, L. Lian, *J. Hazard. Mater.*, Vol. 187, pp. 318–323, 2011.
90. M.P.E. Alez, J. Mattusch, W.D. Einicke, and R. Wennrich, *Chem. Eng. J.*, Vol. 81, Issue 1–3, pp. 187–195. 2001.
91. M.P. Elizalde-González, J. Mattusch, R. Wennrich, and P. Morgenstern, *Micropor. Mesopor. Mater.*, Vol. 46, No. 2–3, pp. 277–286, 2001.
92. Y.-H. Xu, T. Nakajima, and A. Ohki, *J. Hazard. Mater.*, Vol. 92, pp. 275–287, 2002.
93. M.S. Onyango, H. Matsuda, and T. Ogada, *J. Chem. Eng. Jpn.*, Vol. 36, No. 4, pp. 477–485, 2003.
94. S. Ouvrard, M.O. Simonnot, and M. Sardin, *Ind. Eng. Chem. Res.*, Vol. 41, pp. 2785–2791, 2002.
95. S. Ouvrard, M.O. Simonnot, P. Donat, and M. Sardin, *Ind. Eng. Chem. Res.*, Vol. 41, pp. 6194–6199, 2002.
96. T.S. Singh, and K.K. Pant, *Sep. Purif. Technol.*, Vol. 36, pp. 139–147, 2004.
97. S. Kuriakose, T.S. Singh, and K.K. Pant, *Water Qual. Res. J. Can.*, Vol. 39, No. 3, pp. 258–266, 2004.
98. J. Hlavay, and K. Polyak, *J. Colloid Interf. Sci.*, Vol. 284, Issue 1, pp.71–77, 2005.
99. M.E. Pena, G.P. Korfiatis, M. Patel, L. Lippincott, and X. Meng, *Water Res.*, Vol. 11, pp. 2327–2337, 2005.
100. P.K. Dutta, A.K. Ray, V.K. Sharm, and F.J. Millero, *J. Colloid Interf. Sci.*, Vol. 278, pp. 270–275, 2004.
101. S. Bang, M. Patel, L. Lippincott, and X. Meng, *Chemosphere*, Vol. 60, Issue 3, pp. 389–397, 2005.
102. S. Tokunaga, S.A. Wasay, and S.W. Park, *Water Sci. Technol.*, Vol. 35, No. 7, pp. 71–78, 1997.
103. J.A. Wilkie, J.G. Hering, *Colloid, A: Physiocochem. Eng. Aspects*, Vol.107, pp. 97–110, 1996.
104. O.S. Thirunavukkarasu, T. Viraghavan, and K.S. Suramanian, *Water Air Soil Poll.*, Vol. 142, pp. 95–111, 2003.
105. W. Driehaus, M. Jekel, and U. Hildebrandt, *Water SRT-Aqua*, Vol. 47, No. 1, pp. 30–35, 1998.
106. M. Badruzzaman, P. Westerhoff, and D.R.U. Knappe, *Water Res.*, Vol. 38, Issue 18, pp. 4002–4012, 2004.
107. B.J. Lafferty, and R.H. Loeppert, *Environ. Sci. Technol.*, Vol. 39, No. 7, pp. 2120–2127, 2005.
108. E.A. Deliyanni, D.N. Bakoyannakis, A.I. Zouboulis, and E. Peleka, *Sep. Sci. Technol.*, Vol. 38, No. 16, pp. 3967–3981, 2003.
109. E.A. Deliyanni, D.N. Bakoyannakis, A.I. Zouboulis, and K.A. Matis, *Chemosphere*, Vol. 50, Issue 1, pp. 155–163, 2003.
110. P.M. Solozhenkin, E.A. Deliyanni, V.N. Bakoyannakis, A.I. Zouboulis, and K.A. Matis, *J. Mining Sci.*, Vol. 39, No. 3, pp. 287–296, 2003.

111. D. Pokhrel, and T. Viraraghavan, *J. Hazard. Mater.*, Vol. 150, pp. 818–825, 2008.
112. C.L. Brierley, *Geomicrobiol. J.*, Vol. 8, pp. 201–223, 1990.
113. E.A. Deliyanni, D.N. Bakoyannakis, A.I. Zouboulis, and K.A. Matis, *Chemosphere*, Vol. 50, Issue 1, pp. 155–163, 2003.
114. B.J. Mcafee, W.D. Gould, J.C. Nedeau, and A.C.A. da Costa, *Sep. Sci. Technol.*, Vol. 36, No. 14, pp. 3207–3222, 2001.
115. C.M. Elson, D.H. Davies, and E.R. Hayes, *Water Res.*, Vol. 14, Issue 9, pp. 1307–1311, 1980.
116. L. Dambies, A. Roze, and E. Guibal, *Adv. Chitin Sci.*, Vol.4, pp. 302–309, 2000.
117. C.T. Kamala, K.H. Chu, N.S. Chary, P.K. Pandey, S.L. Ramesh, A.R.K. Sastry, and K. Chandra Sekhar, *Water Res.*, Vol. 39, Issue 13, pp.2815–2826, 2005.
118. A. Basu, S. Kumar, and S. Mukherjee, *Indian J. Environ. Health*, Vol. 45, No. 2, pp. 143–150, 2003.
119. S. Ridvan, Y. Nalan, and D. Adil, *Sep. Sci. Technol.*, Vol. 38, No. 9, pp. 2039–2053, 2003.
120. R. Say, N. Yilmaz, and A. Denizli, *Sep. Sci. Technol.*, Vol. 38, No. 9, pp. 2039–2053, 2003.
121. M.C. Teixeira, and V.S.T. Ciminelli, *Environ. Sci. Technol.*, Vol. 39, No. 3, pp. 895–900, 2005.
122. S. Sinha, K. Pandey, D. Mohan, and K.P. Singh, *Ind. Eng. Chem. Res.*, Vol. 42, No. 26, pp. 6911–6918, 2003.
123. F.E. Chigbo, R.W. Smith, and F.L. Shore, *Environ. Poll. Ser. A: Ecol. Biol.*, Vol. 27, No. 1, pp. 31–36, 1982.
124. K.S. Low, and C.K. Lee, *Pertanika*, Vol. 13, No. 1, pp. 129–131, 1990.
125. N.M. Wasiuddin, M. Tango, and M.R. Islam, *Energy Sources*, Vol. 24, pp. 1031–1041, 2002.
126. C. Li, W. Xu, D. Jia, and X. Liu, *J. Chem. Eng. Data*, Vol. 58, pp. 427–435, 2013.
127. D. Vu, X. Li, Z. Li, and C. Wang, *J. Chem. Eng. Data*, Vol. 58, pp. 71–77, 2013.
128. A. Gupta, M. Yunus, and N. Sankararamakrishnan, *Ind. Eng. Chem. Res.*, Vol. 52, pp. 2066–2072, 2013.
129. S. Rijith, T.S. Anirudhan, and T. Shripathi, *Ind. Eng. Chem. Res.*, Vol. 51, pp. 10682–10694, 2012.
130. W. Yan, R. Vasic, A.I. Frenkel, and B.E. Koel, *Environ. Sci. Technol.*, Vol. 46, pp. 7018–7026, 2012.
131. R. Prucek, J. Tuček, J. Kolařík, J. Filip, Z. Marušák, V.K. Sharma, and R. Zbořil, *Environ. Sci. Technol.*, Vol. 47, pp. 3283–3292, 2013.
132. X. Yu, S. Tong, M. Ge, J. Zuo, C. Cao, and W. Song, *J. Mater. Chem. A*, Vol. 1, pp. 959–965, 2013.
133. M. Leist, R.J. Casey, and D. Caridi, *J. Hazard. Mater.*, Vol. 76, No. 1, pp. 125–138, 2000.

134. R.K. Dhar, B.K. Biswas, and G. Samanta, *Curr. Sci.*, Vol. 73, No. 1, pp. 48–59, 1997.
135. D. Zhang, Y. Jia, J. Ma, and Z. Li, *J. Hazard. Mater.*, Vol. 195, pp. 398–404, 2011.
136. M.F. Ahmed, S. Ahuja, M. Alauddin, S.J. Hug, J.R. Lloyd, A. Pfaff, T. Pichler, C. Saltikov, M. Stute, and A. Van Geen, *Water Bangladesh*, Vol. 314, No. 5806, pp. 1687–1688, 2006.

4

Prediction and Optimization of Heavy Clay Products Quality

Milica Arsenović[1,*], Lato Pezo[2], Lidija Mančić[3] and Zagorka Radojević[1]

[1]*Institute for testing of materials IMS, Belgrade, Serbia*
[2]*University of Belgrade, Institute of general and physical chemistry, Belgrade, Serbia*
[3]*Institute of Technical Sciences of SASA, Belgrade, Serbia*

Abstract
The effects of chemical composition, firing temperature (800–1100 °C), and several shape formats of laboratory brick samples on the final product quality were investigated. Prediction of the final laboratory products parameters was evaluated by second order polynomial regression models (SOPs) and artificial neural networks (ANNs), and afterwards compared to experimental results. SOPs showed high r^2 values (0.897–0.913 for compressive strength models, 0.942–0.962 for water absorption, 0.928 for firing shrinkage, 0.988–0.991 for water loss during firing, and 0.941 for volume mass of cubes models). An ANN model, coupled with sensitivity analysis, was obtained with high prediction accuracy: 0.866–0.939 for compressive strength models, 0.954–0.974 for water absorption, 0.882 for firing shrinkage, 0.982–0.988 for water loss during firing, and 0.920 for volume mass of cubes models. The optimal samples' chemical composition and firing temperature were chosen depending on a final usage of the raw material in heavy clay brick industry.

Keywords: Heavy clay products, prediction, optimization

4.1 Introduction

Heavy clays are used as the main raw material in the brick-making industry due to their specific properties. Final properties of clay brick are

*Corresponding author: milica.arsenovic@institutims.rs

Ashutosh Tiwari and Mikael Syväjärvi (eds.) Advanced Materials for Agriculture, Food, and Environmental Safety, (87–120) 2014 © Scrivener Publishing LLC

influenced by many factors due to the heterogeneous nature of used raw materials. Mineral transformations during firing are influenced by various factors: chemical and mineralogical composition, grain-size distribution, maximum heating temperature, heating rate, duration of firing, and kiln atmosphere. The knowledge of the chemical composition of heavy clays is essential for defining suitable compositions required for brick production. Important parameters of fired products that help in evaluation of the heavy clay deposits are compressive strength (*CS*), water absorption (*WA*), firing shrinkage (*FS*), weight loss during firing (*WLF*) and volume mass of cubes (*VMC*). These parameters can be also indicative for final products' durability and quality. Defining the mutual relationship among characteristics of heavy clay, processing conditions, and properties of the final products is a topic that has been extensively studied in previous years [1–7]. Various reports have been published, but only a few of them are related to the application of the mathematical modeling for better understanding of the overall system behavior and the quality of final products. Recently, mathematical modeling has been increasingly used for the study of the given systems and composition to final product properties relations [1–4, 8, 9]. Developed empirical models give a reasonable fit to experimental data and successfully predict technological parameters. In the case of brick production, nonlinear models are found to be more suitable due to variability in chemical composition of raw material and nonlinear behavior during firing of shaped products. The heterogeneous nature of the heavy clays implied a complex material-process-product relationship. Therefore, second order polynomial (SOP) and artificial neural network (ANN) models have gained momentum for modeling and control of brick processing [1–4].

Artificial neural network (ANN) models are recognized as a good modeling tool since they provide the solution to the problems from a set of experimental data, and are capable of handling complex systems with nonlinearities and interactions between decision variables [4]. Sensitivity analysis was used for ANN outputs testing, by changing one input variable at a time, for +2.241% or -2.241% of its nominal value, while all the other variables were fixed to their baseline values. Input variables' effects were calculated with reference to the same central point in the input space, thus increasing the comparability of the results obtained [1, 10].

The specific objective in this study was to investigate the effect of chemical composition and firing temperature on compressive strength, water absorption, firing shrinkage, and weight loss during firing and volume mass of cubes, for the brick production process. The performance of ANN was compared with the performance of second order polynomial models

(SOPs). The focus of this study was to determine the optimal material composition and firing temperature, depending on a final usage of the raw material in heavy clay brick industry. This approach enables shortening of all the necessary analyses when exploring brick clay deposits, by predicting the final products' features on the basis of raw material chemical characteristics. The optimization algorithm used in this study is concerned with finding the maxima of objective functions, subject to constraints. Described mathematical models, applied together, will be extremely helpful in preserving the raw natural resources, and lowering the industrial waste and energy consumption.

4.2 Materials and Methods

4.2.1 Raw Materials and Samples

Representative 139 heavy clays tested are collected in Serbia from the new opened brick factory deposits. Prior to sample preparation, all the raw samples were oven-dried at 105°C and then milled following the usual practice in ceramic laboratories [4]. To simulate industrial processing conditions, the heavy clays were moistened, mixed, and milled to the particle size of 2 mm until homogeneous mass was obtained. Moisture in the samples was uniformed by resting for 24 hours in sealed nylon bags. A laboratory vacuum extruder (Händle) was used to produce tiles (120mm x 50mm x 14mm), hollow blocks with vertical voids (55.3mm x 36mm x 36mm), and cubes (30mm x 30mm x 30mm). Afterwards, samples were slightly dried in the air and in a laboratory dryer at 105±5 °C [3]. The firing step was done in the computer-controlled oxygen atmosphere furnace, by choosing the slow firing regime: up to 610°C, the average heating rate was 1.4°C/min, while at higher temperatures it was increased to 2.5°C/min until the final temperature was reached at which the samples were treated for 2 hours. All tested samples were fired under the same regime up to the final temperatures of 800, 820, 850, 870, 900, 930, 950, 1050, and 1100°C. Afterwards the samples were furnace-cooled to room temperature.

4.2.2 Chemical and Technological Features

Heavy clay chemical composition was tested by classical silicate analysis [11], and the results are presented as average values determined by three measurements.

Water absorption of differently shaped samples was evaluated in fired clay pieces by soaking in distilled water for 24 hours, according to standard procedure [12].

Compressive strength was determined in laboratory hydraulic press Alfred Amsler, CHD, with a measuring range of 100/200/500/1,000 kN and resolution of 0.1/0.2/0.5/1.0 kN. Three specimens for each combination of sample shape (blocks and cubes) and firing temperature were tested [3]. The strength results reported were the average of three specimens with a variation of no more than 10%.

4.2.3 Second Order Polynomial Model and Analysis of Variance

Before ANN modeling, it is recommended to make statistical analysis of the available data, like analysis of variance (ANOVA), in order to check the significant effect of the input variables over the output and to justify the later use of ANN model by coefficient of determination (r^2). The SOP models were obtained for each dependent variable, where factors were rejected when their significance level was $p>0.05$. All the analysis and mathematical modeling was performed using StatSoft Statistica 10.0 software.

The SOP model estimated the main effect of the process variables on final products' properties during the production of heavy clay bricks. The independent variables were: the content of major oxides (SiO_2, Al_2O_3, Fe_2O_3, CaO, MgO, Na_2O, K_2O, MnO, and TiO_2) and firing temperature (in the range of 800–1100°C). The influences of chemical composition and firing temperature on some of the SOP models were published before - compressive strength (*CSB* for blocks and *CSC* for cubes); water absorption (for tiles—*WAT*, for blocks—*WAB*, and for cubes—*WAC* [3], so here the impact of independent variables on: firing shrinkage *FS*; weight loss during firing of tiles—*WLFT*, blocks—*WLFB* and cubes—*WLFC*; and volume mass of cubes—*VMC* is presented. All SOP models were fitted to data collected by experimental measurements. The models of the following form were developed to relate five dependent outputs (Y) to nine process variables (X):

$$Y_k = \beta_{k0} + \sum_{i=1}^{10} \beta_{ki} X_i + \sum_{i=1}^{10} \beta_{kii} X_i^2 + \sum_{i=1}^{10} \sum_{j=j+1}^{10} \beta_{kij} X_i X_j \tag{4.1}$$

where β_{kn} are constant regression coefficients; Y_k, either *CSB*, *CSC*, *WAT*, *WAB*, *WAC*, *FS*, *WLFT*, *WLFB*, *WLFC*, or *VMC*; while X_k are either metal oxides content or production temperature. The significant terms in the model were found using ANOVA for each dependent variable.

4.2.4 Artificial Neural Network Modeling

According to StatSoft Statistica's recommendations, the database is randomly divided into: training data (60%), cross-validation (20%) and testing data (20%). The cross-validation data set was used to test the performance of the network while training was in progress as an indicator of the level of generalization and the time at which the network has begun to over train. Testing data set was used to examine the network generalization capability.

To improve the behavior of the ANN, both input and output data were normalized according to Eq. (4.2):

$$q_{inorm} = \frac{q_i - \min(q_i)}{\max(q_i) - \min(q_i)} \tag{4.2}$$

where q_i is i-th case, with measured compressive strength (CS), water absorption (WA), firing shrinkage (FS), weight loss during firing (WLF) and volume mass of cubes (VMC). Normalized variables gained values in the range of 0 to 1, and have no physical meaning.

In order to obtain a good network behavior, it was necessary to make a trial and error procedure and also to choose the number of hidden layers, and the number of neurons in hidden layer(s). The use of only one layer is advisable, because more layers exacerbates the problem of local minima [4, 13].

A multi-layer perceptron model (MLP) consisted of three layers (input, hidden, and output), which is the most common, flexible, and general-purpose kind of ANN. Such a model has been proven as quite capable of approximating nonlinear functions [4, 13], which is the reason for choosing it in this study. The network consists of one layer of linear output neurons and one hidden layer of nonlinear neurons. The MLP neural network learns using an algorithm called "backpropagation." The Levenberg–Marquardt algorithm is proven to be the fastest and particularly adapted for networks of moderate size. During this iterative process, input data are repeatedly presented to the network [14].

The first estimation of the number of neurons can be obtained from the following equation [15, 16]:

$$m = n\cdot(x+1) + y\cdot(n+1) \tag{4.3}$$

where x and y represent the number of input and output neurons, respectively, n is the number of neurons in the hidden layer and m is the number of weights (connections between layers) in the neural network. m can be taken as the number of training exemplars divided by 10. Some suggestions regarding the number of hidden neurons are as follows: this number should be between the sizes of the input and output layers, it should be 2/3 the size of the input layer, plus the size of the output layer, or less than twice the size of the input layer [16].

In this work, the ANN procedure of StatSoft Statistica was used to model the ANN, and the number of hidden neurons varied from $n = 5$ to 13 (proposed by the program used). There were $x = 10$ inputs, $y = 10$ outputs, and m = 115 to 283 weight coefficients (depending on n). Broyden–Fletcher–Goldfarb–Shanno (BFGS) algorithm, implemented in StatSoft Statistica's evaluation routine, was used for ANN modeling. The information is transferred, between the layers through a "transfer" or "activation" function. This function is typically nonlinear for hidden layers and linear for the output layer. Most common nonlinear activation functions, used in StatSoft Statistica ANN calculation, are logistic, sigmoid, hyperbolic, and tangent functions (also exponential, sine, softmax, Gausian). In most applications, hyperbolic tangent function behaves better as compared to the other functions [13].

Coefficients associated with the hidden layer (both weights and biases) are grouped in matrices W_1 and B_1. Similarly, coefficients associated with the output layer are grouped in matrices W_2 and B_2. If Y is the matrix of the output variables, f_1 and f_2 are transfer functions in the hidden and output layers, respectively, and X is the matrix of input variables, it is possible to represent the neural network, by using matrix notation, as follows [17]:

$$Y = f_1(W_2 \cdot f_2(W_1 \cdot X + B_1) + B_2) \tag{4.4}$$

Weights (elements of matrices W_1 and W_2) are determined during the training step, which updates them using optimization procedures to minimize the error function between network and experimental outputs [15, 16], evaluated according to the sum of squares (SOS) and BFGS algorithm, used to speed up and stabilize convergence [18].

4.2.4.1 *Training, Testing and System Implementation*

After defining the architecture of ANN, the training step is initiated. The training process was repeated several times in order to get the best performance of the ANN, due to a high degree of variability of parameters. It was accepted that successful training was achieved when learning and cross-validation curves (SOS vs. training cycles) approached zero. Testing was carried out with the best weights stored during the training step. Correlation coefficient *r* and *SOS* were used as parameters to check the performance (i.e., the accuracy) of the obtained ANNs.

After the best-behaved ANN is chosen, the model is implemented using an algebraic system of equations to predict *CS*, *WA*, *FS*, *WLF*, and *VMC*, by substitution of the corresponding weights and coefficients matrices in eq. (4.4). This step can be easily achieved in some spreadsheet calculus (Microsoft Office Excel, for instance).

4.2.4.2 Sensitivity Analysis

Sensitivity analysis is a sophisticated technique that is necessary to use for studying the effects of observed input variables and also the uncertainties in obtained models and general network behavior [13, 19]. Neural networks were tested using sensitivity analysis, to determine whether and under what circumstances the combination of obtained models and the expected actual training data might result in an ill-conditioned system [13]. On the basis of developed ANN models, sensitivity analysis is performed in order to more precisely define the influence of major oxides content on the observed outputs. The white noise signals were incorporated by adding or subtracting the Gaussian error of standard deviation (SD = 5%) and zero mean with 98% probability, i.e., ±2.241·SD to each input variable [4, 20]. The white noise was normally distributed with a constant intensity and frequency. It was used to test the model sensitivity and measurement errors.

4.2.5 Fuzzy Synthetic Optimization

Optimization procedure was performed using Fuzzy Synthetic Evaluation (FSE) algorithm implemented in Microsoft Excel 2007, in order to determine the workable optimum conditions for the thermal processing of heavy clay bricks [3]. FSE method was implemented, using the results of models proposed to represent *CS*, *WA*, *FS*, *WLF*, and *VMC*, using eq. (4.5). FSE is a commonly used technique to solve problems with constraints involving nonlinear functions. The method aims to solve a sequence of simple problems whose solutions converge to the solution of the original problem.

Trapezoidal membership function used in this calculation could be written as:

$$A(x,a,m,n,b)=\begin{cases} a\leq x<m, & \dfrac{x-a}{m-a} \\ m\leq x<n, & 1 \\ n\leq x<b, & 1-\dfrac{x-n}{b-n} \end{cases} \quad (4.5)$$

Where x is whether *CS*, *WA*, *FS*, *WLF*, and *VMC*, and the values of a, b, m, and n are function parameters. Interval a - b represents the range in which measured values occurred, while range m - n is the expected optimal value range for output variables, chosen for certain product groups.

4.3 Results and Discussions

Experimental data obtained (chemical composition and physical-mechanical characteristics) were presented using basic descriptive statistics of obtained data, in table 4.1. Both chemical content and final products characteristics varied significantly, implying that fitting of the experimental data could be performed using SOP and ANN modeling.

The 139 heavy clay samples showed the expected typical chemical compositions, rich in silica and aluminum, accompanied by relatively significant amounts of iron oxides, with satisfying content of potash, and minor contents of titan. Harmful constituents that can cause efflorescence were not detected in most of the samples [21–23]. Large differences among technological characteristics indicated that the tested samples contained various contents of clay and other minerals.

4.3.1 Correlation Analysis

The correlation coefficients between independent and dependent parameters in this research are presented in table 4.2. Most of the coefficients were statistically significant ($p<0.01$), due to the large number of measurements. The analysis is done to test the parameters prior to further investigation of the models. Since correlations presume linear relationships [24], the obtained coefficients were not very high, and the non-linearity of the systems was indicated. The highest positive correlations were observed in the

Table 4.1 Descriptive statistics of experimental data for the SOP and ANN modeling.

	SiO_2	Al_2O_3	Fe_2O_3	CaO	MgO	Na_2O	K_2O	MnO	TiO_2	
min	36.39	8.76	3.00	0.43	0.12	0.01	0.40	0.00	0.01	
max	68.28	21.54	13.19	16.57	6.26	2.59	3.87	0.24	1.68	
average	56.18	15.14	5.57	6.36	2.41	1.08	2.18	0.09	0.64	
	CSB	**CSC**	**WAT**	**WAB**	**WAC**	**FS**	**WLFT**	**WLFB**	**WLFC**	**VMC**
min	4.95	8.46	1.11	0.65	4.76	2.64	3.86	3.98	4.32	1.38
max	137.18	180.74	30.95	31.38	27.20	17.17	19.55	19.79	19.65	2.42
average	28.94	54.63	17.14	17.08	13.68	0.19	9.14	9.38	9.72	1.80

Table 4.2 Correlation analysis.

	CSB	CSC	WAT	WAB	WAC	FS	WLFT	WLFB	WLFC	VMC
SiO_2	0.364^{+}	0.308^{+}	-0.685^{+}	-0.681^{+}	-0.672^{+}	-0.193^{+}	-0.810^{+}	-0.819^{+}	-0.814^{+}	0.614^{+}
Al_2O_3	0.530^{+}	0.473^{+}	-0.561^{+}	-0.556^{+}	-0.548^{+}	0.030^{ns}	-0.589^{+}	-0.601^{+}	-0.594^{+}	0.585^{+}
Fe_2O_3	0.459^{+}	0.469^{+}	-0.640^{+}	-0.643^{+}	-0.618^{+}	0.035^{ns}	-0.622^{+}	-0.616^{+}	-0.608^{+}	0.655^{+}
CaO	-0.594^{+}	-0.527^{+}	0.883^{+}	0.877^{+}	0.860^{+}	0.072^{+}	0.934^{+}	0.942^{+}	0.928^{+}	-0.818^{+}
MgO	-0.422^{+}	-0.347^{+}	0.591^{+}	0.598^{+}	0.595^{+}	0.086^{+}	0.618^{+}	0.633^{+}	0.631^{+}	-0.555^{+}
Na_2O	-0.148^{+}	-0.134^{+}	-0.069^{*}	-0.062^{*}	-0.049^{**}	-0.052^{**}	-0.252^{+}	-0.230^{+}	-0.232^{+}	0.032^{ns}
K_2O	0.170^{+}	0.195^{+}	-0.245^{+}	-0.245^{+}	-0.221^{+}	0.080^{+}	-0.219^{+}	-0.223^{+}	-0.244^{+}	0.196^{+}
MnO	0.044^{ns}	0.045^{**}	-0.104^{+}	-0.103^{+}	-0.101^{+}	-0.100^{+}	-0.137^{+}	-0.157^{+}	-0.181^{+}	0.087^{+}
TiO_2	0.152^{+}	0.150^{+}	-0.320^{+}	-0.321^{+}	-0.308^{+}	0.063^{*}	-0.251^{+}	-0.244^{+}	-0.239^{+}	0.291^{+}

$^{+}$Significant at $p<0.01\%$ level, *Significant at $p<0.05\%$ level, **Significant at $p<0.10$ level, 95% confidence limit, ^{ns}Not significant case of CaO content to *WLF* (0.928–0.942) and *WA* (0.860–0.883). MgO content is mostly influenced by the same dependent parameters, but in a lower extent then CaO (0.591–0.633). This is because of the fact that CaO and MgO are mainly hosted with carbonates, which lower *CS* and increase *WA* and *WLF*. The next significant influence observed is relation of SiO_2 to *WLF* (around –0.81) and *WA* (around –0.68). Actually, SiO_2 is present within the clay minerals, but also as free quartz, which is mostly fine-grained since it had a positive effect on *CS* [2, 25, 26]. Relatively high and negative correlations are obtained for Fe_2O_3 – *WA* (from -0.618 to –0.643) and Fe_2O_3 – *WLF* (around -0.62) due to the increase in *CS*. Taking all this into account *CS* is mostly influenced by CaO (negative impact) and Al_2O_3 content (positive impact).

4.3.2 Analysis of Variance and SOP Models

Analysis of variance (ANOVA) was conducted for obtained SOP models, and outputs were tested against the impact of input variables, in table 4.3. ANOVA was performed to demonstrate the influences of the inputs on the outputs, and these are quantified by the sum of squares (SOS). Results concerning *WA* and *CS* were shown elsewhere [3], while here testing of the SOP models for *FS*, *WLF*, and *VMC* is presented. All the variables that entered into the analysis showed a significant effect on the outputs, either as linear or square members, or in the form of a product of two variables. *CS* and *WA* were the most sensitive to the chemical composition, compared to all other physico-mechanical parameters that characterize the final products. Furthermore, the coefficients of determination (r^2) were used to explore the confidence level of the models. SOPs showed high r^2 values (0.897–0.913 for compressive strength models, 0.942–0.962 for water absorption, 0.928 for firing shrinkage, 0.988–0.991 for water loss during firing, and 0.941 for volume mass of cubes models). All variables considered in the ANOVA analysis were also used for the ANN modeling (*See sections 4.3.3 and 4.3.4*).

The parameters that most influence the compressive strength, statistically significant at $p<0.05$ (confidence level of 95%), were the quadratic terms of temperature and CaO contents, whereas the value of CSB was the most sensitive to the content of CaO and CSC to the firing temperature. The third-largest impact on CS in both cases was SiO_2xCaO term. ANOVA, for most of the observed parameters, showed higher influence in the case of CSC than CSB, confirming that the compressive strength of cubes changes more drastically, especially with the firing temperature. The highest effect on water absorption showed the content of CaO (square

Table 4.3 Analysis of variance for the SOP models.

			Sum of squares				
Term	**Source**	**dF**	***FS***	***WLFT***	***WLFB***	***WLFC***	***VMC***
Linear	SiO_2	1	8.60^{*}	6.21^{*}	20.42^{*}	33.14^{*}	0.06^{*}
	Al_2O_3	1	87.00^{*}	11.03^{*}	18.85^{*}	21.47^{*}	0.00^{ns}
	Fe_2O_3	1	10.62^{*}	5.80^{*}	14.13^{*}	20.36^{*}	0.04^{*}
	CaO	1	0.99^{*}	0.57^{*}	6.28^{*}	7.51^{*}	0.02^{*}
	MgO	1	0.16^{ns}	2.24^{*}	6.65^{*}	30.90^{*}	0.01^{*}
	Na_2O	1	0.01^{ns}	68.22^{*}	100.60^{*}	39.91^{*}	0.01^{*}
	K_2O	1	13.82^{*}	1.34^{*}	1.22^{*}	4.11^{*}	0.00^{ns}
	MnO	1	0.09^{ns}	14.98^{*}	16.04^{*}	6.00^{*}	0.00^{ns}
	TiO_2	1	81.21^{*}	7.96^{*}	10.39^{*}	0.00^{ns}	0.00^{ns}
	Temp	1	0.06^{ns}	14.28^{*}	16.37^{*}	8.97^{*}	0.03^{*}
Quadratic	SiO_2	1	25.05^{*}	4.90^{*}	8.91^{*}	9.50^{*}	0.04^{*}
	Al_2O_3	1	13.85^{*}	5.58^{*}	3.05^{*}	1.18^{*}	0.00^{ns}
	Fe_2O_3	1	52.81^{*}	0.68^{*}	0.41^{**}	3.70^{*}	0.02^{*}
	CaO	1	0.29^{ns}	2.09^{*}	5.86^{*}	1.63^{*}	0.20^{*}
	MgO	1	4.72^{*}	3.15^{*}	4.26^{*}	24.52^{*}	0.01^{*}
	Na_2O	1	17.80^{*}	0.98^{*}	10.07^{*}	1.90^{*}	0.01^{*}
	K_2O	1	0.72^{*}	0.02^{*}	4.50^{*}	5.31^{*}	0.13^{*}
	MnO	1	0.97^{*}	13.35^{*}	32.29^{*}	18.95^{*}	0.02^{*}
	TiO_2	1	2.18^{*}	0.63^{*}	0.12^{ns}	0.02^{ns}	0.13^{*}
	Temp	1	3.31^{*}	20.48^{*}	11.27^{*}	3.92^{*}	0.00^{ns}
Interchange	SiO_2 x Al_2O_3	1	14.67^{*}	10.61^{*}	7.73^{*}	3.04^{*}	0.00^{ns}
	SiO_2 x Fe_2O_3	1	32.79^{*}	2.89^{*}	0.04^{ns}	2.14^{*}	0.02^{*}
	SiO_2 x CaO	1	14.01^{*}	3.43^{*}	9.28^{*}	7.23^{*}	0.14^{*}
	SiO_2 x MgO	1	4.52^{*}	3.69^{*}	2.67^{*}	15.99^{*}	0.01^{*}
	SiO_2 x Na_2O	1	20.76^{*}	2.86^{*}	19.91^{*}	0.51^{*}	0.00^{ns}
	SiO_2 x K_2O	1	12.54^{*}	10.46^{*}	2.58^{*}	5.64^{*}	0.10^{*}
	SiO_2 x MnO	1	0.00^{ns}	6.59^{*}	2.98^{*}	0.29^{**}	0.04^{*}
	SiO_2 x TiO_2	1	0.60^{**}	13.06^{*}	12.08^{*}	0.00^{ns}	0.01^{*}
	SiO_2 x Temp	1	1.15^{*}	8.20^{*}	13.68^{*}	6.37^{*}	0.25^{*}
	Al_2O_3 x Fe_2O_3	1	116.80^{*}	0.47^{**}	0.20^{ns}	0.26^{ns}	0.03^{*}
	Al_2O_3 x CaO	1	30.24^{*}	11.16^{*}	6.75^{*}	1.21^{*}	0.00^{ns}

Sum of squares							
Term	Source	dF	*FS*	*WLFT*	*WLFB*	*WLFC*	*VMC*
Interchange	Al_2O_3 x MgO	1	0.39 ns	2.64*	0.61*	8.81*	0.01*
	Al_2O_3 x Na_2O	1	W15.74*	0.93*	13.60*	1.95*	0.22*
	Al_2O_3 x K_2O	1	24.40*	3.13*	0.39**	0.05 ns	0.07*
	Al_2O_3 x MnO	1	0.06 ns	0.32 ns	1.05*	3.42*	0.03*
	Al_2O_3 x TiO_2	1	14.28*	4.88*	4.49*	0.04 ns	0.02*
	Al_2O_3 x Temp	1	4.02*	16.30*	9.92*	5.85*	0.31*
	Fe_2O_3 x CaO	1	1.81*	1.76*	0.13 ns	1.24*	0.01*
	Fe_2O_3 x MgO	1	0.49**	0.08	1.32*	10.34*	0.00 ns
	Fe_2O_3 x Na2O	1	0.66**	32.73*	45.21*	10.85*	0.00 ns
	Fe_2O_3 x K_2O	1	60.02*	1.95*	1.78*	0.59*	0.01*
	Fe_2O_3 x MnO	1	9.33*	22.36*	14.20*	3.79*	0.00 ns
	Fe_2O_3 x TiO_2	1	110.42*	3.88*	5.15*	0.84*	0.03*
	Fe_2O_3 x Temp	1	20.12*	0.30 ns	0.71*	0.01 ns	0.00 ns
	CaO x MgO	1	12.42*	2.51*	1.72*	14.85*	0.00 ns
	CaO x Na_2O	1	6.12*	28.94*	67.17*	16.79*	0.00 ns
	CaO x K_2O	1	0.87*	1.13*	0.03 ns	0.27 ns	0.09*
	CaO x MnO	1	1.22*	3.72*	0.76 ns	3.10*	0.02*
	CaO x TiO_2	1	3.77*	15.79*	13.98*	0.64*	0.01*
	CaO x Temp	1	4.44*	13.48*	13.96*	3.24*	0.38*
	MgO x Na_2O	1	9.95*	25.30*	7.97*	7.19*	0.02*
	MgO x K_2O	1	15.56*	4.31*	1.28*	0.39**	0.03*
	MgO x MnO	1	0.19 ns	0.10 ns	5.64*	8.82*	0.01*
	MgO x TiO_2	1	4.60*	0.41 ns	0.41*	2.19*	0.00 ns
	MgO x Temp	1	0.03 ns	6.06*	2.99*	11.62*	0.06*
	Na_2O x K_2O	1	13.30*	0.65*	1.02*	0.19 ns	0.04*

(Continued)

Table 4.3 (*Cont.*)

			Sum of squares				
Term	Source	dF	*FS*	*WLFT*	*WLFB*	*WLFC*	*VMC*
Interchange	Na_2O x MnO	1	0.24 ns	7.38*	7.16*	8.82*	0.01*
	Na_2O x TiO_2	1	3.81*	2.26*	8.68*	2.08*	0.00 ns
	Na_2O x Temp	1	43.69*	0.81*	2.15*	0.00 ns	0.01*
	K_2O x MnO	1	2.71*	1.84*	0.04 ns	15.15*	0.06*
	K_2O x TiO_2	1	70.20*	16.34*	20.64*	0.26 ns	0.00 ns
	K_2O x Temp	1	3.93*	2.29*	4.75*	0.23 ns	0.17*
	MnO x TiO_2	1	27.73*	1.49*	2.00*	7.50*	0.03*
	MnO x Temp	1	2.58*	1.11*	1.41*	3.86*	0.06*
	TiO_2 x Temp	1	42.51*	12.34*	9.10*	10.40*	0.01*
	Error	1310	231.57ns	191.20ns	143.47ns	140.466ns	1.892ns
	r^2		0.928	0.988	0.991	0.991	0.941

*Significant at p<0.05% level, **Significant at p<0.10 level, 95% confidence limit, nsNot significant, dF – degrees of freedom

member), though the values were ten to thirty times lower than in the case of CS. The influence of firing temperature was also statistically significant, but much less than in CS [3].

Unlike the *CS* and *WA*, most of the output parameters presented in table 4.3 were at higher impact of other major oxides contents, and not as much by firing temperature. The highest influence can be seen in *FS* in the case of $Al_2O_3xFe_2O_3$ and $Fe_2O_3xTiO_2$ terms. The next significant influence on *FS* showed Al_2O_3 and TiO_2 linear terms. Thus, *FS* is to be the most affected by biotite and chlorite content, but also hematite/rutile presence. Kaolinite and smectite were present in most of the samples, although to a low quantity. Their significant influence is reflected in $SiO_2xAl_2O_3$ term. Calcite significantly reduces *FS*, especially when present in the form of large concretions [2]. CaO influence in the case of *FS*, according to ANOVA, was the most pronounced in interchange parameters Al_2O_3 x CaO, SiO_2 x CaO and CaO x MgO, and thus showed the dependance of *FS* on anorthite, montmorillonite, and dolomite contents. Calcite can react with quartz, or Al_2O_3 and SiO_2 from clay mineral dehydroxylation to form calcium-sillicates, which increase *FS* [3, 27–29]. According to the SOP model developed, it is concluded that fine-grained calcite reacted to SiO_2 in most of the samples, since the terms SiO_2xCaO and Al_2O_3xCaO showed strong influence on *FS*. CaOxMgO (dolomite) influenced materials densification, which was observed by the influence in ANOVA calculation. Firing temperature mostly influenced *FS* in terms conjuncted to Na_2O, TiO_2, and Fe_2O_3.

Alkali oxides facilitate the creation of a liquid phase, which promotes densification in reaction with silicon and aluminum [30]. The highest impact on *WLF* parameters showed Na_2O linear term, mostly to *WLFB*. Na_2O i K_2O also induced a significant influence on *FS*, *WLFT*, and *WLFB* in the terms containing another flux, Fe_2O_3 [26]. In the case of *WLF* parameters, CaO was more influential in relation to Na_2O, TiO_2, MgO, Temp and SiO_2, but as the linear term. CaO content and some of the other fluxes gained strong influences on *WLF*, and also SiO_2 from quartz and/or clay minerals showed powerful impact, thus confirming calcium silicate formation. Firing temperature influenced *WLFT*, following the order of importance: quadratic term of Temp, Al_2O_3xTemp, and linear term of Temp. Laboratory tiles showed the highest sensitivity to firing temperature, concerning weight loss, since the greatest specific surface is exposed to furnace conditions, above all sample shapes. Firing temperature mostly influenced *WLFB* following the order of importance: linear term of Temp, CaOxTemp, and SiO_2xTemp. Weight loss depended on combined effects of firing temperature and CaO/SiO_2 contents, and it is being slightly increased with them. Hollow blocks were also sensitive to firing temperature, but less than tiles, and its impact is also combined to the calcite and quartz content. *WLFC* is influenced by firing temperature in the order: MgOxTemp, TiO_2xTemp, and linear term of temperature.

VMC mostly depended on CaOxTemp, Al_2O_3xTemp, and SiO_2xTemp. The fired material density mostly depended on carbonate content and firing temperature applied, but also on clay minerals content and densification abillity.

The residual variance, marked as error in table 4.3, presents the model disagreement with the experimental values (contributions of other members that are not described in the SOP model). The statistically significant discrepancy between experimental data and results obtained using mathematical models, would show inadequacy of the developed model. All the developed models showed statistically insignificant deviation from the experimental values of the model, which confirmed their suitability. The coefficient of determination, r^2 is also a share of output variability of the system, which was calculated based on regression analysis. High r^2 values

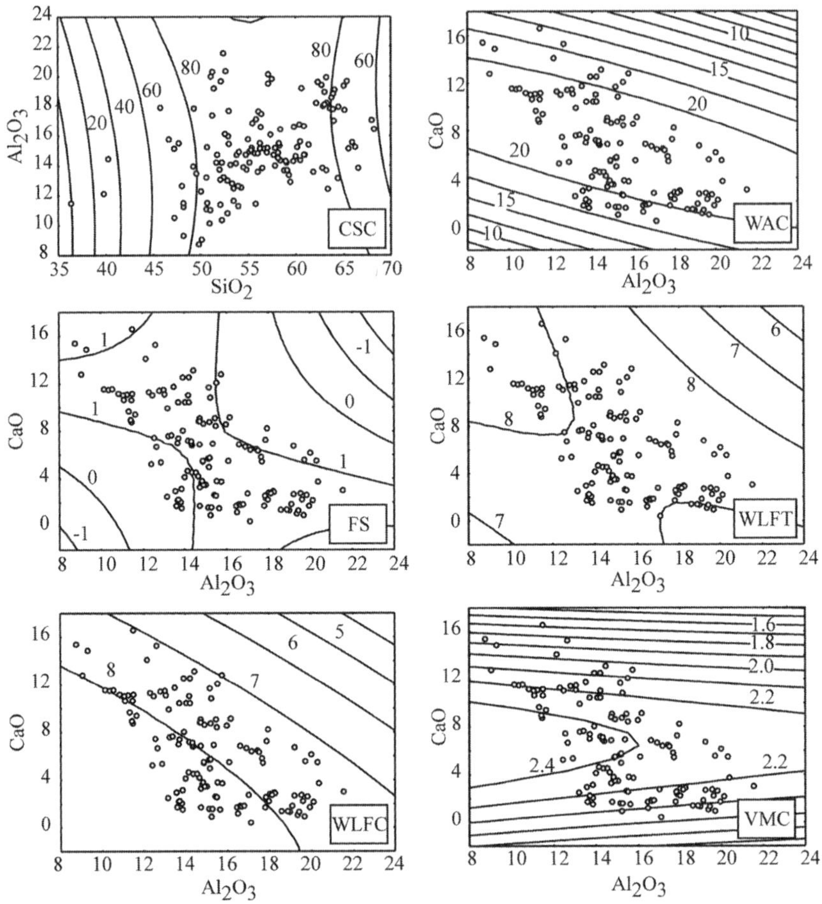

Figure 4.1 Some of the SOP models' most influential parameters

also indicated that the experimental data satisfactorily coincided with the mathematical models.

Figure 4.1 shows graphical dependence of the observed outputs on selected inputs that were the most influential. Analysis revealed that linear, quadratic, as well as interchange terms considerably influenced forming SOP models. It was concluded that all of the obtained models were statistically significant and in agreement with experimental results.

4.3.3 Neurons in the ANN Hidden Layer

Determination of the appropriate number of hidden layers and number of hidden neurons in each layer is one of the most critical tasks in ANN design. The number of neurons in a hidden layer depends on the complexity of the relationship between inputs and outputs. As this relationship becomes more complex, more neurons should be added [13].

The optimum number of hidden neurons was chosen upon minimizing the difference between predicted ANN values and desired outputs, using *SOS* during testing as a performance indicator (table 4.1). Results of *CSB*, *CSC*, *WAT*, *WAB*, *WAC*, *FS*, *WLFT*, *WLFB*, *WLFC*, and *VMC* during testing with five to thirteen neurons in the hidden layer are presented. Used MLPs are marked according to StatSoft Statistica's notation; MLP followed by number of inputs, number of neurons in the hidden layer, and the number of outputs. According to ANN performance, from table 4.4 (sum of r^2 and SOSs for all variables in one ANN), it was noticed that the optimal number of neurons in the hidden layer is thirteen (network MLP 10-13-10, No 4.), when obtaining high values of r^2 and also low values of *SOS*. It was noticed that a greater number of neurons increases the structure complexity, but does not necessarily significantly improve the network behavior [18], (during testing step MLP 10-13-10, No. 4 gained $r^2 = 0.949$, $SOS = 0.010$, while MLP 10-9-10, No. 6 gained $r^2 = 0.943$ and $SOS = 0.011$).

The goodness of fit, between experimental measurements and model-calculated outputs, represented as ANN performance (sum of r^2 between measured and calculated *CS*, *WA*, *FS*, *WLF*, and *VMC* for each ANN) and also the *SOS* between measured and calculated technological parameters, during training, testing and validation steps, are shown in table 4.4. The *SOS* between the experimental and the network predicted values was used as the iteration termination criterion, as StatSoft Statistica's default. As soon as the cross-validation *SOS* starts to increase, the training step is terminated; otherwise, the training step ends after a fixed number of epochs or training cycles [13].

Table 4.4 shows ANN performance data, expressed as the sum of r^2 and *SOS*, for all variables in one ANN, while table 4.5 presents r^2 for each variable (*CSB*, *CSC*, *WAT*, *WAB*, *WAC*, *FS*, *WLFT*, *WLFB*, *WLFC*, and *VMC*) during training, testing, and validation steps.

4.3.4 Simulation of the ANNs

Process outputs *CSB*, *CSC*, *WAT*, *WAB*, *WAC*, *FS*, *WLFT*, *WLFB*, *WLFC*, and *VMC* can be calculated by the eq. (4.4), using matrices W_1 and B_1, and matrices W_2 and B_2, which represent system, incorporating coefficients associated with the hidden layer (both weights and biases). Output variables are calculated by applying transfer functions f_1 and f_2 (from table 4.1)

Table 4.4 ANNs summary.

No.	Network Name	ANN performance			SOS (error)			Activation function	
		Train.	Test.	Valid.	Train.	Test.	Valid.	Hidden	Output
1	MLP 10-5-10	0.896	0.890	0.892	0.019	0.020	0.020	Tanh	Logistic
2	MLP 10-13-10	0.946	0.942	0.944	0.010	0.012	0.012	Tanh	Exponential
3	MLP 10-6-10	0.901	0.899	0.902	0.016	0.017	0.017	Exponential	Logistic
4	MLP 10-13-10	0.949	0.939	0.943	0.010	0.011	0.011	Tanh	Tanh
5	MLP 10-13-10	0.949	0.939	0.944	0.010	0.012	0.011	Logistic	Tanh
6	MLP 10-9-10	0.943	0.938	0.938	0.011	0.012	0.013	Exponential	Identity
7	MLP 10-10-10	0.942	0.936	0.939	0.011	0.013	0.012	Tanh	Identity
8	MLP 10-11-10	0.939	0.936	0.930	0.012	0.013	0.014	Exponential	Tanh
9	MLP 10-7-10	0.923	0.912	0.913	0.015	0.017	0.018	Logistic	Exponential
10	MLP 10-13-10	0.946	0.941	0.941	0.011	0.012	0.012	Tanh	Exponential

Table 4.5 Coefficients of determination, r^2, between experimentally measured and ANN outputs, during train, test, and validation steps.

	ANN name	*CSB*	*CSC*	*WAT*	*WAB*	*WAC*	*FS*	*WLFT*	*WLFB*	*WLFC*	*VMC*
Training											
1	MLP 10-5-10	0.822	0.768	0.925	0.921	0.920	0.774	0.979	0.982	0.976	0.895
2	MLP 10-13-10	0.913	0.895	0.959	0.960	0.965	0.882	0.986	0.988	0.987	0.919
3	MLP 10-6-10	0.884	0.827	0.937	0.938	0.945	0.630	0.982	0.981	0.979	0.903
4	MLP 10-13-10	0.912	0.906	0.962	0.962	0.972	0.881	0.987	0.989	0.989	0.925
5	MLP 10-13-10	0.915	0.909	0.959	0.959	0.965	0.897	0.989	0.991	0.990	0.914
6	MLP 10-9-10	0.909	0.889	0.952	0.953	0.962	0.900	0.983	0.987	0.985	0.912
7	MLP 10-10-10	0.906	0.878	0.950	0.951	0.959	0.898	0.987	0.989	0.986	0.921
8	MLP 10-11-10	0.894	0.882	0.947	0.949	0.954	0.880	0.987	0.990	0.989	0.917
9	MLP 10-7-10	0.870	0.851	0.931	0.935	0.930	0.863	0.981	0.982	0.981	0.906
10	MLP 10-13-10	0.913	0.914	0.955	0.955	0.958	0.876	0.985	0.988	0.987	0.926
Testing											
1	MLP 10-5-10	0.858	0.748	0.928	0.923	0.927	0.697	0.979	0.984	0.980	0.880
2	MLP 10-13-10	0.929	0.866	0.958	0.961	0.974	0.868	0.983	0.987	0.984	0.914
3	MLP 10-6-10	0.913	0.778	0.944	0.945	0.955	0.620	0.980	0.980	0.975	0.897
4	MLP 10-13-10	0.902	0.879	0.963	0.962	0.973	0.839	0.984	0.987	0.987	0.913

(*Continued*)

Table 4.5 (*Cont.*)

5	MLP 10-13-10	0.906	0.880	0.954	0.953	0.968	0.865	0.986	0.989	0.987	0.906
6	MLP 10-9-10	0.912	0.873	0.959	0.960	0.970	0.846	0.982	0.987	0.983	0.913
7	MLP 10-10-10	0.915	0.850	0.947	0.946	0.966	0.872	0.984	0.986	0.983	0.914
8	MLP 10-11-10	0.890	0.861	0.952	0.952	0.961	0.869	0.984	0.990	0.988	0.914
9	MLP 10-7-10	0.893	0.808	0.926	0.932	0.931	0.789	0.978	0.982	0.979	0.901
10	MLP 10-13-10	0.927	0.890	0.957	0.959	0.965	0.833	0.981	0.987	0.986	0.923
Validation											
1	MLP 10-5-10	0.855	0.783	0.925	0.924	0.915	0.710	0.973	0.978	0.971	0.885
2	MLP 10-13-10	0.939	0.889	0.954	0.958	0.965	0.864	0.983	0.986	0.982	0.916
3	MLP 10-6-10	0.880	0.800	0.933	0.936	0.941	0.688	0.981	0.981	0.976	0.903
4	MLP 10-13-10	0.906	0.904	0.967	0.968	0.971	0.847	0.985	0.988	0.987	0.910
5	MLP 10-13-10	0.909	0.916	0.957	0.958	0.962	0.863	0.988	0.990	0.988	0.907
6	MLP 10-9-10	0.915	0.888	0.954	0.957	0.959	0.855	0.979	0.984	0.980	0.909
7	MLP 10-10-10	0.921	0.878	0.947	0.950	0.962	0.865	0.985	0.987	0.984	0.910
8	MLP 10-11-10	0.882	0.880	0.948	0.949	0.945	0.836	0.984	0.989	0.986	0.903
9	MLP 10-7-10	0.877	0.823	0.927	0.932	0.921	0.808	0.979	0.980	0.977	0.901
10	MLP 10-13-10	0.931	0.917	0.946	0.948	0.953	0.837	0.982	0.986	0.983	0.926

in the hidden and output layers, respectively, onto the matrix of input variables *X* using eq. (4.4). The algebraic system of equations is easily evaluated in a spreadsheet (i.e., Microsoft Excel) to predict *CSB*, *CSC*, *WAT*, *WAB*, *WAC*, *FS*, *WLFT*, *WLFB*, *WLFC*, and *VMC*, using eq. (4.4), with shown calculated weights and biases matrices. Table 4.6 presents the elements of matrix W_1 and vector B_1 (presented in "bias" column), and table 4.7 present the elements of matrix W_2 and vector B_2 ("bias" raw) for hidden layer.

ANN models were used to predict experimental variables (*CSB*, *CSC*, *WAT*, *WAB*, *WAC*, *FS*, *WLFT*, *WLFB*, *WLFC*, and *VMC*) and showed the highest match obtained so far for such systems [4]. Figure 4.2 shows simulated curves in comparison with experimental data, for the best found MLP 10-13-10, No. 4 neural network. It can be noted from tables 4.4 and 4.5 that the networks were able to predict reasonably well all process outputs for a broad range of the process variables, shown in table 4.1. The predicted values were very close to the desired values in most cases, although *CSB*, *CSC*, *FS*, and *VMC* prediction was not as good as that for *WAT*, *WAB*, *WAC*, *FS*, *WLFT*, *WLFB*, and *WLFC* in terms of r^2 value, for both SOP and ANN models. *SOS*s obtained with ANN models are of the same order of magnitude as experimental errors for *CSB*, *CSC*, *WAT*, *WAB*, *WAC*, *FS*, *WLFT*, *WLFB*, *WLFC*, and *VMC* reported in the literature [4].

The mean and the standard deviation of residuals have also been analyzed. The mean of residuals were: 0.257 for *CSB*, 0.206 for *CSC*, 0.024 for *WAT*, 0.015 for *WAB*, 0.005 for *WAC*, 0.004 for *FS*, 0.015 for *WLFT*, 0.012 for *WLFB*, 0.002 for *WLFC*, and 0.001 for *VMC*. These results showed a good approximation of a normal distribution around zero with a probability of 95% (2·SD), which means a good generalization ability of ANN model for the range of experimental values of *CSB*, *CSC*, *WAT*, *WAB*, *WAC*, *FS*, *WLFT*, *WLFB*, *WLFC*, and *VMC* shown in table 4.1.

SOP models using StatSoft Statistica 10 were also developed, and sum of squares and r^2 values are presented in table 4.3. It can be seen that these r^2 values are slightly lower than those associated with the ANN model. This agrees with other authors [15, 16, 18]. Although ANN models are more complex (115–283 weights-biases for *CSB*, *CSC*, *WAT*, *WAB*, *WAC*, *FS*, *WLFT*, *WLFB*, *WLFC*, and *VMC* models, for ten different ANNs) than SOPs, ANNs performed better because of the high nonlinearity of the developed system [18]. Tables 4.4, 4.5, and 4.6 showed that ANN models gained much better results, compared to SOPs (table 4.3), regarding the r^2 comparison between experimental and calculated outputs. r^2 between experimental and SOP model outputs, for *CSB*, *CSC*, *WAT*, *WAB*, *WAC*, *FS*, *WLFT*, *WLFB*, *WLFC*, and *VMC* were: 0.913, 0.897, 0.942, 0.946, 0.961, 0.928, 0.988, 0.991, 0.991, and 0.941, respectively, while the best ANN

Table 4.6 Elements of matrix W_1 and vector B_1 (presented in "bias" column).

Hidden neuron	SiO_2	Al_2O_3	Fe_2O_3	CaO	MgO	Na_2O	K_2O	MnO	TiO_2	Temp	Bias
1	-2.081	1.788	0.294	0.178	-0.128	0.525	-0.385	-0.742	0.153	-0.136	4.303
2	-1.617	-0.658	1.230	-0.303	3.332	-0.919	0.110	0.268	0.137	0.191	0.026
3	-1.835	-0.349	0.475	0.292	0.285	0.962	-4.795	-0.696	0.360	0.063	3.423
4	-1.522	-0.397	-0.003	-0.401	-0.067	-0.727	-0.860	-0.438	-0.266	0.951	-2.299
5	-0.509	0.539	1.484	-1.055	-0.027	-0.059	-2.844	0.188	1.330	0.934	0.796
6	0.584	-0.256	1.098	0.238	-0.380	7.791	-4.704	-0.021	-0.104	0.632	-2.439
7	-0.289	-0.079	0.924	-0.239	-0.173	-0.948	0.208	0.456	1.109	1.425	1.029
8	0.153	-0.108	1.480	0.173	0.020	6.644	-0.797	-0.255	0.556	0.156	-3.370
9	-0.770	-2.474	0.435	-0.016	-0.954	4.408	-0.829	0.329	0.300	0.005	5.236
10	-0.010	0.081	0.475	-0.225	-0.555	0.196	-0.020	2.343	0.108	0.364	-0.664
11	-0.337	-5.783	0.102	1.202	0.712	-1.061	-0.975	1.896	0.496	0.098	-5.807
12	-0.012	-0.021	-0.085	0.209	1.496	-1.128	-0.096	1.643	0.450	0.004	-1.333
13	0.931	-0.264	0.270	1.680	-0.835	0.221	-0.190	2.009	0.156	0.626	-2.714

Table 4.7 Elements of matrix W_2 and vector B_2 ("bias" raw).

Hidden neuron	CSB	CSC	WAT	WAB	WAC	FS	WLFT	WLFB	WLFC	VMC
1	−0.091	0.367	−0.349	1.268	−0.470	−0.956	−0.440	0.090	1.730	−0.074
2	−1.190	−0.661	0.573	−0.635	−0.107	−0.290	1.352	−0.703	−0.820	−0.264
3	0.124	1.449	0.178	−2.129	−1.050	0.750	−1.557	−2.785	0.804	−1.060
4	−0.060	2.108	1.457	0.047	−2.503	1.235	−0.690	−1.810	−0.296	−1.593
5	−2.818	0.851	−1.259	−0.051	2.234	1.487	0.207	−2.562	1.269	−0.598
6	−1.642	−0.309	−1.510	−2.525	1.012	−1.446	0.400	2.610	1.404	−0.109
7	−3.055	1.329	−0.263	−1.214	−0.924	−0.079	0.195	−0.161	0.639	−0.109
8	−0.361	−0.186	−1.220	0.195	0.182	−0.713	−0.283	0.003	−0.555	−1.478
9	0.022	0.634	0.127	−0.041	0.756	−1.480	−0.611	−1.151	−0.260	−2.833
10	0.137	−0.663	−1.728	0.026	0.917	0.128	0.032	0.821	−1.338	−0.672
11	−0.767	−0.425	−3.177	0.020	−0.669	−2.022	−0.159	1.464	0.070	−0.427
12	0.972	−1.049	−0.138	−1.288	−0.373	−3.943	0.121	−0.235	0.739	−0.082
13	1.811	−0.172	−0.291	−0.758	0.079	1.233	1.538	−1.610	−0.248	−0.499
Bias	0.628	0.413	0.886	0.860	0.604	0.900	0.777	0.780	0.543	0.674

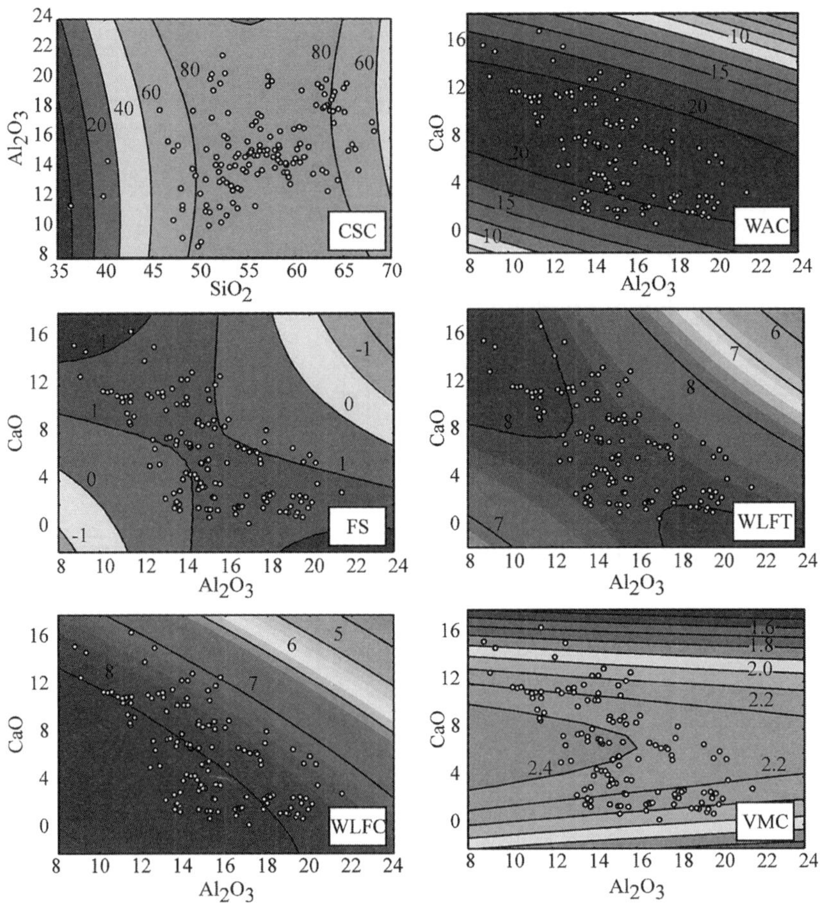

Figure 4.2 ANN prediction performance for *CSC, WAC, FS, WLFT, WLFC* and *VMC* (MLP 10–13-10).

model (MLP 10-13-10, No 4) gained: 0.902, 0.879, 0.963, 0.962, 0.973, 0.839, 0.984, 0.987, 0.987, and 0.913, respectively, during the testing period.

4.3.5 Sensitivity Analysis

In order to access the effect of changes in the output due to the change in the input, a sensitivity analysis was performed. The greater effect observed in the output means greater sensitivity presented with respect to the input [31]. The white noise signals were incorporated by adding or subtracting a Gaussian error of SD = 5% and zero mean with 98% probability [20], i.e., 2.241·SD to each input variable. The white noise was normally distributed and had a constant intensity and frequency. It is used to test the model sensitivity and measurement errors. Due to the large number of combinations (10 input and 10 output variables) a screening experimental design [19] was used for testing the best performance developed ANN model (MLP 10-13-10, table 4.4). A 10-factor screening design, with all inputs as the independent variables, was selected for the study [19]. For this number of input variables number, 21 assays should be made [32]. The coded values for the inputs were −1 and +1, for negative and positive white noise incorporating, and 0 for unchanged experimental values. Table 4.8 shows the final experimental design with twenty-one combinations, which is in diagonal form. The first design contained coded values instead of real, experimentally observed, results (base case), and every next two models present the effect of one input parameter with applying noise in the range of ±SD. That is how each input parameter is tested to show the level of influence on outputs. The complete database (1390 data points) was used, for a total of 1390×21 = 29,190 cases.

The influence of the input variables over the outputs is shown in table 4.9 by calculated SOS values for all twenty-one designs. Obtained SOS values corresponded to the level of experimental errors, and also showed the significance of inputs' influence on outputs. Most of the output variables were affected by firing temperature (which is expected, considering the variance of this variable). *CSB* and *CSC* were found most affected by SiO_2, Al_2O_3 and CaO content. Significant impact of Na_2O, MgO and K_2O content can also be observed, for both *CSB* and *CSC*. Besides firing temperature, the most

Table 4.8 Custom experimental designs for sensitivity analysis.

Assay No.	SiO_2	Al_2O_3	Fe_2O_3	CaO	....	TiO_2	Temp.
1	0	0	0	0	....	0	0
2–3	±1	0	0	0	....	0	0
4–5	0	±1	0	0	....	0	0
6–7	0	0	±1	0	....	0	0
...	...	...	...	....	....	...	...
18–19	0	0	0	0	....	0	0
19–20	0	0	0	0	....	±1	0
20–21	0	0	0	0	....	0	±1

Table 4.9 Sensitivity analyses for *CSB*, *CSC*, *WAT*, *WAB*, *WAC*, *FS*, *WLFT*, *WLFB*, *WLFC* and *VMC*

	CSB	CSC	WAT	WAB	WAC	FS	WLFT	WLFB	WLFC	VMC
+SiO_2	0.638	0.627	1.101	0.133	0.134	1.200	4.069	4.146	0.111	0.427
-SiO_2	2.053	2.027	0.990	1.426	1.423	1.079	6.885	6.801	0.900	0.382
+Al_2O_3	0.582	0.578	1.483	0.452	0.448	10.117	5.829	5.938	0.091	0.601
-Al_2O_3	0.503	0.499	1.528	0.512	0.503	9.671	4.930	4.943	0.087	0.526
Fe_2O_3	0.010	0.010	0.001	0.001	0.001	0.002	0.008	0.008	0.001	0.000
-Fe_2O_3	0.009	0.009	0.001	0.001	0.001	0.002	0.007	0.007	0.001	0.000
+CaO	0.122	0.123	0.309	0.019	0.019	0.362	1.027	1.045	0.007	0.109
-CaO	0.132	0.131	0.299	0.017	0.017	0.380	1.195	1.205	0.006	0.125
+MgO	0.024	0.024	0.011	0.001	0.001	0.016	0.046	0.046	0.001	0.005
-MgO	0.025	0.025	0.011	0.001	0.001	0.017	0.046	0.047	0.001	0.005
+Na_2O	0.052	0.053	0.005	0.000	0.000	0.007	0.018	0.018	0.001	0.002
-Na_2O	0.053	0.052	0.005	0.000	0.000	0.007	0.017	0.017	0.001	0.002
+K_2O	0.039	0.040	0.014	0.012	0.012	0.022	0.085	0.084	0.011	0.007
-K_2O	0.078	0.077	0.013	0.025	0.025	0.021	0.203	0.201	0.020	0.006
+MnO	0.001	0.001	0.000	0.000	0.000	0.000	0.001	0.001	0.000	0.000
-MnO	0.001	0.001	0.000	0.000	0.000	0.000	0.001	0.001	0.000	0.000
+TiO_2	0.003	0.003	0.000	0.000	0.000	0.001	0.002	0.002	0.000	0.000
-TiO_2	0.003	0.003	0.000	0.000	0.000	0.001	0.002	0.002	0.000	0.000
Temp	8.179	8.328	18.561	2.044	2.008	11.858	65.515	65.874	1.005	6.757
-Temp	7.980	7.863	19.39	0.714	0.711	12.503	72.153	72.23	0.234	7.589

effective impact on *WAT*, *WAB* and *WAC* can be observed by Al_2O_3, SiO_2 and CaO content. High influence of K_2O content is observable for *WAC* and *WAT*. *FS* is most affected by Al_2O_3, SiO_2, and CaO content. The influence of MgO, K_2O, and Na_2O is also notable. *WLF*s are most affected by SiO_2 and Al_2O_3. CaO is also very influential. The influences of MgO, Na_2O, and K_2O were also notable. *VMC* is most affected by Al_2O_3 and SiO_2 content. CaO is also very influential. The influences of MgO, Na_2O, and K_2O were also notable.

Table 4.9 shows the influence of the input variables on *CSB*, *CSC*, *WAT*, *WAB*, *WAC*, *FS*, *WLFT*, *WLFB*, *WLFC*, and *VMC*, according to sum of squares, calculated by comparing model predicted values with and without white noise signal applied, according to table 4.8.

Prediction of the final laboratory product parameters evaluated by second order polynomial regression models (SOPs) and optimal artificial neural network (ANN) in conjunction with sensitivity analysis showed quite good agreement with experimental results. SOPs and ANN showed high r^2 values, as well as prediction accuracy, for the observed outputs, which proved them to be useful in predicting final products' quality. Moreover, the results of ANOVA (for SOPs) correspond very well to sensitivity analysis (for ANN model).

4.3.6 Fuzzy Synthetic Optimization

Knowing final product application and optimal values of *CSB*, *CSC*, *WAT*, *WAB*, *WAC*, *FS*, *WLFT*, *WLFB*, *WLFC*, and *VMC* of it, allow ones to find the optimal chemical composition and firing temperature [2, 3, 33]. Fuzzy synthetic optimization of the output variables was accomplished in order to find the content of major oxides (SiO_2, Al_2O_3, Fe_2O_3, CaO, MgO, Na_2O, K_2O, MnO, and TiO_2), and firing temperature that give optimums of *CSB*, *CSC*, *WAT*, *WAB*, *WAC*, *FS*, *WLFT*, *WLFB*, *WLFC*, and *VMC*. Trapezoidal membership function was used, according to eq. 4.5, in which a - b covered the complete interval of obtained output values, and m - n represented the optimal values for observed product group (table 4.10). The optimal parameters, used for FSE evaluation, were given based on our experience with fired laboratory heavy clay products, and published elsewhere [3]. Heavy clay products are divided to three groups, where: Group I is suitable for the production of solid bricks; Group II can be used for hollow bricks and blocks, as well as ceiling elements; and Group III is appropriate for roof tiles and facade elements. The samples belonging to Groups II and III could be also used in a light-weighted bricks, as a primary raw material [2, 3].

Table 4.10 Trapezoidal membership function parameters.

Group	Parameter	*FS*	*WLFT*	*WLFB*	*WLFC*	*VMC*
–	*A*	−2.64	3.86	3.98	4.32	1.38
	B	2.17	19.55	19.79	19.65	2.42
I	*m*	−0.54	10.2	10.2	10.2	1.62
	n	0.11	14.3	14.3	14.3	1.81
II	*m*	0.34	8.2	8.2	8.2	1.76
	n	0.53	11.5	11.5	11.5	2
III	*m*	0.73	7.5	7.5	7.5	1.85
	n	1.11	9	9	9	2.1

The objective function (*F*) is the mathematical function whose maximum would be determined, by summing the FSE results for the ANN model, according to eq. (4.4). Each group of output parameters (*CS*, *WA*, *FS*, *WLF*, and *VMC*) has the equal influence on the function *F*:

$$F(SiO_2, Al_2O_3, Fe_2O_3, CaO, MgO, Na_2O, K_2O, Mno, TiO_2, Temp.) = 0.10.\ (CSB + CSC + WAT + WAB + FS + WAC + WLFT + WLFB + WLFC + VMC) \quad (4.6)$$

The maximum of function *F* represents the optimal major oxide contents and processing temperature, and also the optimum of *CSB*, *CSC*, *WAT*, *WAB*, *WAC*, *FS*, *WLFT*, *WLFB*, *WLFC*, and *VMC*. The values of *F* were determined using eq. (4.6). Values of membership function closer to 1 show the tendency of processing parameters to be highly optimal. Optimized process parameters (inputs and outputs) for all groups were shown in table 4.11. Solid bricks (Group I) are proved to be producible of low quality heavy clays containing the highest free quartz and carbonates content, and subsequently the lowest Fe_2O_3 content [3]. Group II had the the highest Fe_2O_3, Na_2O and K_2O content, and presented highly sinterable heavy clays [34]. Group III contained the most of SiO_2 and the lowest carbonates content. The optimal firing temperatures for Group I were 950–1000°C, and for Groups II and III it was 900°C. Higher clay minerals

Table 4.11 Optimizing parameters.

Optimal inputs										
Group	SiO_2	Al_2O_3	Fe_2O_3	CaO	MgO	Na_2O	K_2O	MnO	TiO_2	Temp
	50.67	11.47	4.73	10.63	3.58	1.26	2.24	0.08	1.08	950
I	52.08	13.04	4.16	11.13	1.65	1.16	1.64	0.20	0.39	950
	54.49	13.91	5.09	8.05	3.7	1.14	1.7	0.08	0.46	1000
II	57.07	19.54	7.13	2.78	1.87	1.65	2.59	0.12	0.38	900
III	67.99	17.07	5.41	0.43	1.53	0.70	1.57	0.01	0.78	900
Optimal outputs										
Group	CSB	CSC	WAT	WAB	WAC	FS	WLFT	WLFB	WLFC	VMC
	16.91	33.33	20.14	20.18	19.47	0.31	12.60	12.95	13.09	1.687
I	16.58	32.39	21.44	21.63	20.96	–0.21	12.88	12.91	12.74	1.702
	15.00	29.56	19.79	19.85	19.26	–0.09	10.44	10.53	10.58	1.728
II	27.72	53.02	15.13	15.46	14.95	0.35	6.62	6.85	6.64	1.780
III	50.60	79.21	10.40	11.15	10.69	–0.18	5.58	5.31	6.11	1.790

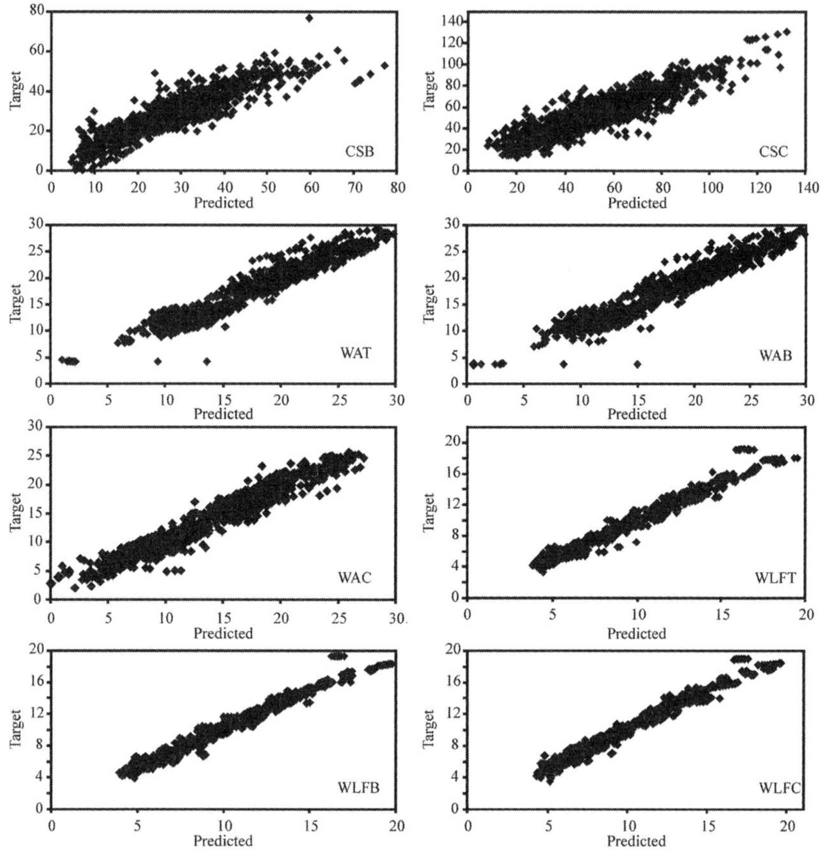

Figure 4.3 Objective functions for solid bricks optimization.

and lower carbonates contents can be held responsible for lower optimal firing temperature.

Comparing the outputs for the three heavy clays products groups, it was concluded that the highest *CS* and *VMC*, and the lowest *WA* and *WLF* were optimal for roofing tiles production.

The objective function, regarding $SiO_2/Al_2O_3/Fe_2O_3/CaO$ content were shown in the surface plots, for Group I (figure 4.3). Objective function gained values between 0.6 and 0.7 on these plots, and the obtained quadratic surface varied depending on the variables used from "peak-type" (for SiO_2 vs. Al_2O_3 graph), "saddle-type" (for SiO_2 vs. Fe_2O_3), and "rising ridge" (for SiO_2 vs. CaO, Al_2O_3 vs. Fe_2O_3, Al_2O_3 vs. CaO, and Fe_2O_3 vs. CaO).

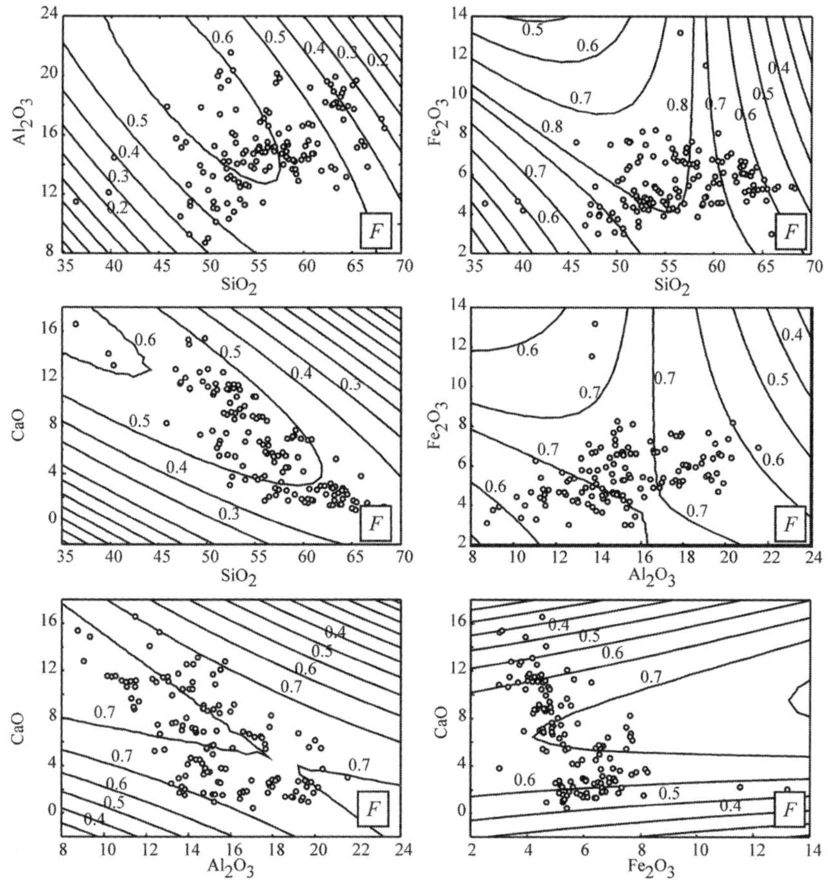

Figure 4.4 The effects of most important parameters on optimization function.

4.4 Conclusions

The ANN-based model was developed for prediction of *CSB*, *CSC*, *WAT*, *WAB*, *WAC*, *FS*, *WLFT*, *WLFB*, *WLFC*, and *VMC* for a wide range of experimental conditions. The model was able to predict experimental data successfully, with ease of implementing it for design and control of the processes and also the effective use for predictive modeling and optimization. As compared to SOP models, ANN models yield a better fit of experimental data, according to r^2 and SOS of both models.

Taking into account that a considerable amount and wide variety of data were used in the present work to obtain the ANN model, and considering that the model turned out to yield a sufficiently good representation of these data, this ANN model can be expected to be very useful in practice for the design and control of brick production process. The model could even be improved by increasing the laboratory experiments database or by making industrial analysis in as many as possible brick factories.

Acknowledgement

The authors are grateful to the Serbian Ministry of Science and Technological Development for the financial support of this investigation included in the projects III 450080 and TR 31055, project cycle 2011–2014.

References

1. M. Arsenović, L. Pezo, S. Stanković, Z. Radojević, Sensitivity analysis of mathematical models for final product properties: Link to DTG curve, *Ceramics International*, http://dx.doi.org/10.1016/j.ceramint.2013.01.049
2. M. Arsenović, S. Stanković, L. Pezo, L. Mančić, Z. Radojević, "Optimization of the production process through response surface method: Bricks made of loess," *Ceramics International*, Vol. 39, No. 3, pp. 3065–3075, 2013.
3. M. Arsenović, S. Stanković, Z. Radojević, L. Pezo, "Prediction and fuzzy synthetic optimization of process parameters in heavy clay brick production," *Ceramics International*, Vol. 39, No. 2, pp. 2013–2022, 2013.
4. M. Arsenović, Z. Radojević, S. Stanković, Ž. Lalić, L. Pezo, "What to expect from heavy clay?" *Ceramics International*, Vol. 39, No. 2, pp. 1667–1675, 2013.
5. M. Arsenović, Z. Radojević, S. Stanković, "Removal of toxic metals from industrial sludge by fixing in brick structure," *Construction and Building Materials*, Vol. 37, pp. 7–14, 2012.
6. M. Arsenović, L. Pezo, Z. Radojević, S. Stanković, "Serbian heavy clays behavior: application in rouch ceramics," *Hemijska industrija*, DOI:10.2298/HEMIND121123006A, 2013.
7. Ž.Lalić, M. Arsenović, Đ. Janaćković, M. Vasić, Z. Radojević, "Influence of increased temperature on clay fast drying process," *Romanian Journal of Materials*, Vol. 39, No. 3, pp. 175–179, 2009.
8. S. Meseguer, T. Sanfeliu, M.M. Jordan, "Classification and statistical analysis of mine spoils chemical composition from Oliete basin (Teruel, NE Spain)," *Environmental Geology*, Vol. 56, pp. 1461–1466, 2009.
9. M. Arsenović, L. Pezo, L. Mančić, Z. Radojević, "Thermal and mineralogical characterization of loess heavy clays for potential use in brick industry", *Thermochimica Acta* 580 (2014) 38–45.

10. X. Qua, C.F. Je-Wub, "One-factor-at-a-time designs of resolution V," *Journal of Statistical Planning and Inference*, Vol. 131, pp. 407–416, 2005.
11. S.L. Rettig, J.W. Marinenko, H.N. Khoury, B.F. Jones, "Comparison of rapid methods for chemical analysis of milligram samples of ultrafine clays," *Clays and Clay Minerals*, Vol. 31, pp. 440–446, 1983.
12. N. Koukouzas, C. Ketikidis, G. Itskos, X. Spiliotis, V. Karayannis, G. Papapolymerou, "Synthesis of CFB-coal fly ash clay bricks and their characterization," *Waste and Biomass Valorization*, Vol. 2, pp. 87–94, 2011.
13. L. Pezo, M. Arsenović, Z. Radojević, "ANN model of brick properties using LPNORM calculation of mineral content", Ceramics International, DOI 10.1016/j.ceramint.2014.02.044.
14. S. Grieua, O. Faugeroux, A. Traoré, B. Claudet, J.-L. Bodnar, "Artificial intelligence tools and inverse methods for estimating the thermal diffusivity of building materials," *Energy and Buildings*, Vol. 43, pp. 543–554, 2011.
15. I.C. Trelea, A.L. Raoult-Wack, G. Trystram, "Note: Application of neural network modeling for the control of dewatering and impregnation soaking process (osmotic dehydration)," Food Sci. Technol. Int., Vol. 3, pp. 459–465, 1997.
16. L.L. Pezo, B.Lj. Ćurčić, V.S. Filipović, M.R. Nićetin, G.B. Koprivica, N.M. Mišljenović, Lj.B. Lević, "Artificial neural network model of pork meat cubes osmotic dehydratation," *Hemijska industrija*, DOI:10.2298/HEMIND120529082P, 2013.
17. T. Kollo, D. von Rosen, *Advanced Multivariate Statistics with Matrices, Mathematics and its Applications*, Vol. 579, Springer, Netherlands, http://dx.doi.org/10.1007/1-4020-3419-9, 2005.
18. L.A. Basheer, M. Hajmeer, "Artificial neural networks: Fundamentals, computing, design and application," *J. Microbiol. Meth.*, Vol. 43, p. 3–31, 2000.
19. G.E.P. Box, D.W. Behnken, "Some new three level designs for the study of quantitative variables," *Technometrics*, Vol. 2, pp. 455–475, 1960.
20. S.S. Sablani, A. Kacimov, J. Pret, A.S. Mujumdar, A. Campo, "Noniterative estimation of heat transfer coefficients using artificial neural network models," *International Journal of Heat and Mass Transfer*, Vol. 48, pp. 665–679, 2005.
21. M. Dondi, G. Guarini, M. Raimondo, C. Zanelli, D.D. Fabbriche, A. Agostini, "Recycling the insoluble residue from titania slag dissolution (tionite) in clay bricks," *Ceramics International*, Vol. 36, pp. 2461–2467, 2010.
22. E. Furlani, G. Tonello, E. Aneggi, S. Maschio, "Preparation and characterization of sintered ceramics made with spent foundry olivine sand and clay," *Ceramics International*, Vol. 38, pp. 2619–2625, 2012.
23. L.C.S. Herek, C.E. Hori, M.H.M. Reis, N.D. Mora, C.R.G. Tavares, R. Bergamasco, "Characterization of ceramic bricks incorporated with textile laundry sludge," *Ceramics International*, Vol. 38, pp. 951–959, 2012.
24. D.L. Fuks, A.E. Kiv, Correlation between the composition and the rate of ionic transport in perovskites, *Advanced Materials Letters*, Vol. 4 (5), p. 328–331, 2013.

25. Z. Radojević, Ž. Lalić, R. Vasić, A. Mitrović, "Brick raw materials quality in Serbia and the influence for the products quality and to the production technology—loess sediments," *Izgradnja*, Vol. 59, pp. 45–49, 2005.
26. F. H. Clews, *Heavy clay technology*, The British ceramic reasearch association, Stoke-on-Trent, London, 1955.
27. J. García-Ten, M.J. Orts, A. Saburit, G. Silva, "Thermal coductivity of traditional ceramics. Part II: Influence of mineralogical composition," *Ceramics International*, Vol. 36, pp. 2017–2024, 2010.
28. M.M. Jordán, M.A. Montero, S. Meseguer, T. Sanfeliu, "Influence of firing temperature and mineralogical composition on bending strength and porosity of ceramic tile bodies," *Applied Clay Science*, Vol. 42, pp. 266–271, 2008.
29. F. Pardo, S. Meseguer, M.M. Jordán, T. Sanfeliu, I. González, "Firing transformations of Chilean clays for the manufacture of ceramic tile bodies," *Applied Clay Science*, Vol. 51, pp. 147–150, 2011.
30. P. Bormans, *Ceramics are More Than Clay Alone*, Cambridge International Science Publishing, Cambridge, UK, 2004.
31. J.J. Montano, A. Palmer, "Numeric sensitivity analysis applied to feedforward neural networks," *Neural Computing and Applications,* Vol. 12, pp. 119–125, 2003.
32. S.W. Cheng, C.F.J. Wu, "Factor screening and response surface exploration," *Statistica Sinica*, Vol. 11, pp. 553–604, 2001.
33. M. Arsenović, L. Pezo, Z. Radojević, "Response surface method as a tool for heavy clay firing process optimization: Roofing tiles," *Processing and Application of Ceramics*, Vol. 6 (4), pp. 209–214, 2012.
34. V. Kornilov, "Reasons for the different effects of calcareous clays on strength properties of ceramics," *Glass and Ceramics*, Vol. 62, UDC:663.3:666.3.325.4:539.4, 2005.

5

Enhancement of Physical and Mechanical Properties of Sugar Palm Fiber via Vacuum Resin Impregnation

M.R. Ishak[1,4,7], Z. Leman[2], S.M. Sapuan[2,3,4,*], M.Z.A. Rahman[5] and U.M.K. Anwar[4,6]

[1]Department of Aerospace Engineering, Universiti Putra Malaysia, Selangor, Malaysia
[2]Department of Mechanical and Manufacturing Engineering, Universiti Putra Malaysia, Selangor, Malaysia
[3]Institute of Advanced Technology (ITMA), Universiti Putra Malaysia, Selangor, Malaysia
[4]Institute of Tropical Forestry and Forest Products (INTROP), Universiti Putra Malaysia, Selangor, Malaysia
[5]Centre of Foundation Studies for Agricultural Science, Universiti Putra Malaysia, Selangor, Malaysia
[6]Wood Finishing Laboratory, Forest Research Institute of Malaysia (FRIM), Kepong, Kuala Lumpur, Malaysia
[7]Aerospace Manufacturing Research Centre, Universiti Putra Malaysia, Selangor, Malaysia

Abstract

A study on the effects of elevating impregnation pressure on the physical and mechanical properties of sugar palm fiber with phenol formaldehyde (PF) and unsaturated polyester (UP) with a constant impregnation time of 5 minutes are presented in this chapter. A significant increase in the weight percentage gain of sugar palm fibers was observed after impregnation under pressure with PF and UP in the ranges of 8.59 to 38.59% and 7.16 to 22.63%, respectively. Other findings showed that the fiber moisture content dropped from 8.19% to between 0.75 and 0.46% for PF and between 0.87 and 0.44% for UP, while water absorption was reduced from 116.82% to between 61.64 and 22.52% for PF and between 63.49 and 23.31% for UP. The mechanical properties of impregnated fibers showed

**Corresponding authors*: sapuan@eng.upm.edu.my; mohdridzwan@upm.edu.my

Ashutosh Tiwari and Mikael Syväjärvi (eds.) Advanced Materials for Agriculture, Food, and Environmental Safety, (121–144) 2014 © Scrivener Publishing LLC

significant improvement ($p \leq 0.05$) with the increase in pressure. The tensile strength increased from 243.77 MPa to between 251.39 and 297.67 MPa for PF and between 287.84 and 344.71 MPa for UP. The tensile modulus increased from 3.07 GPa to between 3.1 and 3.98 GPa for PF and between 3.15 and 3.98 GPa for UP, while the elongation at break decreased from 25.16% to 17.66 and 6.26% for PF and between 24.24 and 19.05% for UP. Low tensile strength and elongation at break of PF-impregnated fibres showed that their fibers are inferior in toughness (48.51 MJ/m^3 to between 33.08 and 12.29 MJ/m^3) compared to UP-impregnated fibres (48.51 MJ/m^3 to between 57.56 and 45.21 MJ/m^3). The outcome of this study is a new technique for the enhancement of physical and mechanical properties of natural fibers and their composites.

Keywords: Sugar palm fiber, *Arenga pinnata,* impregnation modification, vacuum resin impregnation, impregnation pressure

5.1 Introduction

Natural fibers offer numerous advantages; they are abundantly available, have high specific mechanical properties, are low in cost, have low density, are safe to handle, are not a threat to the environment, and are renewable and biodegradable [1–17]. Despite these advantages, they suffer from some drawbacks due to the presence of hydroxyl groups (OH) in the fibre. When natural fibres are exposed to high humidity, these hydroxyl groups attract and hold water molecules through hydrogen bonding, which leads to fiber swelling and causes the presence of voids at the composite interface, resulting in deterioration in mechanical properties [18–20] and loss in dimensional stability [21] of its fibers and composites. This was proven by the limited use of natural fiber composite products in indoor environments (where humidity is lower) compared to outdoor environments. In composite processing, the addition of hydrophilic natural fibers to hydrophobic plastics results in a composite with poor mechanical properties due to nonuniform fiber dispersion in the matrix and inferior fiber-matrix interphase [22, 23]. This is due to the fact that natural fibers are hydrophilic in nature and generally contain large numbers of hydroxyl groups which make them polar (A_xOH: molecules with an -OH at one end), while most hydrophobic polymer matrices are non-polar (C_xA_y: most carbon compounds). Incorporation of these polar and non-polar characteristics creates compounding difficulties that lead to nonuniform dispersion of fibers within the matrix, which impairs the efficiency of the composite [20]. Besides contributing to the high moisture absorption of the fiber, the polar nature also results in high moisture sorption in composites that leads to voids in the fiber matrix interphase, which would apparently reduce the mechanical properties of their composites. If moisture were not removed from natural fibers prior to compounding by drying, the products would become porous [23]. Also, absorption of moisture causes natural fibers to be susceptible to rotting and to be less resistant to microbial attack [24], yielding to poor fiber mechanical properties. Thus, the high hygroscopicity of natural fibers pose serious problems to their properties whether they are being stored or are being processed. The problems may also affect their composites' properties or their composite service life. In order to reduce hygroscopicity, a number of studies were carried out to modify the hydrophilic property of natural fibers. One of the successful treatments to reduce hygroscopicity is by reacting the hydroxyl groups of natural fibers with chemical reagents. Natural fibers can also be treated thermally by heating the fiber and inducing chemical changes to its hydroxyl groups. As a result of the reaction of chemical reagents or the damages to macromolecular constituents after heating, the treated hydroxyl groups become less effective in absorbing moisture compared to the untreated fiber.

Another new fiber treatment to be developed includes physical modification of the fiber in which the fiber is impregnated with thermosetting polymer via a vacuum resin impregnation process. This enhances the properties of sugar palm fiber before it is used for composite reinforcement. The principle idea behind this process is to impregnate the fiber cell lumen and cell wall and to enclose the fiber surface with hydrophobic thermosetting polymer in order to prevent the absorption of water molecules to hydrophilic onto the fiber.

The main objectives of this study were to compare the differences between the two impregnation agents, phenol formaldehyde (PF) and unsaturated polyester (UP), and to ascertain the effects of elevating impregnation pressures on the physical and tensile properties of sugar palm fibers.

5.2 Experimental

5.2.1 Materials

5.2.1.1 Preparation of Fibers

Sugar palm (*Arenga pinnata*) fibers were from Kampung Kuala Jempol, Negeri Sembilan, Malaysia. The fibers were obtained from matured sugar palm trees, aged more than 6 years and with a height of 20 m. For optimum properties of sugar palm fibers [17], the fibers were obtained within the live palm fronds. The fibers were washed and air-dried for 24 h before being dried in a Mermert oven at 80°C for 24 h. Phenol formaldehyde (PF) with molecular weight of 600 and commercial grade unsaturated polyester (UP) type isophtalic with molecular weight of 2000 were used as impregnation agents, supplied by Pultrution Innovative Sdn. Bhd., Seremban, Negeri Sembilan, Malaysia.

5.2.2 Methods

5.2.2.1 Impregnation of the Fibers

The sugar palm fibers were submerged in a 1000 mL beaker filled with the impregnation resin agents (PF and UP) and placed inside a glass vessel. The glass vessel was evacuated at various impregnation pressures: 1000, 900, 800, 700, 600, and 500 mmHg (133.32, 119.99, 106.66, 93.33, 79.99, and 66.66 kPa) for 5 min before the vacuum vessel was slowly released within 30 s. The impregnated fibers were removed and excess resin was then drained for 5 min before the fibers were heated for polymerization at 140°C for 30 min. The fibers were weighed using an analytical balance with the capability of reading up to 0.0001 g.

5.2.2.2 Determination of Physical Properties

The determination of weight percentage gain, density, specific gravity, moisture content, and water absorption of the fibers were determined using the following equations:

$$\text{Weight percentage gain, } WPG\ (\%) = (m_b - m_a) / m_a \ x\ 100 \quad (5.1)$$

where m_a = mass of fiber before impregnation (g), m_b = mass of fiber after impregnation (g).

$$\text{Density, } \rho\ (\text{g/cm}^3) = m_o / V_g \quad (5.2)$$

where m_o = mass of oven dry fiber (g), V_g = green volume of fiber (cm^3).

$$\text{Specific gravity, } SG = \rho_f / \rho_w \quad (5.3)$$

where ρ_f = density of fibers (g/cm^3); ρ_w = density of water (g/cm^3).

$$\text{Moisture content, } MC\ (\%) = (m_a - m_o) / m_o \ x\ 100 \quad (5.4)$$

where m_a = mass of air dry fiber (g), m_o = mass of oven dry fiber (g).

$$\text{Water absorption, } WA\ (\%) = (m_2 - m_1) / m_1 \ x\ 100 \quad (5.5)$$

where m_1 = mass of fiber before soaking (g), m_2 = mass of fiber after soaking in distilled water for 24 h.

A scanning electron microscope (SEM) and an optical microscope model Lieca MS 5 (with constant magnification of 400x) were used to observe the surface and cross-sectional views of the sugar palm fibers after impregnation.

5.2.2.3 Determination of Tensile Properties

The control impregnation agent along with the PF- and UP-impregnated fibers at various impregnation pressures underwent single fiber tensile testing. A total of 150 specimens were prepared and tested at room temperature of 23°C and a relative humidity of 50% using an Instron type universal testing machine with a load capacity of 5 kN.

5.2.2.4 Method of Result Analysis

The mean of the samples was used to show the changes in physical and tensile properties of the impregnated fibers, while Standard error was used to show the significant difference between the variables. The results were analyzed using the Statistical Package for the Social Sciences (SPSS) software. A one-way analysis of variance (ANOVA), Duncan's multiple range test, was used to determine whether increasing impregnation pressure would show any significant effects on the physical and tensile properties of the fiber. The mean followed with the different letters (a, b, c, d, e, and f) for each series (1000, 900, 800, 700, 600, and 500 mmHg) in impregnation agents was significantly different. For the analysis, an alpha level of 0.05 for all statistical tests was used. Linear regression analysis was used to show a trend line in the series data. It revealed whether a particular data set of sugar palm fiber increased or decreased in physical and mechanical properties after an increase in impregnation pressure. The trend lines were used to describe the relationship strength between the two variables (impregnation pressure and impregnation agent), to assess whether these variables have any relationship at all, and to determine if these variables contain redundant information.

5.3 Results and Discussion

5.3.1 Physical Properties of Impregnated Fiber

The weight percentage gain (WPG) of impregnated sugar palm fibers at various impregnation pressures is shown in Figure 5.1. The results showed that the weight of resin absorbed in the fiber (after polymerizing) was due

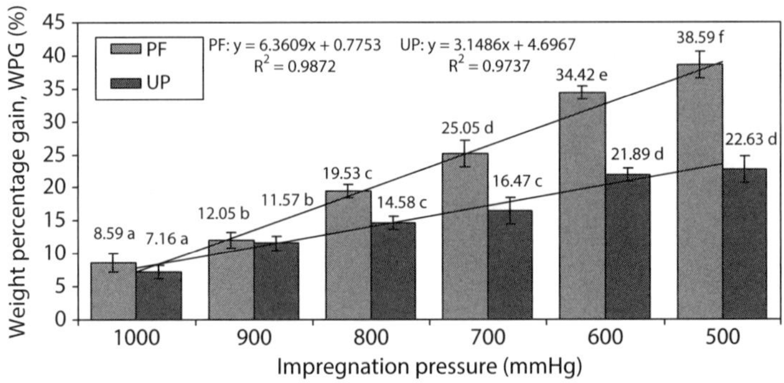

Figure 5.1 WPG of PF- and UP-impregnated sugar palm fibers at various impregnation pressures.

to impregnation. The WPG of the fiber significantly increased linearly as impregnation pressure increased. The highest WPG was found with the fiber impregnated with pressure of 500 mmHg for PF and 600 mmHg for UP. No significant increase was seen when the pressure increased to 500 mmHg. It was also observed that the increasing rate of WPG for fiber impregnated with PF was higher than that of the fiber impregnated with UP, as shown in the regression trend lines.

The different increasing rates can be explained by the difference in molecular weights of the impregnation agents. Due to the low molecular weight of the PF resin that was specially modified to be 500, it has shorter branching in the polymer unit where it has smaller resin molecules that allow the resin to penetrate into fiber cell wall. This is in contrast with UP, which has a higher molecular weight of 2000 [25, 26]. In order to have better penetration of the resin, the molecule size of the resin must be sufficiently small to penetrate deep into the cell wall, requiring the resin to be unpolymerized or only slightly polymerized [27]. The penetration of resin into the fiber cell wall depends on several important factors such as the size of the oligomer or polymer [28], sufficient time of immersion in resin, and the impregnation pressure. It is also believed that higher molecular weight resins might penetrate only into the cell lumen and not into the cell wall. Bearing in mind that resin with a lower molecular weight has the tendency to leach out from the cell wall, which can cause loss of cell wall bulking effect [29], proper control should be taken on the impregnated fiber. Such measures to prevent the resin from leaching out from the cell wall and cell lumen of fiber include immediate polymerization of the resin after impregnation, or prepolymerization of the resin.

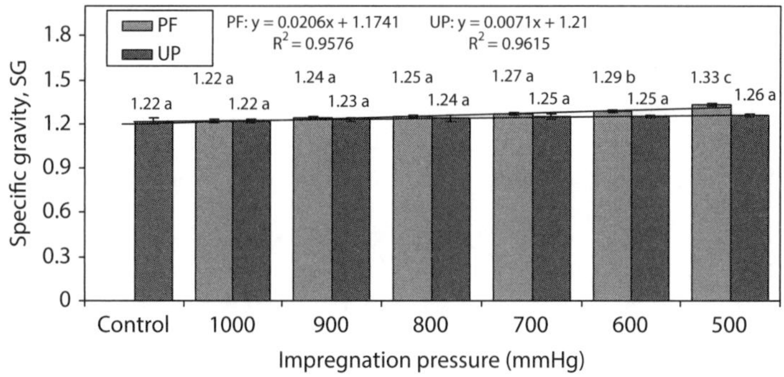

Figure 5.2 SG of PF- and UP-impregnated sugar palm fibers at various impregnation pressures.

The better penetration of PF than UP resin into the fiber could be related to the slight increase of its specific gravity (SG). It is shown in Figure 5.2 that the SG of fibers impregnated with PF statistically increased slightly from 1.22 to 1.29 and 1.33 after impregnation with a pressure of 600 and 500 mmHg, respectively. The slight increase in SG of PF-impregnated fibers was expected since it absorbed more resin, which also increased the WPG. On the other hand, no significant increase in SG was observed for those fibers impregnated with UP, as shown by the low value of WPG. According to Deka *et al.* [30], in many cases of impregnation modification, impregnation with low WPG usually results in no changes to SG. Nevertheless, both results are acceptable since impregnation with these agents showed that there was little or no weight gain resulting from impregnation. In the case of fiber impregnated with UP, no weight gain was seen as shown in the regression trends (where both show a horizontal slope). Since lightweight material enhances low fuel consumptions, the results can be extremely helpful when designing and producing a lightweight material that can contribute to low fuel consumption.

The relationship between moisture absorption rate and impregnation pressure is shown in Figure 5.3. It can be seen that there is a significant difference in the rate of moisture absorption between the control samples and the impregnated fibers in which the impregnated fibers show significantly lower values of moisture content (MC) than those of the control samples. This indicates significant improvement in physical properties as a result of MC reduction following impregnation with polymer resin. The MC of impregnated fibers was found to be significantly reduced after impregnation at pressures of 1000, 900, and 800 mmHg for both resin types (0.41 to 0.46% for PF; 0.44 to 0.53% for UP), while no reduction in MC was

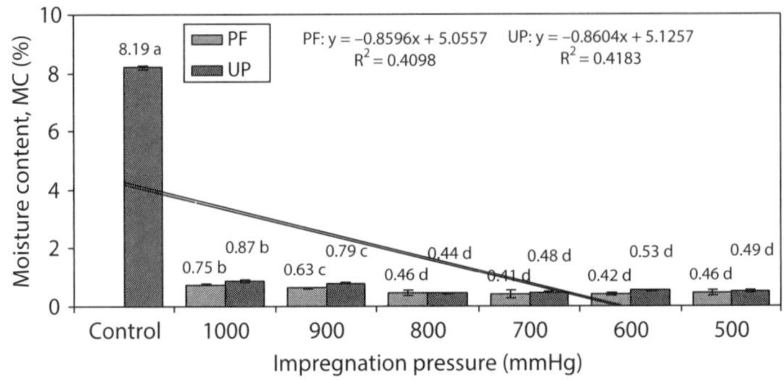

Figure 5.3 MC of PF- and UP-impregnated sugar palm fibers at various impregnation pressures.

observed when pressure increased from 700 and 500 mmHg. It was also observed that the rate of moisture absorption was not much different between PF-impregnated fibers and UP-impregnated fibers, as shown by the slope in the respective regression trend lines.

It was expected that the fiber without impregnation would have much higher MC due to absorption of moisture into its cell wall. This can be attributed to the hydrophilic property of natural fibers where hydrogen bonding is formed between the hydroxyl groups and water molecules [31–34]. This is due to the fact that natural fiber contains hydroxyl or OH groups, which are present throughout its structure and attract and hold water molecules by the mechanism of chemical bonding called hydrogen bonding. When natural fibers are exposed to a surrounding atmosphere containing high humidity, initially water molecules are absorbed directly onto the hydrophilic groups of the fiber, followed by other water molecules attracting to other hydrophilic groups; they also may form further layers on top of the water molecules that have already been absorbed [35]. This absorption continues until the equilibrium moisture content (EMC) with the surrounding air is attained. This study has successfully proven that the penetration of moisture and the formation of hydrogen bonds are inhibited by enclosing the fiber with hydrophobic materials of polymer resin via a fiber-resin impregnation process. With impregnation and polymerization, the natural fibers became moisture resistant as they were enclosed with cured resin that would expel moisture from the fiber. This can be better understood by observing the optical micrographs (400x) (Figs. 5.4a–g) that show the fibers enclosed with the resin after being impregnated. The surface of the control sample (Fig. 5.4a) was significantly different than the surface of the UP-impregnated fibers (Figs. 5.4b–g).

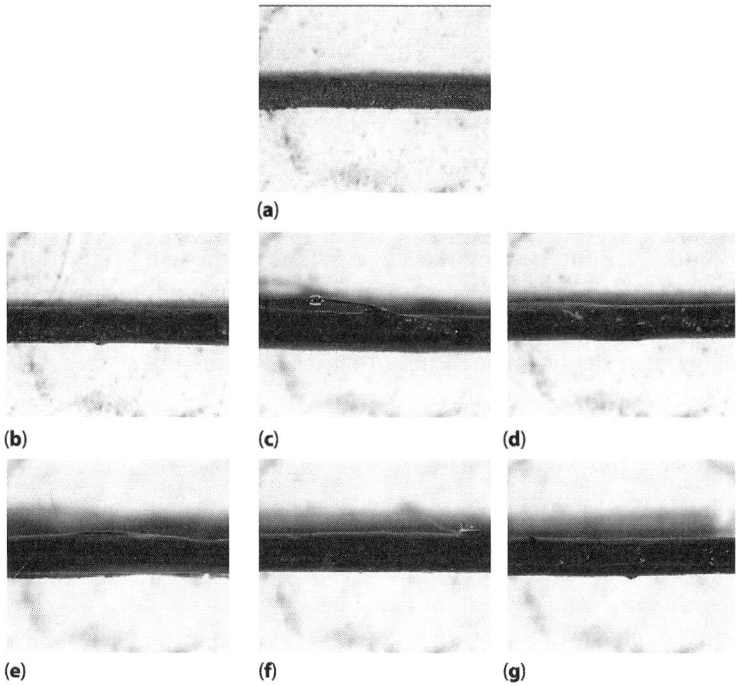

Figure 5.4 (a) Optical micrograph of control sample; (b) Optical micrograph of UP-impregnated fibers (1000 mmHg); (c) Optical micrograph of UP-impregnated fibers (900 mmHg); (d) Optical micrograph of UP-impregnated fibers (800 mmHg); (e) Optical micrograph of UP-impregnated fibers (700 mmHg); (f) Optical micrograph of UP-impregnated fibers (600 mmHg); (g) Optical micrograph of UP-impregnated fibers (500 mmHg).

Figure 5.4a clearly shows that the natural surface of the control sample is different from the impregnated fibers where the latter were enclosed with the UP resin (Figs. 5.4b–g). It also shows that the UP-impregnated fiber with an impregnation pressure of 500 mmHg (Fig. 5.4g) was fully enclosed with the resin, while the fiber with an impregnation pressure of 1000 mmHg (Fig. 5.4b) seemed to be partially enclosed, where only a small change in the fiber surface was observed. A higher degree of encapsulation of this fiber (Fig. 5.4g) is also accompanied by the increase in impregnation pressure that gives deeper penetration onto the fiber surface compared to the lower pressure (Fig. 5.4b). This allows resin molecules to be uniformly distributed onto the fiber surface and more resin molecules are believed to be embedded into the micropores (small holes that exist at the inter-microfibrillar region along the fiber). This is because with the increase in the number of resin molecules being embedded in the fiber surface, the mechanical interlocking force between the fiber and the matrix is larger. It encourages the uniformity of the polymeric structure, resulting in an enhancement of the bonding strength between the polymer on the fiber surface and the polymer matrix of the composite. In other words, the impregnation modification improved the stress transfer between the fiber and the matrix, which further improved bonding and mechanical properties of the composite. Enclosing the fiber surface has other benefits such as reducing defects and healing the surface flaws of the natural fiber, thereby reducing the degree of anisotropy of the natural fiber and its composites.

Besides improvement in physical and mechanical properties, another advantage of resin impregnation is that it increases fiber biological degradation resistance. Without this resistance, prolonged water molecules located in the cell wall can initiate degradation of the fiber cell wall. Low MC is desirable to ensure natural fiber durability as opposed to a fiber with higher MC, which promotes fiber decay. It is a well-known fact that ambient MC, chemicals, and radiation often cause changes which result in a slow drift in the physical and mechanical properties of fibers and their composites [36–44].

Further evidence of improvement in physical properties via resin impregnation was found when a significant reduction in water absorption (WA) was seen after the impregnated fibers were submerged into water for 24 h, as shown in Figure 5.5. According to Chen *et al.* [19], at high humidity levels, liquid water may be held by the forces of surface tension in capillary spaces available in the material. With impregnation modification, the resin effectively served as the fiber's barrier from diffusion of water molecules into cell wall and capillary spaces (lumen), or at least it reduced the permeability

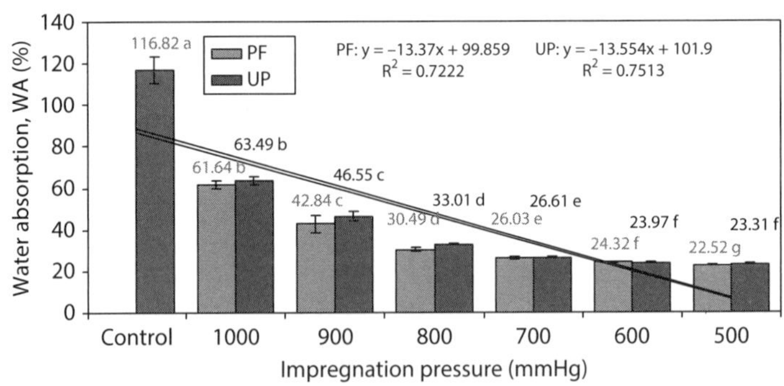

Figure 5.5 WA of PF- and UP-impregnated sugar palm fibers at various impregnation pressures.

behavior of the fiber once the micropores were blocked with the resin. Although water molecules are absorbed into the cell wall, the fiber can only swell by an additional amount of water which is dependent upon the bulking of the cell wall due to the volume occupied by the impregnation agent. Thus, it is concluded that the moisture absorption of fibers is reduced due to the physical blocking of cell wall OH groups, which also contributes to the improvement in the dimensional stability of the fiber and its composites. It was also observed that the rate of absorption of the fiber impregnated with PF was sligthly lower than that of UP, as seen in the regression trends. It was expected that a higher WPG of PF imparted a slightly lower WA than UP. The resin molecules were forced into the cell wall of the fiber as the pressure increased, resulting in deep penetration into the fiber, and extensive penetration was believed to occur due to the smaller resin molecules of PF. The WA was found to reduce as pressure increased, of which the lowest WA was observed for the fibers impregnated at pressures of 500 mmHg (22.52%) for PF and 600 mmHg (23.97%) for UP, while no significant increase was observed at the higher pressure of 500 mmHg (23.31%).

The phenomena can well be understood by comparing the scanning electron microscope (SEM) cross-section views of fractured surfaces of the control, PF-, and UP-impregnated fibers (at a pressure of 500 mmHg). Figure 5.6a shows that the fiber has an open lumen, which is believed to be a mechanism for water to fill before diffusing into the cell wall. During impregnation, the lumen and the cell wall were impregnated with resin, which immediately prevented the penetration of water. The resin was deposited extensively into the fiber cell wall and it enclosed the hydrophilic property of the fiber, which caused the fiber to become highly water resistant. This is shown in Figure 5.6b where the lumen of the fiber was partially impregnated with PF resin and is similar to Figure 5.6c, which shows the lumen partially filled with UP resin.

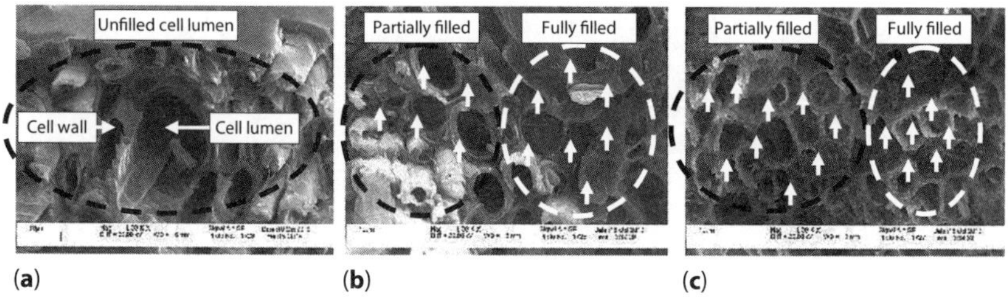

Figure 5.6 (a) Surface of control sample; (b) Surface of PF-impregnated fiber; (c) Surface of UP-impregnated fiber.

In addition to considering the quantitative results of improvement in physical properties (Figs. 5.3 and 5.5), these photographs have been taken as evidence that the impregnation agents are located in the cell lumen and might also be within the cell wall. It is important to emphasize that in this study mere lumen filling (without diffusion of resin into the cell wall) would not be expected to give satisfactory improvement in the dimensional stability of fibers and their composites. Polymerization of an impregnation agent within the cell wall is still required, where it results in a lower degree of swelling when the impregnated fiber or its composite is exposed to moisture. If diffusion of resin did not occur in the cell wall, the water molecules would still be absorbed into the cell wall when the fiber is exposed to water, even though the lumen is fully filled with the resin. This is due to the attractive force of hydrogen bonding in the polymeric constituent of fiber cell walls. This causes the cell wall to swell and the swelling occurs continuously until the cell wall is fully saturated with water molecules. This is called the fiber saturation point (FSP), which indicates that the cell wall cannot increase its moisture content any further. The water molecules that are absorbed into the fiber beyond this point become located in the lumen, or in other macrovoids [29], preventing further swelling of the cell wall.

5.3.2 Tensile Properties of Impregnated Fibre

Figure 5.7 shows the distinct tensile stress strain behaviors of impregnated fibers after impregnation with PF and UP. It shows that as the impregnation pressure of sugar palm fiber increased, the tensile strength and tensile modulus also increased and the elongation at break decreased.

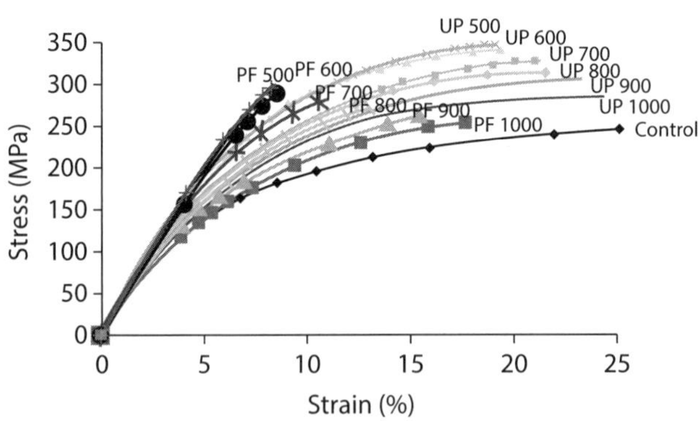

Figure 5.7 Stress-strain behavior of PF- and UP-impregnated sugar palm fiber at various impregnation pressures.

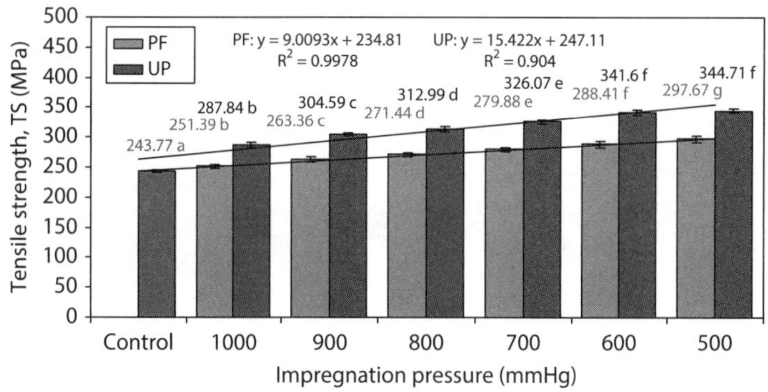

Figure 5.8 Tensile strength of PF- and UP-impregnated sugar palm fiber at various impregnation pressures.

It is shown in Figure 5.8 that the tensile strength of PF- and UP-impregnated fibers significantly increased as the impregnation pressure increased. It is more pronounced in the UP-impregnated fibers, as seen in the regression trends. The highest tensile strength of PF-impregnated fibers was found at an impregnation pressure of 500 mmHg (297.67 MPa), while for the UP-impregnated fibers it was at 600 mmHg (341.6 MPa) after no significant increase at 500 mmHg (344.71 MPa). It shows that the resin that was enclosed and impregnated into the fiber cell wall not only acted to improve physical properties, but at the same time it also increased the strength of the fiber. This is due to the fact that resin absorbed into the fiber gave a reliable measure of the amount of substance in the fiber, which reacted to stress. Since more added substances (such as PF and UP resins) in the fiber provide more area to resist stress, a greater amount of added substances resulted in a stronger impregnated fiber compared to the control sample; therefore, an increase in tensile strength is another important advantage of fibers impregnated with PF and UP resins. Similar findings were reported by many researchers [45–47] that show that resin impregnation improved the strength of several lignocellulosic materials.

The increase in tensile strength in fibers can also be explained by the reduction of moisture content. When the fiber absorbed moisture and diffused it into the cell wall, the water molecules started to form hydrogen bonds with the hydroxyl groups of the fiber, and the cellulose chains moved apart, which resulted in an increase in microfibril size. This caused the change in the size of the cell, which in turn caused the fiber to swell. As the celluloses inside the fiber moved away from each other due to fiber swelling, large gaps between the cellulose chains resulted and weakened

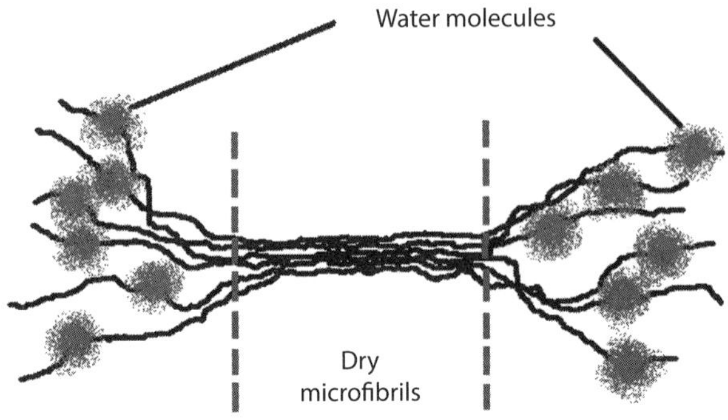

Figure 5.9 Illustration of wet and dry microfibrils of natural fibers.

intermolecular and intramolecular bonding in the fiber. On the other hand, if the water molecules were removed, the microfibrils would shrink and decrease in size, causing the cellulose chains to become closer and form stronger intermolecular bonding among the microfibrils as well as intramolecular bonding within the microfibrils. Fibers are stronger as the cellulose chains form strong bonding among themselves. This is illustrated in Figure 5.9 where there are portions of wet microfibrils (at both ends of the microfibrils) due to the presence of moisture, while the dry portion of the microfibrils shrink (the middle portion). The decrease in the strength of natural fiber due to the presence of moisture has also been reported by several researchers [48–60, 18] found an interesting result in that the strength of pineapple leaf fiber (PALF) decreased by 50% when tested under wet conditions.

Figure 5.10 indicates that the strength of the natural fiber increased exponentially with the reduction of MC. It also shows the changes in strength and shrinkage that took place at a MC below the FSP value. This explains the point at which water molecules are present in the cell wall and that no free water molecules (whether liquid or vapor) are present in the cell lumen. As the number of water molecules in the cell wall (referred to as bound water) is reduced, the fiber shrinks linearly and swells when moisture is absorbed until the moisture content reaches the FSP. As a consequence, when the MC is higher than the FSP, water molecules (vapor or liquid) are present in the cell lumen and fiber swelling stops despite further increase in fiber moisture content. Thus, if the strength and swelling are determined as a function of fiber MC, it exhibits a constant value above the FSP [25, 26].

The resin that was locked in the fiber caused the fiber to become more rigid and stiff where it approached the behavior of brittle PF and UP resins. It can be observed in Figure 5.11 that there is an increase in tensile

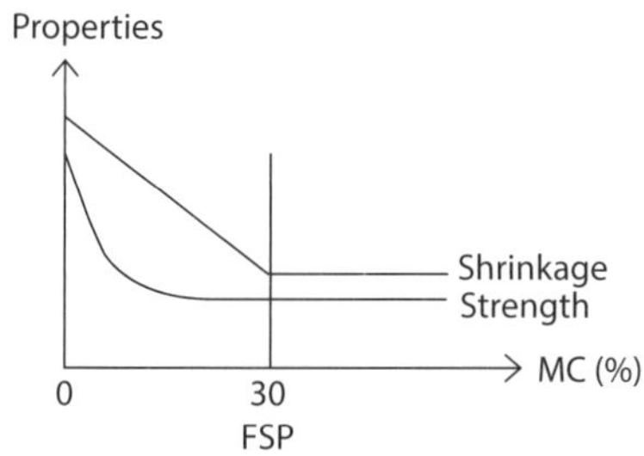

Figure 5.10 Relation of shrinkage and strength of natural fiber upon FSP.

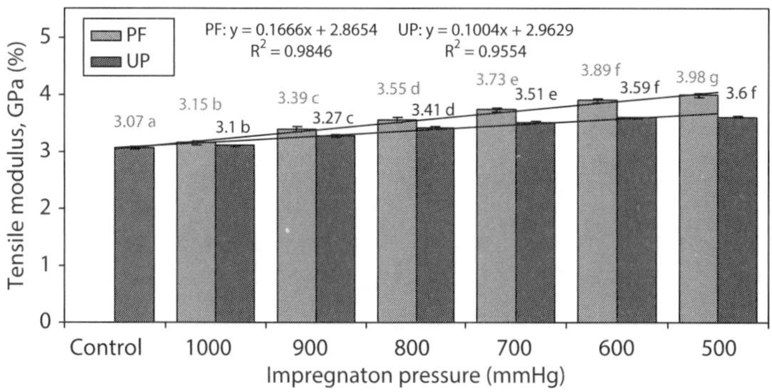

Figure 5.11 Tensile modulus of PF- and UP-impregnated sugar palm fibers at various impregnation pressures.

modulus as the pressure increases. The impregnated fibers became brittle as impregnation pressure increased. The opposite trend was obtained for tensile strength where the PF-impregnated fibers showed a significantly higher tensile modulus than UP-impregnated fibers, as observed in their regression trends. It is also observed that the fibers reached a significant maximum tensile modulus when impregnated at a pressure of 500 mmHg: 3.98 GPa (PF) and 3.60 GPa (UP). There was also no significant difference in the tensile modulus of UP-impregnated fiber between pressures of 600 and 500 mmHg. The increase in brittleness, especially for the fiber impregnated with PF, explains that less energy was absorbed before rupture, which means a reduction in fiber toughness. This information is vital in designing materials used in structural applications where it is important to produce a tough material as opposed to a brittle material.

As observed earlier in the stress strain curve, the elongation at break of fiber after impregnation was reduced considerably as the impregnation pressure increased, as shown in Figure 5.12. The figure shows that the

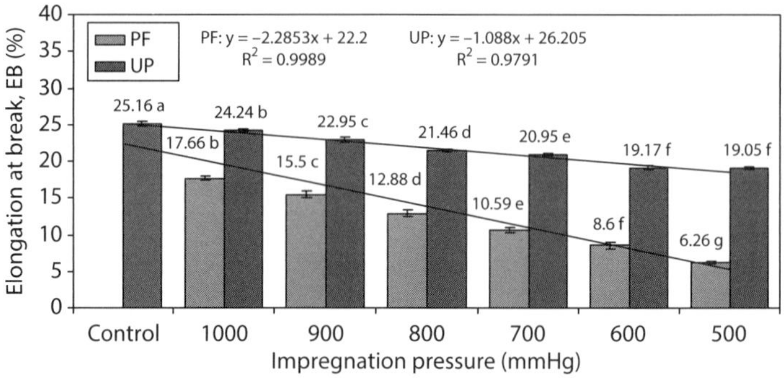

Figure 5.12 Elongation at break of PF- and UP-impregnated sugar palm fibers at various impregnation pressures.

lowest elongation at break for both fibers is at a pressure of 500 mmHg and that the rate of reduction is more apparent in the fiber impregnated with PF as seen by the slope in their regression trends. It is observed at a pressure of 500 mmHg that the elongation at break of the fiber impregnated with PF (6.26%) and UP (19.05%) was significantly different. This was a similar trend found with tensile modulus; no significant difference in elongation at break was observed between impregnation pressures of 600 mmHg (19.17%) and 500 mmHg (19.05%). It was expected that the higher the pressure, the more resin that would be absorbed into the fiber. This was proven by the increase in its WPG, resulting in the resin having more of an influence on the fiber properties. The improvement in strength and modulus after impregnation modification with resin has also been reported by previous recearchers [30, 46, 47, 61, 25, 26]. It is seen in Figure 5.7 (stress strain curve) that the impregnated fiber became brittle and had low elongation at break compared to the control sample (Fig. 5.12).

An important issue to be highlighted in this research is that while the impregnated fibers increased in stress values, they showed significant decreases in elongation at break, which may result in the reduction of fiber toughness. Based on the area under the stress strain curve, the fiber toughness can be measured to represent the total energy the material can absorb before rupture. It is observed in Figure 5.13 that the toughness of PF-impregnated fibers experiences a significant drop at high impregnation pressure, as shown by the regression trend lines (with a slope of -4.2373 compared to -1.3385 for UP).

Statistically ($p \leq 0.05$), no significant decrease in toughness was observed for the fiber impregnated with UP compared to the control sample with an increase in impregnation pressure. This is because a material is known to

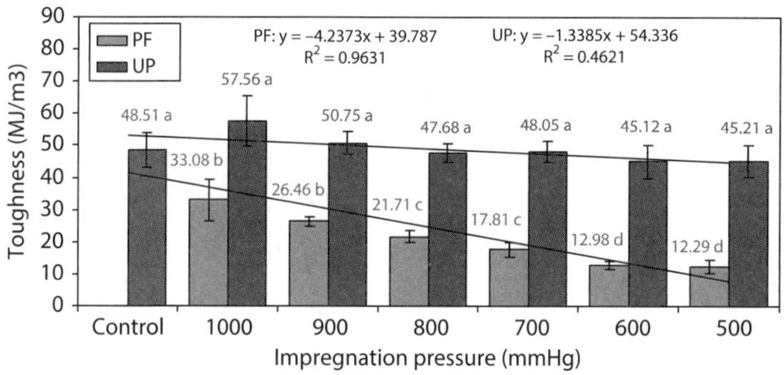

Figure 5.13 Toughness of PF- and UP-impregnated sugar palm fibers at various impregnation pressures

be strong and tough if it ruptures at high stress and exhibits high strain values. In contrast, for a material low in toughness, it may be strong (has high ultimate stress), but also has low strain values as found in brittle materials. High toughness translates to high stress and strain values and this is a desirable property of fibers produced especially for composite reinforcement. Toughness is an important criterion for engineers to consider when selecting, modifying, or designing new material that requires high durability. It is also important during an impregnation process because most impregnation modification, especially when using PF, results in an increase in material brittleness. Although it is good to have materials with high modulus that can resist deformation, most of the time it is a lot better to have materials that bend instead of break, and if they bend, stretch, or deform in some way, they do not break [62, 63]; therefore, it is important to not focus solely on producing a fiber with a high modulus (as found in PF-impregnated fibres) and high strength. It is also worth considering producing a strong fiber that has high toughness.

In addition to toughness, it is also important to consider weight gain of materials after undergoing modification, as it can increase the final weight of the end product (composite). The PF-impregnated fibers show significant increases in SG (Fig. 5.2) as a result of a higher WPG (Fig. 5.1), while no significant weight gain is shown in UP-impregnated fibers.

With regard to human safety and the environment, it is well known that the formaldehyde emission of PF can have severe effects compared to UP. It causes allergic reactions such as watery eyes, burning sensations in the eyes, and many other side effects to human health, not to mention further problems for the disposal of formaldehyde waste. Although the lower molecular weight of PF gives better penetration of resin molecules into the fiber cell wall, it exhibits lower bonding capability and deterioration of mechanical properties [64]. A longer time is required for polymerization of the resin with problems of leachable resin from fiber, and worse, producing higher formaldehyde emission compared to the high molecular weight PF.

Impregnation with the PF was better than that of the UP; however, after considering the Duncan's multiple range test results, not much significant difference in physical properties was observed between the PF and UP resins where the maximum reduction of MC (at the significant highest pressure) was found to be 94.66 and 94.08%, respectively, while for WA, the results were found to be 80.72 and 79.76%, respectively. This less significant effect of using different impregnation agents can also be seen in their regression trends. Contrary to mechanical properties, it is found that the fiber impregnated with UP was significantly different than with PF, as shown in the regression trends. The maximum tensile strength for PF and UP was found to be 22.11 and 40.77%, respectively, while elongation at break was found to be 24.05 and 75.12%, respectively. A significantly lower tensile strength and elongation at break of PF gives a significant reduction in fiber toughness by 73.94% for fibers impregnated at a pressure of 500 mmHg compared to UP and the control sample. The use of fiber with such properties should be avoided because it is unreliable in sustaining the durability of its composites during service conditions. According to statistical analysis on UP-impregnated fibers, for physical and mechanical properties, the optimum impregnation pressure was found to be 600 mmHg after no significant increase was seen when the pressure increased to 500 mmHg. After considering the results obtained for both physical and mechanical properties, and after taking into account human health, environmental factors, and processability, it is worth stating that in order to enhance the properties of sugar palm fiber, the fiber should be impregnated with UP resin at an impregnation pressure of 600 mmHg for 5 min.

5.4 Conclusions

A study on impregnation modification via vacuum resin impregnation at various impregnation pressures and with different impregnation resins on physical and mechanical properties of sugar palm fiber were carried out. The finding showed that:

1. There was significant improvement in physical and mechanical properties when the impregnation pressure was increased to 500 mmHg from 1000 mmHg, which indicated

better penetration of impregnation agents (PF and UP) into the cells of the sugar palm fiber.
2. As the pressure increased, the SG of the PF-impregnated fibers statistically increased, while no significant increase was seen for those fibers impregnated with UP.
3. The moisture content and water absorption of PF- and UP-impregnated fibers were found to decrease significantly after being impregnated.
4. As the impregnation pressure increased, the mechanical properties of PF- and UP-sugar palm fibers showed significant improvement ($p \leq 0.05$) where it was seen that the tensile strength and tensile modulus had significantly increased, while the elongation at break decreased.
5. It was also observed that significantly low tensile strength and low elongation at break of PF-impregnated fibers resulted in the fiber being less superior in toughness compared to the UP-impregnated fibers.
6. It was discovered that besides having less side effects to human health and the environment and not contributing to an increase in SG, better property enhancement was seen for UP-impregnated fibers, especially in mechanical properties, compared to PF-impregnated fibers.

Acknowledgments

The authors would like to acknowledge the Ministry of Agriculture and Agro-Based Industry (MOA) of Malaysia for funding the research grant (Science Fund, project number: 05-01-04-SF1114). Part of this paper was presented in the Eighth International Conference on Composite Science and Technology (ICCST/8 2011) held in Kuala Lumpur.

References

1. Z. Leman, S.M. Sapuan, A.M. Saifol, M.A. Maleque, and M.M.H.M. Ahmad, Moisture absorption of sugar palm fibre reinforced epoxy composites, *Mater. Design*, Vol. 29, pp. 1666–1670, 2008.
2. A. Aidy, A.B. Sanuddin, and Saifuliwan Ezzeddin, The effect of aging on *Arenga pinnata* fiber-reinforced epoxy composite, *Mater. Design*, Vol. 31, No. 7, pp. 3550–3554, 2010.

3. D. Bachtiar, S.M. Sapuan, E.S. Zainudin, A. Khalina, and K.Z.H.M. Dahlan, The effect of alkali treatment and compatibilising agent on tensile properties of short sugar palm fibre reinforced high impact polystyrene composites, *Bioresources*, Vol. 6, pp. 4815–4823, 2011.
4. Y.A. El-Shekeil, S.M. Sapuan, A. Khalina, and E.S. Zainudin, Development of a new kenaf bast fiber-reinforced thermoplastic polyurethane composite, *BioResources*, Vol. 6, pp. 4662–4672, 2011.
5. M.J. Suriani, M.M.H.M. Hamdan, H.Y. Sastra, and S.M. Sapuan, Study of interfacial adhesion of tensile specimens of *Arenga pinnata* fiber reinforced composites, *Multidiscip. Model. Mater. Struct.*, Vol. 3, pp. 213–224, 2006.
6. S.M. Sapuan, and M.A. Maleque, Design and fabrication of natural woven fabric reinforced epoxy composite for household telephone stand, *Mater. Design*, Vol. 26, pp. 65–71, 2005.
7. H.Y. Sastra, J.P. Siregar, S.M. Sapuan, Z. Leman, and M.M.H.M. Hamdan, Flexural properties of *Arenga Pinnata* fibre reinforced epoxy composites, *Am. J. Appl. Sci.*, Special Issue, pp. 21–24, 2005.
8. M.R. Ishak, A.M.M. Edeerozey, I.S. Othman, Z. Leman, and S.M. Sapuan, Mechanical properties of kenaf bast and core fibre reinforced unsaturated polyester composites, *IOP Conf. Series: Mater. Sci. Eng., Vol.* 11, No. 1, 012006, 2010.
9. M.Y.M. Zuhri, S.M. Sapuan, N. Ismail, and R. Wirawan, Mechanical properties of short random oil palm fibre reinforced epoxy composites, *Sains Malays.*, Vol. 39, pp. 87–92, 2009.
10. M.A. Maleque, F.Y. Belal, and S.M. Sapuan, Mechanical properties study of pseudo-stem banana fiber reinforced epoxy composite, *Arab. J. Sci. Eng.*, Vol. 32, pp. 359–364, 2007.
11. H.Y. Sastra, J.P. Siregar, S.M. Sapuan, Z. Leman, and M.M.H.M. Hamdan, Tensile properties of *Arenga pinnata* fibre-reinforced epoxy composites, *Polym. Plast. Technol. Eng.*, Vol. 45, pp. 1–8, 2006.
12. J. Sahari, S.M. Sapuan, Z.N. Ismarrubie, and M.Z.A. Rahman, Comparative study of physical properties based on different parts of sugar palm fibre reinforced unsaturated polyester composites, *Key Eng. Mat.*, Vol. 471–472, pp. 502–506, 2011a.
13. D. Bachtiar, S.M. Sapuan, and M.M. Hamdan, The effect of alkaline treatment on tensile properties of sugar palm fibre reinforced epoxy composites, *Mater. Design*, Vol. 29, pp. 1285–1290, 2008.
14. M.R. Ishak, Z. Leman, S.M. Sapuan, M.Y. Salleh, and S. Misri, The effect of sea water treatment on the impact and flexural strength of sugar palm fibre reinforced epoxy composites, *Int. J. Mech. Mater. Eng.*, Vol. 4, pp. 316–320, 2009.
15. J. Sahari, S.M. Sapuan, Z.N. Ismarrubie, and M.Z.A. Rahman, Investigation on bending strength and stiffness of sugar palm fibre from different parts reinforced unsaturated polyester composites, *Key Eng. Mat.*, Vol. 471–472, pp. 455–460, 2011b.

16. Z. Leman, S.M. Sapuan, M.R. Ishak, and M.M.H.M. Ahmad, Pre-treatment by water retting to improve the interfacial bonding strength of sugar palm fibre reinforced epoxy composite, *Polym. Renew. Re.*, Vol. 1, pp. 1–12, 2010.
17. M.R. Ishak, S.M. Sapuan, Z. Leman, M.Z.A. Rahman, and U.M.K. Anwar, Characterization of sugar palm (*Arenga pinnata*) fibres: Tensile and thermal properties, *J. Therm. Anal. Calorim.*, DOI: 10.1007/s10973-011-1785-1, 2011.
18. S. Mishra, M. Misra, S.S. Tripathy, S.K. Nayak, and A.K. Mohanty, Potentiality of pineapple leaf fibre as reinforcement in PALF-polyester composite: Surface modification and mechanical performance, *J. Reinf. Plast. Compos.*, Vol. 20, pp. 322–334, 2001.
19. H. Chen, M. Miao, and X. Ding, Influence of moisture absorption on the interfacial strength of bamboo/vinyl ester composites, *Comp. Part A*, Vol. 40, pp. 2013–2019, 2009.
20. M.J. John, and S. Thomas, Biofibres and biocomposites, *Carbohydrate Polymers*, Vol. 71, pp. 343–364, 2008.
21. M.M. Rahman, and M.A. Khan, Surface treatment of coir (*Cocos nucifera*) fibers and its influence on the fibers' physico-mechanical properties, *Comp. Sci. Technol.*, Vol. 67, pp. 2369–2376, 2007.
22. W.L. Bradley, and T.S. Grant, The effect of the moisture absorption on the interfacial strength of polymeric matrix composites, *J. Mater. Sci.*, Vol. 30, pp. 5537–5542, 1995.
23. H.M. Akil, M.F. Omar, A.A.M. Mazuki, S. Safiee, Z.A.M. Ishak, and A. Abu Bakar, Kenaf fiber reinforced composites: A review, *Mater. Design*, Vol. 32, pp. 4107–4121, 2011.
24. S. Ramarad, Preparation and properties of kenaf bast fiber filled (plasticized) poly(lactic acid) composites, MSc Thesis, Universiti Sains Malaysia, 2008.
25. C.A.S. Hill, "Impregnation modification," in: *Wood Modification: Chemical, Thermal and Other Processes*, C.A.S. Hill, ed., John Wiley & Sons Ltd., West Sussex, pp. 149–173, 2006.
26. C.A.S. Hill, "Modifying the properties of wood," in: *Wood Modification: Chemical, Thermal and Other Processes*, C.A.S. Hill, ed., John Wiley & Sons Ltd., West Sussex, pp. 19–44, 2006.
27. J.S. Robbins, W.D. Mockel, and C.J. Lyons, Average molecular weight determination of a polyester by NIR spectroscopy, *ISA Transactions*, Vol. 33, pp. 377–380, 1994.
28. A.J. Stamm, and R.M. Seborg, Resin treated wood – impreg, USDA Forest Service, Forest Products Laboratory, Report No. 1360, 1962.
29. *R.R. Ibach, and W.D. Ellis,* "Lumen modifications," in: *Handbook of Wood Chemistry and Wood Composites*, W.R.M. Rowell, ed., CRC Press, Boca Raton, pp. 421–445, 2005.
30. M. Deka, P. Das, and C.N. Saikia, Studies on dimensional stability, thermal degradation and termite resistant properties of bamboo (*Bambusa tulda* Roxb.) treated with thermosetting resins, *Bamboo Rattan*, Vol. 2, pp. 29–41, 2003.

31. R.H. Hu, M.Y. Sun, and J.K. Lim, Moisture absorption, tensile strength and microstructure evolution of short jute fiber/polylactide composite in hygrothermal environment, *Mater. Design*, Vol. 31, pp. 3167–3173, 2010.
32. H.D. Rozman, R.N. Kumar, H.P.S. Khalil, A. Abusamah, and R. Abu, Fibre activation with glycidyl methacrylate and subsequent copolymerization with diallyl phthalate, *Eur. Polym. J.*, Vol. 33, pp. 1213–1218, 1997.
33. H.D. Rozman, R.N. Kumar, A. Abusamah, and M.J. Saad, Rubberwood–polymer composites based on glycidyl methacrylate and diallyl phthalate, *Appl. Polym. Sci.*, Vol. 67, pp. 1221–1226, 1998.
34. M.S. Islam, S. Hamdan, M.R. Rahman, I. Jusoh, and A.S. Ahmed, The effect of crosslinker on mechanical and morphological properties of tropical wood material composites, *Mater. Design*, Vol. 32, pp. 2221–2227, 2011.
35. W.E. Morton, and J.W.S. Hearle, *Physical Properties of Textile Fibres*, Manchester, The Textile Institute, 1993.
36. G. Pritchard, "Environmental testing of organic matrix composites," in: *Mechanical Testing of Advanced Fibre Composites*, J.K. Hodgkinson, ed., Boca Raton, CRC Press, pp. 269–291, 2000.
37. H. Abral, M. Fitra Gafar, H. Andriyanto, Ilhamdi, S.M. Sapuan, M.R. Ishak, and Evitayani, Alkali treatment of *Pandanus Odoratissimus* fibers and its effect on unsaturated polyester composite, *Polym.-Plast. Technol. Eng.*, Vol. 51, pp. 12–18, 2012.
38. G. Ramgopal, R. Ramani, P. Ramachandra, and C. Ranganathaiah, UV degradation of bivoltine silk fiber: A positron annihilation study, *Eur. Polym. J.*, Vol. 34, pp. 1423–1427, 1998.
39. F. Gouanvé, S. Marais, A. Bessadok, D. Langevin, and M. Métayer, Kinetics of water sorption in flax and PET fibers, *Eur. Polym. J.*, Vol. 43, pp. 586–598, 2007.
40. H.P.S.A. Khalil, H. Ismail, H.D. Rozman, and M.N. Ahmad, The effect of acetylation on interfacial shear strength between plant fibres and various matrices, *Eur. Polym. J.*, Vol. 37, pp. 1037–1045, 2001.
41. B.D. Agarwal, and G.S. Giare, Influence of the properties of the matrix material on the fracture toughness of short-fibre composites, *Mater. Sci. Eng.*, Vol. 52, pp. 139–145, 1982.
42. T.W. Chou, and A. Kelly, Fibre composites, *Mater. Sci. Eng.*, Vol. 25, pp. 35–40, 1976.
43. R.C. Wetherhold, and M.A. Mahmoud, Tensile strength of notched composite materials, *Mater. Sci. Eng.*, Vol. 79, pp. 55–65, 1986.
44. H.D. Rozman, G.S. Tay, R.N. Kumar, A. Abusamah, H. Ismail, and Z.A. Mohd Ishak, Polypropylene-oil palm empty fruit bunch-glass fibre hybrid composites: A preliminary study on the flexural and tensile properties, *Eur. Polym. J.*, Vol. 37, pp. 1283–1291, 2001.
45. M. Deka, and C.N. Saikia, Chemical modification of wood with thermosetting resin: Effect on dimensional stability and strength property, *Bioresour. Technol.*, Vol. 73, pp. 179–181, 2000.

46. M.I. Shams, H. Yano, and K. Endou, Compressive deformation of wood impregnated with low molecular weight phenol formaldehyde (PF) resin I: Effects of pressing pressure and pressure holding, *Wood Sci.*, Vol. 50, pp. 337–342, 2004.
47. U.M.K. Anwar, M.T. Paridah, H. Hamdan, S.M. Sapuan, and E.S. Bakar, Effect of curing time on physical and mechanical properties of phenolic-treated bamboo strips, *Ind. Crops Prod.*, Vol. 29, pp. 214–219, 2009.
48. J.M. Dinwoodie, *Timber: Its Nature and Behaviour*, London, Spon Press, 2000.
49. K.F. Chung, and W.K. Yu, Mechanical properties of structural bamboo for bamboo scaffoldings, Eng. Struct., Vol. 24, pp. 429–442, 2002.
50. V.S. Godbole, and S.C. Lakkad, Effect of water absorption on the mechanical properties of bamboo, J. Mater. Sci. Lett., Vol. 5, pp. 303–304, 1986.
51. A. Stamboulis, C.A. Baillie, and T. Peijs, Effects of environmental conditions on mechanical and physical properties of flax fiber, Composites Part A, Vol. 32, pp. 1105–1115, 2001.
52. A.C. Karmaker, Effect of water absorption on dimensional stability and impact energy of jute fiber reinforced polypropylene, J. Mater. Sci. Lett., Vol. 16, pp. 462–464, 1998.
53. A. Espert, F. Vilaplana, and S. Karlsson, Comparison of water absorption in natural cellulosic fibers from wood and one-year crops in polypropylene composites and its influence on their mechanical properties, Composites Part A, Vol. 35, pp. 1267–1276, 2004.
54. A.K. Bledzki, and J. Gassan, Composites reinforced with cellulose based fibers, Prog. Polym. Sci., Vol. 24, pp. 221–274, 1999.
55. D. Nabi Saheb, and J.P. Jog, Natural fiber polymer composites: A review, Adv. Polym. Technol., Vol. 18, pp. 351–363, 1999.
56. M. Mizanur Rahman, UV-cured henequen fibers as polymeric matrix reinforcement: Studies of physico-mechanical and degradable properties, *Mater. Design*, Vol. 30, pp. 2191–2197, 2009.
57. Y. Diamant, G. Marom, and L.J. Broutman, The effect of network structure on moisture absorption of epoxy resins, J. Appl. Polym. Sci., Vol. 26, pp. 3015–3025, 1981.
58. H.N. Dhakal, Z.Y. Zhang, and M.O.W. Richardson, Effect of water absorption on the mechanical properties of hemp fiber reinforced unsaturated polyester composites, Compos. Sci. Technol., Vol. 67, pp. 1674–1683, 2007.
59. N. Sombastsompop, and K. Chaochanchaikul, Effect of moisture content on mechanical properties, thermal and structural stability and extruded texture of poly(vinyl chloride)/wood sawdust composites, Polym. Int., Vol. 53, pp. 1210–1218, 2004.
60. P.V. Joseph, M.S. Rabello, L.H.C. Mattoso, K. Joseph, and S. Thomas, Environmental effects on the degradation behavior of sisal fiber reinforced polypropylene composites, Comp. Sci. Technol., Vol. 62, pp. 1357–1372, 2002.

61. Y.F. Loh, M.T. Paridah, Y.B. Hoong, and A.C.C. Yoong, Effects of treatment with low molecular weight phenol formaldehyde resin on the surface characteristics of oil palm (*Elaeis quineensis*) stem veneer, *Mater. Design*, Vol. 32, pp. 2277–2283, 2011.
62. B.Z. Jang, *Advanced Polymer Composites: Principles and Applications*, Materials Park, OH, USA, ASM International, 1994.
63. G. Odian, *Principles of Polymerization*, New York, John Wiley and Sons, 1991.
64. D.N.S. Hon, "Analysis of adhesives," in: *Handbook of Adhesive Technology*, A. Pizzi and K.L. Mittal, eds., Marcel Dekker, New York, pp. 293–324, 2003.

6

Environmentally-Friendly Acrylates-Based Polymer Latices

Sweta Shukla*,[1] and J.S.P. Rai[2]

[1]*Assistant Professor, KIET Ghaziabad, U.P. India*
[2]*Director, Harcourt Butler Technological Institute, Nawabganj, Kanpur, U.P. India*

Abstract

Acrylic copolymers have achieved prime importance in various avenues of industrial application such as adhesive in packaging and construction, binder and stabilizer in textiles, base coat in leather, topcoat in floor polish, pigment binder in paper industries, etc. Acrylic resins are generally known to have superior exterior durability and weather resistance compared to other organic binders/film-formers commonly used in surface coatings. The emulsion polymerization of acrylic monomers provides a good combination of performance properties required in film formation. These properties depend on the copolymer composition, particle size, molecular weight, glass transition temperature (Tg), morphology of the latex particles, etc.

In this chapter, the reaction components and polymerization process conditions that affect the inherent characteristics of the polymers are presented. There is a discussion of the incorporation of comonomer, which would lead to an improvement in the flexibility of films produced along with adhesion, gloss, hardness, solvent resistance, water resistance, etc. Also, there is a recently developed successful approach for achieving balanced properties which involves the copolymerization of film-forming monomers like acrylate and methacrylates with crosslinkable monomers.

Keywords: Acrylate, coating, crosslinker, waterborne

**Corresponding author*: sweta_hbti@rediffmail.com

Ashutosh Tiwari and Mikael Syväjärvi (eds.) Advanced Materials for Agriculture, Food, and Environmental Safety, (145–176) 2014 © Scrivener Publishing LLC

6.1 Introduction

About the time of the American Revolution, most of the paints used were based on natural resins, but with the increasing demand for coatings, natural resins became insufficient. Moreover, the paints based on natural resins do not exhibit satisfactory film properties. As a result, today the paint industry has become dependent mainly on synthetic resins because of their much superior properties, easy availability to suit the requirements and also their cost effectiveness. Synthetic polymers as binders play a key role in the process of developing suitable coatings. Since the use of latex as a coating binder was first examined in 1946, synthetic polymers have become more and more important. Latex dispersions form films at ambient temperatures by coalescence of relatively soft particles containing solid polymer. The selection of binder for formulating a coating is based on the desired film properties for the specified end use. Industrial coatings differ from those used in most household paint applications. They are used on larger-scale industrial buildings and equipment. They have a higher glass transition temperature so that they are tougher than household paints, and have better barrier properties through copolymerizsation with more hydrophobic monomers. Industrial coatings and an important class of binders used in surface coatings are often acrylic copolymers, such as polyurethane, epoxy, alkyd resin latexes, etc.

Monomers are the building block of the polymer, and their selection is mainly based on the desired end-product properties (adhesive strength, water resistance, minimum film-forming temperature, etc.). These applications require film-forming latex with high clarity, good stability and high mechanical strength. The desired properties of the emulsions depend on the glass transition temperature (Tg), visco-elastic modulus of the polymer which, in turn, depends upon the type of monomer, sequence length distribution of the monomer, and the molecular weight of the polymer. For film-forming behavior, Tg must be well below the application temperature to allow high molecular mobility for interdiffusion and entanglement.

6.1.1 Alkyds

Alkyd resins are essentially short-branched polyesters derived as the reaction products of vegetable oils triglycerides, polyols and dibasic acids or their anhydrides. Alkyd resins have been the workhorse for the coating industry over the last half century. At present, they hold the majority share of the world market for non-aqueous binders. They are used extensively because of their low cost and versatility, but for high performance/challenging areas, they are not suitable [1].

6.1.2 Urethanes

Urethanes and polyurethanes are essentially polymeric esters of carbamic acid or derivatives of carbamic acid with general formula R-NH-COOR [2]. They are prepared by reaction of diisocyanate with polyfunctional compound containing active hydrogen atoms such as water, alcohols, amines, etc. The main advantage of the coatings based on urethanes are their high chemical and abrasion resistance, low temperature cure, and wide range of flexibility coupled with toughness. However, toxicity and yellowing tendency in outdoor applications limit their use as binders. Urethane paints are toxic and should be used with caution. Because they contain iscocyanantes, airborne compounds that enter the lungs or skin, investigating proper precautions and aerating techniques is important when working with urethane.

6.1.3 Epoxies

Epoxy resins are polyether resins containing more than one epoxy group capable of being converted into the thermoset form. These resins, on curing, do not create volatile products in spite of the presence of a volatile solvent. The large groups of polyepoxides or epoxide resins [3] which are commonly used in surface coatings are based upon the reaction products of epichlorohydrin and bisphenol A. Because of high chemical and corrosion resistance, excellent adhesion, low shrinkage, toughness, and good abrasion resistance properties, epoxy resins find ample application in the industrial maintenance field.

The coatings industry is the biggest consumer of epoxy resins. These resins are used mostly as chemical and special purpose coatings. Epoxy resins provide thin-layer durable coatings having mechanical strength and good adhesion to a variety of substrates. They are resistant to chemicals, corrosion, and solutions. They find applications in washing machines and appliances, ships and bridges, pipelines and chemical plants, automobiles, farm implements, containers, and floor coatings. Epoxy coating formulations are available as liquid resins, solid resins, high molecular weight thermoplastic resins, multifunctional resins, radiation curable resins, and special purpose resins. Epoxy baking finishes are obtained by high molecular weight epoxy resins crosslinked by phenolic or amino resins. These resins are used as lining for tanks, cars, drums, pails, pipes, downhole oilfield tubing, and food cans. Epoxy acrylic systems provide excellent coatings for appliances, kitchen cabinets, outdoor furniture, aluminum siding, and other metal products. Waterborne coatings are made by dispersing or emulsifying the resins with surfactants. Such coatings also have been based on emulsified liquid epoxy resins cured with emulsified polyamide resins. These formulations are used in anionic electrodeposited coatings. They provide exterior and interior coatings for underground pipes, and electrical equipment appliances reinforcement. The main drawbacks of epoxy resins for industrial use are their brittleness and high cost.

6.1.4 Acrylics

The term acrylic resin is generally applied to the polymer and copolymers of esters of acrylic and methacrylic acids. Acrylics were first prepared in 1880 as acrylate by Otto Rohm (1876–1939). He patented it in 1915, and its suggested use was as a substitute for drying oils in industrial paints and lacquers. Based on the continuing work in Rohm's laboratory, the first limited production of acrylates began in 1927 by Rohm and Haas Company in Darmstadt, Germany. The use of various acrylates could not be explored for a variety of applications due to their cost till 1950. The cost of the acrylates was reduced by Rohm and Haas in 1950, which resulted in extensive use of these monomers, particularly in the field of coating and printing. The polymers based on acrylates contribute to almost 50% of all the polymers used for coating applications. These are the most common class of resins used as binders in surface coatings. Acrylic resins may be used alone or as blends with other resins to form a suitable binder system for coatings [4]. Paints containing acrylic resins as binders have been known since 1930s. They are now one of the largest product classes in the paint and coating sector. Acrylates and methacrylates are extremely versatile building blocks used as monomers to synthesize homo- or copolymers.

Acrylics can be formulated as thermoplastic resins, thermosetting resins, and as a water emulsion latex. The resins are formed from polymers of acrylate esters, predominantly polymethyl methacrylate and polyethyl acrylate. The acrylate resins do not contain tertiary hydrogens attached directly to the polymer backbone chain and, as a result, are exceptionally stable to oxygen and ultraviolet light deterioration. The repeating units of the acrylic backbone are joined to make long polymer chains. A wide range of monomers is available for use in designing a specific acrylic system. Typically, mixtures of monomers are chosen for the properties they impart to the polymer. The glass transition temperature of the polymer can be varied by selecting the proper monomers. This permits a varied area of application.

These monomers readily polymerize themselves to form homopolymers, or copolymerize with a variety of other monomers to yield acrylic copolymers. The properties of the resulting acrylate polymers are related not only to the chemical nature of the monomers but also to the extent and distribution of monomers, crystallinity and crystallite size distribution, molecular weight distribution, etc. These factors influence the hardness, flammability, weatherability, chemical resistance, softening point, electrical properties, etc. The copolymers based on acrylate have been found to possess excellent adhesive, solubility and other properties desired for general coating and adhesive purposes. One of the major applications of acrylate polymers is in the area of waterborne coatings.

Literature concerned with the free radical polymerization of acrylic resin has been reported by many workers in the field [5–8]. The relative ease of polymerization and the wide range of properties among the acrylic resins have led to the commercial production of many different resins suitable for a broad variety of applications [9]. For example, these resins play a prominent role in the paint industry due to the unique combination of their properties [10].

Monomeric methacrylic acid esters copolymerize with one another and often with several other vinyl monomers to increase the range of available polymer properties. The properties of the synthesized resins are dependent not only on their average molecular weight but also on their chemical composition. Therefore, by judicious adjustments of the amount of each type of monomer and synthesis conditions, polymers with desired properties can be prepared. Polymers of this class are noted for their water-clear color and retention of their properties upon aging under severe service conditions. A small amount of functionalized monomers are often copolymerized with methacrylic monomers to modify the properties of the polymer directly or by providing sites for further reactions.

6.1.4.1 *Homopolymers of Acrylic Esters*

Polymers that contain only a single type of repeat unit are known as homopolymers, while polymers containing a mixture of repeat units are known as copolymers. Copolymerization is one of the most powerful techniques for effecting systematic changes in polymer properties and is widely used in the production of various commercial polymers. It is defined as polymerization in which two or more structurally distinct monomers are polymerized so as to incorporate these into the same polymer chain. There has been enormous commercial success in the synthesis of various copolymers. This technique can be used to vary melting point, glass transition temperature

(Tg), crystallinity, solubility, elasticity, permeability, and chemical reactivity of existing polymers within limits [1]. Intense research activity continues in this area of polymer science where a combination of monomers results in a product of interest.

Acrylic monomers are versatile because of their low cost and good combination of properties. Acrylate and methacrylate monomers are unsymmetrically substituted ethylenes having moderate to high boiling points depending on the identity and nature of the group attached. These may be represented by the generic formula:

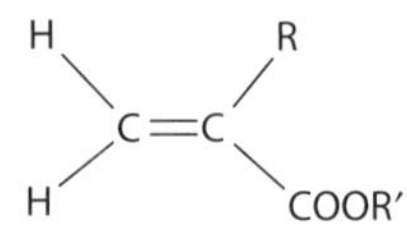

When R is a hydrogen atom, the monomer is referred to as an acrylate and when R is a methyl group, the monomer is called methacrylate. The acrylate or methacrylate monomers are derived from the corresponding acrylic and methacrylic acids (where R' = H) by a simple esterification reaction with an alcohol. However, in most of the cases the acrylate and methacrylate monomers are synthesized using primary and secondary alcohols, as the tertiary alcohol yields esters with limited stability and commercial acceptability.

Acrylic monomers include not only the monomeric alkyl ester of acrylic and methacrylic acid but also the free acids and their derivatives such as amides, nitriles and the esters bearing the functional group on the side chain, for example, fluoroacrylates, cyanoacrylates, etc.

The important uses of the acrylic resins/polymers include automotive finishes, clear lacquers for polished metals, finishes for aluminum sidings, enamels for washing machines, exterior and interior water-thinned coatings for masonry, wood, plaster, wallboard, etc. The important properties of acrylic resins as film formers imparted to the coating films are that they form brilliantly clear and durable (UV resistant) films with a varied range of hardness and flexibility. They have excellent resistance to chemical attack and are versatile in bulk, solution, dispersion and emulsion polymerization for their use in lacquers and enamels. They have variability in physical and chemical properties due to choice of monomers.

A number of homopolymers were synthesized by different techniques for achieving the desired properties. The properties of poly(alkyl acrylate) s depend on the nature of the alkyl group. For example, poly(methyl acrylate) is rubber-like, whereas poly(methyl methacrylate) is a hard material. The Tg decreases with an increase in the alkyl side-chain length in poly(n-alkyl (meth) acrylate)s. Increasing the bulkiness of the alkyl group also restricts the motion of backbone polymer chain, and hence the Tg within a series of isomers increases. The Tg of polymer is important for attainment of properties required for specific applications. Control of the Tg is achieved through copolymerization of various combinations of monomers. The selection of monomer is based on the lowest cost commensurate with achieving the properties for end-use applications.

J.K. Fink [11] gave an overview of acrylic and methacrylic ester polymers. A large variety of monomers is known and discussed because of the possibility of esterifying the acrylic acid and methacrylic acid with various alcohols. Acrylic resins are appreciated for their exceptional clarity, self-extinguishing, transparency, and superior design adaptability. Special formulations are also described in detail.

Poly(methyl methacrylate) (PMMA) is type of thermoplastic used throughout the world in applications such as transparent all-weather sheets, electrical insulation, bathroom units, automotive parts, surface coating and ion exchange resins, etc. Hamielec *et al.* [12] studied the free radical polymerization of MMA initiated by an ultraviolet (UV) light. They found that the termination rate constant decreases at high conversions.

Wunderlich [13] investigated polymerization of butyl acrylate in benzene and butyl propionate at 30, 50, and 70°C as a function of the concentration of initiator and monomer. The rate of polymerization and the molecular weight distribution of the formed polymers were measured at conversions less than 3%. The results deviate from the ideal kinetic model of radical polymerization. The reaction rate constants were obtained based on the conversion and molecular weight data.

Scott and Senogles [14] investigated the polymerization kinetics of n-lauryl acrylate (LA) in ethyl acetate and n-heptane at 40°C. They found that primary radical terminations were unlikely to play any significant role at higher monomer concentration, and autoacceleration effects do not occur during the high conversion polymerization of LA.

The homopolymers of soft monomers like alkyl acrylates, having 4–17 carbon atoms in the alkyl group, such as 2-ethylhexyl acrylate (2-EHA), BA and EA have Tg below 0°C. These monomers possess inherent tackiness and lack of sufficient cohesive strength. Since homopolymers are produced from a single monomeric species, the properties of such latexes are predetermined, and cannot be modified without the incorporation of additional chemical components. For this reason, most commercial emulsions are comprised of a comonomer system (i.e., a combination of different monomers) rather than a single monomer. The properties (e.g., glass transition temperature, mechanical properties, and solvent resistance) of the resulting copolymer can be controlled by choosing an appropriate comonomer system, such as BMA, MMA, styrene, etc.

6.1.4.2 *Copolymers of Acrylic Esters*

Acrylic copolymers are the copolymers obtained from the polymerization of esters derived from monohydric alcohols with acrylic acid, methacrylic acid, or lower alkyl substituted acrylic acids. Copolymerization of MMA with monomers such as ethyl-, butyl-, isobutyl-, 2-ethylhexyl-, 2-hydroxyethyl-, dodecyl- and cetyl-methacrylates has been investigated in the past with an aim to study the effect of structure on the properties of copolymers [15–22].

Bulk polymerizations of MMA with n-hexyl acrylate (HA) or EHA were carried out using benzoyl peroxide as initiator by Varma *et al.* [23]. Polymer samples were prepared using HA or EHA from 0 to 1 mole fraction in the initial feed. Molecular characterization of the copolymers and homopolymers was done by gel permeation chromatography (GPC) and intrinsic viscosity measurements. To determine the copolymer compositions, ^{1}H-NMR spectroscopy was used.

Li *et al.* [24] made an experimental investigation of the kinetics of bulk polymerization of MMA and ethylene glycol dimethacrylate (EGDMA) initiated with Azobisisobutyronitrile (AIBN) at 70°C. They covered a wide range of divinyl-vinyl monomer ratios and determined the effect of chain transfer agent concentration on the gel point conversion.

Rouhallah and Ali-Asghar [25] prepared a series of acrylic copolymer solutions by copolymerizing MMA with BA in toluene to high conversion using dibenzoyl peroxide as initiator. Various reaction parameters, such as MMA concentration and mode of addition of reactants to the reaction vessel, were changed to obtain an acceptable quality of the resin for paint. The properties of the resin and the effect of MMA concentration on properties (viscosity, drying time, hardness and adhesion) were measured according to ASTM standard tests. A linear relation between final percentage conversion and MMA concentration was observed. Drying time, hardness, and Tg of the samples increased with MMA concentration; adhesion of the sample remained constant up to 50 wt% of MMA and then decreased significantly with further increase in MMA concentration.

Jang and Kim [26] synthesized copolymers of styrene with EHA, LA, lauryl methacrylate (LMA), and stearyl acrylate (SA) by suspension polymerization technique using benzoyl peroxide (BPO) as an initiator with a varying monomer feed ratio. The copolymers were characterized by FTIR, ^{1}H-NMR, DSC, and a solubility test, and they were random in nature having similar composition as in the monomer feed. Hwa [27] synthesized MMA and methylene dimethacrylate copolymer by suspension polymerization at 60°C in the presence of a free radical catalyst.

Statistical copolymer films that consist of two types of monomers, a sticky and a glassy monomer, were investigated by Diethert *et al.* [28]. These model systems of pressure-sensitive adhesive films were probed with X-ray reflectivity and mechanical tack measurements. The influence of the type of monomers being copolymerized, the monomer ratio, and the sample age on the near-surface composition as well as the resulting adhesive performance of the films were analyzed. The copolymer contains EHA as the major component and styrene, maleic acid anhydride, or MMA as a minor component. The minimization of surface free energy resulted in an internal reorganization and the component of the statistical copolymer with the lower surface tension was enriched at the free surface.

Zhang *et al.* [29] synthesized an acrylic emulsion of MMA, BA, and acrylic acid with high hydroxyl content. In their work ammonium persulfate was used as initiator; sodium lauryl sulfate (SLS) mixed with p-octyl polyethylene glycol phenyl ether (OP) was used as emulsifier. It was concluded that emulsion with high hydroxyl content is useful in the coating industry.

Moraes *et al.* [30] investigated MMA, BA, and EHA as alternatives to EA in the composition of methacrylic acid (MAA)–EA copolymers for the production of hydrophobically-modified alkali-soluble emulsions. The EHA gave reasonable results as a replacement comonomer for the synthesis and applications of alkali-soluble emulsion.

Latex properties of poly(urethane acrylate) (Pu-A) and MMA copolymer have been studied by Zhang *et al.* [31] They found that the cast film was significantly influenced by the amount of 2-hydoxypropyl acrylate, dimethylpropionic acid and MMA.

Bakhshi *et al.* [32] used emulsion polymerization for the synthesis of poly(butyl acrylate-co-glycidyl methacrylate) by selecting different monomer feed ratio at 75°C with potassium persulfate (KPS) as initiator. Solubility, gel content, and copolymer compositions were determined. The composition of copolymer was determined by the ^{1}H-NMR analysis.

H.Y. Erbill [33] copolymerized vinyl acetate and BA monomer by emulsion polymerization using the controlled monomer and initiator addition method. Particle size distribution, viscosity average molecular weights, Tg and molar composition of the copolymers were determined. Surface tension components of the copolymer films were calculated from contact angle data of various liquids by using van Oss–Good methodology. A decrease of average particle size and Tg was found by the increase of BA content. Surface tension was also affected by the BA content in the monomer feed.

6.2 Polymerization Techniques

Acrylic resins have been widely used in the organic coatings/adhesive industry. Depending on the method of polymerization, the copolymers find different applications. Bulk polymerized methacrylates are widely used as materials in dentistry [34], but when polymerized in solution, they can be used as varnishes [35], and suspension polymerization results in acrylic beads [36]. Emulsion polymerization has been an important and widely used process for the manufacture of polymer products, e.g., paints, adhesives, coatings, and binders [37–54]. One of the advantages of this process is the possibility of obtaining polymers of high molecular weight at a reasonable rate of reaction. The kinetics and mechanism of conventional emulsion polymerization have been investigated extensively. From an industrial perspective, one of the major objectives in the operation of emulsion polymerization processes is that of a faster and safer operation with consistent quality. The polymerization reaction proceeds as a classical double bond addition reaction initiated via a free-radical mechanism [37–39]. In order to obtain well-defined latex, it is important to optimize the polymerization process with regard to emulsifiers, initiator, and monomers. However, particle size, Tg, surface charge density of latex particles, and type of monomers change the properties of polymers synthesized by emulsion polymerization.

The advantages of the emulsion polymerization process are numerous and mainly arise from the absence of organic solvents and the compartmentalization of the reaction. Water constitutes an inert and harmless continuous phase, which acts to maintain a relatively low viscosity of the end product and provide good heat transfer. An additional difference between emulsion polymerization [55], bulk and solution polymerization is that, in emulsion polymerization, the size of the gel is limited by the size of the polymer particle. Monomer-swollen micelles may also exist in the reaction system when the concentration of surfactant in the aqueous phase is above its critical micelle concentration (CMC). Only a small fraction of the relatively hydrophobic monomer is present in the micelles (if present) or dissolved in the aqueous phase. The polymerization proceeds by decomposition of a water-soluble initiator producing the free radicals which propagate in the aqueous phase to form oligomeric radicals by adding a small fraction of the monomers present in the aqueous phase.

Emulsion polymerization is applicable to a number of hydrophobic monomers such as butadiene, styrene, MMA, BA, etc. The emulsion technique was adopted by several scientists [56–59] to synthesize the crosslinked copolymers because, in emulsion polymerization, crosslinkers allow the control of particle morphology and enhance the mechanical properties of latexes used for paints and coatings. The kinetic and mechanistic features of emulsion polymerization are strongly reflected in molecular weight and its distribution, chemical composition, and product properties. Many studies have focused on the effect of various reaction parameters on kinetics of seeded emulsion polymerization.

6.2.1 Component of Emulsion Polymerization

In emulsion polymerization, it is important to optimize the emulsifier, stabilizer, initiator, chain transfer agent (CTA), temperature, the monomer feed ratio, etc. These parameters affect the kinetics of reaction and composition of polymer, including its inherent characteristics which play a significant role in developing properties, and definitely the film properties such as wettability, adhesion, swelling, etc., when used in coating compositions.

6.2.1.1 Emulsifier

An emulsifier is a substance which stabilizes an emulsion by increasing its kinetic stability. Emulsifiers are also known as surface active agents or surfactants. The main functions of emulsifier in the emulsion polymerizations are to facilitate particle nucleation and to enhance the colloidal stability of the growing polymer particle. It controls the particle size and particle size distribution of latex products.

There are four types of surfactants used in polymerization such as anionic, nonionic, cationic and amphoteric or zwitterionic. Out of these, anionic and nonionic surfactants are commonly used in emulsion polymerization because of their enhanced compatibility with negatively charged latex particles. In addition, a mixture of surfactants is often used together in a synergistic manner to control the particle size and to impart colloidal stability to the latex. The effect of the nature and type of surfactant on the emulsion polymerization of vinyl monomer is well documented in the literature [60–62]. Anionic surfactants can provide repulsive force between two similarly charged electric double layers to the latex particles. By contrast, nonionic surfactants can impart two approaching particles with the steric stabilization mechanism. Reactive surfactants, which are surface-active molecules with an active vinyl group, are also used in order to bind the surfactant chemically to the surface of the particles, with the advantages of reduced desorption during film formation and reduced water sensitivity of the latex film.

Huang *et al.* [63] studied the emulsion copolymerization of methyl methacrylate and octyl acrylate by using ammonium sulfate allyloxy nonylphenoxy poly(ethyleneoxy) (10) ether (DNS-86) as a a reactive surfactant, and a conventional surfactant, sodium dodecylbenzene sulfonate (DBS), with a similar structure as a comparison sample. A series of latex samples were prepared with two kinds of surfactants, and their properties characterized and compared. Analysis by ^{1}H-NMR proves that the reactive surfactant was incorporated into the resulting copolymers. The atomic force microscopy (AFM) proves that the reactive surfactant DNS-86 migrates to the surface of the latex film to a much less degree than the conventional surfactant DBS. The stability and water-resistance of the latex films prepared by reactive surfactant DNS-86 were found to be better than those prepared by the conventional surfactant DBS.

Cationic surfactants are important as corrosion inhibitors, fuel and lubricating oil additives, germicides and hair conditioners. Important applications of cationic surfactants in textiles include their use as fabric softeners, fixatives for anionic dyes and dyeing rate retarders for cationic dyes. Cationic and anionic surfactants are usually incompatible. Cationic surfactants are compatible with nonionics and zwitterionics. Usage of cationic surfactants is small compared to anionics and nonionics. The common types of cationic surfactants are long-chain amines and quarternary amine salts. The amphoteric/zwitterionic surfactants are very mild, making them particularly suited for use in personal care and household cleaning products. They can be anionic (negatively charged), cationic (positively charged) or nonionic (no charge) in solution, depending on the acidity or pH of the water. They are compatible with all other classes of surfactants and are soluble and effective in the presence of high concentrations of electrolytes, acids and alkalis. These surfactants have excellent dermatological properties. They are frequently used in shampoos and other cosmetic products, and also in hand dishwashing liquids because of their high foaming properties. An example of an amphoteric/zwitterionic surfactant is alkyl betaine.

The most commonly used anionic surfactants are alkyl sulphates, alkyl ethoxylate sulphates, and soaps. Apart from this, the use of other anionic surfactants such as alkylated disulphonated diphenyl oxide surfactant having C_6–C_{16} hydrophobes (Dowfax series), sodium dioctyl sulphosuccinate (Aerosol series) and some nonionic surfactants such as octyl and nonyl phenol polyethoxylate are also reported [64, 65]. A detailed analysis of the performance of Dowfax 2AI as an emulsifier in styrene/butadiene was reported by Vanderhoff *et al.* [66].

The presence of surfactant influences the kinetics of polymerization and ultimately the bulk film properties. Freeney *et al.* [67] explained the role of certain surfactants in reaction kinetics in getting monodispersed latex. Emelie *et al.* [68, 69] studied the batch emulsion copolymerization of MMA/BA using anionic (SLS) and nonionic (polyethylene oxides) surfactants. Butler *et al.* [70] studied the effect of three different types of surfactants (ionic, polymeric and electrosteric) on the water sensitivity of P(BA-co-MMA) latex films. The water sensitivity of the synthesized polymer was greatly affected by factors such as surfactant mobility, crystallinity and surfactant/polymer polarity.

Forcada and Unzueta [71] studied the effect of mixed surfactant (anionic and nonionic) in emulsion polymerization of MMA and BA. The use of nonionic surfactant caused the colloidal destabilization of the system and led to a broader particle size distribution with a larger particle diameter. The polymerization stabilized by the nonionic surfactant alone resulted in a slower rate of polymerization and a larger latex particle size. At low anionic surfactant concentration, the final number of particles per unit volume of water increased with increasing the total surfactant concentration for the polymerization stabilized by mixed anionic and nonionic surfactants. On the other hand, at high anionic surfactant concentration and a ratio of anionic surfactant to nonionic surfactant between 1:1 and 1:3, a smaller population of particles was produced. Furthermore, latex products with larger particle sizes and narrower particle size distributions were obtained from polymerizations stabilized by mixed anionic and nonionic surfactants compared to the polymerization stabilized by anionic surfactant alone.

Yu *et al.* [72] reported the emulsion polymerization of MMA, BA, dimethylaminoethyl methacrylate (DMAEMA) using a combination of emulsifiers and found that the particle size and surface tension of the latexes decreased with an increase in emulsifier concentration. Sundardi and Zubir [73] prepared the emulsions with monomers having epoxy and carboxyl groups using radiation emulsion polymerization and investigated the influence of the irradiation dose rate and emulsifiers. Xu *et al.* [74] prepared the emulsions with similar monomers by the seeded emulsion polymerization technique. These copolymer emulsions possessed self-crosslinking property. Glycidyl methacrylate (GMA) copolymer emulsions without carboxyl groups were prepared by Zurkova *et al.* [75] and Okubo *et al.* [76].

The PBMA/PMMA polymer networks were synthesized by emulsion polymerization with SLS and polyoxyethylene nonylphenol ether as the emulsifier, distilled water as the continuous medium, and KPS as the initiator by Zhu [77]. The effect of emulsifier concentration, initiator concentration, and polymerization temperature on monomer conversion and polymerization rate was investigated. Experimental data indicated that the polymerization rate and monomer conversion increased with the increase of emulsifier concentration, initiator concentration, and reaction temperature.

J.I. Amalvy [78] investigated the effect of SLS in the synthesis of carboxylic acrylic latexes of MMA, EA and methacylic acid (MAA) by emulsion polymerization method. It was found that the surfactant concentration had an important effect on the final properties of the latexes. Depending on the final use of latex, an optimal range of surfactant concentration was found for which the particle size, water uptake and mechanical properties are appropriate.

Urretabizkaia and Asua [79] studied the effect of reaction parameters, such as feed flow rate, surfactant concentration, distribution of surfactant in the feed, initiator concentration and solid content, on the monomer conversion, terpolymer composition, and total number of polymer particles on the kinetics of high solid content terpolymerization of VA, BA, and MMA. Polymerization rate was reported to be independent of the number of particles. A high value of an average number of radicals per particle of the system suggested that the radical termination in aqueous phase was negligible.

Yang *et al.* [80] prepared an emulsion polymer by using monomers MMA, BA, 2-EHA, acrylic acid, and 2-hydroxyethylmethacrylate (2-HEMA) in the presence of KPS and SLS as initiator and emulsifier, respectively. The effects of the ratio of emulsifiers and their concentration, the amount of initiator, the functional monomers and ratio of monomers on the emulsion polymerization and the film performance were studied. The effect of the type and amount of surfactant on the overall polymerization features and final product properties for the seeded semicontinuous emulsion terpolymerization of 2-EHA, styrene, and methacrylic acid were investigated by Masa *et al.* [81]. They found that the type and amount of surfactant had no effect on the polymerization rate, and the viscosity of the latexes decreased as the amount of emulsifier increased.

6.2.1.2 Initiator

In emulsion polymerization, the reaction is initiated by addition of either water-soluble (thermal or redox) or oil-soluble initiator. Water-soluble initiator produces the free radicals in aqueous phase, whereas oil-soluble initiator produces the free-radicals in monomer-swollen micelle, monomer droplets, monomer-swollen polymer particles and to some extent in aqueous phase. So the loci of polymerization with different initiator systems bring in different kinetics and mechanism of polymerization. Mostly, the water-soluble thermal initiators are sodium, potassium and ammonium salt of persulphates, which are generally employed between 50–90°C. The decomposition of persulphate is accelerated at acidic pH. The initiator efficiency is reduced at pH < 3. The use of a redox initiator (combination of oxidizing and reducing agents), e.g., tert-butyl hydroperoxide-sodium formaldehyde sulfoxylate; persulphate–bisulphite is also useful at low temperature for preparing high molar mass polymer with a low level of branching [82]. Monomer conversion during the polymerization provides knowledge about the polymerization rate and kinetics of the polymerization process, which can be measured gravimetrically and by gas chromatography.

Lee *et al.* [83, 84] studied the kinetics of emulsion polymerization of BA and MMA. The results showed an increase in polymerization rate and number of polymer particles with an increase in initiator concentration, but a decrease in molecular weight and particle size. Schneider *et al.* [85] studied the effect of initiators in high solid content BA/MMA/AA emulsions. They found an increase in the reaction rate with the oil-soluble initiator and a better stability of nucleated particles with the water-soluble initiator. Capek *et al.* [86] also studied the effect of both water-soluble and oil-soluble initiators on the emulsion polymerization of BA and found that the polymerization rate was faster with the ammonium persulphate (APS) initiator.

The emulsion copolymerization of EHMA and MMA was carried out with the use of a bifunctional initiator 1,4-butylene glycol di(2-bromoisobutyrate) by Eslami and Zhu [87]. The effects of initiator concentration and temperature profile on the polymerization kinetics and latex stability were examined. Both EHMA homopolymerization and successive copolymerization with MMA proceeded in a living manner and gave good control over the polymer molecular weights. The polymer molecular weights increased linearly with the monomer conversion with polydispersities lower than 1.2. A low-temperature prepolymerization step was found to be helpful in stabilizing the latex systems, whereas further polymerization at an elevated temperature ensured high conversion rates. The EHMA polymers were effective as macroinitiators for initiating the block polymerization of MMA. Triblock poly(methyl methacrylate–2-ethylhexyl methacrylate–methyl methacrylate) samples with various block lengths were synthesized. The MMA and EHMA reactivity ratios determined by a nonlinear least-square method were ~0.903 and 0.930, respectively, at 70°C.

You *et al.* [88] studied the kinetics of emulsion copolymerization of EMA/LMA. They found that the overall rate of polymerization was only slightly dependent on monomer and stabilizer concentrations and independent on initiator concentration. Naghash *et al.* [89] synthesized vinyl acetate-acrylic emulsion copolymer in the presence of APS as initiator. They found that the polymerization rate increases with increasing its concentration.

6.2.1.3 Monomer(s)

The emulsion polymerization of MMA and BA in the presence of a small amount of sodium methacrylate with KPS initiator at 70°C was investigated by Zhang *et al.* [90]. They found the effect of BA concentration on the polymer properties. Increasing the BA in the feed caused the decrease of particle diameter, surface tension, and viscosity of the latexes and an increase in particle number, polymerization rate, surface charge density and average molecular weight of polymers. The effect of the monomer ratio on the microstructure of BA/MMA emulsion copolymer was investigated by Gonzalez *et al.* [91]. They observed that the addition of MMA led to a decrease in conversion rate and gel content.

Vail *et al.* [92] synthesized random MMA/BMA copolymers in the presence of a chain transfer agent. The reactivity ratios for the monomer pair and Tg of the polymers were determined. Laureau *et al.* [93] studied the effect of monomer composition of EHA/MMA (constant positive and negative gradient) latex particles and found that the homogeneity and heterogeneity of copolymer compositions affected the adhesive strength (tack, peel and shear).

Tigli and Evren [94] synthesized acrylic resins by emulsion polymerization using MA, EA, and BA monomers. The film structures of homo- and copolymers were investigated and three of them [P(MMA/BA) 1:1, P(MMA/EA) 1:1.5, P(MMA/MA) 1:3] were indicated as appropriate binders for paint production. The films were characterized by their mechanical properties like hardness, flexibility, adhesion, gloss, and UV resistance.

Yuan *et al.* [95] prepared the MMA-BA-MAA copolymer emulsions as seed latexes and the seeded emulsion polymerization of MMA-MAA-DVB was used to prepare carboxylated core particles. The hydrophobic shell was then synthesized onto the core using styrene, acrylonitrile, and DVB as comonomers. The hollow latex particles were obtained by alkalization treatment of the core-shell latex particles. The effects of the feeding rate of monomer mixture, contents of emulsifier, sodium dodecylbenzene sulfonate (SDBS) and crosslinking agent DVB, and ratio of the monomers during the core stage and shell stage on the morphology and volume expansion of the latex particles were investigated. The results indicated the formation of monodispersed hollow latex particles of large size when the feeding rate was 0.1 g/min, SDBS content was 0.15 and 0.2 wt% during the core stage and shell stage, respectively. DVB contents were 1% during the preparation of shell copolymers, and the monomer ratio of the core particle to shell layer was 1:8.

Prior *et al.* [96] studied the effect of comonomer, i.e., vinyl versatate (VV), BA, and EHA, on the colloid-stabilized VA polymers to assess their impact on latex coating properties. The VV showed advantages in scrub resistance, gloss, hydrophobicity, and higher Tg; BA developed better wet adhesion efficiency, gel content and hiding efficiency; EHA offered performance comparable to BA in hiding efficiency, lowering copolymer Tg, and gloss development. A combination of these monomers, in many instances, afforded a better balance of performance properties than individual monomers alone. In this study, a simplex-centroid design was utilized to statistically map polymer compositions for determining the effect of comonomer composition on latex and paint film properties.

Similarly, Gower and Shank [97] studied the effect of varied monomer composition on adhesive performance and peeling master curves for acrylic pressure sensitive adhesive (PSA). The synthesis and characterization of a structured latex particle of acrylic copolymers and their peel adhesion behavior were studied by Mayer *et al.* [98]. Bhabe *et al.* [99] studied the synthesis and characterization of copolymer of MMA with other acrylate comonomers such as EA or BA or 2-EHA and discussed its performance as a PSA. They found that an increase in MMA concentration increases the cohesive strength of the latex.

The effect of functional monomer on the adhesive strength was also reported by Ghosh and coworkers [100]. They studied the use of GMA for developing crosslinkable PSA and reported that the peel strength increased with an increase in the percentage of GMA in the copolymer composition. Pedraza *et al.* [101] studied the effect of functional monomers, i.e., 2-HEMA and MAA, on the film properties of the polymer latex and found that the mechanical properties of the polymer films increased with the increase in carboxyl and hydroxyl functionality.

6.2.1.4 *Crosslinking Monomer(s)*

Multifunctional monomers, often known as crosslinkers, are usually employed in free-radical polymerizations to produce polymer networks which find applications in medicine, pharmaceuticals, paints, column packing, optics, polymer additives, ion-exchange resins and rubbers. When used in emulsion polymerization, crosslinkers allow control of particle morphology and enhancement of the mechanical properties of latexes used for paints and coatings [55]. The crosslinking process involves the linking of polymer chains to obtain a network which results in an increase in film hardness, and resistance to solvents, chemicals and detergents. Polymer chains are viscoelastic in nature and crosslinking will increase the rigidity of chains, which retards the segmental motion of chains. Modification of the monomer that leads to a crosslinkable monomer has been receiving great attention in recent years due to their potential for the development of products such as thermosetting molding compounds, coatings, superabsorbent and ion-exchange resins [102].

The crosslinking reaction increases the brittleness and adhesion of the coatings. Present technology on thermosetting latexes has shown the successful crosslinking of latexes with different crosslinking agents, either external or copolymerized within the polymer backbone. However, the incorporation of crosslinking technology into common thermoplastic latexes involves additional variables that affect the processes of latex production and property development on films. Besides the types of chemistry involved in the crosslinking reactions, one of the most important variables is the incorporation of functional monomers and crosslinking agents into the dispersions. The type of functional groups and crosslinkers, as well as the addition and localization, would be expected to influence processes of synthesis, film formation and mechanical behavior of latex films. It is anticipated that the presence of increased functionality of monomers during synthesis influences the physical properties of the particles and the structure of the film. On the other hand, the overall mechanical properties and solvent resistance of latex films may not be enhanced if the polymer is not sufficiently crosslinked. Thus, functionality and crosslinker levels must be enough to provide solvent resistance and cohesive strength, without disrupting other desired properties mandated by the requirements of specific applications such as flexibility or impact resistance.

The distribution of functional groups within a crosslinked latex film is considered as a controlling factor for structure-property relationships if it is believed that the enhancement of strength through crosslinking is achieved primarily when the crosslinker is located in the proximities of functional groups. Ideally, systems with a specially controlled functionality location would allow specific crosslinked structures with similar or even improved mechanical behavior compared to homogeneous crosslinked systems. Such control of functionality location could be synergistically combined with packing optimization through the use of different particle sizes within a single latex dispersion (e.g., bimodal latex dispersions). However, as a result of the incorporation of functional groups within the polymer backbone and the addition of crosslinking agents, packing and distribution of particles during film formation could present dissimilar features compared to traditional monomodal thermoplastic latexes.

In some thermosetting latex dispersions, crosslinking reactions are benefitted by the presence of strong or weak acid groups. For instance, crosslinking reactions of hydroxyl and/or carboxyl groups in acrylic latex dispersions with melamine-formaldehyde-based resins or cycloaliphatic epoxide resins are usually enhanced by the addition of strong acid catalysts such as sulfonic acids. The presence of such strong acids along with weak acids (as reactive groups) on the surface of latex particles gives an acidic character to the dispersion. In this way, weak bases are typically used for pH control, further stabilization of particles, and additionally as blocking agents for reactive and catalyst acid sites. In this way, acid-base interactions between catalysts, blocking agents and reacting groups are expected to affect film forming processes and crosslinking reactions.

During the last decade, polymers with hydrophilic crosslinking agents have received much attention [103]. The hydrophilicity and flexibility imparted by the crosslinking agents improved the salvation characteristics and enhanced the reactivity of the polymer-bound functional groups [104]. Ethylene glycol dimethacrylate (EGDMA) has been widely used as a crosslinking agent in the synthesis of hydrophilic or hydrophobic copolymer networks [105]. Some aspects of the copolymerization of MMA with EGDM have been investigated since 1945 [106–116]. Some authors estimated the efficiency of crosslinking [108, 114] and determined the Tg of the copolymers [109]. The fast transient fluorescence technique was used to study the sol–gel phase transition in free-radical crosslinking monomers.

Since the 1970s several scientists [117–123] have investigated the behaviors of crosslinked PMMA. Crosslinked PMMA is used in a variety of commercial materials such as membranes [124], dental fillings [125], dentures, latex coatings [126], computer-to-sensor data transmission links [127], composites [128], optically clear sheeting, limb prostheses, bathtubs, bathroom sinks and shower surrounds, etc., where crosslinking plays a vital role in both processing and ultimate properties.

Crosslinked copolymers can be synthesized by various polymerization techniques such as bulk, solution, suspension and emulsion. A number of scientists studied the crosslinked copolymers synthesized by bulk technique. Huang *et al.* [123] used DSC to determine monomer conversion, rate of polymerization, reaction rate constants, and reaction order during the copolymerization of EGDMA and HEMA. This detailed research revealed trends in reaction parameters that were dependent on both EGDMA concentration and temperature. Copolymers of MMA with different glycol dimethacrylates were synthesized by Loshack and Fox

[129]. They presented a method for determination of residual unsaturations in copolymers and indicated that the efficiency of crosslinking in a monovinyl-divinyl copolymerization will be less than 100%. A kinetic study of the bulk copolymerization of HEMA with EGDMA was investigated by Ajzenberg and Ricard [130].

Ramelow and Pingili [131] synthesized EGDMA-MMA copolymer by using photochemical initiator, UV radiation. Infrared spectroscopy was used to calculate reactivity ratios and to identify the type of copolymerization. The reactivity ratios of EGDMA and MMA were calculated as 0.6993 and 1.8635, respectively, and it was found that as the MMA percentage in the monomer feed increases, the percent conversion decreases.

Copolymer dispersions useful as aqueous automotive topcoats were prepared by solution polymerization by Andre *et al.* [132]. Polymer of AA-BA-HEMA-MMA was dispersed in aqueous amino-methylpropanol and crosslinked with melamine resins to give a 23-μm-thick coating. A paint film with outstanding finished appearance, and acid, water, adhesion and bending resistances was obtained by using a solvent-based thermosetting base finishing coat. Melamine resins were used as a crosslinker in the above coating [133].

Hosoya *et al.* [134] synthesized monodispersed P(MMA-EGDMA) packing materials by a multi-step swelling and polymerization method with cyclohexanol or toluene as porogenic solvent. When toluene was applied as a porogen, the seed polymer severely affected the porous structure, while no effect was observed with cyclohexanols compared chromatographically with those prepared by the corresponding suspension polymerization methods.

Crosslinked copolymers of acrylamide were obtained by the aqueous suspension polymerization method by Dragan *et al.* [135]. In this method, DVB and N,N-methylene-bis-acrylamide (MBAA) were used as crosslinking agents and EA as a third comonomer. The real acrylamide content in the crosslinked copolymers was calculated taking into account the nitrogen content after the removal of the soluble fractions. The water uptake dependence on the copolymer structure gave information on the hydrophilicity of these copolymers. Crosslinked copolymers of AA and EA and some ion exchangers derived from them containing either primary amine groups or carboxylic groups were studied by Dragan *et al.* [136], where DVB and MBAA were used as crosslinkers. The copolymers and the corresponding ion exchangers were characterized by IR spectroscopy, swelling behavior and thermogravimetric analysis. The ion-exchange properties were correlated with the crosslinker nature and the chemical reactions performed on the acrylamide copolymers.

Bryjak [137] evaluated BA copolymerizd with EGDMA as potential chromatographic packings. It was found that slightly crosslinked copolymers (up to 30 wt% EGDMA) did not provide porous material, while for matrices which exceeded 40 wt% of EGDMA the porous structure remained unchangeable. The increase of crosslinker content greatly affected the surface hydrophobicity.

Copolymer systems based on various molar ratios of MMA/BA and different mass content of N-methylolacrylamide (NMA) as a cross-linking agent were investigated by Leskovac *et al.* [138]. The changes in mechanical properties of the copolymers were found to be connected with the crosslinking and/or degradation mechanism. It was concluded that a higher content of MMA units can lead to preferential depolymerization; whereas an opposite effect was observed when BA contents increased. The observed changes in mechanical properties were reduction of yield point, strength, elongation, and increase in brittleness, and may be attributed to the presence of various ratios of constitutive sequences and crosslinked structures before and after degradation.

Copolymers of BA-MMA-AA and intraparticle crosslinking agents containing NMA and EGDMA were prepared by emulsion copolymerization by Yoo *et al.* [58] The films were prepared from the mixture of copolymers and the interparticle crosslinking agents which were prepared from hexamethylene diisocyanate and aziridine ethanol. The film was cast using mixtures of the copolymer and the interparticle crosslinking agent, and crosslinked in a convection oven. The effects of the contents of the intra/interparticle crosslinking agents were evaluated. By increasing the contents of EGDMA, the roughness of the films was increased because of the effects of EGDMA which acted as an intraparticle crosslinking agent. By increasing the contents of the interparticle crosslinking agent, roughness was also increased by the reaction between the copolymers and interparticle crosslinking agent. Tensile strength, water and chemical resistance of the films were increased, whereas elongation of the film was decreased by increasing the content of interparticle crosslinking agents.

To determine the difference in tensile properties and swelling behavior, a series of crosslinking monomers such as macromonomer crosslinker (Mac), EGDMA and aliphatic urethane acrylate were copolymerized with BMA by emulsion technique by El-Aasser *et al.* [59]. They found that copolymers prepared with Mac were tough in comparison with copolymers made with EGDMA. The factors responsible for the difference were the presence of longer linear or lightly crosslinked PBMA chains and the looseness of the crosslinked network structure in the PBMA-co-Mac copolymers.

Hidalgo *et al.* [139] prepared latex with a styrene/BA/amide functional monomer of varied hydrophilicity through emulsion polymerization. The polymerization kinetics, size and morphology of latex particles and the location of the functional groups in the final latexes were studied. It was shown that increasing the hydrophobicity led to better homogeneity in the copolymer formed during the polymerization, while the more hydrophilic functional monomer partly homopolymerized in a separate phase.

Duarte *et al.* [140] studied the equilibrium solubility of carbon dioxide in P(MMA-co-EHA-co-EGDMA) by gravimetric method at different temperatures and pressures. The crosslinked copolymer showed the Fickian behavior and Fick's diffusion model was applied to determine the amount of carbon dioxide and the diffusion coefficients.

The effect of crosslinking on intercellular polymer diffusion in P(BMA-co-BA-co-EGDMA) latex films containing 0.1–4 mol% EGDMA as a crosslinking agent was monitored by fluorescent energy-transfer measurements and by atomic force microscopy in a study by Tamai *et al.* [141]. The polymer diffusion decreased with an increase in the levels of crosslinking.

A film was prepared from BMA/ethylene urea (EU) emulsion copolymer by Okubo *et al.* [142]. The wet adhesion of the emulsion film on an alkyd resin was significantly improved by copolymerization with a small amount of EU (0.5–1.0 mol%). The localization of EU and the cleanliness at the particle surface were also key factors in the improvement of the wet adhesion of the polymer emulsion film.

A reactive monomer, acetoacetoxyethyl methacrylate (AAEM), has long been recognized for its ability to provide a versatile crosslinking site in thermoset coatings [143]. An important feature of this crosslinking monomer was reported to be a better crosslinking agent for the epoxy resins in comparison to aziridine and isocyanates. It has relatively low toxicity, which makes it attractive for high performance, environmentally-friendly, non-hazardous coatings. The AAEM-containing waterborne coatings have gained such high interest as evidenced by recent patents [144–147] and literature reports [148, 149]. Noomen [150] synthesized acid-functional resins based on MMA-BA-AAEM and methacrylic acid in a ratio of 21:37:30:12, by weight using emulsion technique. The mechanism of drying and crosslinking of acetoacetoxy functional latexes was also investigated.

The effects of divinyl monomers (AMA and BDA) on the kinetics and mechanical properties (i.e., Tg, storage modules and molecular weight between entanglements) of the film were studied by Bouvier-Fontes *et al.* [55]. The results showed that the addition of the crosslinkers did not significantly affect the kinetics. Surprisingly, the most reactive crosslinker, BDA, produced the least crosslinked latex, while AMA formed a highly crosslinked latex. This was explained by the fact that intramolecular chain transfer to the polymer was favored by low monomer concentrations.

Polymers of EGDM, diethylene glycol dimethacrylate, triethylene glycol dimethacrylate and tetraethylene glycol dimethacrylate were prepared by photoinitiated polymerization in the presence of 2,2-dimethoxy-2-phenyl-acetophenone at 27°C by Bowman and Peppas [150]. These systems were used predominantly as crosslinking agents during the copolymerization of hydrophilic monomers for the production of important biomedical polymers [151].

A series of poly(propylene glycol-diacrylate) (PPGDA) having molecular weight in the range of 300–3000 were synthesized by Malucelli *et al.* [152]. They studied the preparation and characterization of PPGDA oligomers. Moreover, mixtures of the PPGDA oligomers with a typical epoxy acrylic resin were examined in order to evaluate the flexibility properties of these oligomers introduced in a rigid network.

Water-soluble polymers [153] were synthesized by free radical copolymerization of MMA, α-methyl-ω-methacroyl-poly(ethylene glycol) and 2-HEMA. The methacrylate containing hydrogel precursors were obtained by the reaction of free hydroxyl groups of HEMA in the polymer with methacrylic anhydride in the presence of catalyst. γ-Irradiation was used as an elegant method for crosslinking and simultaneous sterilization.

Formulations for waterborne printing inks based on self-crosslinking ketone functional acrylate emulsions were prepared by de Krom *et al.* [154–156]. Results of measurements of gloss, drying, adhesion and various resistance properties of the coatings on nonabsorbent substrates have been discussed. These inks dried very rapidly and have good adhesion, resistance to solvents and water, with high gloss.

A self-curing thermosetting water-dispersed polymer consisting of styrene, ethyl acrylate, methacrylic acid, N-isobutoxymethyl acrylamide and methacryloyloxyethyl phosphate has been used as an interior coating for cans by Martino [157]. It has been claimed that a coating consisting of an acrylic polymer latex and aminoplast curing agent has good adhesion and gloss, and is resistant to water, solvent and chemicals. The baking schedule has been reported to be 26°C for 42 seconds by Kuhn *et al.* [158].

Novel self-crosslinkable graft copolymers with a pendant blocked isocyanate in the first segment, and a hydroxyl group in the second segment, were developed by Yukawa *et al.* [159]. M-isopropenyl-α, α-dimethylbenzyl isocyanate were copolymerized with butyl acrylate, 2-hydroxyethyl acrylate to form the graft copolymer, by incorporating neutralizable functionalities such as carboxyl or tertiary amino groups to the second segment of the graft copolymer, to prepare self-dispersible (self-stabilized) aqueous coating vehicles.

Fan *et al.* [160] studied the process of self-curing of water-based acrylic adhesives. The composition contained acrylic ester emulsion, which could be self-cured by introducing epoxy and acrylamide groups. The adhesives showed excellent adhesion and solder resistance when the ratio of the two self-curing monomers was 7:6. The self-curing reaction was illustrated by IR spectra.

6.2.2 Applications of Acrylic Polymers

Aqueous emulsion polymers [161] are environmentally-friendly alternatives to traditional solvent-based polymers. The technology of water-based polymer binders (or dry powders) play a key role with regard to sustainability in many market segments. The acrylic latexes prepared by the emulsion polymerization were used in various fields. Latexes are aqueous dispersions of solid polymeric particles obtained through an emulsion polymerization technique. Due to its environmentally-friendly attributes, as well as the special features of the synthesis and its structure-property interrelated characteristics, emulsion polymerization has been the subject of extensive research during the past sixty years. Polymers obtained by emulsion polymerization have the advantage of reaching high molecular weights without developing the increased viscosity characteristic of other polymers. In this way, a wide variety of homopolymers and copolymers obtained by emulsion polymerization have proven applications in fields such as coatings, adhesives, synthetic rubbers, floor polishes and additives amongst others. The first line of waterborne acrylic emulsion paints called "Liquitex" was produced in 1954 [162].

Butyl acrylate is one of the principal acrylic monomers included in emulsions intended for decorative coatings. In many cases it is copolymerized with MMA to obtain a film of the requisite degree of hardness [163]. Yoshida [164] proposed weather-resistant lacquer compositions based on a copolymer of BA and MMA. A solution of this copolymer and 4-methacryloyloxy-2,2,6,6-tetramethyl-piperdine, when sprayed on a urethane enamel-coated surface, gave coatings with good weather, water and alcohol resistances. El-Aasser and Vanderhoff *et al.* have systematically investigated different effects of emulsion copolymerization of vinyl acetate (VA) and BA on latex particle size, molecular weight, latex stability, surface characteristics and the morphological, mechanical properties of their films [165–169]. Copolymer [170] synthesized using BA-acrylonitrile-acrylic acid was used as peelable coatings.

Grubert *et al.* [171] synthesized aqueous polymer dispersions of a halogen-containing monomer (vinyl dichloride) by copolymerization with BA. The polymer gave a film without discoloration. The MMA-co-BA [25] polymer also fulfilled the requirement of good quality traffic paint. Water-thinned coatings [172] from BuA-2-HEMA-2-isocynatoethyl methacrylate synthesized in the presence of $(NH_4)S_2O_8$ were transparent with good water resistance.

A patent [173] claimed dispersions for the weather-resistant coatings by the monomer mixtures containing MMA-Bu acrylate-acrylic acid. The films were dried at 80°C and aged at 40°C to give a clear, transparent, peelable, water-thinned, and good solvent resistant coating. In another patent [174], aqueous acrylic polymer emulsions for sprayable clear coatings have been synthesized by polymerization of acrylic acid-BA-tert. BA-hydroxypropyl acrylate-MMA-Styrene. Emulsion of this polymer was sprayed on primed steel, dried at room temperature and baked for 30 min at 130°C to give a 42-μm-thick film with 82–84% gloss.

The 2-EHA is also incorporated with other monomers in latex coatings. A combination of 2-EHMA with St and MMA is beneficial in multilayer coatings with good appearance. Polyisocynate was used as crosslinking agent in this combination [175]. The compounds useful for building exteriors contain emulsions of polymers having Tg of −10 to +50°C [176]. A copolymer combination of 2-EHA-acrylonitrile-AA (Tg = −53°C), useful in flooring coating, was proposed by Schwerzel *et al.* [177]. A copolymer from MMA, 2-EHA and methacrylic acid was synthesized by Takeuchi *et al.* [178]. When this copolymer was neutralized with ethanolamine, coated on metal and dried, it gave a glossy, and water-resistant coating.

A water-thinned metallic coating for automobile bodies was developed by Sugiura [179] containing a thermosetting acrylic resin, a metallic pigment, nonmeltable crosslinked resin particles and a metal silicate. The metallic coating was applied on a substrate, and a clear coating was applied on the resulting uncured metallic coating, and both of the coatings were heated for cure at once to give a surface with high gloss.

The latex paints were considered free from any pollutants as they were waterborne and the derived paint films were waterproof. The paint was easily formulated from an aqueous latex and a dispersion of pigment in water. The film-forming latexes are also used in floor coatings (polishes), printing inks, adhesives, paper overprint varnishes, carpet backing and paper making.

Although a vast amount of literature is available on various types of acrylic coatings, a systematic data is lacking for understanding the polymerization kinetics of the effect of higher acrylates [180–181] on crosslinkable coatings based on methyl methacrylate and their performance on various substrates. Since these coatings are widely used in industries, there is a broad scope for further opportunities of new findings within the framework of crosslinkable coatings. This needs effort and further investigation.

References

1. Anon, Fast-dry, high solids coating compositions based on acetoacetate-functionalized alkyd resins, US Patent: 6794049, 2004.
2. P. Thomas, *Polyurethanes: Waterborne and Solvent-based Surface Coating Resins and Their Applications - Vol 3*, John Wiley and Sons, Chichester, pp. 7–11, 1998.
3. L.T. Gangotri, and N. Dhaneshwar, *Paint India*, Vol. LVII, No. 2, p. 83, 2007.
4. P.K.T. Oldring, and G. Hayward, *A Manual of Resins for Surface Coatings, 2nd Ed.*, Vol. 2, SITA Technology, London, p. 121, 1987.
5. B. Sirnivasulu, P. Paghunath Rao, and E.V. Sundaram, *J. Appl. Polym. Sci.*, Vol. 43, p. 1521, 1991.
6. I. Czajilik, T. Foldes-Berezsnich, F. Tudos, and S. Szakacs, *Eur. Polym. J.*, Vol. 17, No.2, p. 131, 1981.
7. P. Bitler, British Patent 20,097, 409A.
8. T.F. McKenna, C. Graillat, and J. Guillot, *Polym. Bull.*, Vol. 34, p. 361, 1995.
9. B.B. Kine, and R.W. Novak, *Acrylic and Methacrylic Ester Polymers Encyclopedia of Polymer Science and Engineering, 2nd Ed.*, Vol. 1, Wiley, New York, p. 234, 1986.
10. C.R. Martens, *Technology of Paints, Varnishes and Lacquers*, Renhold, New York, p. 111, 1968.
11. J.K. Fink, *Reactive Polymers Fundamentals and Application, 2nd Ed.*, p. 235, 2013.
12. S. Zhu, Y. Tian, and A.E. Hamielec, *Macromolecules*, Vol. 23, p. 1144, 1990.
13. W. Wunderlich, *Macromol. Chem.*, Vol. 177, No.4, p. 973, 1976.
14. G.E. Scott, and E. Senogles, *J. Macromol. Sci.-Chem.*, Vol. 4, No. 5, p. 1105, 1970.
15. I.K. Varma, and S. Patnaik, *Eur. Polym. J.*, Vol. 12, p. 259, 1976.
16. I.K. Varma, M.V. Nair, and V.K. Karan, *J. Thermal Anal.*, Vol. 35, p. 989, 1989.
17. I.K. Varma, M.V. Nair, V.K. Karan, and D.S. Varma, *Thermochim. Acta*, Vol. 142, No. 2, p. 189, 1989.
18. M.S. Choudhary, and I.K. Varma, *J. Macromol. Sci. Chem.*, Vol. A20, p. 771, 1983.
19. M. Patnaik, V. Choudhary, and I.K. Varma, *Eur. Polym. J.*, Vol. 28, p. 1433, 1992.
20. M.S. Choudhary, and I.K. Varma, *Eur. Polym. J.*, Vol. 15, No. 10, p. 957, 1979.
21. I.K. Varma, and S. Patnaik, *Eur. Polym. J.*, Vol. 13, p. 175, 1977.
22. Idemistsu, Jpn. Kokai Tokkyo Koho JP 58, 109, 556, cf. CA Vol. 100, p. 52557q, 1984.

23. M. Patnaik, V. Choudhary, and I.K. Varma, *Eur. Polym. J.*, Vol. 28, No. 11, p. 1433, 1992.
24. W.H. Li, A.E. Hamielec, and C.M. Crowe, *Polymer*, Vol. 30, p. 1513, 1989.
25. B. Rouhallah, and M. Ali-Asghar, *J. Appl. Polym. Sci.*, Vol. 70, p. 367, 1998.
26. J. Jang, and B.S. Kim, *J. Appl. Polym. Sci.*, Vol. 77, p. 903, 2000.
27. J.C.H. Hwa, *J. Polym. Sci.*, Vol. 58, p. 715, 2003.
28. A. Diethert, Y. Peykova, N. Willenbacher, and P.M. Buschbaum, *Applied Material Interfaces*, Vol. 2, No. 7, p. 2060, 2010.
29. F. Zhang, Y. Wang, L. Yuan, and C. Chai, *J. Macromol. Sci. A*, Vol. 41, No.1, p. 15, 2004.
30. R.P. Moraes, C. Graillat, G. Jeanson, S. Haw, C. Favero, and F.L. Timothy, *J. Colloid Interf. Sci.*, Vol. 352, p. 19, 2010.
31. H.T. Zhang, R. Guan, Z.H. Yin, and L.L. Lin, *J. Appl. Polym. Sci.*, Vol. 82 No.4, p. 941, 2001.
32. H. Bakhshi, M.J. Zohuriaan-Mehr, H. Bouhendi, and K. Kabiri, *Polymer Testing*, Vol. 28, p. 730, 2009.
33. H. Yildirim Erbil, *Polymer*, Vol. 37, No. 24, p. 5483, 1996.
34. S. Matsukawa, T. Hayakawa, and K. Nemoto, *Dent. Mater.*, Vol. 10, p. 343 1994.
35. I. Lewinstein, N. Fuhrer, and Y. Ganor, *J. Prosthet. Dent.*, Vol. 89, p. 70, 2003.
36. P.J. Dowding, and B. Vincent, *Colloid Surface A: Physicochem. Eng. Aspects*, Vol. 161, p. 259, 2000.
37. M. Jain, R.A. Vora, and U.S. Satpathy, *Eur. Polym. J.*, Vol. 39, p. 2069, 2003.
38. S. Arora, R. Gesthuisen, and S. Engell, *Comp. Chem. Eng.*, Vol. 31, p. 552, 2007.
39. J.E. Jonsson, O.J. Karlsson, H. Hassander, and B. Tornell, *Eur. Polym. J.*, Vol. 43, No. 4, p. 1322, 2007.
40. C.S. Chen, J.J. Lin, Y.L. Lin, and S.Z. Lai, *Eur. Polym. J.*, Vol. 42, No. 5, p. 1033, 2006.
41. L.L. Arbina, L.M. Gugliotta, M.J. Barandiaran, and J.M. Asua, *Polymer*, Vol. 39, No. 17, p. 4047, 1998.
42. X. Cheng, Z.X. Chen, T.S. Shi, and H.Y. Wang, *Colloid. Surface. A*, Vol. 292, No. 2–3, p. 119, 2007.
43. J.E. Jonsson, H. Hassander, and B. Tornell, *Macromolecules*, Vol. 27, No. 7, p. 1932, 1994.
44. J. Stubbs, O. Karlsson, J.E. Jonsson, E. Sundberg, Y. Durant, and D. Sundberg, *J. Colloid. Surface. A*, Vol. 153, p. 255, 1999.
45. L.E. Karlsson, O.J. Karlsson, and D.C. Sundberg, *J. Appl. Polym. Sci.*, Vol. 90, No. 4, p. 905, 2003.
46. O.J. Karlsson, J.M. Stubbs, R.H. Carrier, and D.C. Sundberg, *Polym. React. Eng.*, Vol. 11, p. 589, 2003.
47. S. Lee, and A. Rudin, *J. Polym. Sci. Poly. Chem.*, Vol. 30, p. 2211, 1992.
48. C.L. Winzor, and D.C. Sundberg, *Polymer*, Vol. 33, No. 18, p. 3797, 1992.
49. O. Karlsson, H. Hassander, and B. Wesslen, *J. Appl. Polym. Sci.*, Vol. 63 No. 12, p. 1543, 1997.

50. O.J. Karlsson, H. Hassander, and B. Wesslen, *J. Appl. Polym. Sci.*, Vol. 77, p. 297, 2000.
51. J.E. Jonsson, O.J. Karlsson, H. Hassander, and B. Tornell, *Macromolecules*, Vol. 34, No. 5, p. 1512, 2001.
52. N.M. Ahmad, F. Healtey, and P.A. Lovell, *Macromolecules*, Vol. 31, No. 9, p. 2822, 1998.
53. I.A. Maxwell, B.R. Morrison, D.H. Napper, and R.G. Gilbert, *Macromolecules*, Vol. 24, No.7, p. 1629, 1991.
54. V.N. Kislenko, *J. Colloid. Surface. A*, Vol. 152, No.1–2, p.199, 1999.
55. L. Bouvier-Fontes, R. Pirri, J.M. Asua, and J.R. Leiza, *Macromolecules*, Vol. 38, p.1164, 2005.
56. J. Tang, T. Ding, E.S. Daniels, V.L. Damonie, M.S. El-Aasser, and A. Klein, *J. Appl. Polym. Sci.*, Vol. 88, No. 1, p. 88, 2003.
57. T. Ding, E.S. Daniels, M.S. El-Aasser, and A. Klein, *J. Appl. Polym. Sci.*, Vol. 97, No.1, p. 248, 2005.
58. Y. Yoo, G.H. Hong, S.R. Hur, Y.S. Kim, S.G. Lee, H.J. Kim, and J.H. Lee, *J. Appl. Polym. Sci.*, Vol. 112, pp. 1587–1594, 2009.
59. M.H. Ghazaly, E.S. Daniels, V.L. Dimonie, A. Klein, L.H. Sperling, and M.S. El-Aasser, *J. Appl. Polym. Sci.*, Vol. 88, No. 1, p. 42, 2003.
60. H. Chu, and I. Piirma, *Polym. Bull.*, Vol. 21, p. 301, 1989.
61. H. Wang, and H. Chu, *Polym. Bull.*, Vol. 24, p. 207, 1990.
62. M.E. Woods, J.S. Dodge, and I.M. Krieger, *J. Paint. Technol.*, Vol. 40, No. 527, p. 541, 1968.
63. H. Huang, D. Lu, L. Shen, and R. Guan, *J. Macromol. Sci. Pure Appl. Chem.*, Vol. 45, pp. 242–247, 2008.
64. Dow Chemical, Technical Bulletin on Dowfax Surfactants, 1986.
65. Cytec Industries, Technical Bulletin on Surfactants, 1983.
66. J.W. Vanderhoff, V.L. Dimonie, and M.S. El-Aasser, *J. Appl. Polym. Chem.*, Vol. 41, p. 1549, 1990.
67. P.J. Freeney, D.H. Napper, and R.G. Gillbert, *J. Colloid Interf. Sci.*, Vol. 118, No. 2, p. 493, 1987.
68. B. Emelie, C. Pichot, and J. Guillot, *Macromol. Chem. Suppl.*, Vol. 10/11, p. 43, 1985.
69. B. Emelie, C. Pichot, and J. Guillot, *Macromol. Chem.*, Vol. 192, p. 1629, 1991.
70. L.N. Butler, C.M. Fellows, and R.G. Gilbert, *J. Appl. Polym. Sci.*, Vol. 92, No. 3, p. 1813, 2004.
71. E. Unzueta, and J. Forcada, Polymer, Vol. 36, No. 5, p. 1045, 1995.
72. Z.Q. Yu, B.G. Li, B.F. Li, and Z.R. Pan, *Colloid. Surface. A: Physicochem. Eng. Asp.*, Vol. 153, p. 31, 1999.
73. F. Sundardi, and A. Zubir, *J. Macromol. Sci. Chem. A*, Vol. 22, No. 12, p. 1647, 1985.
74. Z. Xu, G. Lu, S.Cheng, and J. Li, *J. Appl. Polym. Sci.*, Vol. 56, p. 575, 1995.
75. E. Zurkova, K. Bouchal, D. Zdenkova, Z. Pelzbauer, F. Svec, and J. Kalal, *Polym. Sci. Polym. Chem. Ed.*, Vol. 21, p. 2949, 1983.

76. M. Okubo, Y. Nakamura, and T. Matsumoto, *J. Polym. Sci. Polym. Chem. Ed.*, Vol. 18, p. 2451, 1980.
77. G. Zhu, *Fibers and Polymers*, Vol. 9, No. 2, pp. 107–112, 2008.
78. J.I. Amalvy, *Pigment and Resin Technology*, Vol. 27, No. 1, pp. 20–27, 1998.
79. A. Urretabizkaia, and J.M. Asua, *J. Polym. Sci. Part A: Polym. Chem.*, Vol. 32, No. 9, p. 1761, 1994.
80. R. Yang, T. Zhau, E. Chen, and W. Jiong, *Tuliao Gongye*, Vol. 32, p. 4, 2002.
81. J.A. Masa, J. Forcada, and J.M. Asua, *Polymer*, Vol. 34, No. 13, p. 2853, 1993.
82. A. Zosel, and B. Shuler, *J. Adhesion Sci.*, Vol. 70, No. 1, p. 179, 1999.
83. C.F. Lee, K.R. Lin, and W.Y. Chiu, *J. Appl. Polym. Sci.*, Vol. 51, No. 9, p. 1621, 1994.
84. C.F. Lee, and W.Y. Chiu, *J. Appl. Polym. Sci.*, Vol. 56, No. 10, p. 1263, 1995.
85. M. Schneider, A. Graillat, A. Guyot, and T.F. McKenna, *J. Appl. Polym. Sci.*, Vol. 84, No. 10, p. 1916, 2002.
86. I. Capek, and P. Potisk, *Eur. Polym. J.*, Vol. 31, No. 12, p.1269, 1995.
87. H. Eslami, and S. Zhu, *J. Polym. Sci.: Part A: Polym. Chem.*, Vol. 44, p. 1914, 2006.
88. X. You, V.L. Dimonie, and A. Klein, *J. Appl. Polym. Sci.*, Vol. 82, p. 1691, 2004.
89. H.J. Naghash, S. Mallakpour, and N. Mokhtarian, *Prog. Org. Coat.*, Vol. 55, p. 375, 2006.
90. M.G. Zang, Z.X. Weng, Z.M. Huang, and Z.R. Pan, *Eur. Polym. J.*, Vol. 34, No. 9, p. 1243, 1998.
91. I. Gonzalez, J.M. Asua, and J.R. Leiza, *Polymer*, Vol. 48, p. 2542, 2007.
92. N.K. Vail, J.W. Barlow, J.J. Beaman, H.L. Marcus, and D.L. Bourell, *J. Appl. Polym. Sci.*, Vol. 52, p. 789, 1994.
93. C. Laureau, M. Vicente, M.J. Barandiaran, J.R. Leiza, and J.M. Asua, *J. Appl. Polym. Chem.*, Vol. 81, p. 1258, 2004.
94. R.S. Tigli, and V. Evren, *Prog. Org. Coat.*, Vol. 52, p. 144, 2005.
95. C.D. Yuan, A.H. Miao, J.W. Cao, Y.S. Xu, and T.Y. Cao, *J. Appl. Polym. Sci.*, Vol. 98, p. 1505, 2005.
96. R.A. Prior, W.R. Hinson, O.W. Smith, and D.R. Bassett, *Prog. Org. Coat.*, Vol. 29, p. 209, 1996.
97. M.D. Gower, and R.A. Shank, *J. Appl. Polym. Sci.*, Vol. 93, No. 6, p. 2909, 2004.
98. A. Mayer, T. Pith, G.H. Hu, and M. Lambla, *J. Polym. Sci. Polym. Phys.*, Vol. 33, p. 1781, 1995.
99. M.D. Bhabhe, P.S. Galvankar, V.M. Desai, and V.D. Athawale, *J. Appl. Polym. Sci.*, Vol. 56, No. 4, p. 485, 1995.
100. S. Ghosh, and N. Krishnamurti, *Eur. Polym. J.*, Vol. 36, No. 10, p. 2125, 2000.
101. E.P. Pedraza, and M.D. Soucek, *Polymer*, Vol. 46, No. 24, p. 1174, 2005.
102. S. Zhu, Y. Tian, A.E. Hamielec, and D.R. Eaton, Polymer, Vol. 31, No. 1, p. 154, 1990.
103. M. Kempe, and G. Barany, J. Am. Chem. Soc., Vol. 118, p. 7083, 1996.

104. P.K. Ajikumar, and K.S. Devaky, *Proc. Indian Acad. Sci. (Chem. Sci.)*, Vol. 112, p. 465, 2000.
105. O. Oguz, and G. Cigdem, *J. Appl. Polym. Sci.*, Vol. 46, No. 3, p. 421, 1992.
106. C. Walling, *J. Am. Chem. Soc.*, Vol. 67, p. 441, 1945.
107. S. Loshack, and T.G. Fox, *J. Am. Chem. Soc.*, Vol. 75, p. 3544, 1953.
108. S. Loshack, *J. Polym. Sci.*, Vol. 15, p. 391, 1955.
109. M. Gordon, and R.J. Roe, *J. Polym. Sci.*, Vol. 21, p. 27, 1956.
110. M. Gordon, and R.J. Roe, *J. Polym. Sci.*, Vol.21, p. 39, 1956.
111. M. Gordon, and R.J. Roe, *J. Polym. Sci.*, Vol. 21, p. 57, 1956.
112. M. Gordon, and R.J. Roe, *J. Polym. Sci.*, Vol. 21, pp.75, 1956.
113. A.R. Shultz, J. Am., Chem. Soc. Vol. 80, pp.1854, 1958.
114. P. Hayden, and H. Melville, *J. Polym. Sci.*, Vol. 43, p. 215, 1960.
115. K. Horie, A. Otagawa, M. Muraoka, and I. Mita, *J. Polym. Sci., Polym. Chem. Ed.*, Vol. 13, p. 445, 1975.
116. R.S. Whitney, and W. Burchard, *Makromol. Chem.*, Vol. 181, p. 869, 1980.
117. T.D. Landin, and W.C. Macosko, *Macromolecules*, Vol. 21, No. 3, p. 846, 1988.
118. S. Zhu, Y. Tian, A.E. Hamielec, and D.R. Eaton, *Polymer*, Vol. 31, p. 154, 1990.
119. T.Y. Guo, C. Xi, G.J. Hao, M.D. Song, and B.H. Zhang, *Advances in Polymer Technology*, Vol. 24, No. 4, p. 298, 2005.
120. G.C. Eastmond, and J.E. Richardson, *Macromolecules*, Vol. 24, No. 11, p. 3189, 1991.
121. B. Shaughnessy, and J. Yu, *Macromolecules*, Vol. 27, No. 18, p. 5079, 1994.
122. K. Anseth, C.N. Bowman, and N.A. Peppas, *J. Polym. Sci. Part A: Polym. Chem.*, Vol. 32, p. 139, 1994.
123. C.W. Huang, Y.M. Sun, and W.F. Huang, *Polymer*, Vol. 38, p. 1873, 1997.
124. J. Spevacek, and B. Schneider, *Adv. Colloid Interface Sci.*, Vol. 27, p. 81, 1987.
125. E.M. Woliff, *Aust. Dent. J.*, Vol. 7, p.439, 1962.
126. K.S. Anseth, L.M. Kline, T.A. Walker, K.J. Anderson, and C.N. Bowman, *Macromolecules*, Vol. 28, No. 7, p. 2491, 1995.
127. Y. Takezawa, S. Tanno, N. Taketani, S. Ohara, and H. Asano, *J. Appl. Polym. Sci.*, Vol. 42, No. 10, p. 2811, 1991.
128. C.C.M. Ma, and C.H. Chen, *Polym. Eng. Sci.*, Vol. 31, p. 1086, 1991.
129. S. Loshack, and T.G. Fox, *J. Am. Chem. Soc.*, Vol. 75, p. 3544, 1953.
130. N. Ajzenberg, and A. Ricard, *J. Appl. Polym. Sci.*, Vol. 80, No. 8, p. 1220, 2001.
131. U.S. Ramelow, and S. Pingili, *Polymers*, Vol. 2, p. 265, 2010.
132. O.L.P. Andre, C.H.P. Geerts, H.P.H. Schelten, and M.M.J.M. Van Kerckhove, European Patent, p. 462, 677, 1997.
133. K. Furuuchi, and S. Mori, Patent WO 9722420, 1997.
134. K. Hosoya, Y. Kishii, N. Tanaka, K. Kimata, S. Maruya, T. Araki, and J.M.J. Fréchet, *Chem. Lett.*, p.1145, 1992.
135. S. Dragan, and C.D. Vlad, *Macromol. Symp.*, Vol. 181, p. 47, 2002.
136. S. Dragan, M.V. Dinu, and C.D. Vlad, *J. Appl. Polym. Sci.*, Vol. 89, No.10, p. 2701, 2003.

137. M. Bryjak, *Angew. Makromol. Chem.*, Vol. 215, p. 129, 2003.
138. M. Leskovac, V. Kovacevic, D. Stanojevic, and M. Bravar, *J. Appl. Polym. Sci.*, Vol. 53, No. 13, p. 1717, 1994.
139. M. Hidalgo, J.Y. Cavaille, J. Cuillot, A. Guyot, J. Perez, and R. Vassoille, *J. Polym. Sci. B: Polym. Phys.*, Vol. 33, p. 1559, 1995.
140. A.R. Duarte, L.E. Anderson, C.M. Duarte, S.G. Kazarian, *J. Supercrit. Fluids*, Vol. 36, p. 160, 2005.
141. T. Tamai, P. Pineng, and M.A. Winnik, *Macromolecules*, Vol. 32, No. 19, p. 6102, 1999.
142. M. Okubo, M. Matsuda, A. Terada, Y. Kagawa, and S. Kondo, *J. Appl. Polym. Sci.*, Vol. 90, Issue 7, p. 1825, 2003.
143. F. Del Rector, W.W. Blount, and D.R. Leonard, Higher-Solids and Powder Coatings Symposium, Feb 3–5. New Orleans, LA, 1988.
144. J.G. Carey, G.C. Overbeek, M.P.J. Heuts, and Y.W. Smak, US Patent 5,002,998, 1991.
145. A.C. Lavoie, D.A. Bors, and W.T. Brown, US Patent 5,494,975, 1996.
146. C.L. Mao, N. Chen, and C.F. Tien, US Patent 5,536,784, 1996.
147. J.W. Taylor, and J.R. Salisbury, US Patent 5,663,266 1997.
148. A. Noomen, *Prog. Org. Coat.*, Vol. 23, p. 325, 1994.
149. Geurink, *21st Int. Conf. Org. Coat. Sci.*, Athens, Greece, p. 161, 1995.
150. C.N. Bowman, and N.A. Peppas, *J. Appl. Polym. Sci.*, Vol. 42, p. 201 1991.
151. N.A. Peppas, *Hydrogels in Medicine and Pharmacy*, CRC Press, Boca Raton, FL, 1987.
152. G. Malucelli, G. Gozzelino, F. Ferrero, R. Bongiovanni, and A. Priola, *J. Appl. Polym. Sci.*, Vol. 65, p. 491, 1997.
153. G. Snellings, W. Mondelaus, and E. Schacht, *Polym. Int.*, Vol. 52, p.1653, 2003.
154. A. de Krom, H. Mulder, and D. Mestach, *Farbe Lack*, Vol. 107, p. 97, 2001.
155. A. de Krom, H. Mulder, and D. Mestach, *Farbe Lack*, Vol. 107, p. 100, 2001.
156. A. de Krom, H. Mulder, and D. Mestach, *Farbe Lack*, Vol. 107, p. 104, 2001.
157. P.C. Martino, and P.P. Winner, US Patent, US 4487859, 1984.
158. T. Kuhn, and G. Schlenk, US Patent, US 4476286 1984.
159. Y. Yukawa, M. Yabuta, and Y. Nakao European Patent, EP 519513, cf CA 1119–73808e, 1993.
160. H. Fan, J. Yu, Z. Chen, W. Li, and W. Chen, *Huaxue yu Nianhe*, Vol. 1, pp. 1–3, 13, cf CA 130–297510s, 1999.
161. H. Lutz, H.P. Weitzel, W. Huster, *Polymer Science: A Comprehensive Reference*, Vol. 10, pp. 479–518, 2012.
162. C. Decker, and K. Zahouily, *Polym. Mater. Sci. Eng.*, Vol. 68, p. 70, 1993.
163. H. Warson, and C.A. Finch, *Applications of Synthetic Resin Latices*, Wiley, Chichester, p. 758, 2001.
164. M. Yoshida, Japan Patent 03124776, 1991.
165. J.W. Vanderhoff, *J. Polym. Sci. Polym. Symp.*, Vol. 72, p. 161, 1985.
166. S.C. Misra, C. Pichot, M.S. El-Aasser, and J.W. Vanderhoff, *J. Polym. Sci., Polym. Chem. Ed.*, Vol. 21, p. 2383, 1983.

167. M.S. El-Aasser, T. Makgawinta, J.W. Vanderhoff, and C.Pichot, *J. Polym. Sci. Polym. Chem. Ed.*, Vol. 21, p. 2363, 1983.
168. M.S. El-Aasser, T. Makgawinta, S.C. Misra, J.W. Vanderhoff, C. Pichot, and M.F. Llanro, *Applied Science*, London, p. 215, 1981.
169. S.C. Misra, C. Pichot, M.S. El-Aasser, and J.W. Vanderhoff, *Polym. Lett.*, Vol. 17, p. 567, 1979.
170. N. Okameto, and T. Tatsuho, Japan Patent 03259966, 1991.
171. H. Grubert, P. Dersch, J. Neutzner, R. Hummerich, and P.J. Claassen, European Patent 524 513, 1993.
172. M. Nikami, S. Matsuhira, and N. Misu, Japan Patent 11116658, 1997.
173. H.P. Mueller, H. Gruttmann, J. Petzoldt, H. Mueller, and C. Irle, European Patent 1132413 A1 2001.
174. S. Schwarte, M. Dangschet, M. Reimann, and C. Wastermann, German Patent 4,009,931, 1991.
175. A. Shiota, H. Yoshihara, H. Isaka, M. Yoshizumi, and A. Aida, Japan Patent 02035676, 2002.
176. S. Makineura, and T. Nakamura, Japan Patent 03201440, 2003.
177. T. Schwerzel, P. Fickeisen, and H.J. Neumann, European Patent, Vol. 490, p. 191, 1992.
178. H. Takeuchi, T. Morioka, K. Tomijima, and F. Tsukiyana, Japan Patent 03153782, 1991.
179. K. Sugiura, H. Onoda, T. Okuda, and Y. Matsuno, Japanese Patent JP 2003226843, 2003.
180. S. Bajpai, J.S.P. Rai, I. Nigam, *J. Appl. Polym. Sci.*, Vol. 122, Issue 1, p. 676, 2011.
181. S. Bajpai, J.S.P. Rai, I. Nigam, *J. Appl. Polym. Sci.*, Vol. 112, Issue 4, p. 2374, 2009.

Part 2

INVENTIVE NANOTECHNOLOGY

7

Nanoparticles for Trace Analysis of Toxins: Present and Future Scenario

Anupreet Kaur* **and Shivender Singh Saini**

Department of Chemistry, Punjabi University-Patiala, Punjab, India

Abstract

Nanotechnology is a multidisciplinary field that has gained significant momentum in recent years. Nanoparticles are the key players that promised many benefits through their nano-enabled applications in multiple sectors. The success of the technique in field conditions is a factor of the interdisciplinary work that is involved. Nanoremediation has the potential to clean up large contaminated sites *in situ*, reducing clean up time and eliminating the need for removal of contaminants, and hence reducing the contaminant concentration to near zero. The explosive growth in nanotechnology research has opened the doors to new strategies using nanometallic particles for extraction of pollutants. This chapter briefly deals with recent advances and applications of nanotechnology for removal of environmental pollutants. Under the nanotechnology umbrella, a number of new procedures for producing nanomaterials ultimately used for treatment of wastewater are presented. Research advances for the use of metals, bimetallic nanoparticles, mixed oxides and carbon nanomaterials in environmental remediation are also reviewed.

Keywords: Silica nanoparticles, preconcentration, SPNE, APTZ, NNI, MWCNTs, FET, AFB1, AuNPs, VB, QD, AOPs, GFH

7.1 Introduction

"Nanotechnology is the understanding and control of matter at dimensions between approximately 1 and 100 nanometers, where unique phenomena

**Corresponding author*: anu_chem06@yahoo.co.in

Ashutosh Tiwari and Mikael Syväjärvi (eds.) Advanced Materials for Agriculture, Food, and Environmental Safety, (177–196) 2014 © Scrivener Publishing LLC

enable novel applications" (National Nanotechnology Initiative [NNI] 2008). Encompassing nanoscale science, engineering, and technology, nanotechnology involves imaging, measuring, modeling, and manipulating matter at this length scale. Although industrial sectors involving semiconductors; memory and storage technologies; display, optical, and photonic technologies; energy; biotechnology; and health care produce the most products containing nanomaterials, there are increasing efforts to use nanotechnology as an environmental technology to protect the environment through pollution prevention, treatment, and clean up of long-term problems such as hazardous waste sites. The technology could be a beneficial replacement of current practices for site remediation. However, potential risks are poorly understood and might lead to unintended consequences.

Nanoremediation has the potential not only to reduce the overall costs of cleaning up large-scale contaminated sites, but also to reduce clean-up time, eliminate the need for treatment and disposal of contaminated soil, and reduce some contaminant concentrations to near zero—all *in situ*. Proper evaluation of nanoremediation, particularly full-scale ecosystem-wide studies, needs to be conducted to prevent any potential adverse environmental impacts. In addition to remediating pollution, nanoparticles can be used as sensors to monitor toxins, heavy metals and organic contaminants in land, air and water environments, and have been found to be more sensitive and selective than conventional sensors. In this chapter, we present a background and overview of current practice, research findings related to nanotechnology, issues surrounding the use of nanotechnology for environmental remediation, and future directions.

7.2 Nanoremediation Using TiO_2 Nanoparticles

Tiny particles of titanium dioxide are found as key ingredients in wall paints, sunscreens, and toothpaste; they act as reflectors of light or as abrasives. However with decreasing particle size and a corresponding change in their surface-to-volume ratio, their properties change so that crystalline titanium dioxide nanoparticles acquire catalytic ability; activated by the UV component in sunlight, they break down toxins or catalyze other relevant reactions.

Titanium oxide photocatalyts have been widely studied for solar energy conversion and environmental applications in the past several decades because of their high chemical stability, good photoactivity, relatively low cost and nontoxicity. However, the photocatalytic capability of TiO_2 is limited to only ultraviolet light; to overcome this problem both chemical and physical modification approaches were developed to extend the absorption band-edge of TiO_2 into visible region [1–5].

In the photocatalytic oxidation process, organic pollutants are destroyed in the presence of semiconductor photocatalyts, an energetic light source, and an oxidizing agent such as oxygen or air. Only photons with energies greater than the band gap energy (ΔE) can result in the excitation of valence band (VB) electrons which then promote the possible reactions. The absorption of photons with energy lower than ΔE or longer wavelength usually causes energy dissipation in the form of heat.

The illumination of the photocatalytic surface with surface energy leads to the formation of positive hole (h^+) in the valence band and an electron (e^-) in the conduction band. The positive hole oxidizes either pollutants directly or water to produce $\cdot OH$ radicals, whereas the electron in the conduction band reduces the oxygen adsorbed on the photocatalyt (TiO_2). The activation of TiO_2 by UV light can be:

Overall reaction:

$$\mathbf{TiO_2 + hv \rightarrow e^- + h^+ \; e^- + O_2 \rightarrow O_2^-} \tag{7.1}$$

Oxidative reaction:

$$\mathbf{h^+ + organic\ moiety \rightarrow CO_2} \tag{7.2}$$

$$\mathbf{h^+ + H_2O \rightarrow \cdot OH + H^+} \tag{7.3}$$

Reductive reaction:

$$\mathbf{OH + organic\ moiety \rightarrow CO_2} \tag{7.4}$$

In recent years, advanced oxidation processes (AOPs) using titanium dioxide (TiO_2) have been effectively used to detoxify recalcitrant pollutants present in industrial wastewater. Titanium dioxide has singular characteristics that have made it an extremely attractive photocatalyst: high photochemical reactivity, high photocatalytic activity, low cost, stability in aquatic systems and low environmental toxicity. The general detailed mechanism of dye degradation upon irradiation is described below.

$$\text{Dye} + hv \rightarrow \text{Dye}^* \tag{7.5}$$

$$\text{Dye}^* + TiO_2 \rightarrow \text{Dye}\bullet + + TiO_2(e) \tag{7.6}$$

$$TiO_2\,(e) + O_2 \rightarrow TiO_2 + O_2^{\bullet -} \tag{7.7}$$

$$O_2^{\bullet -} + TiO_2\,(e) + 2H^+ \rightarrow H_2O_2 \tag{7.8}$$

$$H_2O_2 + TiO_2\,(e) \rightarrow \bullet OH + OH^- \quad (7.9)$$

$$Dye\bullet^+ + O_2\ (\text{or } O_2^{\bullet-} \text{ or } \bullet OH) \rightarrow \text{peroxylated or hydroxylated intermediates} \rightarrow \text{degraded or mineralized products} \quad (7.10)$$

Photocatalytic degradation of rhodamine 6G (R-6G), methyl red, malachite, 4-nitrophenol, yellow 27, yellow 50, violet 51 and bisphenol-A. Photocatalytic degradation of the diuron by TiO_2 and by Pt/TiO_2, also on Au/TiO_2 for diuron as well as its didemethylatedproduct,3,4-dichlorophenylurea, has been reported.The photocatalytic degradation of s-triazines herbicides (atrazine, simazine, trietazine, prometon and prometryn) was first studied by Pelizzetti *et al.* [6]. Several studies were published about the solar TiO_2 photocatalyzed oxidation of s-triazines (2-chloro-, 2-methoxy and 2-methylthio-s-triazines) and the mechanistic pathways of the observed photoproducts [7–13]. Recently, other three herbicides belonging to the triazine group, irgarol [2-methylthio-4-(tert-butylamino)-6-(cyclopropylamino)-s-triazine], cyromazine [*N*-cyclopropyl- 1,3,5-triazine-2,4,6-triamine] and metamitron {4-amino-6-phenyl-3-methyl-1,2,4-triazin-5(4*H*)-one}, were studied and similar results were observed.

The photocatalytic degradation of four representative compounds of anilide and amide herbicides (3,4-dichloropropioamide, propanil, alachlor, propachlor) have been studied [14–17].

p-Chloraniline, tetrachlorohydroquinone, H_2O_2, tetrachlorocatechol and *o*-chloraniline were identified as the principal intermediates in the photocatalytic destruction of PCP in the presence of TiO_2 [18–21]. The major PCP intermediates detected were 2,3,5,6-tetrachloro-1,4-benzoquinone, 2,3,5,6-tetrachloro-1,4-hydroquinone and 2,3,5,6-tetrachlorophenol [22]. The pesticide permethrin can be easily photodecomposed into CO_2 and Cl^- ions in a fluoro surfactant/TiO_2 aqueous dispersion [23].

Atrazine *{*2-chloro-4-ethyl-amino-6-isopropylamino-1,3,5-triazine*}* is one of the most common pesticides found in groundwater sources and drinking water supplies. McMurray *et al.* [24] studied the photocatalytic decomposition of this herbicide by TiO_2 films in industrial water.

TiO_2 nanoparticles were used for preconcentration of trace arsenite and arsenate in natural water. Adsorption of Pb, Cd, Cu, Zn, and Ni to titanium dioxide nanoparticles was studied by Engates *et al.* [25]. TiO_2 had also been used for the preconcentration of metal ion such as Cu, Cr, Zn, Cd, Se, Ho, Au, Nd, Tm, La, Y, Tb, Eu, Dy from various environmental samples such as wastewater, sediments, coal ash, vehicle exhaust particulates and geological samples; and also physical modification with diethyldithiocarbamate (DDTC), 1-(2-pyridylazo)-2-naphthol (PAN), dithizone and 8-hydroxyquinolone used for the separation and extraction of Cr,Cu, Pb, Zn,Fe, Al, Y, Yb from natural, waste and environmental water samples, biological samples and also in food samples [26–34].

Since the expression of super-hydrophilicity with TiO_2 photocatalyst was presented in 1997 by Hashimoto *et al.* research into this feature has been active. Consequently, it has been increasingly applied to fog-proofing and self-cleaning applications for mirrors, including road mirrors (curve mirrors) and door mirrors on cars, as well as window glass panels. Since the hydrophilicity of TiO_2 photocatalysts is more positively maintained by the addition of SiO_2 or a more porous structure of TiO_2 particles, improvement in the composition and layer-forming method for TiO_2 photocatalysts is now underway. Air purification is one example of the most advanced application of TiO_2 photocatalysts. For example, photocatalysts are used in deodorizing filters in air-purifiers, incorporating UV lamps to eliminate aldehyde or VOC in indoor air. Since then, there has been a report concerning the successful elimination of low-concentration NOx in outdoor environments. A TiO_2 photocatalyst oxidizes NO into NO_2 and eventually into NO_3^-, hence removing NO from the air [35]. Nanogold supported on TiO_2-coated glass fiber was used for removing toxic CO gas from air. The outstanding catalytic activities of nanogold for oxidizing CO at low temperature, and various reactions of nanogold catalysts have been studied. These include CO oxidation, preferential oxidation of CO in the presence of excess hydrogen (PROX), water-gas shift reaction (WGSR), hydrogenation and oxidation.

7.3 Gold Nanoparticles for Nanoremediation

Gold nanoparticles (AuNPs), one of the wide varieties of core materials available, coupled with tunable surface properties in the form of inorganic or inorganic-organic hybrid, have been reported as an excellent platform for a broad range of analytical methods. The modification of the Au surface with appropriate chemical species can improve the separation and preconcentration efficiency, analytical selectivity, and method reliability. Because of their high surface-to-volume ratio, easy surface modification, and simple synthesis methods, gold nanoparticles (AuNPs) are becoming an attractive material as an alternative to conventional solvent extraction and solid-phase extraction. Through covalent bond formation (Au-S bonds), electrostatic attraction, hydrophobic adsorption, and molecular recognition, AuNPs have been applied successfully to the extraction/removal of a variety of compounds, peptides, proteins, heavy metal ions, and polycyclic aromatic hydrocarbons. Nanocomposite of gold and aluminium nanoparticles has been used for the preconcentration of Hg(II) from natural water.

Gold nanoparticle loaded on activated carbon (AuNP-AC) with 1-((6-(-(2,4-dihydroxybenzylideneamino))hexylimino) methyl)benzene-2,4-diol (DHBAHMB) has been applied for enrichment and preconcentration of trace amounts of Cu(II), Fe(III) and Zn(II) ions in real samples [36]. Modified citrate-stabilized gold nanoparticles (AuNPs) have been used for the enrichment and preconcentration of endocrine disruptors in real samples [37]. In environmental samples, 2,5-dimercapto-1,3,4-thiadiazole-stabilized gold nanoparticles (DMT-AuNPs) were used for the determination of picogram Hg(II).

N-1-(2-Mercaptoethyl)thymine modification of gold nanoparticles is a highly selective and sensitive colorimetric chemosensor for Hg(II) [38, 39]. A novel alternative approach using the so-called solid-phase nanoextraction (SPNE) for the preconcentration of polycyclic aromatic hydrocarbons (PAHs) from drinking water has been proposed by Wang *et al.* [40, 41]. Alkanethiol-modified AuNPs coated on silica gel were used for the SPNE of steroids (progesterone and testosterone propionate) by Liu *et al.* [43]. Qu *et al.* used AuNPs for the SPNE of nine compounds (ethanol, benzene, 1-butanol, chlorobenzene, 1-pentanol, anisole, phenol, methyl benzoate, benzyl alcohol) [44]. Aromatic compounds (benzene, naphthalene, phenanthrene, anthracene) were preconcentrated by immobilization of noctadecanethiol-modified Au-coated polystyrene particles on capillary by Kobayashi *et al.* Immobilization of BSA-modified AuNPs on capillary was used for the SPNE of Dansyl-norvaline by Liu *et al.* Neutral compounds (thiourea, benzophenone, biphenyl and pyrethroid pesticides) and drug substances (propiophenone, benzoin and warfarin) and PAHs were preconcentrated by dodecanethiol-modified AuNPs by Glenonn's group. Aflatoxin B1 detection by AuNPs also gave good results. Gold nanoparticles were also employed for an immunoassay for the detection of aflatoxin B1 (AFB1) in foods.

7.4 Zero-Valent Iron Nanoparticles

Iron is one of the most abundant elements on earth. Elemental iron has been used as an ideal candidate for remediation because it is inexpensive, abundant, and easy to prepare and apply to a variety of systems, and devoid of any known toxicity induced by its usage. The concept of using metals, such as iron, as remediation agents is based on reduction–oxidation or "redox" reactions, in which a neutral electron donor (a metal) chemically reduces an electron acceptor (a contaminant). Nanoscale iron particles have surface areas significantly greater than larger-sized powders or granular iron, which leads to enhanced reactivity for the redox process. As a result, iron nanoparticles have been extensively investigated for the decomposition of halogenated hydrocarbons to benign hydrocarbons and the remediation of many other contaminants, including anions and heavy metals [44]. Zero-valent iron nanoparticles are highly reactive and react rapidly with surrounding media in the subsurface [45]. A significant loss of reactivity can occur before the particles are able to reach the target contaminant. In addition, zero-valent iron nanoparticles tend to flocculate when added to water, resulting in a reduction in effective surface area of the metal. Therefore, the effectiveness of a remediation depends on the accessibility of the contaminants to the nanoparticles; and the maximum efficiency of remediation will be achieved only if the metal nanoparticles can effectively migrate without oxidation to the contaminant or the water/contaminant interface. To overcome such difficulties, a commonly used strategy is to incorporate iron nanoparticles within support materials, such as polymers, porous carbon, and polyelectrolytes [46, 47]. Zero-valent iron removes aqueous contaminants by reductive dechlorination, in the case of chlorinated solvents, or by reduction to an insoluble form, in the case of aqueous metal ions [48, 49] (Figure 7.1). Zero-valent iron removes aqueous contaminants by the reduction of nitrate compounds (Figure 7.2). Increasing the surface area of zero-valent iron nanoparticles results in an increased rate of remediation. In general, chlorinated organics (C x H y Cl z) and iron in aqueous solutions can be expressed by equation:

$$C_xH_yCl_z + zH^+ + zFe^0 \rightarrow C_xH_{y+z} + zFe^{2+} + zCl^- \quad (7.11)$$

Iron undergoes classical electrochemical/corrosion redox reactions in which iron is oxidized from exposure to oxygen and water:

$$2Fe^0(s) + O_2(g) + 2H_2O \rightarrow 2Fe^{2+}(aq) + 4OH^-(aq) \quad (7.12)$$

$$Fe^0(s) + 2H_2O(g) \rightarrow Fe^{2+}(aq) + H_2(g) + 2OH^-(aq) \quad (7.13)$$

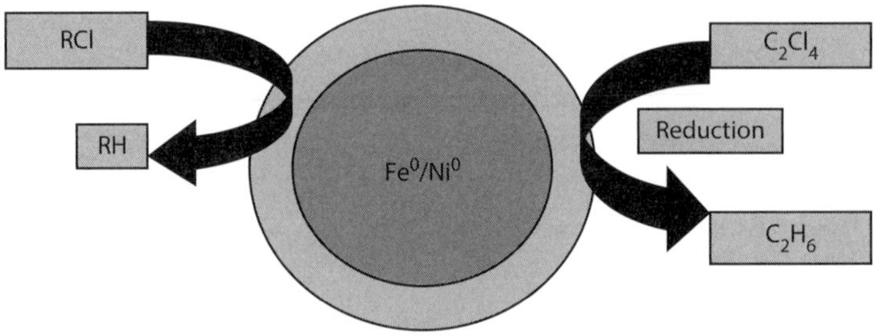

Figure 7.1 Dechlorination by zero-valent iron nanoparticles.

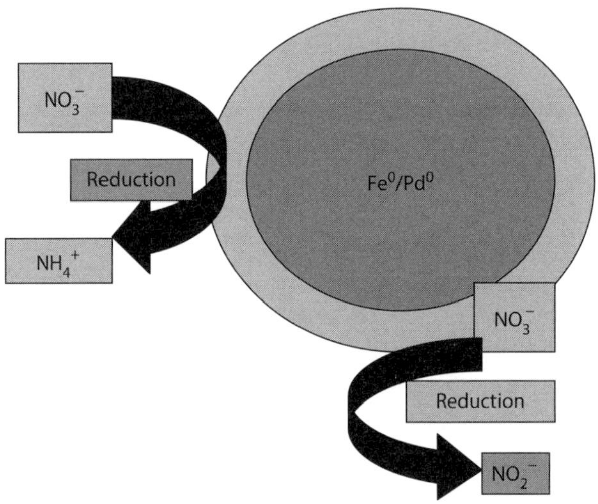

Figure 7.2 Nanoremediation of nitrate by zero-valent iron nanoparticles.

Fe(II) reacts to give magnetite (Fe_3O_4), ferrous oxide [$Fe(OH)_2$] and ferric hydroxide [$Fe(OH)_3$] depending on redox conditions and pH. For example, chromium(VI) can be reduced by Fe(II) to the generic scheme shown in Reaction 7.14 [50].

$$Cr^{6+} + 3Fe^{2+} \rightarrow Cr^{3+} + 3Fe^{3+} \tag{7.14}$$

Iron nanoparticles have been used for the separation of As(III) [51]. Phosphates are a growth nutrient for microorganism in water. As a result of increased phosphorus concentration, an excessive growth of photosynthetic aquatic micro- and macro-organisms occurs and ultimately becomes a major cause of eutrophication, or extensive algae growth. All parameters being equal, HAIX containing hydrated iron oxide nanoparticles was compared with a granular ferric hydroxide (GFH) without any ion exchange material. The HAIX provided significantly greater phosphate removal capacity [52]. Phosphate breakthrough with HAIX occurred after nearly four thousand bed volumes, while the commercially available GFH column from the US Filter Corp. showed a breakthrough after one thousand bed volumes. Iron nanoparticles have been encapsulated with silica in order to increase stability and prevent aggregation. Iron oxide has been shown to retard the proliferation of bacteria. The incorporation of iron oxide catalyzed ozonation technology increases the retention of bacteria to the surface of membranes, resulting in improved remediation of water. Iron oxide catalyzed ozonation and membrane filtration combine to improve inactivation and/or removal of bacteria [53]. There is fast adsorption of methylene blue on polyacrylic acid-bound iron oxide magnetic nanoparticles [54].

7.5 Silicon Oxide Nanoparticles for Nanoremediation

Silica nanoparticles are promising materials as a solid-phase extractant because of their large surface area, high adsorption capacity, low temperature modification, lesser degree of unsaturation and low electrophilicity (Figure 7.3). The sequence of reactivity is expressed as follows:

$$Zr(OR)_4, Al(OR)_4 > Ti(OR)_4 > Sn(OR)_4 >> Si(OR)_4 \quad (7.15)$$

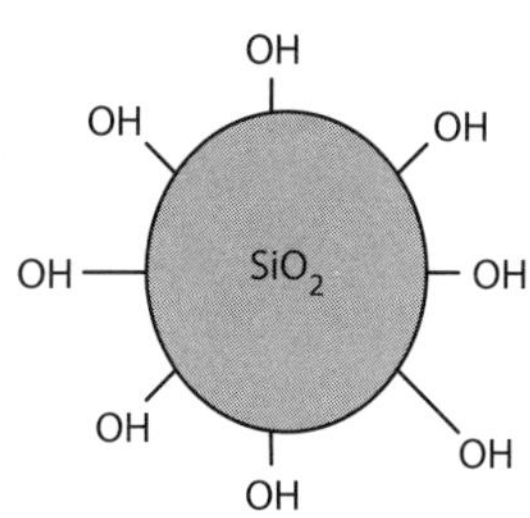

Figure 7.3 Silica nanoparticle.

Some pollutants are poorly adsorbed on nanoparticles. To overcome this problem, physical or chemical modification of the surface of these nanoparticles with certain functional groups containing some donor atoms such as oxygen, nitrogen, sulfur and phosphorus is necessary. The most often used method is to load a kind of specific chelating reagent by physical or chemical procedure. The formal method is simple but the loaded reagent is prone to leaking out from the sorbent, while the chemically bonded material is more stable and can be used repeatedly. The modification of nanometer-sized materials is usually required in order to prevent a conglomeration of particles and to improve its consistency in relation to other materials. Also, the modification of nanometer-sized materials can improve their selectivity toward pollutants. Selectivity of suitable specific functional groups towards metal ions depends on certain factors such as: (1) size of the modifiers, (2) activity of loaded group, and also on (3) the basis of the concept of hard-soft acids and bases. Chemisorption of nanoparticles provides immobility, mechanical stability and water insolubility, thereby increasing the efficiency, sensitivity and selectivity.

Chemical modification is a process that leads to a change in chemical characteristics of the surface of nanoparticles. By the modification, the adsorption properties are significantly affected. Chemisorption of chelating molecules on nanoparticle surface provides immobility, mechanical stability and water insolubility, thereby increasing the efficiency, sensitivity and selectivity of nanoparticles for the analytical application. Chemical modification of nanoparticles by the silylation procedure using different silylating agents such as 3-aminopropyltriethoxysilane, 3-chloropropyltriethoxysilane and 3-mercaptopropyltriethoxysilane provides immobility, mechanical stability and water insolubility. N-[3-(trimethoxysilyl)propyl]ethylenediammine-modified SiO_2 nanoparticles have been used for the preconcentration of some toxic heavy metal ions such as Hg(II), Cu(II) and Zn(II) [55]. Modified silica nanoparticles have also been used for the preconcentration of drugs and pesticides. Silylation of silica nanoparticles followed by their chemical modification using 4-(2-pyridylazo)-resorcinol [56] and these modified SiO_2 nanoparticles have been used for the selective preconcentration of Hg(II). The SiO_2 nanoparticles have also been modified with acetylsalicylic acid, p-dimethylaminobenzaldehyde and 5-sulfonylsalicylic acid for use in the preconcentration of Cr(III), Fe(III), Pb(II) and Cu(II) [57–59]. Chemical modification of silica nanoparticles are shown in Figure 7.4.

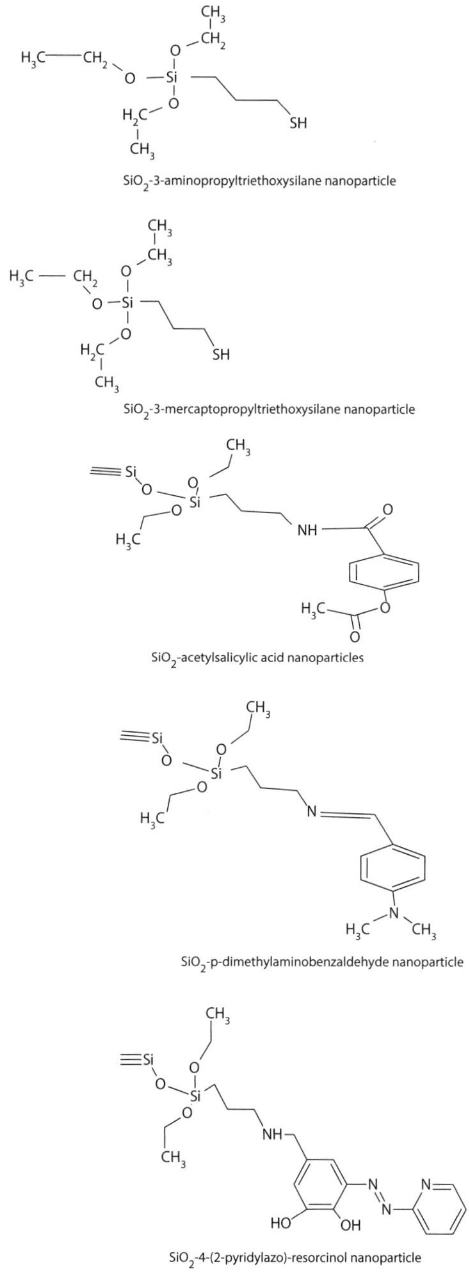

Figure 7.4 Chemical modification of silica nanoparticles with various silanes and ligands.

7.6 Other Materials for Nanoremediation

Biosensors and affinity sensor devices have been shown to have the ability to provide rapid, cost-effective, specific and reliable quantitative and qualitative analysis. To date, developments in nanomaterial and biosensor fabrication technology are moving forward rapidly, with new and novel nanobiorecognition materials being developed which can be applied as the sensing receptors for analysis of mycotoxins. Biosensors as tools have proven to be able to provide rapid, sensitive, robust and cost-effective quantitative methods for on-site testing. The development of biosensor devices for different mycotoxins has attracted much research interest in recent years, with a range of devices being developed and reported in the scientific literature. However, with the advent of nanotechnology and its impact on developing ultrasensitive devices, mycotoxin analysis is also benefiting from the advances taking place in applying nanomaterials in the development of sensors. The application of nanotechnology in biosensors can range from the transducer device, the recognition ligand, the label and the running systems (e.g., instruments). Their application in sensor development has been due to the excellent advantages offered by these materials in miniaturization of the devices and signal enhancements which result in high precision and accuracy, and also amplification of signal by the use of nanoparticles as labels. The high surface-to-volume ratio offered by nanomaterials makes these devices very sensitive and can allow a single molecule detection, which is very attractive in contaminant monitoring, such as for toxins. The development of micro/nanosensor devices for toxins analysis is increasing due to their extremely attractive characteristics for this application.

The development of micro/nanosensor devices for toxins analysis is increasing due to their extremely attractive characteristics for this application. In principle these devices are miniature transducers fabricated using conventional thin- and thick-film technology. Their novel electron transport properties make them highly sensitive for low-level detection. Multi-toxins detection (e.g., mycotoxins) in foods can be conducted using single micro/nanoelectrode array chip with high sensitivity and rapid analysis time. The use of micro/nanoarrays for analysis applications in foods can produce highly sensitive sensors. Multi-mycotoxins detection has also been reported in the literature using different sensor platforms combined with multi-ELISA assays. Therefore, multi-toxins can be detected on a single microelectrode array chip with multi-array working electrode, where different antibody is immobilised to detect a specific mycotoxin. Micro/nanoelectrode arrays have unique properties, including small capacitive charging current and faster diffusion of electroactive species, which will result in an improved response time and greater sensitivity. The use of lab-on–a-chip is expanding in all areas of analysis due to the advantages of using small samples to analyze several markers/toxins, i.e., offer high throughput analysis. These types of devices will be attractive for mycotoxin analysis since several toxins may exist in the same food or feed sample.

A range of sensors are being developed for mycotoxins based on the above technologies which can be applied on the farm or in the factory and can be operated by unskilled personnel. Current trends to produce chip-based micro/nanoarrays for multi-mycotoxins analysis are challenging but possible, and will have a significant impact on risk assessment testing. The use of nanoparticles such as gold, silver, metal oxides and quantum dots assay developments will enhance the capability of the biosensor technology for mycotoxins analysis. Early and sensitive detection will aid in eliminating these toxins from entering the food chain and preventing ill health and protecting life.

The development of biosensors for the rapid, reliable and low-cost determination of mycotoxins in foodstuffs has received considerable attention in recent years, and various types of assays have already been devised for several of the major groups of mycotoxins. One format uses the phenomenon of *surface plasmon resonance* (SPR) to detect the change in mass that occurs when mycotoxin-specific antibodies attach to a mycotoxin that has been covalently bonded to the surface of a sensor chip. A recent application developed and optimized for measuring deoxynivalenol in wheat extracts gave results that were in good agreement with LC–MS data. Moreover, SPR sensor chips with immobilized deoxynivalenol could be reused more than 500 times without significant loss of activity. Because the instrumentation is now commercially available, this format could find widespread application in future mycotoxin analysis. A second format using *fiber-optic probes* can be adapted for continuous monitoring of mycotoxin levels. This sensor uses the evanescent wave of light that can form around the surface of an optical fiber. Antibodies attached to the surface of the fiber trap fluorescent mycotoxins (e.g., aflatoxins) or fluorescent analogues of mycotoxins (e.g., derivatized fumonisins) with the evanescent zone, permitting their detection. Two different benchtop devices have been designed for the fumonisins and aflatoxins. Unfortunately, most of the SPR and fiber-optic biosensor procedures for mycotoxin analysis still require some form of sample clean-up/preconcentration in order to be truly effective in the analysis of real samples and to achieve adequate sensitivity. Moreover, the majority of these devices lack the ability to perform simultaneous analyses of multiple samples. Recently, array biosensors have been developed and demonstrated for a variety of applications. The ability of array biosensors to analyze multiple samples simultaneously for multiple analytes offers a significant advantage over other types of biosensors. In particular, a rapid, multianalyte array biosensor developed by Ngundi *et al.* [60] at the Naval Research Laboratory of Washington D.C., USA, has demonstrated the potential to be used as a screening and monitoring device for clinical, food and environmental samples. The device, which is portable and fully automated, can be used with different immunoassay formats. One interesting application is the development of a competitive immunoassay for the detection and quantification of ochratoxin A in a variety of spiked food and beverage samples. A simple extraction procedure was employed with no need for clean-up or preconcentration of the sample extract. This is the first demonstration that a rapid biosensor can be used in a competitive assay format to detect a mycotoxin in extracts of relevant foods. However, further work aimed at developing a dual-analyte assay for deoxynivalenol and ochratoxin A showed that improvements are still necessary to reduce the analysis time and increase the sensitivity.

Carbon nanotubes were used for the determination of zearalenone in urine samples by Andres *et al.* [61]. Multiwalled carbon nanotubes were modified with an enzyme, aflatoxin detoxifizyme. The MWCNTs were used for enzyme immobilization of aflatoxins detoxifizyem (APTZ) and for the determination of sterigmatocystin [62] and also carbon nanotubes field effect transistors (FET) that had been functionalized with protein G and IgG to detect *Aspergillus flavus* in contaminated milled rice [63]. Optical sensors based on nanomaterials have been applied much less for the detection of analytes of interest in the food industry. Quantum dots (QD) are practically the only nanomaterial. They are nanocrystals of inorganic semiconductors that are somewhat restricted to a spherical shape of around 2 to 8 nm diameter [64]. Their fluorescent properties are size-dependent and, therefore, they can be tuned to emit at desired wavelengths (between 400 and 2000 nm) if synthesized in different composition and size. In this way, QDs of different sizes can be excited with a single wavelength and emission controlled at different wavelengths, thus providing for simultaneous detection. This, together with their highly robust emission properties, makes them more advantageous for labeling and optical detection than conventional organic dyes [65]. Their high quantum yields and their narrow emission bands produce sharper colors, leading to higher sensitivity and the possibility of multiplexing of analysis [66, 67]. Costa *et al.* [68] have reviewed the progress in exploiting these novel probes in optical sensing, as well as their still unexploited sensing capabilities. In the analytical chemistry field their major application has been as fluorescent labels, while an application for food analysis has been, up to now, unexploited. Goldman *et al.* [69] have used QDs for fluoroimmuno assays of toxins. They detected four toxins simultaneously, three of which are naturally responsible for food or waterborne sickness. The CdSe-ZnS core-shell QDs were capped with dihydrolipoic acid and bioconjugates with the appropriate antibodies were prepared. A sandwich immunoassay was performed in microtitre plates, where the toxins and different QDs were incubated for an hour. Fluorescence was measured at adequate wavelengths and, although there was spectral overlap, deconvolution of spectra revealed the fluorescence contribution of all toxins. Signals increased with toxin concentration in different ranges according to the particular toxin. No LODs were reported. Although the authors treated the bioconjugate QDs as fluororeagents, they can be considered as "chemosensing devices." An ultrasensitive densitometry method was used for the detection of cytokines by nanoparticle-modified aptamers [70, 71]. Gold nanoparticles were also employed for an immunoassay for the detection of aflatoxin B1 (AFB1) in foods [72].

7.7 Conclusion

Decontamination is the reduction or removal of chemical and biological agents by means of physical means or chemical neutralization or detoxification techniques. Nanotechnology has shown huge potential in areas as diverse as drug development, water decontamination, information and communication technologies, and the production of stronger, lighter materials and human health care. Water and air are two vital components of life on earth; the existence of life on earth is made possible largely because of their importance to metabolic processes within the body. Clean and fresh water and air are essential for the existence of life. The recent developments in nanotechnology have raised the possibility of environmental decontamination through several nanomaterials, processes and tools. This chapter summarizes the expertise that various decontamination approaches can bring to the successful realization of environmental remediation.

References

1. M.R. Hoffmann, S.T. Martin, W. Choi, and D.W. Bahnemann, *Chemical Reviews*, Vol. 95, p. 69, 1995.
2. F.L. Toma, G. Bertrand, D. Klein, C. Coddet, *Environ. Chem. Lett.*, Vol. 2, p. 117, 2004.

3. J.C. Sin, S.M. Lam, A.R. Mohamed, K.T. Lee, *Int. J. Photoenergy*, Vol. 2012, pp. 1–23, Doi:10.1155/2012/185159.
4. W.T. Tsai, M.K. Lee, T.Y. Su, and Y.M. Chang, *J. Hazard. Mater.*, Vol. 168, p. 269, 2009.
5. H. Katsumata, M. Sada, Y. Nakaoka, S. Kaneco, T. Suzuki, K. Ohta, *J. Hazard. Mater.*, Vol. 171, p. 1081, 2009.
6. E. Pelizzetti, V. Maurino, C. Minero, V. Carlin, E. Pramauro, O. Zerbinati, M.L. Tosato, *Environ. Sci. Technol.*, Vol. 24, p. 1559, 1990.
7. L. Muszkat, L. Bir, L. Feigelson, *J. Photochem. Photobiol. A: Chem.*, Vol. 87, p. 85, 1995.
8. I.K. Konstantinou, T.M. Sakellarides, V.A. Sakkas, T.A. Albanis, *Environ. Sci. Technol.*, Vol. 35, p. 398, 2001.
9. E. Pelizzetti, C. Minero, V. Carlin, M. Vincenti, E. Pramauro, *Chemosphere*, Vol. 24, p. 891, 1992.
10. O. Borio, B.M. Gawlik, I.R. Bellobono, H. Muntau, *Chemosphere*, Vol. 37, p. 975, 1998.
11. E. Pelizzetti, V. Carlin, C. Minero, E. Pramauro, M. Vincenti, *Sci. Total Environ.*, Vol. 123, p. 161, 1992.
12. Y. Sanlaville, S. Guittonneau, M. Mansour, E.A. Feicht, P. Meallier, A. Kettrup, *Chemosphere*, Vol. 33, p. 353, 1996.
13. I. Textier, C. Giannotti, S. Malato, C. Richter, J. Delaire, *Catal. Today*, Vol. 54, p. 297, 1999.
14. I.K. Konstantinou, V.A. Sakkas, T.A. Albanis, *Appl. Catal. B: Environ.*, Vol. 34, p. 227, 2001.
15. H.M.K.K. Pathirana, R.A. Maithreepala, *J. Photochem. Photobiol. A: Chem.*, Vol. 102, p. 273, 1997.
16. L. Muszkat, M. Halmann, D. Raucher, L. Bir, *J. Photochem. Photobiol. A: Chem.*, Vol. 65, p. 109, 1992.
17. G.A. Peñuela, D. Barceló, *J. Chromatogr. A*, Vol. 754, p. 187, 1996.
18. E. Vulliet, C. Emmelin, J.M. Chovelon, C. Guillard, J.M. Herrmann, *Appl. Catal. B: Environ.*, Vol. 38, p. 127, 2002.
19. C. Minero, E. Pelizzetti, S. Malato, J. Blanco, *Sol. Energy*, Vol. 56, p. 42, 1996.
20. G. Mills, M.R. Hoffmann, *Environ. Sci. Technol.*, Vol. 27, p. 1681, 1993.
21. C. Minero, E. Pelizzetti, S. Malato, J. Blanco, *Chemosphere*, Vol. 26, p. 2103, 1993.
22. W.F. Jardim, S.G. Moraes, M.M.K. Takiyama, *Water Res.*, Vol. 31, p. 1728, 1997.
23. H. Hidaka, H. Jou, K. Nohara, J. Zhao, *Chemosphere*, Vol. 25, p. 1589, 1992.
24. T.A. McMurray, P.S.M. Dunlop, J.A. Byrne, *J. Photochem. Photobiol. A: Chem.*, Vol. 182, p. 43, 2006.
25. K.E. Engates, H.J. Shipley, *Environ. Sci. Polllut.Res.*, Vol. 18, p. 386, 2011.
26. C.N. Kuo, H.F. Chen, J.N. Lin, B.Z. Wan, *Catal. Today*, Vol. 122, p. 270, 2007.
27. P. Li, Y. Liu, and L. Guo, *J. Anal. At. Spectrom.*, Vol. 19, p. 1006, 2004.
28. P. Liang, Y. Qin, B. Hu, and T. Peng, *Anal. Chim. Acta*, Vol. 440, p. 207, 2001.

29. L. Shunxin, Q. Shahua, H. Ganquan, and H. Fei, *Freseniu's J. Anal. Chem.*, Vol. 365, p. 469, 1999.
30. Y. Hang, Y. Qin, and J. Shen, *J. Sep. Sci.*, Vol. 26, p. 957, 2003.
31. P. Liang, B. Hu, Z. Jiang, Y. Qin, and T. Peng, *J. Anal. At. Spectrom.*, Vol. 16, p. 863, 2001.
32. P. Liang, T. Shi, H. Lu, Z. Jiang, and B. Hu, *Spectrochim. Acta B*, Vol. 58, p. 1709, 2003.
33. H. Zheng, X. Chang, N. Lian, S. Wang, Y. Cui, and Y. Zhai, *Intern. J. Environ. Anal. Chem.*, Vol. 86, p. 431, 2004.
34. P. Liang, L. Yang, B. Hu, and Z. Jiang, *Anal. Sci.*, Vol. 19, p. 1167, 2003.
35. N. Lian, X. Change, H. Zheng, S. Wang, Y. Cui, and Y. Zhai, *Microchim. Acta*, Vol. 151, p. 81, 2005.
36. L. Yang, B. Hu, Z. Jiang, and H. Pan, *Microchim. Acta*, Vol. 144, p. 227, 2004.
37. F. Marahel, M. Ghaedi, M. Montazerozohori, S. Khodadoust, *Ind. J. Sci. Technol.*, Vol. 4, p. 1234, 2011.
38. H.B. Noh, K.S. Lee, B.S. Lim, S.J. Kim, Y.B. Shim, *Electrophoresis*, Vol. 31, p. 3053, 2010.
39. N. Vasimalai, S.A. John, *Analyst*, Vol. 37, p. 3349, 2012.
40. L. Chen, T. Lou, C. Yu, Q. Kang, *Analyst*, Vol. 136, p. 4770, 2011.
41. H.Y. Wang, A.D. Campiglia, *Anal. Chem.*, Vol. 80, p. 8202, 2008.
42. H.Y. Wang, S.J. Yu, A.D. Campiglia, *Anal Biochem.*, Vol. 385, p. 249, 2009.
43. F.K. Liu, *J. Chin. Chem. Soc.*, Vol. 55, p. 69, 2008.
44. Q.S. Qu, F. Shen, M. Shen, X.Y. Hu, G.J. Yang, C.Y. Wang, C. Yan, Y.K. Zhang, *Anal. Chim. Acta*, Vol. 609, p. 76, 2008.
45. W. Yantasee, C.L. Warner, T. Sangvanich, R.S. Addleman, T.G. Carter, R.J. Wiacek, G.E. Fryxell, C. Timchalk, and M.G. Warner, *Environ. Sci. Technol.*, Vol. 41, p. 5114, 2007.
46. D. Maity, and D.C. Agrawal, *J. Magne. Magnetic Mater.*, Vol. 308, p. 46, 2007.
47. S. Laurent, D. Forge, M. Port, A. Roch, C. Robic, L. Vander Elst, and R.N. Muller, *Chem. Rev.*, Vol. 108, p. 2064, 2008.
48. W.X. Zhang, D.W. Elliot, *Remediation J.*, Vol. 16, p. 7, 2006.
49. J. Quinn, C. Geiger, C. Clausen, C. Brooks, K. Coon, *Environ Sci Technol.*, Vol. 39, p. 1309, 2005.
50. B. Schrick, J. Blough, A. Jones, and T.E. Mallouk, *Chem. Mater.*, Vol. 14, p. 5140, 2002.
51. S. Ponder, J. Darab, T. Mallouk, *Environ. Sci. Technol.*, Vol. 34, p. 2564, 2000.
52. I.N. Savina, C.J. English, R.L.D. Whitby, Y. Zheng, A. Leistner, S.V. Mikhalovsky, A.B. Cundy, *J. Hazard. Mater.*, Vol. 192, p. 1002, 2011.
53. G. Zelmanov, R. Semiat, *J. Environ. Eng. Manag.*, Vol. 10, p. 1923, 2011.
54. S.Y. Mak, D.H. Chen, *Dyes and Pigments*, Vol. 61, p. 93, 2004.
55. Y.F. Huang, Y.F. Wang, X. P. Yan, *Environ. Sci. Technol.*, Vol. 44, p. 7908, 2010.
56. W.C. Shu, F.K. Li, F.H. Ko, *Anal. Bioanal. Chem.*, Vol. 399, p. 103, 2011.
57. J.H. Lin, T.W. Lung, *Reviews. Anal. Chem.*, Vol. 31, p. 153, 2012.

58. F. Marahel, M. Ghaedi, M. Montazerozohori, and S. Khodadoust, *Indian J. Sci. Technol.*, Vol. 4, p. 1234, 2011.
59. H.B. Noh, K.S. Lee, B.S. Lim, S.J. Kim, Y.B. Shim, *Analyst*, Vol. 137, p. 3349, 2012.
60. N. Vasimalai, S.A. John, *Analyst*, Vol. 137, p. 3349, 2012.
61. L. Chen, T. Lou, C. Yu, Q. Kang, *Analyst*, Vol. 136, p. 4770, 2011.
62. H.Y. Wang, A.D. Campiglia, *Anal Chem.*, Vol. 80, p. 8202, 2008.
63. H.Y. Wang, S.J. Yu, A.D. Campiglia, *Anal Biochem.*, Vol. 385, p. 249, 2009.
64. F.K. Liu, *J. Chin. Chem. Soc.*, Vol. 55, p. 69, 2008.
65. Q.S. Qu, F. Shen, M. Shen, X.Y. Hu, G.J. Yang, C.Y. Wang, C. Yan, Y.K. Zhang, *Anal. Chim. Acta*, Vol. 609, p. 76, 2008.
66. K. Kobayashi, S. Kitagawa, H. Ohtani, *J. Chromatogr. A*, Vol. 1110, p. 95, 2006.
67. F.K. Liu, G.T. Wei, F.C. Cheng, *J. Chin. Chem. Soc.*, Vol. 50, p. 931, 2003.
68. T. O'Mahony, V.P. Owens, J.P. Murrihy, E. Guihen, J.D. Holmes, J.D. Glennon, *J. Chromatogr. A*, Vol. 1004, p. 181, 2003.
69. L. Yang, E. Guihen, J.D. Holmes, M. Loughran, G.P. O'Sullivan, J.D. Glennon, *Anal. Chem.*, Vol. 77, p. 1840, 2005.
70. L. Yang, E. Guihen, J.D. Glennon, *J. Sep. Sci.*, Vol. 28, p. 757, 2005.
71. K.S. Abhijith, M.S. Thakur, *Anal. Methods*, Vol. 4, p. 4250, 2012.
72. S. Xiulan, *Food Control*, Vol. 17, p. 256, 2006.
73. V.A. Tertykh, V.V. Yanishpolskii, O. Yu. Panova, *J. Therm. Anal. Calorim.*, Vol. 62, p. 545, 2000.

8

Recent Developments in Gold Nanomaterial Catalysts for Oxidation Reaction through Green and Sustainable Routes

Biswajit Chowdhury*, Chiranjit Santra, Sandip Mandal and Rawesh Kumar

Department of Applied Chemistry, Indian School of Mines, Dhanbad, India

Abstract

The breakthrough discovery of gold nanoparticles as a catalyst towards various chemical reactions has opened a new horizon in surface science. Among the different reactions, oxidation reactions have demanded special attention as selective oxidation of organic compounds is an important technique for conversion of biomass to useful fine chemical substances. Oxidation by molecular O_2 is a greener route for producing oxygenates, improving the atom economy of the process. It is found that 2–5 nm-sized gold nanoparticles can participate as catalyst for many important reactions, e.g., propylene epoxidation, CO oxidation, alcohol oxidation, glucose oxidation, etc. An extensive study encompasses the fact that the nature of support controls the activity of gold nanoparticles. Among different oxide supports, the mesoporous silicious and non-silicious oxides demonstrated high catalytic activity of gold nanoparticles in the oxidation reaction. During the course of research and development in the liquid-phase reaction, the supports of interest have shifted from activated carbon and metal oxides to organic polymers as a stabilizing agent or a solid support for gold nanoparticles. While investigating the reaction mechanism it is found that gold nanoclusters act as primary active sites as well as secondary sites depending on the nature of support, type of reaction and reaction condition. The development of mesoporous mixed oxide material triggered its use as a support matrix not only because of its thermal stability

**Corresponding author*: biswajit_chem2003@yahoo.com

Ashutosh Tiwari and Mikael Syväjärvi (eds.) Advanced Materials for Agriculture, Food, and Environmental Safety, (197–242) 2014 © Scrivener Publishing LLC

but also exploitation of the redox chemistry and acid-base bifunctionality on the catalyst surface. In this chapter, an attempt has been made to explore the chemical property of supported gold nanoparticles, which might be helpful to apply these advanced materials for physical and biological purposes.

Keywords: Gold nanoparticles, catalyst, green and sustainable methods

8.1 Introduction

In recent years, the science of chemistry has mostly focused on the problems facing the environment. An emphasis on the emerging area of green chemistry is needed in the design and attainment of sustainable development. The international union of pure and applied chemistry has been playing a role in advancing and promoting the continuing emergence and impact of green chemistry. The term green chemistry, as adopted by the Working Party, is defined as: "The invention, design and application of chemical products and processes to reduce or to eliminate the use and generation of hazardous substances." The adoption of green chemistry as one of the primary methods of pollution prevention is a fairly recent phenomenon. Paul Anastas coined the term "green chemistry" in order to focus attention on an area of research and development that was undergoing rapid expansion and that was increasingly characterized by the emergence of distinctive objectives and principles [1]. Despite remarkable scientific achievements and successful industrial applications of several technologies, some processes still produce large amounts of coproduct, which is not desirable from an economical and environmental point of view. Oxidation reaction is an important reaction for the production of huge quantities of ingredients and monomers which are important in diverse areas of human life. Beside this, it has become very clear that the impact of this production on the environment might have been much greater than it is if a considerable effort had not been spent on the continuous improvement of the oxidation technologies—by replacing toxic or dangerous reactants, by developing more efficient systems for heat recovery and for energy integration in the plant, and by minimizing waste streams. To reach the ultimate goal of a sustainable oxidation process, the scientific community has been focusing on oxidation reaction which must be carried out by maintaining the following conditions [2]:

- The reaction must be single-step and/or solvent-free;
- mild reaction conditions must be used, ideally involving benign reactant and benign catalyst;

- the stoichiometric process must be replaced by a catalytic process; and
- durable heterogeneous catalyst of very high activity and desired product selectivity must be designed.

Design of reaction-specific catalyst is a key technology for these aspects. It not only reduces the environmental burden but also increases the reactant conversion and desired product selectivity, which is the main purpose of sustainable chemistry. Different types of industrially demanded chemicals, e.g., ethylene oxide, propylene oxide, aldehyde, ε-caprolactam, phenol, hydrogen peroxide, etc., are produced by oxidation reaction.

In 1857, Michael Faraday delivered the Bakerian Lecture to the Royal Society of London, entitled "Experimental Relations of Gold (and Other Metals) to Light" [3]. He reasoned that there was merit in observing the action of light on metal particles which were small compared to the wavelength of light; gold sprang to mind because "known phenomena appeared to indicate that a mere variation in the size of its particles gave rise to a variety of resultant colours". The physical and chemical properties of solid depend on the type of motions that its electrons may execute, which depends on the space available for their motion, i.e., their surface confinement.

Bulk and nanoparticles of palladium and platinum show many similarities in catalytic activity [4, 5]. However, gold nanoparticles show an exceptional behavior as reported in literature. Prof. Haruta's [6] pioneering discovery of the catalytic activity of gold nanoparticles has opened up new horizons in the chemistry of gold nanoparticles, which is supported by many interesting literature reports [7]. Nowadays, gold nanoparticles are widely used as a catalyst for different types of oxidation reactions in gas phase like propylene epoxidation, ehanol oxidation, and in liquid phase for oxidation of glucose, hydrogenation, dehydrogenation, N-formylation, [8] and many other types of chemical reactions [9–12].

Gold (Au) is the most noble metal of the periodic table; it has the highest electron affinity, with several known compounds that show Au with a negative oxidation number; its Pauling electronegativity has a value of 2.54, which is the largest among all metals [13]. Gold is resistant to any kind of corrosion: Au_2O_3 is the only oxide that has a positive formation enthalpy (ΔH_f° = + 19.3 kJ/mol) [14]. From different literature reviews and experimental observations, the origin of the catalytic activity of gold nanoparticles was considered to be dependent on:

- Quantum size effects;
- the presence of low-coordinated gold sites;

- charge transfer between the support and gold; and
- the perimeter interface between gold and the oxide support.

Supported gold represents a second generation of catalysts for alcohol and carbohydrate oxidation: they show a great improvement in selectivity and stability and thus represent challenging substitutes for platinum group metals. From a commercial point of view, gold is more suitable than Pt, Pd, etc., and involves many organic reactions. The kinetic aspects of gold catalysis are that the turnover frequency is strongly related to the size of metallic gold particles morphology and the metal–support interaction. Considering other fundamental peculiarities, such as biocompatibility, availability and easy recovery, gold appears to be an exciting catalyst for sustainable processes based on the use of clean reagents, particularly O_2, often in aqueous solution or in the absence of solvent, under mild conditions.

8.1.1 Quantum Size Effects

Different literature reviews and experimental results, have observed that the catalytic activity of gold nanoparticles supported on different materials is strongly dependent on the particle size of gold and nature of the support material [12, 15, 16–21] and type of reactions [22, 16–21]. This was indicated by the valence band spectra measured by ultraviolet photoelectron spectroscopy (UPS) [23]. As the gold size diminished, the peak at 2 eV in the UPS spectra disappeared. A similar effect could be observed in the core level X-ray photoelectron spectra, indicated by a shift in the binding energies towards the higher energies. It is important to note that by ascribing the dependence of the catalytic activity on the Au particle size to a higher density of active sites (corner atoms), the turnover frequency (per corner atom) does not solely depend on the particle size; it is the same in both small and large particles. This is fundamentally different from the quantum size effect [24–27], which ascribes the activity of small Au particles to a particle-size-dependent turnover frequency, arising from changes in electronic structure as the particle size decreases. To prepare highly active gold catalysts, it is crucial to minimize the size of the gold particles to diameters >10 nm and to deposit them as hemispherical particles, which maximize the perimeter lengths. Scheme 8.1 shows there are several preparation methods, e.g., impregnation, co-precipitation, deposition-precipitation, etc. Recently, Haruta *et al.* [28, 29] demonstrated solid grinding as an efficient tool for preparation of highly active gold nanoclusters on oxide support.

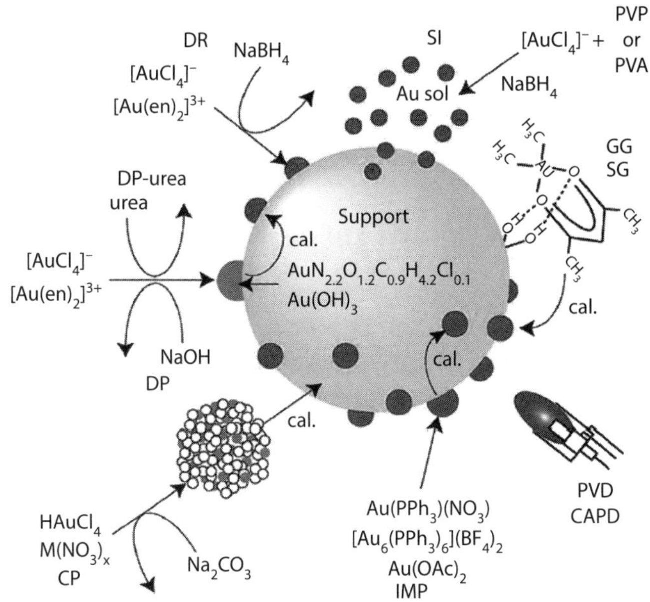

Scheme 8.1 Illustration of various preparation methods for supported gold catalysts (cal., calcinations). Reproduced with permission from [30].

8.1.2 Charge Transfer between Gold and Metal Oxide Support

The interaction between gold nanoparticles and the oxide support is an important phenomenon because it modifies the electronic structure of gold and these gold nanoparticles may play an important role in different types of chemical reaction [10, 31–36]. For example, Campbell [37] and Goodman *et al.* [38, 39] found that metallic gold was indispensable for the genesis of catalytic activity on Au/TiO_2(110) model catalysts, whereas Carrettin *et al.* [40], Schwartz *et al.* [41], Yang *et al.* [42] and Calla *et al.* [43] concluded that metallic gold nanoparticles are responsible for CO oxidation reaction. The cationic gold species in supported gold catalyst are present in the form of AuO_x or AuO(OH) [33, 36, 44, 45–48]. Bond and Thompson [49] proposed that cationic gold naoparticles are responsible for the activation of dioxygen in the catalytic process. The positive charge accumulated on the gold can promote adsorption of some reactants, such as CO and hydrocarbons [48, 50]. In the case of negatively charged gold nanoparticles, an extra electron from the gold readily transfers to the antibonding $2\pi^*$ orbital of the adsorbed O_2, which weakens the O–O bond and activates oxygen molecule for further catalytic reaction

[51, 52]. It was demonstrated that the catalytic activity of gold nanoparticles strongly depends on the type of the support material [32, 53, 54]. Often an electron transfer from the support to the gold cluster promotes its catalytic activity because it opens a route for an extra charge transfer to the adsorbed O_2.

8.1.3 Formation of Reactive Gold–Metal Oxide Perimeter Interfaces

According to Prof. Haruta, the perimeter interface sites between gold nanoparticles and oxide supports act as a reactive site for high surface area gold catalysts [32, 53, 55]. In the case of oxidation reaction by using molecular O_2, the perimeter interface of gold nanoparticles is important. It was assumed that the perimeter interface of gold nanoparticles activate/dissociate the molecular oxygen to give reactive oxygen species for oxidation reaction [56, 57]. In the case of CO oxidation reaction, it is considered that the CO is adsorbed on the gold surfaces, most likely at the edges and corners, and oxygen molecule is activated at the perimeter interfaces [58, 53, 59–60]. The fractions of edges and corners and of perimeter interface changes with a decrease in diameter of gold nanoparticles [55]. However, in the cases of glucose oxidation and alcohol oxidation, small gold nanoclusters play a crucial role.

8.2 Propylene Epoxidation Reaction

The selective addition of oxygen across a carbon-carbon double bond to form an epoxide or oxirane function embodies a class of chemical transformation of great importance, not only in the production of industrial organic chemicals, but also in organic synthesis. Although heterogeneous epoxidation of ethylene over supported silver catalyst has been practiced since 1937, the process is a singularity in several respects. An important characteristic is its use of the simplest oxidant, O_2, either as air or in purified form. However, lack of success for using silver catalyst for higher homologue of ethylene such as propylene could trigger the development of other catalyst systems. At present there are many processes that operate to produce propylene epoxide commercially by different processes. The synthesis of ethane oxide and propene oxide using the chlorohydrin route was first described in 1859 by Wurtz. In this reaction, the alkene reacts with hypochlorous acid (HOCl) to produce the chlorohydrin. The hypochlorous acid is produced *in situ* by an equilibrium reaction of the acid with water and chlorine. The chlorohydrin is thereafter dehydrochlorinated, using aqueous potassium hydroxide to produce the epoxide. This route has long been the main process for producing both ethene oxide and propene oxide. In the 1940s, the process began to be phased out for the ethane epoxidation because of the development of a more-efficient direct epoxidation process using a silver catalyst [61].

Hydroperoxide processes are based on the peroxidation of an alkane to an alkyl hydroperoxide. These alkyl hydroperoxides then react with propene, producing propene oxide and an alcohol. A characteristic of these processes is that, besides propene oxide, a coproduct is produced in a fixed ratio, usually 2–4 times the amount of propene oxide produced. Currently, two variants of this process are applied commercially (Figure 8.1).

However, a disadvantage of the hydroperoxide process is the production, in a fixed ratio, of a coproduct (either styrene or tert-butyl alcohol, depending on which variant of the hydroperoxide process is applied).

The energy required to activate O_2 by splitting it into oxygen atoms is 497 kJ mol-1, even larger than the bonding energy (431 kJ mol^{-1}) of the C–H bond in the most stable hydrocarbon, methane [17, 63]. When such a large amount of energy is put into the system, it is very difficult to control the reactivity of atomically dissociated species of oxygen, and, most likely,

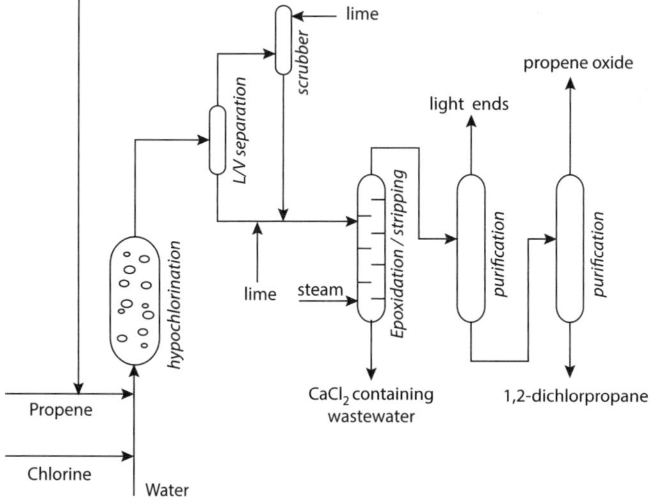

Figure 8.1 Schematic representation of the chlorohydrin process for the production of propene oxide. Reproduced with permission from [62].

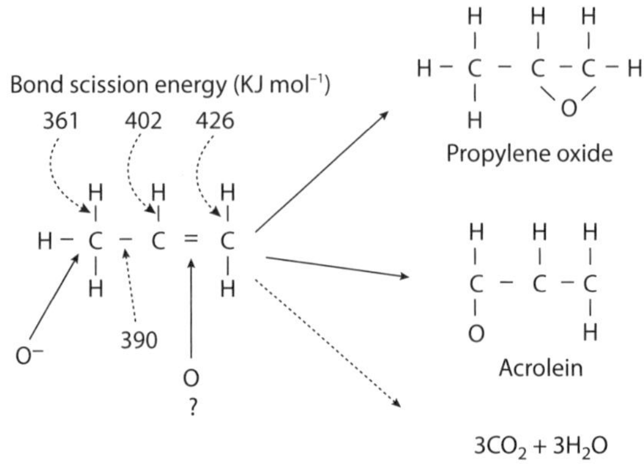

Figure 8.2 Three routes for propylene oxidation and bond scission energies in the propylene molecule. Reproduced with permission from [31].

negatively charged oxygen radicals (O^-) are formed. The O^- radicals will preferentially attack the allylic C–H bond of propene (Figure 8.2) to produce acrolein [64].

The discovery of Haruta *et al.* showing the unique property of gold nanoparticle for generating H_2O_2 from molecular H_2 and O_2, thereby demonstrating synthesis of PO in gas phase, opened a new horizon in catalysis research [65, 66, 67]. A target of 10% conversion of propylene, > 90% selectivity of PO and hydrogen efficiency of 50% was set to develop the catalyst to reach the production of PO at a commercial level (Figure 8.3).

It is observed that the formation of propylene epoxide from propylene over supported gold catalyst depends on the nature of support as well as size and shape of gold nanoparticles. It is found that 2–5 nm-sized gold nanoparticles of hemispherical shape give the highest selectivity for propylene epoxide [32, 68] (Figure 8.4). The effect of the titanium phase on reactivity to PO formation can be pieced together from its coordination environment, the distance between two neighboring titanium atoms (Figure 8.5), hydrophobicity and porosity. Among different supports, anatase TiO_2, TS-1, Ti-MCM-41, Ti-MCM-48, Ti-HMM, Ti-SiO_2, and Ti-TUD-1 are found to be very effective for Au/Ti interactions which are found most effective towards propylene epoxidation reaction.

While scrutinizing the role of titanium phase during propylene epoxidation reaction, it was found that the existence of isolate Ti^{4+} species and porosity of the Ti-containing mesoporous silica plays a crucial role. The importance of the isolated tetrahedral Ti species was emphasized [69] by

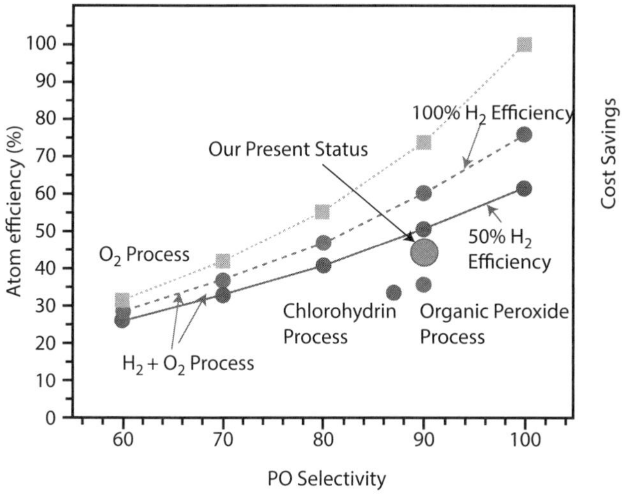

Figure 8.3 Status of propylene epoxide production till 2006 (claimed by Prof. Haruta *et al.* [67]).

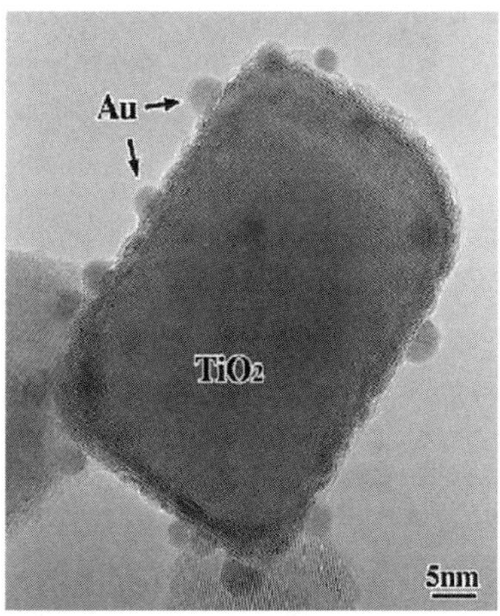

Figure 8.4 HRTEM of Au/TiO_2 catalyst. Reproduced with permission from [32].

Haruta *et al.*, where Au deposited on Ti-doped nonporous silica was used as catalyst (Figure 8.6).

The Au-Ti systems of nanogold particles deposited on anatase TiO_2 possess two inherent weaknesses: faster deactivation and low PO yields (< 2%). It is only selective for PO at temperature below 100°C [70]. To improve the conversion by raising the temperature could produce only PO combustion reaction. The improvement in PO yield by using Ti-MCM-41/48,

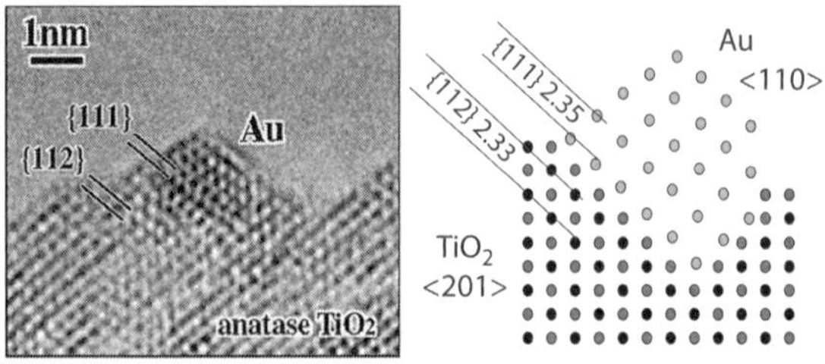

Figure 8.5 (Left) Transmission electron micrograph for Au/TiO2 contact interface and (Right) its schematic diagram. Reproduced with permission from [68].

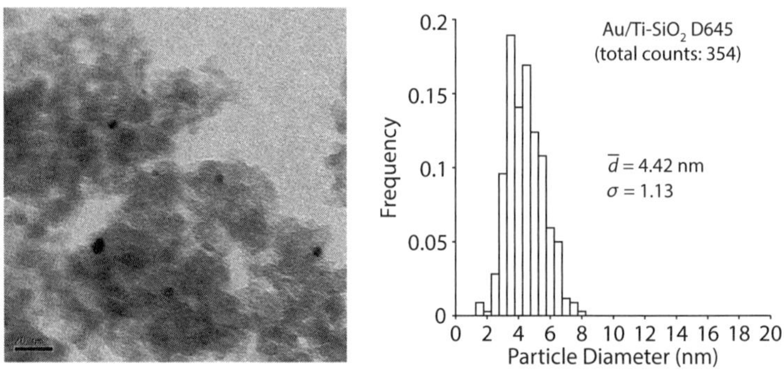

Figure 8.6 HRTEM photograph of Au/Ti-SiO_2 catalyst and particle size distribution of gold. (Obtained from presentation of X. Nijhuis, Eindhoven University of Technology).

Ti-HMM, Ti-SiO_2 was up to 4–5%. It is found that titanium-incorporated conventional mesoporous silica was very effective in producing PO with high selectivity. Among the different parameters, the catalyst preparation (i.e., gold deposition), washing with water and calcination temperature had a significant effect on the activity of the catalysts for propylene epoxidation. The silylated three-dimensional mesoporous titanosilicate material was very promising for producing PO up to 8% yield. This probably suggests that the distance between two neighboring Ti cations in Ti-O-Ti entities may determine the optimal reaction temperature to some extent. The mesoporous titanosilicate discovered by Sinha *et al.* [80] could raise the PO yield up to 9% after silylation and using $Ba(NO_3)_2$ promoter.

The major problem associated with a Ti-incorporated mesoporous silica system is the fast deactivation of the catalyst. Uphade *et al.* [71] extracted the organic compound from used Ti-MCM-48 catalyst and by GC-MS studies and found many oligomarized products in the catalyst surface

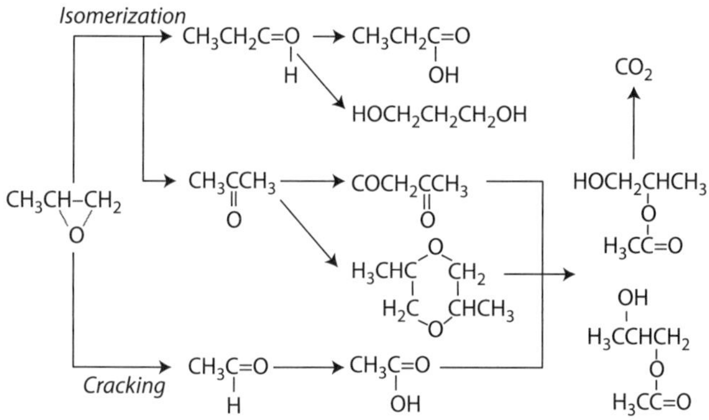

Figure 8.7 Probable side reactions occurring on Au/Ti-MCM-48 catalyst during direct propylene epoxidation reaction. (Obtained from presentation of M. Haruta, Tokyo Metropolitan University).

(Figure 8.7). This might be due to the presence of acidic sites in the catalyst which form oligomerized byproducts, thereby covering the catalyst surface, leading to deactivation. Sinha *et al.* [80] speculated that an increase in hydrophobicity in the catalyst surface might control the adsorption of PO on the catalyst surface, improving the PO yield steady state in time over stream (Figure 8.8). Even though they could improve the catalyst performance with respect to propylene conversion, PO selectivity and hydrogen efficiency, they failed to control the deactivation. Later Chowdhury *et al.* [72] made an attempt to use a small amount of trimethylamine as a gaseous promoter on the promoted mesoporous titanosilicate catalyst and were successful in keeping the catalyst activity steady for up to a 10 hour run. It is found that there is an optimum concentration of trimethylamine in the range of 10–15 ppm present during pretreatment and cofeed to find the best effect on the catalyst performance (Figure 8.9).

Also, they observed that regenerated catalyst gave almost equal activity to that of the fresh catalyst, which was quite interesting for recycling catalyst for successive runs (Figure 8.10). Attempts to improve the PO yield by changing the reaction parameters like pretreatment temperature and condition, reaction temperature, space velocity and feed composition are well documented in the literature. They found that catalyst pretreatment at 250°C in the presence of molecular H_2 and O_2, reaction temperature at 150°C, space velocity of 4000 h^{-1} and Ar:Py:H_2:O_2 (70:10:10:10) is effective.

Based on the experimental evidence described above, TS-1, which is a naturally hydrophobic microporous material, and supported nanogold catalysts are supposed to give the best results. Recent results from the

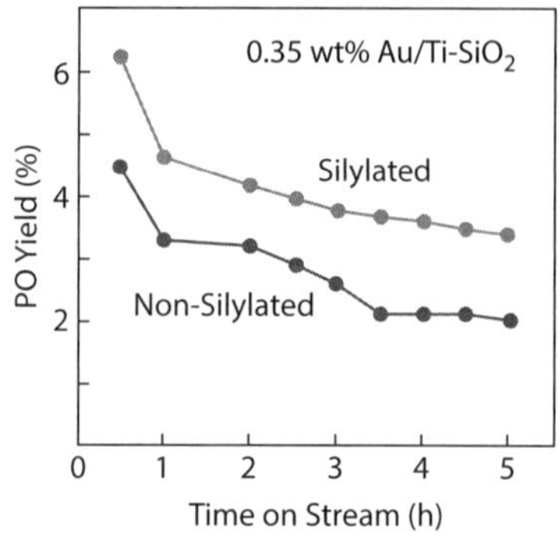

Figure 8.8 Improved performance of Au-Ti-SiO_2 catalysts during direct propylene epoxidation reaction.

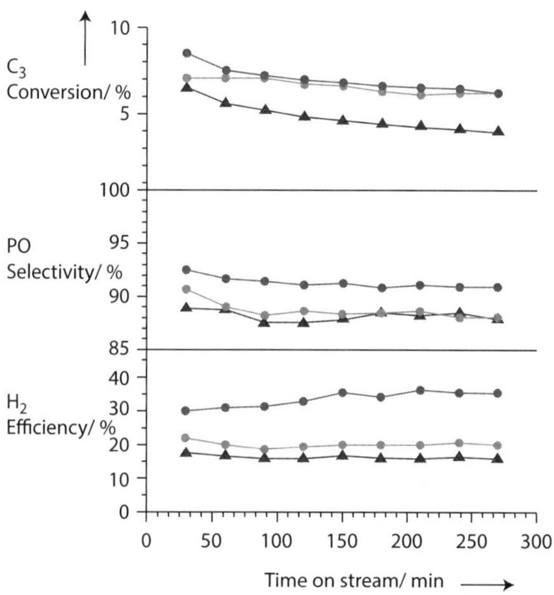

Figure 8.9 Controlled deactivation of Au/Ti-SiO_2 catalysts in the presence of trimethylamine in the reaction feed. Reproduced with permission from [67].

Prof. Delgass group indeed demonstrated that these are the most active Au/TS-1 catalysts reported, where a stable PO production is observed at temperatures ranging from 140–200°C [73–74, 75].

Another important point to be highlighted is the hydrogen efficiency during direct propylene epoxidation using molecular H_2 and O_2 over

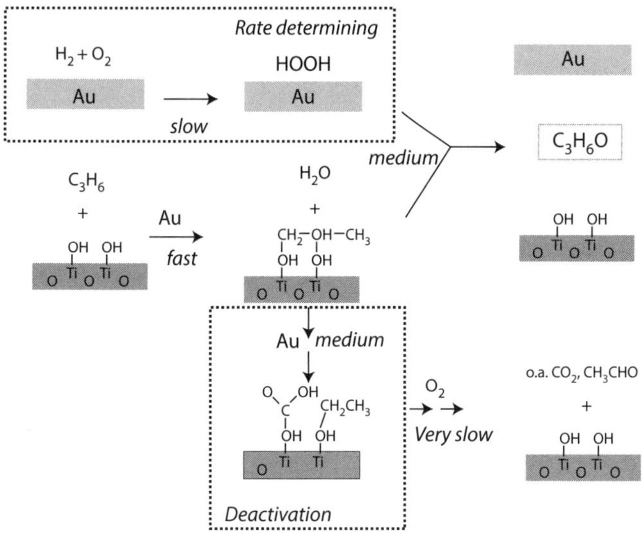

Figure 8.10 Schematic representation of the model of reaction and deactivation for the propene epoxidation over gold-titania catalysts. Reproduced with permission from [76].

supported gold nanoparticle catalyst. The hydrogen efficiency is defined as "moles of PO formed/moles of Hydrogen converted." The higher hydrogen efficiency demands more attention as it shows the green route for propylene epoxidation reaction. A target for 50% hydrogen efficiency was set for commercialization of the PO production process. Uphade *et al.* used CsCl promoter in the Au/Ti-MCM-48 catalyst to suppress the hydrogen combustion for water formation (Figures 8.11 and 8.12).

It is found by Chowdhury *et al.* that a small amount of trimethylamine can cover some small portion of bare gold surface, thereby improving the hydrogen efficiency. It has been investigated that the local heat produced in the catalyst bed is responsible for hydrogen combustion reaction. So use of thermal conducting material like silicon carbide would be very effective for decreasing hydrogen combustion reaction.

Though it is a dream path, the synthesis of PO by direct epoxidation of propylene using molecular O_2 is a green and sustainable route; however, until recently the initiated work has not been successful. Quite recently, a breakthrough discovery by Haruta *et al.* using gold nanocluster catalyst in the size range of less than 2 nm was able to produce propylene epoxide in the presence of molecular O_2 [78]. It was found that the deposition-precipitation method was more effective in preparing gold nanoparticles of 2–5 nm size. However. the gold nanocluster

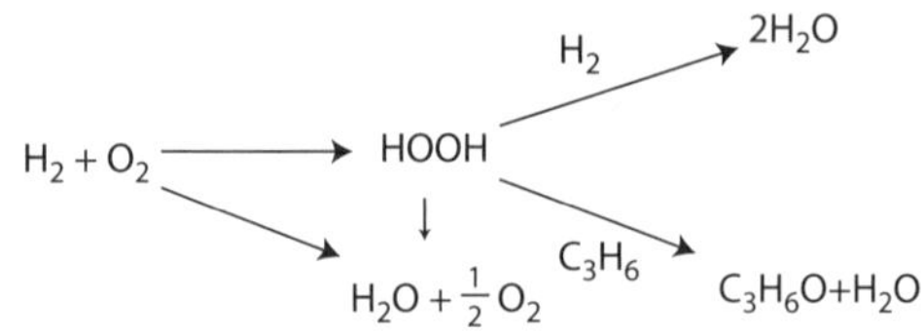

Figure 8.11 Routes for water formation in propene epoxidation using hydrogen and oxygen [77].

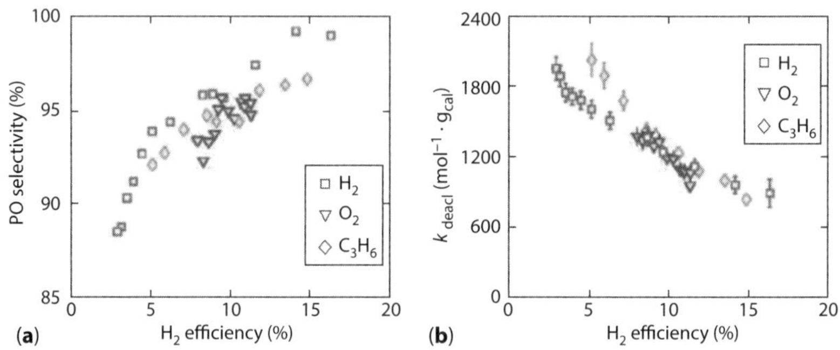

Figure 8.12 Selectivity towards propene oxide (a) and deactivation rate constant (b) as a function of hydrogen efficiency at different reactant concentrations (1 wt% Au on Ti-SiO_2, 403K, GHSV 10000 mL $g_{cat}^{-1}h^{-1}$; the concentration of one reactant is varied while the other two are fixed at 10 vol%) [77].

catalyst prepared by solid-state grinding method was very much effective in dispersing gold nanoclusters on a variety of supports, irrespective of the isoelectric pH of the oxides (Figure 8.13). However, since there is still controversy about whether deposition-precipitation or solid-state grinding is the best method to prepare gold nanocluster, further research is needed.

The usage of a microreactor system for the direct epoxidation of propene over a gold-titania-based catalyst system using a mixture of hydrogen, oxygen and propene allows for the safe operation of the reaction in the explosive regime [73, 75–76]. The propene concentration does not influence the propene oxide formation rate; however, higher propene concentrations significantly reduce the catalyst deactivation rate. Hydrogen increases the rate of the epoxidation reaction, while it only has a minor influence on the rate of deactivation and reactivation. Oxygen has a beneficial effect on the epoxidation reaction, slightly decreases the deactivation rate, and is

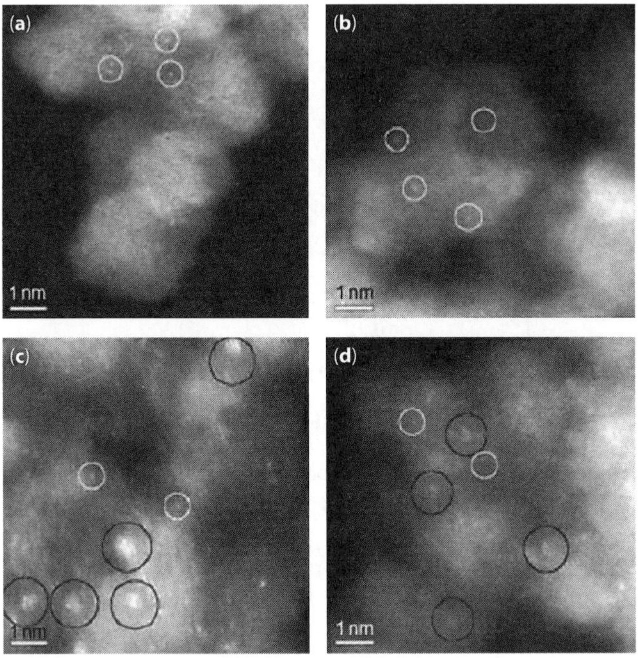

Figure 8.13 Aberration-corrected HAADF-STEM images of Au atoms and gold clusters in (A and B) the inactive and (C and D) the active Au/FeO_x catalysts; the images were acquired by aberration-corrected STEM. Reproduced with permission from [79].

beneficial for the catalyst reactivation. It can be concluded that it is highly beneficial to perform the direct propene epoxidation in a microreactor system at the highest possible feed concentrations for each of the reactants, as this results in the highest propene oxide productivity and the lowest deactivation rate.

8.3 Reaction Mechanism

Several studies have been done to elucidate the reaction mechanism of direct propylene epoxidation over titanosilicate supported gold nanoparticle catalyst (Figure 8.14). Formation of hydroperoxo species seems to be indispensable for producing PO by direct epoxidation of propylene in the presence of H_2 and O_2 over gold surface, as revealed by the results of Haruta *et al.* [80]. For many years the formation, characterization and reactivity of

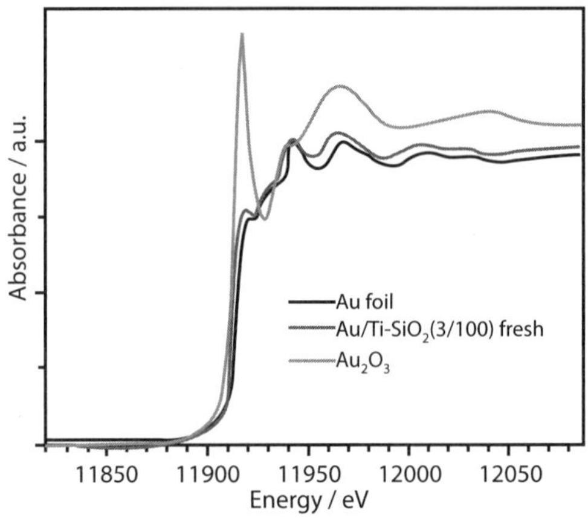

Figure 8.14 XANES spectra of Au L_3-edge spectra of the Au/Ti-SiO_2 catalyst.

these hydroperoxo intermediates have been studied in depth by a plethora of experimental techniques [81–84]. Recently, spectroscopic studies have been found in the literature showing the formation of Ti-peroxo/hydroperoxide species for Pd supported over different types of Ti-containing zeolites [85], and also for Ti-SBA-15 material [86] in the presence of liquid H_2O_2. It has also been found that organic hydroperoxides in liquid phase form a peroxo complex analogous to the hydroperoxo species formed by hydrogen peroxide [36].

Based on experimental results and literature reports a mechanistic model for the formation of superoxo/hydroperoxo species on gold surface is shown in Figure 8.15. The potential ability of Au to form peroxo-type species has been postulated [87] and experimentally verified [88–91]. As per this model, formation of hydrogen peroxide from H_2 and O_2 is initiated on gold surface.

Once H_2O_2 is generated it can be transferred into hydroperoxo species on tetrahedrally coordinated Ti cation sites, and then react with propylene adsorbed on SiO_2 surfaces to form PO. The Ti-hydroperoxo species, which is UV active, and Ti-OO$^-$ species, which is ESR active, can be detected directly as reported by Chowdhury *et al.* [72, 92] and shown in Figure 8.16. The presence of both the species in large amounts in the presence of gold emphasizes the vital role of gold on Ti-hydroperoxo species formation, which is indispensable for production of PO from propylene (Figure 8.17). The work by Bravo *et al.* [93] shows the formation of Ti-hydroperoxo species by *in-situ* XANES spectroscopy.

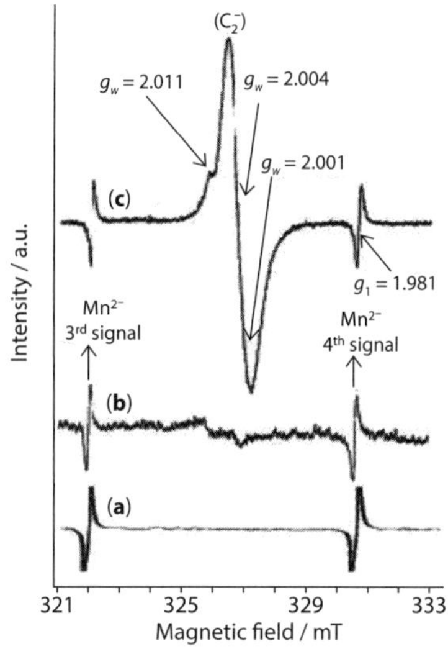

Figure 8.15 EPR spectra showing the role of gold on the formation of paramagnetic peroxo species on Au/Ti-SiO_2 catalyst surface: (a) Au/ Ti-SiO_2 fresh sample before propylene epoxidation, (b) Ti-SiO_2 after propylene epoxidation under $C_3H_6/O_2/H_2/$ Ar (1/1/1/7 in volume) for 270 min, (c) Au/Ti-SiO_2 after propylene epoxidation under $C_3H_6/O_2/$ H_2/Ar (1/1/1/7 in volume) for 270 min. Mn^{2+} is used as an internal standard. Reproduced with permission from [92].

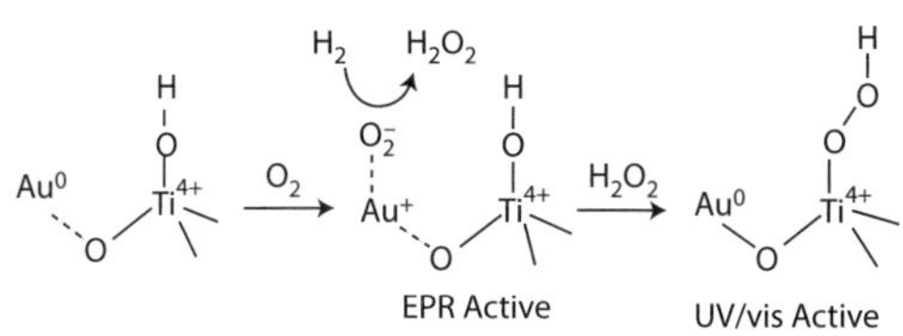

Figure 8.16 Possible mechanistic model for the formation of hydroperoxo species on Au/ titanosilicate materials. Reproduced with permission from [92].

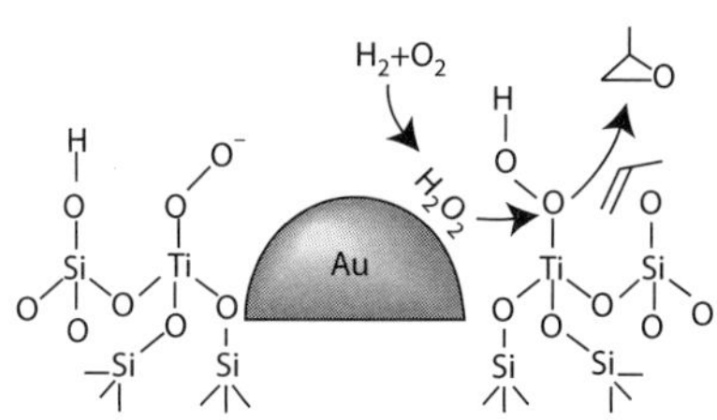

Figure 8.17 Probable reaction mechanisms for propylene epoxidation.

8.4 Glucose Oxidation

Selective oxidation of natural resources is a task of key importance for producing oxygenates to be employed as building blocks in chemical processes that range from kilogram-scale applications in pharmaceuticals to thousands of tone-scales in chemicals. Oxidation of glucose to gluconic acid with heterogeneous catalyst is a typical example of these green chemistry principles (Scheme 8.2). Gluconic acid, used as an intermediate in the food and pharmaceutical industry and is a biodegradable chelating agent, is produced by biochemical oxidation of glucose as the only industrial route [94].

The worldwide availability of carbohydrates has for a long time attracted interest in their oxidative transformation. According to the expected trend, catalytic aldose oxidation occurs at the aldehydic group, leading to carboxylic acid or carboxylate. In this area, efforts have been mainly devoted to the improvement of the catalyst efficiency and process operation. The great interest in the valorization of renewable biological resources has led to extensive investigation on the catalytic oxidation of glucose and other carbohydrates since 2002. Free gluconic acid is presently manufactured starting from calcium gluconate and sulfuric acid, thus producing a large amount of $CaSO_4$ as a byproduct. This is due to the fact that at low pH values the enzymatic catalysis is inhibited. Applying catalytic methods, it has been found that Pd, Pt, and Au catalysts are scarcely active (TOFs 2, 60 and 50 h^{-1}, respectively) in the presence of free gluconic acid [95].

There have been several reports on supported palladium or platinum catalysts which have intensively investigated the aerobic oxidation of glucose. However, the major drawbacks to these catalysts are that they have low catalyst durability and relatively low selectivity [96]. Figure 8.18 shows the conversion-time plot; naked gold particles having a mean diameter of 3.6 nm behave as an active catalyst allowing 21% glucose conversion in the first 200 s. Under similar conditions, Cu, Ag, Pd, and Pt colloidal particles of similar dimension (3–5 nm) were scarcely active [97]. Huiming *et al.* reported that unsupported nanoporous gold (NPG) catalysts exhibited effective catalytic activity and high gluconic acid selectivity (~99%) [98].

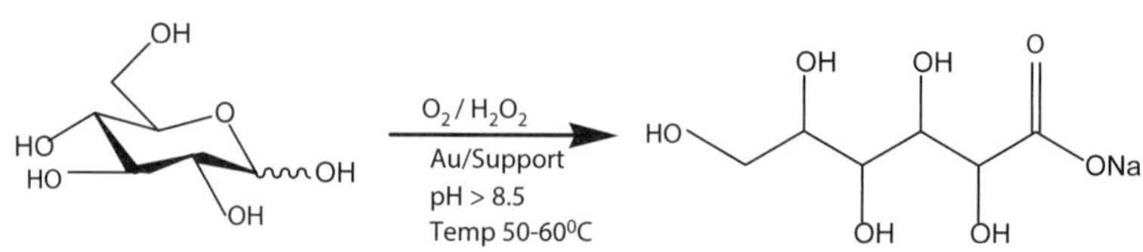

Scheme 8.2 Catalytic oxidation of glucose to gluconic acid.

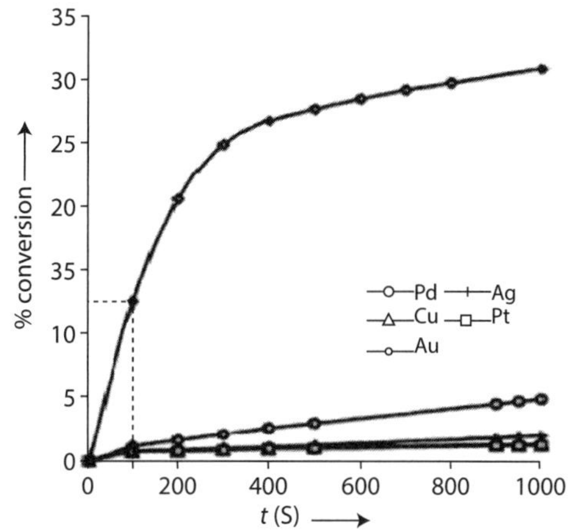

Figure 8.18 The activity of different metal particles in glucose oxidation. [Metal] = 10^{-4} M, [glucose] = 0.4 M, T = 303 K. Reproduced with permission from [98].

Baatz *et al.* demonstrate that the preparation of gold catalyst (size < 2 nm) for liquid phase glucose oxidation by the incipient wetness method has some advantages over the DP urea method, even at high gold loading [99]. A traditional method was optimized by Pruesse *et al.* [100] for preparing an efficient catalyst for glucose oxidation, based on the impregnation of the active component onto the support (0.3% Au/Al_2O_3).

Early reports using heterogeneous catalysts demonstrated the use of platinum on carbon as a catalyst for selective oxidation of glucose reaction; however, it was found that these catalysts suffered from deactivations, which were decreased when the reaction was carried out at high pH, but catalyst deactivation could not eliminated. Maintaining the pH at 9 eliminated the deactivation, suggesting that the deactivation observed could be due to a poisoning effect from competitive adsorption of one of the reaction components. However, the use of bimetallic catalysts for the glucose oxidation reaction was first reported by Besson *et al.* They used Bi-Pd supported on carbon and got the conversion close to 100%. Wenkin *et al.* carried out an extensive study into the preparation of carbon-supported bismuth-promoted palladium catalysts for glucose oxidation, but they found in each case that leaching of bismuth into the reaction solution was detected for all the catalysts tested.

Biella *et al.* proposed the use of gold supported on carbon prepared by the sol-immobilization method for this oxidation. It was found that gold catalysts were more active than platinum, palladium-bismuth or platinum-palladium-bismuth trimetallic catalysts. The gold was also

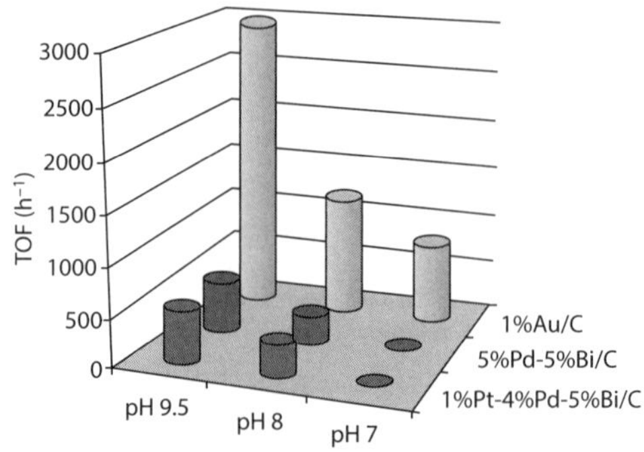

Figure 8.19 Catalytic performance of different catalysts in glucose oxidation. Reproduced with permission from [95].

reported to maintain activity at other pH values than the optimized pH 9.5 (Figure 8.19). Upon recycling, the gold catalyst did display a reduction in its activity, and this was considered to be due to metal leaching.

In a further study by the same group, "naked" gold nanoparticles, formed by the colloidal method, were used for the oxidation of glucose. Comotti *et al.* investigated the combination of gold, platinum, palladium and rhodium as bimetallic catalyst (Table 8.1). They observed that the gold platinum catalysts were the most active [101]. At higher pH 9.5, bimetallic colloidal particles appeared more stable towards agglomeration than monometallic gold particles, allowing higher conversions [102].

One of the most important points of heterogeneous catalysis is the so-called metal-support interaction. The size of nanoparticles was directed much more by the nature of support and also the preparation technique followed. Many studies have revealed that the catalytic properties of noble metal nanocrystals strongly depend on their size, shape, composition and structure [103, 104]. Pina *et al.* [95] reported that as the catalytic conversion is related to the nanoparticles average diameter, there will be a correlation between activity and particle size, expressed by the simple model $v = k/r$, where "v" means the catalyst controlled rate and "r" is the nanoparticle radius. This model follows the kinetic tests with a constant mass of metal of a given density and uniform distribution of nanoparticles having spherical shape. To verify their proposed model, they prepared a series of colloidal particles of radius 3–10 nm by varying the gold precursor concentration from 50 to 600 mg l^{-1}. The glucose oxidation reactions were tested using these catalysts following the standard experimental conditions {[glucose] = 0.38 M; [Au] = 3.2×10^{-5} M: glucose : Au = 12 000, T = 303 K, pH = 9.5, pO_2 = 1 bar}. Figure 8.20 shows catalytic activity inversely proportional to

Table 8.1 Oxidation with mono- and bimetallic catalysts at pH 9.5. Glucose/Au = 20,000; T = 323 K; reaction time = 2.5 h. Reproduced with permission from [102].

Metals	Type of catalyst	Conversion %	TOF (h^{-1})
Au	Supported	43	17200
Au–Pt (2:1)	Supported	44	17600
Au	UnSupported	12	4600
Pt	UnSupported	<2	<500
Pd	UnSupported	5	2000
Au–Pt (2:1)	UnSupported	26	10500
Au–Pd (2:1)	UnSupported	28	11600

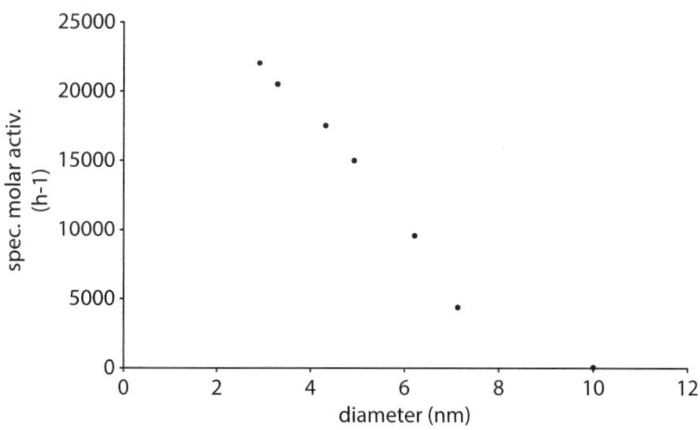

Figure 8.20 Correlation between activity and particle size of unsupported gold nanoparticles during the aerobic oxidation of glucose. Experimental conditions: [glucose] = 0.38 M; [Au] = 3.2×10^{-5} M; glucose : Au = 12 000, T = 303 K, pH = 9.5, pO_2 = 1 bar. Reproduced with permission from [105].

the diameter of the nanoparticles in the range of 2.5–4.5 nm but a deviation from linearity for particles larger than 6 nm, as well as the inactiveness for those larger than 10 nm.

The rapid decrease of activity for particles larger than 6 nm indicates a discontinuity in the catalytic behavior, which is not only due to geometrical factors, but likely due to electronic factors influencing the electrophilic properties of the particles. By the way, other properties show a sudden

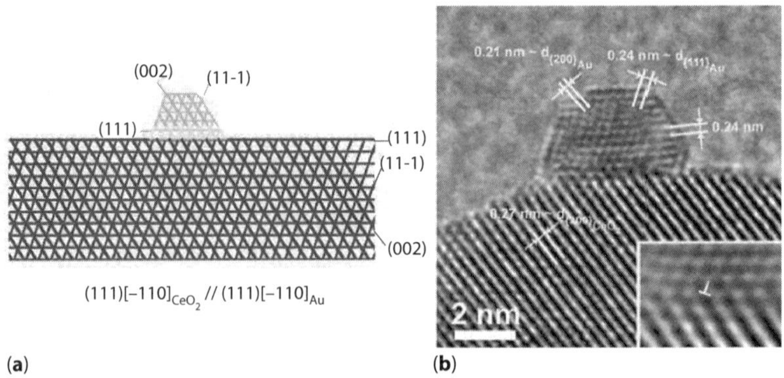

Figure 8.21 HRTEM micrographs of gold islands deposited on CeO_2 nanoparticles following the deposition-precipitation and coprecipitation methods respectively. Dislocation networks are highlighted in the insets. Reproduced with permission from [109].

discontinuity in the region close to the metal-nonmetal transitions in gold, as the melting point collapses.

The dependency of catalytic activity on noble metal catalysis mainly originates from the different crystal planes exposed on the nanocrystal surface (Figure 8.21). There are two types of crystal planes: (i) high-index facets, low catalytic activity during catalytic cycle due to high surface energy, leading to possible structural instability [106], and (ii) low-index facets, including {111}, {100} and {110} planes with both high activity and good stability [107, 108].

Very recently, Wang *et al.* [110] reported the shape-dependent catalytic activities of Au nanocrystals toward electrooxidation of glucose in alkaline medium. They found that {100}-bounded cubic Au nanocrystals are significantly more active than the {110}-bounded rhombic dodecahedral and {111}-bounded octahedral Au nanocrystals.

It is generally accepted that AuOH sites on the Au surfaces act as the active species for glucose oxidation (11) and the oxidation of glucose is strongly dependent on the number of AuOH sites. Among the three types of Au nanocrystals (Figure 8.22, left), the cubic ones have the highest current densities at all three oxidation peaks (Figure 8.23, right). The current densities at -0.40 V (oxidation of glucose to gluconolactone and a very limited number of AuOH sites on the nanocrystal surfaces due to the reductive nature of the electrodes) for the cubic Au nanocrystals are 6.5 and 2.6 times those of the octahedral and rhombic dodecahedra Au nanocrystals, respectively. A similar comparison shows this to be 1.5 and 1.7 times higher at 0.25 V (further oxidation of gluconolactone and the population

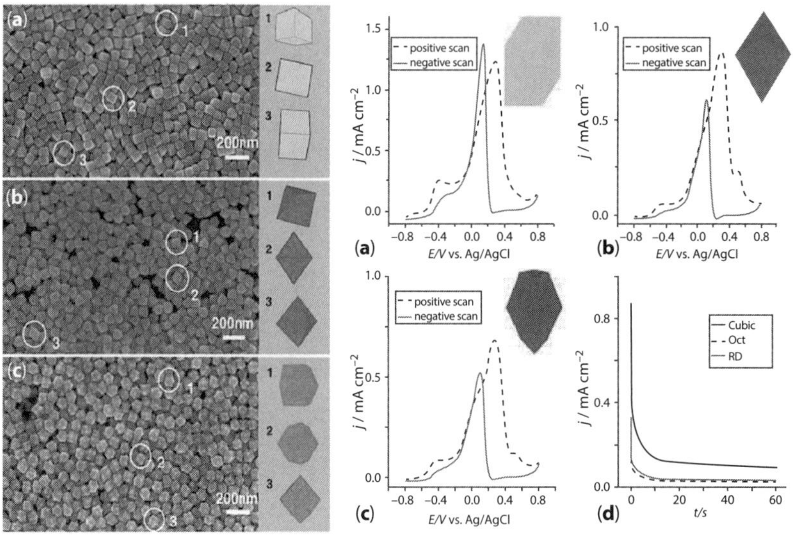

Figure 8.22 (left) SEM images of the differently shaped Au polyhedra: (a) cubes, (b) octahedra, (c) rhombic dodecahedra, and (right) Cyclic voltammograms of the different Au polyhedra on glassy carbon electrodes in a 0.1 M NaOH solution of 10 mM glucose at a scan rate of 50 mV s1: (a) cubes, (b) octahedra, (c) dodecahedra. Dash lines represent positive scanning and gray lines represent negative scanning. (d) Chronoamperometric curves of differently shaped Au polyhedra toward glucose oxidation at 0.4 V vs Ag/AgCl. Reproduced with permission from [110].

of the AuOH sites on the electrode surfaces increases) and 2.2 and 2.6 times higher at 0.10 V (reduction of surface gold oxide and, thus, enough surface active sites are available for the direct oxidation of glucose occurs). Armed with both CV results, we can conclude that the electrocatalytic activities of the different Au polyhedra toward glucose oxidation are as follows: cubes > rhombic dodecahedra ≈ octahedra.

Recently, Haruta *et al.* [111] developed a new preparation method, solid grinding (SG), by using a volatile organogold complex, [Me_2Au(acac)] (acac = acetylacetonate), and succeeded in depositing small (2–4 nm) Au clusters onto several kinds of metal oxides and carbon supports by this SG method, and compared it with the conventional deposition-precipitation (DP) method. This study reflects the importance of support and preparation method for oxidation of glucose having a similar range of nanoparticles size (Figure 8.23).

The highest catalytic activity was observed for Au/ZrO_2 (SG), with a turnover frequency of 45 $mol_{glucose} mol_{surface\ Au}^{-1}\ s^{-1}$ at 50°C and pH 9.0. The catalytic activity of Au/Al_2O_3 was comparable to that of Au/ZrO_2 (DP) and slightly higher than that of Au/carbon at 70°C. However, Au/Al_2O_3

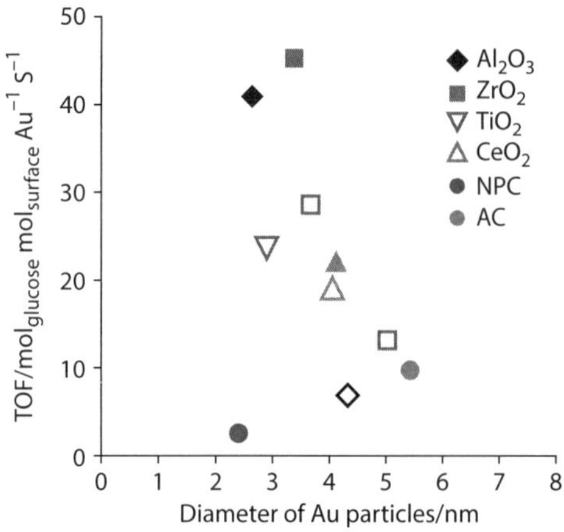

Figure 8.23 Turnover frequencies for glucose oxidation (glucose reacted (mol) per number of surface Au atoms (mol) per second) versus the diameter of the Au particles. Filled and open symbols indicate samples prepared by SG and DP, respectively. Reaction conditions: 5 wt% glucose solution (175 mL), Au catalyst (30 mg), O_2 (120 mLmin^{-1}), 50°C, pH 9.0. Reproduced with permission from [111].

maintained a higher catalytic activity than that of Au/ZrO_2 (DP) at lower temperature. The apparent activation energies were calculated to be 27, 53, 72 and 96 kJmol^{-1} for Au supported on Al_2O_3 (SG), ZrO_2 (DP), NPC (SG) and AC (SG), respectively. The activation energies for glucose oxidation differed according to the nature of the support and tended to be lower for Au on metal oxides, as the activation energy over unsupported Au colloids was reported to be 47 kJ mol^{-1}.

Further investigation was carried out by Baatz *et al.* [112] on the influence of the calcinations temperature on the catalytic activity of glucose oxidation over gold nanoparticles supported on Al_2O_3. Therefore, they investigated with two different catalysts (0.36 wt% Au/Al_2O_3 and 0.60 wt% Au/Al_2O_3) which were calcined at temperatures between 150 and 220°C and between 150 and 300°C, respectively. Figure 8.24 shows the maximum activity depending on the calcination temperature at about 200°C in the case of the catalyst with 0.36 wt% Au content, and a broader maximum between 200 and 250°C for the catalyst with 0.60 wt% gold content. For both catalysts, a significantly lower activity was observed when they were prepared at calcination temperatures below 200°C. This is caused by an incomplete reduction of the catalysts. The decrease in activity at temperatures higher than the optimum temperature could be caused by a sintering of small gold particles to larger agglomerates. Due to these results, 200°C was chosen as the optimum calcination temperature.

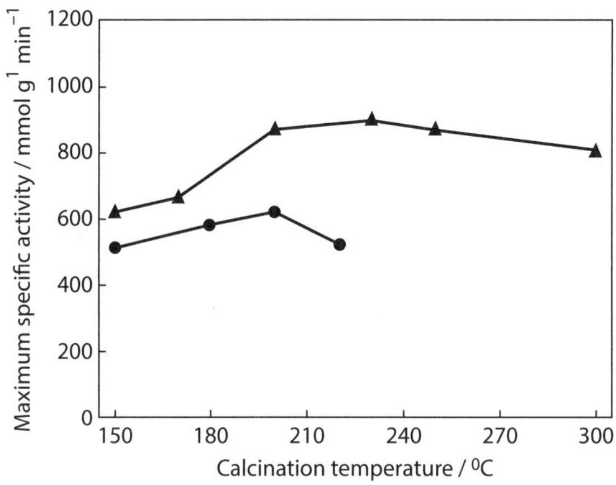

Figure 8.24 Influence of the calcination temperature on the catalytic activity of gold catalysts prepared by DP NaOH (▲ – 0.60 wt% Au; • – 0.33 wt% Au). Reproduced with permission from [111].

Comotti *et al.* [97] reported that during the catalytic oxidation of glucose (Figure 8.25), gold coagulated into larger particles owing to the formation of sodium gluconate. This is very common with other electrolytes; promoted sol coagulation leaves a colorless, inactive solution after about 400 s. The growth of gold crystallites during the reaction has been followed by X-ray diffraction (XRD) analysis at various time intervals after sol immobilization on carbon.

In the time over stream observation of glucose oxidation, the growth of gold nanoparticles would decrease the catalytic performance, as particle size > 6–7 nm is not very effective for glucose oxidation reaction.

The catalytic conversion of glucose, or more generally, the rate of reaction not only depends on the nature of support and nanoparticle morphology but also on the reaction parameters, e.g., catalyst amount, oxygen partial pressure, dissolved oxygen concentration, etc. A recent investigation by Prüße *et al.* [113] shows the influence of the apparent reaction rate on the applied catalyst amount at reaction conditions of 20 wt% initial glucose concentration and 9 bar oxygen partial pressure. Figure 8.26 indicates that oxygen mass transfer does not limit the apparent reaction rate up to a catalyst amount of at least 1200 mg under these conditions.

The curves in Figure 8.27 clearly show the positive effect of a higher oxygen partial pressure on the catalytic activity. Up to a five-fold increase in activity can be achieved by raising the oxygen partial pressure from 1.5 to 9 bar. For each partial pressure a maximum curve is obtained in dependence

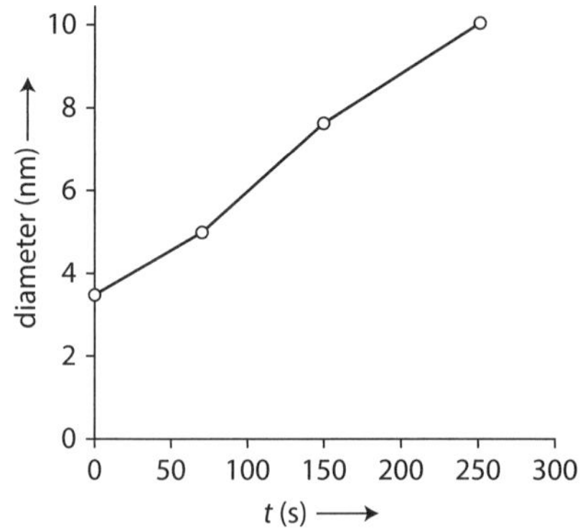

Figure 8.25 Growth of gold particles during glucose oxidation. Reproduced with permission from [97].

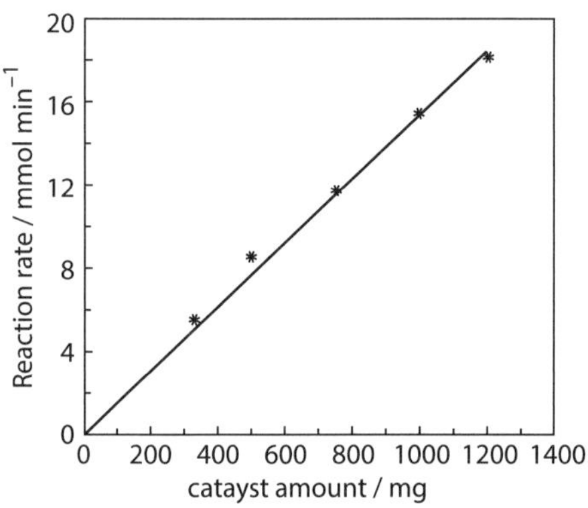

Figure 8.26 Reaction rate vs catalyst amount plot for glucose oxidation. Other reaction conditions: $c_{0,glucose}$ = 20 wt%, 9 bar oxygen, 40°C, pH 9. Reproduced with permission from [114].

of the initial glucose concentration. The maximum activity is obtained at the highest applied oxygen pressures and 20–30 wt% initial glucose concentration. However, the resulting space–time yield of the gold-catalyzed process at those concentrations is considerably different from the fermentation process. At 20 wt% initial glucose concentration and 9 bar oxygen pressure, a maximum space–time yield of above 200 $gl^{-1}h^{-1}$ was achieved with the gold catalyst, which is much higher than the industrial fermentation process, maximum space–time yields of 13 $gl^{-1}h^{-1}$.

Further study was done to understand the role of dissolved oxygen on catalytic conversion. Figure 8.28 shows the activity in dependence of

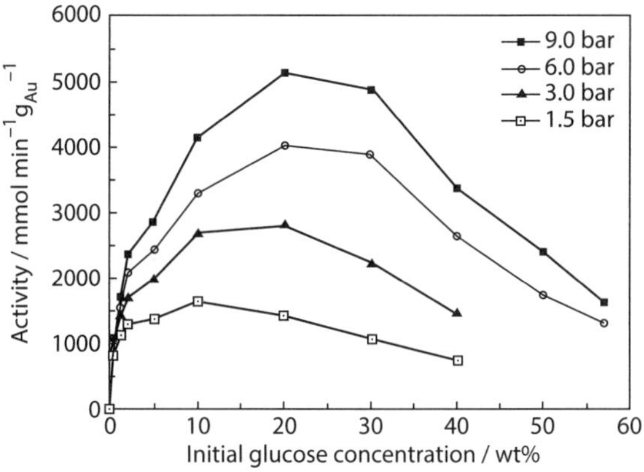

Figure 8.27 Catalyst activity vs initial glucose concentration at 4 different oxygen partial pressures. Other reaction conditions: 1000 mg catalyst, 40°C, pH 9. Reproduced with permission from [114].

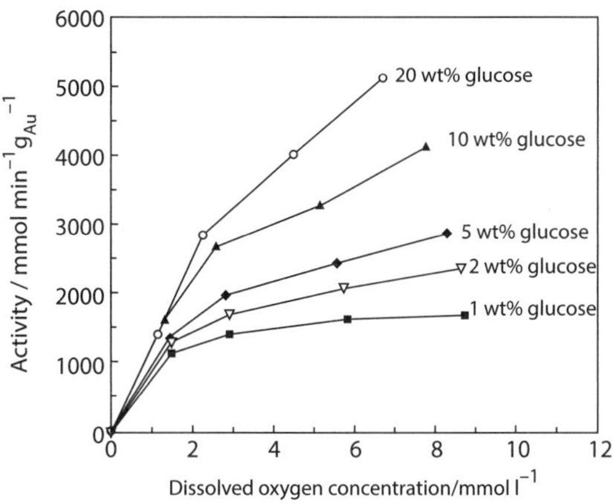

Figure 8.28 Catalyst activity vs calculated dissolved oxygen concentration at different initial glucose concentrations. Other reaction conditions: 1000 mg catalyst, 40°C, pH 9. Reproduced with permission from [114].

the calculated dissolved oxygen concentrations at various initial glucose concentrations. A positive effect of an increased dissolved oxygen concentration on the activity can be noted. The larger the initial glucose concentration is, the higher the activity increase. It was found that by increasing the glucose concentration up to 20 wt% the oxygen reaction order slightly fluctuated (rises), but at high concentration of 30–40 wt% the oxygen

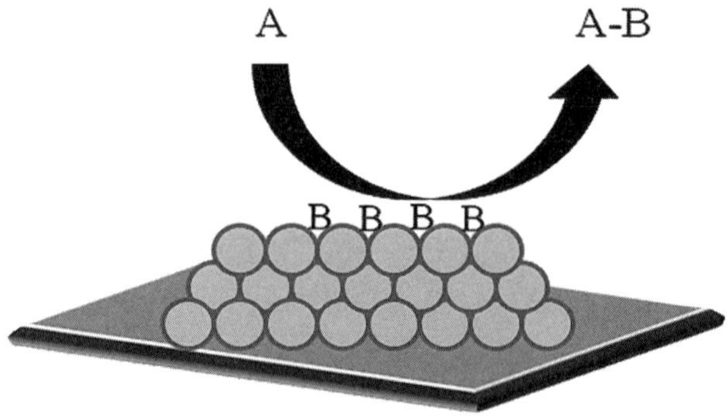

Scheme 8.3 Eley-Rideal mechanism

reaction order remains constant at a value near 0.9. They conclude that with rising glucose concentrations, glucose successfully competes with oxygen for adsorption sites, leading to lower oxygen coverage and, thus, to a higher reaction order for oxygen.

To know the reaction mechanism we have to analyze the stepwise pathway of the reaction. From the literature it was found that Pina *et al.* [95] try to explain the kinetic mechanism of "naked" gold particles as a catalyst for glucose oxidation. They found that low glucose concentrations (< 0.1 M) show a first order reaction; the assumed higher concentration is 0.5 M. The reaction follows the Eley-Rideal mechanism [115] (Scheme 8.3).

This mechanism gives rise to a rate equation that justifies both the first order with respect to oxygen and the decreasing order with respect to glucose, being first order at low concentration and tending to zero for large values.

A similar observation was reported by Comotti *et al.* [97]. Kinetic tests were carried out employing unsupported colloidal gold particles as "naked" particles; it was found that the fundamental detection of hydrogen peroxide, instead of water, was the reduction product of dioxygen. From Scheme 8.4 it can be seen that the rate determining step of the gold-catalyzed glucose oxidation reaction is the adsorbed glucose which is oxidized by dioxygen dissolved in water, according to a first order dependence of the reaction ratevon pO_2.

The fundamental step is the formation of the electron-rich gold species derived by the hydrated glucose anion with gold. This species is supposed to activate oxygen by nucleophilic attack, and the derived dioxo-gold intermediate can then behave as a bridge for the two electrons transfer from glucose to dioxygen.

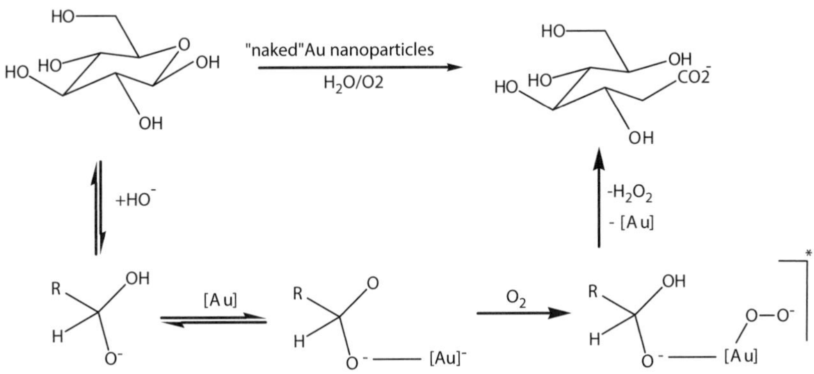

Scheme 8.4 Probable reaction pathway of liquid-phase glucose oxidation reaction.

Considering the reaction mechanism of the gold-catalyzed oxidation of glucose discussed in the above section, the following points are noticeable:

1. Monometallic gold particles are active in the presence of alkali.
2. Unsupported gold nanoparticles have a similar mechanism to that of supported catalyst, however, the supported gold catalyst is stable for a longer time than the "naked" particles.
3. Hydrogen peroxide instead of water is the byproduct of gluconate.

8.5 Alcohol Oxidation

Selective oxidation of alcohol to aldehyde or ketone is an important chemical transformation in the fine chemical industry. The worldwide annual production of carbonyl compound is above 10 million tons [116]. According to Global Industry Analysts, 1.26 million tons of aldehyde were produced in the year 2012. Since application of stoichiometric inorganic and organic oxidizing agents such as chromium (VI), permanganate, peracid, etc., is not acceptable any more from an environmental and economic point of view, there has been a growing demand for alternative green oxidation technology for alcohol oxidation reaction. Molecular O_2 plays an important role as an oxidizing agent in this respect. Its advantages are ready availability and low cost; and O_2 is an eco-friendly gas that produces water as a side product compared with inorganic oxidant which produces toxic waste.

Application of solid catalysts for the vapor-phase oxidation of simple, small-chain alcohols to the corresponding carbonyl compounds is well established [117–119]. The reasonable volatility and thermal stability of reactant and product make it suitable for gas-phase reaction. But in the case of complex molecule synthesis, gas-phase reaction is not suitable because the molecules do not have reasonable volatility and thermal stability. Solid catalysts active in the liquid phase under mild conditions have a much broader application range [120, 121]. A major challenge in liquid-phase oxidation with solid catalysts is to prevent leaching of the active species [122]. Supported noble metal catalysts are active for oxidation of different types of alcohol and polyols. Platinum and palladium metals are most commonly used for alcohol oxidation reaction under mild conditions using molecular O_2 as an oxidant. Different types of promoters such as Bi, Pb, Cd, Co, Cu, Se, etc., are used for alcohol oxidation reaction. Today, various bi- and multimetallic catalysts are applied that are more active, more selective, and less prone to deactivation than monometallic catalysts. The probable redox model for selective oxidation reaction is shown in Figure 8.29.

Oxide and mixed oxide catalysts are potential catalyst for alcohol oxidation, dehydrogenation reaction, etc. These catalysts are mainly used for the gas-phase oxidation reaction by using molecular oxygen, dehydrogenation reaction, etc. Hydrated ruthenium oxide and vanadium pentoxide

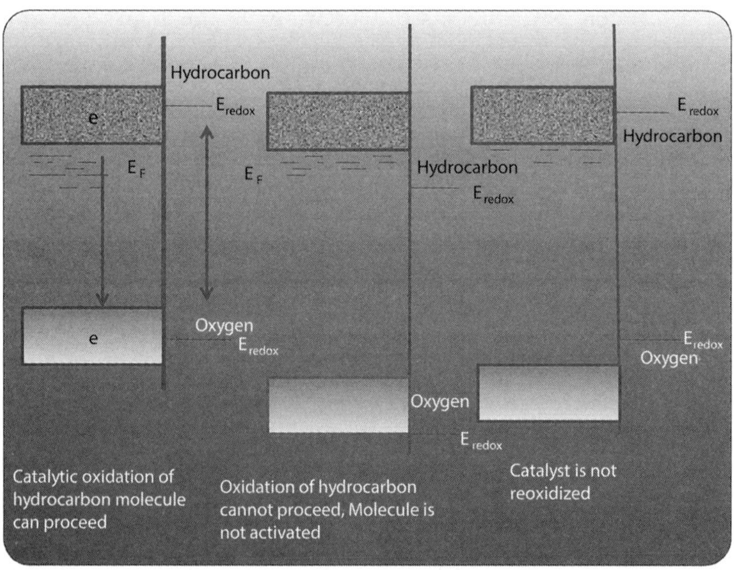

Figure 8.29 Probable redox model for the selective oxidation reaction.

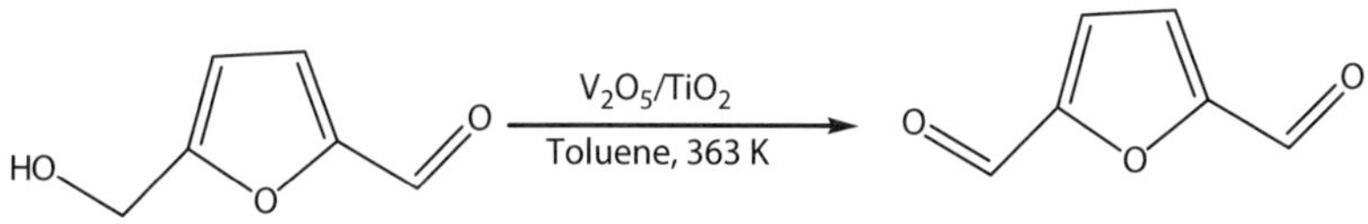

Scheme 8.5 Oxidation of 5-Hydroxymethylfurfural with Air over 15% V_2O_5/TiO_2

are useful for alcohol oxidation reaction. Hydrated ruthenium oxide is a stoichiometric oxidant applied for alcohol oxidation reaction such as cinnamyl alcohol oxidation reaction [123]. The V_2O_5 supported on TiO_2 material is an active catalyst for the oxidation of 5-(hydroxymethyl) furfural to 2,5-furandicarboxaldehyde (Scheme 8.5) [124].

A broad range of oxides and mixed oxides were tested in the synthesis of ethyl pyruvate from ethyl lactate [125]. In the case of SnO_2 and ZrO_2, the catalyst is highly selective when conversion of ethyl lactate is low, but selectivity of ethyl pyruvate dropped with increased conversion of ethyl lactate. The MnO_2, PdO and different types of binary oxides by addition of Co^{2+}, Fe^{3+} and Mn^{3+} are efficient catalysts for different types of alcohol oxidation reaction. Different types of zeolite, hydrotalcite and phosphate materials are used as potential catalysts for alcohol oxidation reaction.

Catalytic activity of the supported nanoparticles can also be promoted by defects in the support material. Defects can trap the metal nanoparticle and enhance charge transfer between the support and the nanoparticle. Yoon *et al.* [126] reported that gold octamers bound to oxygen vacancies (F centers) on MgO (001) are the smallest clusters to catalyze the low-temperature oxidation of CO to CO_2, whereas clusters deposited on close-to-perfect magnesia surfaces remain chemically inert. It has also been shown that the surface F centers on MgO play a critical role in the activation of gold on Au/MgO catalysts [127, 128]. Defects are not the only factor responsible for charging of the supported metal nanoparticles. Recently, it was demonstrated that the charge accumulated on the supported gold nanoparticle can be tuned by varying the thickness of the metal oxide layer deposited on the metal support [129–133]. Thus, the support effects can play an even larger role in gold nanocatalysis than the particle size.

Solid catalysts are widely used for the gas- or vapor-phase oxidation of simple, small-chain alcohols to the corresponding carbonyl compounds [117–119]. For gas-phase oxidation of alcohol important requirements are the reasonable volatility and thermal stability of reactant and products which strongly limit the synthesis of complex molecules. Solid catalysts active in the liquid phase under mild conditions have a much broader application range. A major challenge in liquid-phase oxidation with solid catalysts is to prevent leaching of the active species. The reactants, and particularly the carboxylic acid-type byproducts, are frequently excellent chelating agents and accelerate dissolution and deactivation of the catalysts. The hydroxyl group is the primary functional group in aliphatic alcohols and has greater gas-phase Brönsted acidity than water. The reactions of alcohols on oxygen-covered gold nanoparticles catalysts are initiated by the activation of O-H bonds by adsorbed atomic oxygen to form adsorbed alkoxy and water [134]. The differences in the product distributions of over supported gold nanoparticles catalysts depend on the structure of the alkyl groups present in the alcohol as well as the atomic oxygen coverage over the catalysts.

Recently, gold nanoparticles have been widely applied for alcohol oxidation reaction by maintaining a mild condition in liquid phase. The catalytic activities of gold nanoparticles strongly depend on the size of gold nanoparticles, nature of support material (for supported gold nanoparticles), oxidation state of gold nanoparticles, etc. The application of gold nanoparticles on oxidation reaction is not yet broad, but extensive research is going on in this field. Gold has an advantage compared to other precious metals like Ag, Pt, and Pd because leaching of gold nanoparticles is lower compared to other metals. In order to obtain highly active gold catalysts, a chemical grafting and reduction process was developed in which amino, thiol or diamino groups were grafted onto the surfaces support material and then Au was introduced via a neutralization reaction, followed by a reduction procedure [135–137]. Rossi and Prati [138–150] have shown that supported gold nanoparticles can be very effective catalysts for the oxidation of alcohols, including diols. Gold is highly active and selective in glycerol oxidation reaction [151–153]. Carbon-supported gold nanoparticles are effective for a range of substrates in gas-phase and liquid-phase oxidation reaction. These studies using Au/carbon catalysts were extended by Carrettin *et al.* [26, 154, 155] to show that Au supported on graphite can oxidize glycerol to glycerate with 100% selectivity using dioxygen as the oxidant under relatively mild conditions with yields approaching 60%. It was observed that the selectivity to glyceric acid and the glycerol conversion were dependent upon the glycerol/NaOH ratio. In general, with high concentrations of NaOH, exceptionally high selectivities to glyceric acid can be observed. However, decreasing the concentration of glycerol and increasing the mass of the catalyst and the concentration of oxygen lead to the formation of tartronic acid via consecutive oxidation of glyceric acid. Interestingly, this product is stable with these catalysts. It is apparent that, with careful control of the reaction conditions, 100% selectivity to glyceric acid can be obtained

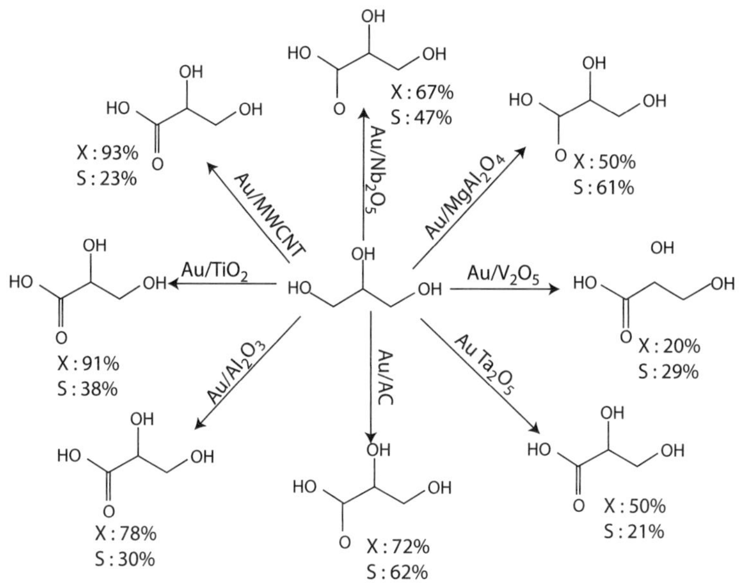

Scheme 8.6 Examples of supported gold catalyst used for glycerol oxidation reaction.

with 1 wt% Au/C. A few examples of supported gold catalysts and their activity (Conversion: X and Selectivity: S) are shown in Scheme 8.6. From a review of different literature [151–153] it is observed that in the case of glycerol oxidation reaction, glyceric acid formed as a main product along with different types of byproduct.

In the case of aromatic alcohol-supported gold nanoparticles catalysts, an exceptionally high activity and selectivity towards the desired product are shown. Some examples of alcohol conversion are shown in Figure 8.30. Baiker *et al.* repoted that Cu-Mg-Al mixed oxide acts as an effective support material for gold nanoparticles, and this Au/Cu-Mg-Al catalyst was effectively used for alcohol oxidation reaction [156]. Corma and coworkers [157–159] showed that an Au/CeO_2 catalyst is active for the selective oxidation of alcohols to aldehydes and ketones and the oxidation of aldehydes to acids. In their studies, the catalysts are active at relatively mild conditions using O_2 as an oxidizing agent. They also carried out the alcohol oxidation reaction without the presence of solvent and any type of base like NaOH. The catalytic activity was ascribed to the Au/CeO_2 catalyst stabilizing a reactive peroxy intermediate from O_2. Subsequently, Hutchings *et al.* [160] showed that Au-Pd alloy supported on TiO_2 enhanced the activity for alcohols under solvent-free conditions by a factor of over 25. Recently, Chowdhury *et al.* also showed that ceria-based mixed oxide can be used

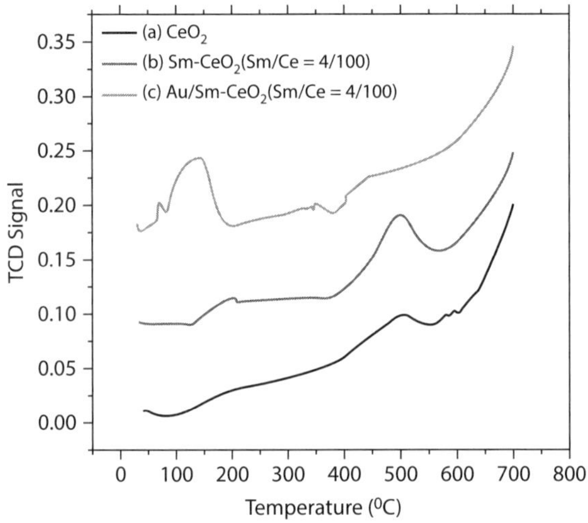

Figure 8.30 H_2-TPR of the catalysts. Reproduced with permission from [161].

as an effective support for gold nanoparticles. They investigated the role of different types of dopant like samarium, manganese and different alkali metal like Ba^{2+}, Ca^{2+} and Mg^{2+} on cerium oxide material [161–163]. In their investigation they showed that gold nanoparticles preferably nucleate on the oxide vacant site and these supported gold nanoparticles which facilitate the easy reducibility of the catalyst.

Alcohol oxidation over Au/CeO_2 catalyst following a complex multistep mechanism was given by Corma *et al.* [158]. According to this mechanism (Scheme 8.7), the interaction between gold and ceria will give rise to an important population of positively charged gold and Ce^{3+} species. The alcohol(s) (Scheme 8.8) or the surface adsorbed alkoxide will then react with the Lewis acid sites of Au/CeO_2 to give a metal alkoxide, which subsequently undergoes a rapid hydride transfer from C-H to Ce^{3+} and Au^+ to give the ketone and Ce-H (indicated as LA-H) and Au-H. Upon admission of oxygen into the system and coordination to the oxygen-deficient sites of ceria, formation of cerium-coordinated superoxide (Ce-OO⊠) species occurs [164]. The increment of oxide vacant site on support material facilitates the formation of cerium-coordinated superoxide (Ce-OO⊠) species. These superoxide species evolve into cerium hydroperoxide by hydrogen abstraction from Au-H, and are responsible for the formation, after reduction of Ce(IV), of the initial Au^+ species. The absence of gold would render this step impossible and lead to a depletion of Ce(III).

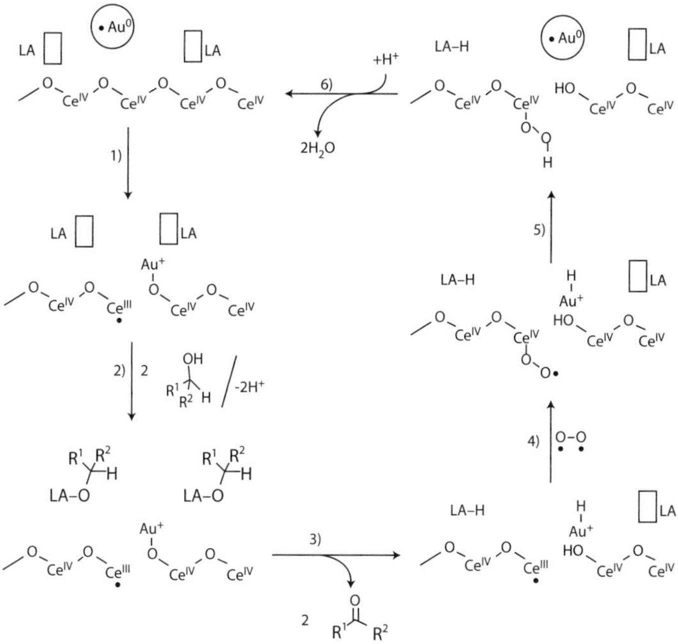

Scheme 8.7 Reaction mechanism of alcohol oxidation reaction over Au/CeO_2 catalyst. Reproduced with permission from [158].

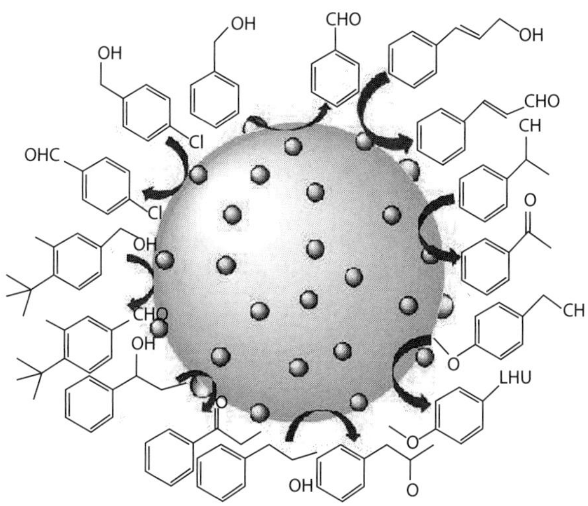

Scheme 8.8 Different types of alcohol oxidation over supported gold catalyst.

Methanol was oxidized on Au(110) via the activation of the O-H bond by adsorbed atomic oxygen to form methoxy at 200 K temperature. This methoxy group further reacted to from methanol, hydrogen, and methyl formate at 250 K and CO_2 at 350 K temperature [165]. The expected product formaldehyde was not observed, which may be due to the fundamental difference in the bonding of methoxy to the gold surface, rendering it more susceptible to attack by the aldehyde [166]. Methoxy decomposed to adsorbed formaldehyde on gold at a lower temperature (250 K), so that its reaction with adsorbed methoxy to form methyl formate was more likely due to a longer surface lifetime during heating. Formate also existed as a stable intermediate on Au(110) through secondary oxidation of formaldehyde [165]. The decomposition temperature of the formate on Au(110) was 350K, lower than that observed on Ag(110) (400 K) and Cu(110) (475 K) [167]. The formation of methyl formate via coupling of formaldehyde with methanol was attributed to the fact that the decomposition temperature of methoxy via β-H elimination is lower than the desorption temperature of the aldehyde formed [168]. In methanol oxidation reaction some groups detected H_2O, CO and CO_2 as the decomposed product of the methoxy group [169, 170], whereas some other research groups claimed that methyl formate, formaldehyde and formic acid are formed as a decomposed product of the methoxy group [171]. The discrepancies in their result may be due to the following factors: (i) the method for preparing oxidized Au(111) surface was different, possibly leading to differences in the bonding environment of the oxygen; (ii) methanol was exposed to the surface at a higher temperature (160 K), close to the temperature where water leaves the surface; (iii) the extent of the esterification reaction appears to be low, so the methyl formate may have gone undetected if the mass spectrometer detector is far away from the sample.

Ethanol oxidation is an important reaction from an industrial perspective. Ethanol was selectively oxidized on Au(111) surface with the formation of acetaldehyde and water as a main product [172]. No other partial oxidation products (i.e., methane, ethane, ethylene, ethanol, acetic acid, ethylene oxide and methyl formate) or C1-containing species (i.e., CO, CO_2, formaldehyde, and formic acid) were detected during the reaction [172]. A trace amount of ethyl acetate was detected during the reaction. When the surface of gold was covered by higher oxygen coverage, CO_2 was also formed in ethanol oxidation reaction. At lower oxygen coverage ethyl acetate is the main product for ethanol oxidation reaction. Liu *et al.* showed that in ethanol oxidation reaction acetaldehyde was the main product; also that ethyl acetate (230 K) and acetic acid (450 and 545 K) were formed on

Au(111) precovered with 0.4 ML atomic oxygen when prepared by decomposition of O_3 at 200 K [173].

8.5.1 Mechanism for Alcohol Oxidation Reaction

The mechanism of gold-catalyzed alcohol oxidation by using molecular oxygen was followed by a multi-step mechanism given by Boronat *et al.* [174] (Scheme 8.9). Initially deprotonation of the hydroxyl group forming a metal-alkoxide intermediate and then β-hydride elimination took place yielding the carbonylic product. Oxygen reacts with the hydrogen atoms to produce H_2O or participates in the first step of the mechanism by assisting the deprotonation of the alcohol. In the case of metal oxide-supported gold nanoparticles catalysts, the alkoxide intermediate is probably formed on the support or at the metal-support interface. But when the reaction is catalyzed by naked gold nanoparticles in solution, stabilized by polymers, or supported on carbon or SiO_2, activity has been related to the presence of low coordinated atoms placed at corner or edge positions [110, 175–177]. However, the effect of chemisorbed basic O atoms is considerably more important and decreases the activation energy for deprotonation of the hydroxyl group by ~20 kcal/mol, while it has no influence on the dissociation of the C-Hβ bond [178]. It can then be concluded that alcohol oxidation is favored by the presence of chemisorbed oxygen atoms able to abstract protons and dissociate the hydroxyl group. These basic oxygen atoms are stabilized on 3D gold nanoparticles and not on small planar clusters.

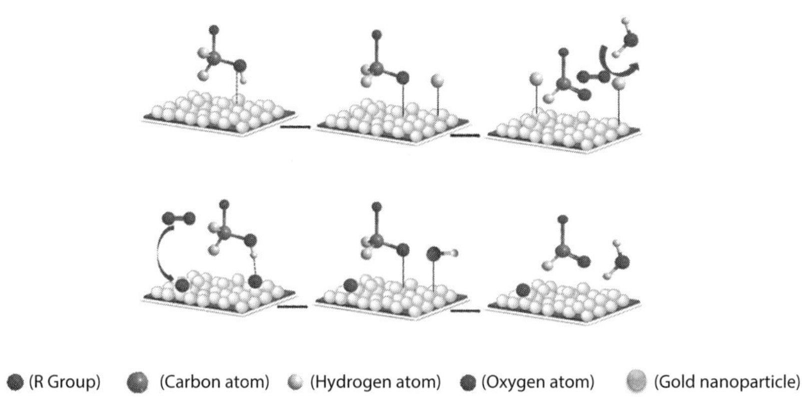

Scheme 8.9 Alcohol oxidation reaction on supported gold nanoparticles catalyst.

8.6 Conclusion

The catalytic activity of gold nanoparticles/nanoclusters plays an immense role in oxidation reactions. By fine tuning gold nanoparticle size, support oxides and reaction conditions, industrially challenging reactions like propylene epoxidation, glucose oxidation and alcohol oxidation can be carried out. The chemistry regarding the role of gold nanoparticles and nanoclusters varies with the nature of the reaction. High thermal stability and less leaching of the gold catalyst compared to other coinage metals have proven the need for extensive exploratory research on catalytic activity using an environmentally benign industrial route for fine chemical synthesis.

References

1. P.T. Anastas, J.C. Warner, *Green Chemistry: Theory and Practice*, Oxford Science Publications, Oxford 1998.
2. J.M. Thomas, R. Raja, *Catal. Today*, Vol. 117, pp. 22–31, 2006.
3. M. Faraday, *Philos. Trans. R. Soc. London*, 147, 1857, pp. 145–1811.
4. B. Hammer, J.K. Nørskov, *Nature*, Vol. 376, pp. 238–240, 1995.
5. A.G. Sault, R.J. Madix, C.T. Campbell, *Surf. Sci.*, Vol. 169, pp. 347–356, 1986.
6. M. Haruta, T. Kobayashi, H. Sano, N. Yamada, *Chem. Lett.*, Vol. 16, pp. 405–408, 1987.
7. P.P. Edwards, J.M. Thomas, *Angew. Chem. Int. Ed.*, Vol. 46, pp. 5480–5486, 2007.
8. R.P. Chambers, M. Boudart, *J. Catal.*, Vol. 5, pp. 517–528, 1966.
9. M.W. Roberts, *Catal. Lett.*, Vol. 67, pp. 1–4, 2000.
10. A.S.K. Hashmi, G.J. Hutchings, *Angew. Chem. Int. Ed.*, Vol. 45, pp. 7896–7936, 2006.
11. D.J. Gorin, B.D. Sherry, F.D. Toste, *Chem. Rev.*, Vol. 108, pp. 3351–3378, 2008.
12. A. Stephen K. Hashmi, *Chem. Rev.*, Vol. 107, pp. 3180–3211, 2007.
13. S.A. Cotton, *Chemistry of Precious Metals*, Springer-Verlag, 1997.
14. G.C. Bond, *Catal. Today*, Vol. 72, pp. 5–9, 2002.
15. Y. Zhang, X. Cui, F. Shi, Y. Deng, *Chem. Rev.*, Vol. 112, pp. 2467–2505, 2012.
16. (a) L. Guczi, A. Beck, Z. Paszti, *Catal. Today*, Vol. 181, pp. 26–32, 2012.
17. (b) S. Panigrahi, S. Basu, S. Praharaj, S. Pande, S. Jana, A. Pal, S.K. Ghosh, T. Pal, *J. Phys. Chem. C*, Vol. 111, pp. 4596–4605, 2007.
18. (c) W.C. Ketchie, Y.L. Fang, M.S. Wong, M. Murayama, R.J. Davis, *J. Catal.*, Vol. 250, pp. 94–101, 2007.
19. (d) Y. Guan, E.J.M. Hensen, *Appl. Catal. A: Gen.*, Vol. 361, pp. 49–56, 2009.
20. (e) J. Huang, Y. Wang, J. Zheng, W.L. Dai, K. Fan, *Appl. Catal. B: Environ.*, Vol. 103, pp. 343–350, 2011.

21. (f) M. Ciobanu, B. Cojocaru, C. Teodorescu, F. Vasiliu, S.M. Coman, W. Leitner, V.I. Parvulescu, *J. Catal.*, Vol. 296, pp. 43–54, 2012.
22. T.V. Choudhary, D.W. Goodman, *Appl. Catal. A: Gen.*, Vol. 291, pp. 32–36, 2005.
23. L. Guczi, D. Horvath, Z. Paszti, L. Toth, Z.E. Horvath, A. Karacs, G. Peto, *J. Phys. Chem. B*, Vol. 104, pp. 3183–3193, 2000.
24. M. Valden, X. Lai, D.W. Goodman, *Science*, Vol. 281, pp. 1647–1650, 1998.
25. M. Valden, S. Pak, X. Lai, D.W. Goodman, *Catal. Lett.*, Vol. 56, pp. 7–10, 1998.
26. S. Carrettin, P. McMorn, P. Johnston, K. Griffin, G.J. Hutchings, *Chem. Commun.*, 2002, pp. 696–697.
27. X. Lai, T.P. Clair, M. Valden, D.W. Goodman, *Prog. Surf. Sci.*, Vol. 59, pp. 25–52, 1998.
28. C. Qi, J. Huang, S. Bao, H. Su, T. Akita, M. Haruta, *J. Catal.*, Vol. 281, pp. 12–20, 2011.
29. J. Huang, E. Lima, T. Akita, A. Guzmán, C. Qi, T. Takei, M. Haruta, *J. Catal.*, Vol. 278, pp. 8–15, 2011.
30. T. Takei, T. Akita, I. Nakamura, T. Fujitani, M. Okumura, K. Okazaki, J. Huang, T. Ishida, M. Haruta, *Adv. Catal.*, Vol. 55, pp. 1–126, 2012.
31. B.K. Min, C.M. Friend, *Chem. Rev.*, Vol. 107, pp. 2709–2724, 2007.
32. M. Haruta, *Catal. Today*, Vol. 36, pp. 153–166, 1997.
33. J. Guzman, B.C. Gates, *Angew. Chem. Int. Ed. Engl.*, Vol. 42, pp. 690–693, 2003.
34. R. Meyer, C. Lemire, S.K. Shaikhutdinov, H.J. Freund, *Gold Bull.*, Vol. 37, pp. 72–124, 2004.
35. M.C. Kung, R.J. Davis, H.H. Kung, *J. Phys. Chem. C*, Vol. 111, pp. 11767–11775, 2007.
36. M.S. Chen, D.W. Goodman, *Chem. Soc. Rev.*, Vol. 37, pp. 1860–1870, 2008.
37. C.T. Campbell, *Science*, Vol. 306, pp. 234–235, 2004.
38. M.S. Chen, D.W. Goodman, *Science*, Vol 306, pp. 252–255, 2004.
39. M. Valden, X. Lai, D.W. Goodman, *Science*, Vol. 281, pp. 1647–1650, 1998.
40. S. Carrettin, Y. Hao, V.A. Guerrero, B.C. Gates, S. Trasobares, J.J. Calvino, A. Corma, *Chem. Eur. J.*, Vol. 13, pp. 7771–7779, 2007.
41. V. Schwartz, D.R. Mullins, W. Yan, B. Chen, S. Dai, S.H. Overbury, *J. Phys. Chem. B*, Vol. 108, pp. 15782–15790, 2004.
42. J.H. Yang, J.D. Henao, M.C. Raphulu, Y. Wang, T. Caputo, A.J. Groszek, M.C. Kung, M.S. Scurrell, J.T. Miller, H.H. Kung, *J. Phys. Chem. B*, Vol. 109, pp. 10319–10326, 2005.
43. J.T. Calla, M.T. Bore, A.K. Datye, R.J.J. Davis, *Catal.*, Vol. 238, pp. 458–467, 2006.
44. J. Guzman, B.C. Gates, *J. Am. Chem. Soc.*, Vol. 126, pp. 2672–2673, 2004.
45. (a) Y. Kobayashi, S. Nasu, S. Tsubota, M. Haruta, *Hyperfine Interact.*, Vol. 126, pp. 95–99, 2000.

46. (b) S.T. Daniells, A.R. Overweg, M. Makkee, J.A. Moulijn, *J. Catal.*, Vol. 230, pp. 52–65, 2005.
47. (c) D. Boyd, S. Golunski, G.R. Hearne, T. Magadzu, K. Mallick, M.C. Raphulu, A. Venugopal, M.S. Scurrell, *Appl. Catal. A: Gen.*, Vol. 292, pp. 76–81, 2005.
48. (d) G.J. Hutchings, M.S. Hall, A.F. Carley, P. Landon, B.E. Solsona, C.J. Kiely, A. Herzing, M. Makkee, J.A. Moulijn, A. Overweg, J.C. Fierro-Gonzalez, J. Guzman, B.C. Gates, *J. Catal.*, Vol. 242, pp. 71–81, 2006.
49. G.C. Bond, D.T. Thompson, *Gold Bull.*, Vol. 33, pp. 41–51, 2000.
50. A. Lyalin, T. Taketsugu, *J. Phys. Chem. C*, Vol. 114, pp. 2484–2493, 2010.
51. A. Lyalin, T. Taketsugu, *J. Phys. Chem. C*, Vol. 113, pp. 12930–12934, 2009.
52. A. Lyalin, T. Taketsugu, *J. Phys. Chem. Lett.*, Vol. 1, pp. 1752–1757, 2010.
53. M. Haruta, *Chem. Rec.*, Vol. 3, pp. 75–87, 2003.
54. M. Okumura, S. Tsubota, M. Haruta, *J. Mol. Catal. A: Chem.*, Vol. 199, pp. 73–84, 2003.
55. M. Haruta, *Faraday Discuss.*, Vol. 152, pp. 11–32, 2011.
56. M.M. Schubert, S. Hackenberg, A.C. van Veen, M. Muhler, V. Plzak, R.J.J. Behm, *J. Catal.*, Vol. 197, pp. 113–122, 2001.
57. M.M. Schubert, M.J. Kahlich, H.A. Gasteiger, R.J.J. Behm, *Power Sour.*, Vol. 84, pp. 175–182, 1999.
58. I.X. Green, W. Tang, M. Neurock, J.T. Yates Jr., *Science*, Vol. 333, pp. 736–739, 2011.
59. (a) U. Landman, B. Yoon, C. Zhang, U. Heiz, M. Arenz, *Top. Catal.*, Vol. 44, pp. 145–158, 2007.
60. (b) C.H.F. Peden, "Surface science of catalysis catalysis," in: *ACS Symp. Ser., Vol. 482*, D.J. Dwyer, F.M. Hoffmann, eds., Amer. Chem. Soc., Washington DC, USA, 1992, p. 143.
61. J.R. Monnier, *Appl. Catal. A: Gen.*, Vol. 221, pp. 73–91, 2001.
62. T.A. Nijhuis, M. Makkee, J.A. Moulijn, B.M. Weckhuysen, *Ind. Eng. Chem. Res.*, Vol. 45, pp. 3447–3459, 2006.
63. W.C. Ketchie, Y.L. Fang, M.S. Wong, M. Murayama, R.J. Davis, *J. Catal.*, Vol. 250, pp. 94–101, 2007.
64. J. Huang, M. Haruta, *Res. Chem. Intermed.*, Vol. 38, pp. 1–24, 2012.
65. M. Haruta, S. Tsubota, T. Hayashi, US Patent 5623090A1, 1997.
66. T. Hayashi, K. Tanaka, M. Haruta, *J. Catal.*, Vol. 178, pp. 566–57, 1998.
67. B. Chowdhury, J.J. Bravo-Suárez, M. Date, S. Tsubota, M. Haruta, *Angew. Chem. Int. Ed.*, Vol. 45, pp. 412–415, 2006.
68. M. Haruta, M, Daté, *Appl. Catal. A: Gen.*, Vol. 222, pp. 427–437, 2001.
69. C. Qi, T. Akita, M. Okumura, M. Haruta, *Appl. Catal. A: Gen.*, Vol. 218, pp. 81–89, 2001.
70. C. Qi, *Gold Bull.*, Vol. 41, pp. 224–234, 2008.
71. B.S. Uphade, T. Akita, T. Nakamura, M. Haruta, *J. Catal.*, Vol. 209, pp. 331–340, 2002.
72. B. Chowdhury, J.J. Bravo-Suárez, M. Date, S. Tsubota, M. Haruta, *Angew. Chem. Int. Ed.*, Vol. 45, pp. 412–415, 2006.

73. (a) T.A. Nijhuis, T. Visser, B.M. Weckhuysen, *J. Phys. Chem. B*, Vol. 109, pp. 19309–19319, 2005.
74. (b) L. Cumaranatunge, W.N. Delgass, *J. Catal.*, Vol. 232, pp. 38–42, 2005.
75. (a) T.A. Nijhuis, T.Q. Gardner, B.M. Weckhuysen, *J. Catal.*, Vol. 236, pp. 153–163, 2005.
76. (b) T.A. Nijhuis, M. Makkee, J.A. Moulijn, B.M. Weckhuysen, *Ind. Eng. Chem. Res.*, Vol. 45, pp. 3447–3459, 2006.
77. [67] J. Chen, S.J.A. Halin, J.C. Schouten, T.A. Nijhuis, *Faraday Discuss.*, Vol. 152, pp. 321–336, 2011.
78. J. Huang, T. Akita, J. Faye, T. Fujitani, T. Takei, M. Haruta, *Angew. Chem. Int. Ed. Engl.*, Vol. 48, pp. 7862–7866, 2009.
79. A.A. Herzing, C.J. Kiely, A.F. Carley, P. Landon, G.J. Hutchings, *Science*, Vol. 321, pp. 1331–1335, 2008.
80. A.K. Sinha, S. Seelan, M. Okumura, T. Akita, S. Tsubota, M. Haruta, *J. Phys. Chem. B*, Vol. 109, pp. 3956–3965, 2005.
81. S. Bordiga, A. Damin, F. Bonino, G. Ricchiardi, A. Zecchina, R. Tagliapietra, C. Lambertiabc, *Phys. Chem. Chem. Phys.*, Vol. 5, pp. 4390–4393, 2003.
82. K.L. Antcliff, D.M. Murphy, E. Griffiths, and E. Giamello, *Phys. Chem. Chem. Phys.*, Vol. 5, pp. 4306–4316, 2003.
83. M. Che, A.J. Tench, *Adv. Catal.*, Vol. 31, pp. 77–133, 1982.
84. M. Che, A.J. Tench, *Adv. Catal.*, Vol. 32, pp. 1–148, 1983.
85. V.N. Shetti, P. Manikandan, D. Srinivas, P. Ratnasamy. *J. Catal.*, Vol. 216, pp. 461–467, 2003.
86. A. Tuel, L.G. Hubert-Pfalzgraf, *J. Catal.*, Vol. 217, pp. 343–453, 2003.
87. P.P. Olivera, E.M. Patrito, H. Sellers, *Surf. Sci.*, Vol. 313, pp. 25–40, 1994.
88. P. Landon, P.J. Collier, A.J. Papworth, C.J. Kiely, G.J. Hutchings, *Chem. Commun.*, Vol. 18, pp. 2058–2059, 2002.
89. M. Okumura, Y. Kitagawa, K. Yamagcuhi, T. Akita, S. Tsubota, M. Haruta, *Chem. Lett.*, Vol. 32, pp. 822–823, 2003.
90. C. Sivadinarayana, T.V. Choudary, L.L. Dacmen, J. Eckert, and D.W. Goodman, *J. Am. Chem. Soc.*, Vol. 126, pp. 38–39, 2004.
91. H. Liu, A.I. Kozlov, A.P. Kozlova, T. Shido, K. Asakura, Y. Iwasawa, *J. Catal.*, Vol. 185, pp. 252–264, 1999.
92. B. Chowdhury, J.J. Bravo-Suàrez, N. Mimura, J. Lu, K.K. Bando, S. Tsubota, M. Haruta, *J. Phys. Chem. B*, Vol. 110, pp. 22995–22999, 2006.
93. J.J. Bravo-Suarez, K.K. Bando, J. Lu, M. Haruta, T. Fujitani, S.T. Oyama, *J. Phys. Chem. C*, Vol. 112, pp. 1115–1123, 2008.
94. X. Liang, C.J. Liu, P. Kuai, *Green Chem.*, Vol. 10, pp. 1318–1322, 2008.
95. C.D. Pina, E. Falletta, L. Prati, M. Rossi, *Chem. Soc. Rev.*, Vol. 37, pp. 2077–2095, 2008.
96. T. Mallat, A. Baiker, *Chem. Rev.*, Vol. 104, pp. 3037–3058, 2004.
97. M. Comotti, C. Della Pina, R. Matarrese, M. Rossi, *Angew. Chem. Int. Ed.*, Vol. 43, pp. 5812–5815, 2004.

98. H. Yin, C. Zhou, C. Xu, P. Liu, X. Xu, Y. Ding, *J. Phys. Chem. C*, Vol. 112, pp. 9673–9678, 2008.
99. C. Baatz, U. Prüße, *J. Catal.*, Vol. 249, pp. 34–40, 2007.
100. C. Baatz, N. Decker, U. Pruesse, *J. Catal.*, Vol. 258, pp. 165–169, 2008.
101. M. Sankar, N. Dimitratos, P.J. Miedziak, P.P. Wells, C J. Kiely, G.J. Hutchings, *Chem. Soc. Rev.*, Vol. 41, pp. 8099–8139, 2012.
102. M. Comotti, C.D. Pina, M. Rossi, *J. Mol. Catal. A: Chem.*, Vol. 251, pp. 89–92, 2006.
103. B. Lim, M. Jiang, P.H.C. Camargo, E.C. Cho, J. Tao, X. Lu, Y. Zhu, and Y. Xia, *Science*, Vol. 324, pp. 1302–1305, 2009.
104. Z. Peng, H. Yang, *J. Am. Chem. Soc.*, Vol. 131, pp. 7542–7543, 2009.
105. C.D. Pina, E. Falletta, *Catal. Sci. Technol.*, Vol. 1, pp. 1564–1571, 2011.
106. A.R. Tao, S. Habas, P. Yang, *Small*, Vol. 4, pp. 310–325, 2008.
107. A. Aldaz, V. Montiel, E. Herrero, *J. Am. Chem. Soc.*, Vol. 132, pp. 5622–5624, 2010.
108. Y.X. Chen, S.P. Chen, Z.Y. Zhou, N. Tian, Y.X. Jiang, S.G. Sun, Y. Ding, Z.L. Wang, *J. Am. Chem. Soc.*, Vol. 131, pp. 10860–10862, 2009.
109. J. Majimel, M. Lamirand-Majimel, I. Moog, C. Feral-Martin, M. Tréguer-Delapierre, *J. Phys. Chem. C*, Vol. 113, pp. 9275–9283, 2009.
110. J. Wang, J. Gong, Y. Xiong, J. Yang, Y. Gao, Y. Liu, X. Lu, Z. Tang, *Chem. Commun.*, Vol. 47, pp. 6894–6896, 2011.
111. T. Ishida, N. Kinoshita, H. Okatsu, T. Akita, T. Takei, M. Haruta, *Angew. Chem. Int. Ed.*, Vol. 47, pp. 9265–9268, 2008.
112. C. Baatz, N. Thielecke, U. Pruße, *Appl. Catal. B: Environ.*, Vol. 70, pp. 653–660, 2007.
113. U. Prüße, M. Herrmann, C. Baatz, N. Decker, *Appl. Catal. A: Gen.*, Vol. 406, pp. 89–93, 2011.
114. N. Toshima, H. Zhang, *Macromol. Symp.*, Vol. 317–318, pp. 149–159, 2012.
115. W.M.M. Kessels, A.H.M. Smets, M.C.M. van de Sanden, *J. Non-Cryst. Solids*, Vol. 338–340, pp. 27–31, 2004.
116. M. Matsumoto, N. Watanabe, *J. Org. Chem.*, Vol. 49, pp. 3432–3435, 1984.
117. (a) M. Muhler, in: *Handbook of Heterogeneous Catalysis*, G. Ertl, H. Knozinger, J. Weitkamp, eds., VCH: Weinheim, Vol. 5, p. 2274, 1997.
118. (b) M. Kraus, in: *Handbook of Heterogeneous Catalysis*, G. Ertl, H. Knozinger, J. Weitkamp, eds., VCH: Weinheim, Vol. 5, p. 2159, 1997.
119. (c) W.F. Holderich, *Catal. Today*, Vol. 62, pp. 115–130, 2000.
120. (a) M. Hudlicky, *Oxidations in Organic Chemistry, ACS Monographs 186*, American Chemical Society: Washington, DC, p. 114, 1990.
121. (b) R.A. Sheldon, I.W.C.E. Arends, A. Dijksman, *Catal. Today*, Vol. 57, pp. 157–166, 2000.
122. R.A. Sheldon, M. Wallau, I.W.C.E. Arends, U. Schuchardt, *Acc. Chem. Res.*, Vol. 31, pp. 485–493, 1998.
123. C. Moreau, R. Durand, C. Pourcheron, D. Tichit, *Stud. Surf. Sci. Catal.*, Vol. 108, pp. 399–406, 1997.

124. T.L. Stuchinskaya, M. Musawir, E.F. Kozhevnikova, I.V. Kozhevnikov, *J. Catal.*, Vol. 231, pp. 41–47, 2005.
125. H. Hayashi, S. Sugiyama, Y. Katayama, K. Sakai, M. Sugino, N. Shigemoto, *J. Mol. Catal.*, Vol. 83, pp. 207–217, 1993.
126. B. Yoon, H. Hakkinen, U. Landman, A.S. Worz, J.M. Antonietti, S. Abbet, K. Judai, U. Heiz, *Science*, Vol. 307, pp. 403–407, 2005.
127. M. Sterrer, M. Yulikov, E. Fischbach, M. Heyde, H.P. Rust, G. Pacchioni, T. Risse, H.J. Freund, *Angew. Chem. Int. Ed. Engl.*, Vol. 45, pp. 2630–2632, 2006.
128. Z. Yan, S. Chinta, A.A. Mohamed, J.P. Fackler, D.W. Goodman, *J. Am. Chem. Soc.*, Vol. 127, pp. 1604–1605, 2005.
129. C. Zhang, B. Yoon, U. Landman, *J. Am. Chem. Soc.*, Vol. 129, pp. 2228–2229, 2007.
130. M. Sterrer, T. Risse, U. Pozzoni, L. Giordano, M. Heyde, H. Rust, G. Pacchioni, H. Freund, *Phys. Rev. Lett.*, Vol. 98, pp. 096107:1–096107:4, 2007.
131. P. Frondelius, H. Hakkinen, K. Honkala, *Phys. Rev. B*, Vol. 76, pp. 073406: 1–073406:4, 2007.
132. K. Honkala, H. Hakkinen, *J. Phys. Chem. C*, Vol. 111, pp. 4319–4327, 2007.
133. C. Harding, V. Habibpour, S. Kunz, A.N.S. Farnbacher, U. Heiz, B. Yoon, U. Landman, *J. Am. Chem. Soc.*, Vol. 131, pp. 538–548, 2009.
134. X.Y. Liu, R.J. Madix, C.M. Friend, *Chem. Soc. Rev.*, Vol. 37, pp. 2243–2261, 2008.
135. J. Gu, J. Lin, G. You, L. Xiong, S. Qian, Z. Hua, H. Chen, *Adv. Mater.*, Vol. 17, pp. 557–560, 2005.
136. K. Zhu, J. Hu, R. Richards, *Catal. Lett.*, Vol. 100, pp. 195–199, 2005.
137. C. Yang, H. Sheu, K. Chao, *Adv. Funct. Mater.*, Vol. 12, pp. 143–148, 2002.
138. (a) L. Prati, M. Rossi, *J. Catal.*, Vol. 176, pp. 552–560, 1998.
139. (b) L. Prati, M. Rossi, *Stud. Surf. Sci. Catal.*, Vol. 110, pp. 509–516, 1997.
140. (c) L. Prati, G. Martra, *Gold Bull.*, Vol. 32, pp. 96–101, 1999.
141. (d) S. Coluccia, G. Martra, F. Porta, L. Prati, M. Rossi, *Catal. Today*, Vol. 61, pp. 165–172, 2000.
142. (e) L. Prati, F. Porta, S. Biella, M. Rossi, *Catal. Lett.*, Vol. 90, pp. 23–29, 2003.
143. (f) F. Porta, L. Prati, M. Rossi, G. Scari, *J. Catal.*, Vol. 211, pp. 464–469, 2002.
144. (g) C. Bianchi, F. Porta, L. Prati, M. Rossi, *Top. Catal.*, Vol. 13, pp. 231–236, 2000.
145. (h) C. Bianchi, S. Biella, A. Gervasini, L. Prati, M. Rossi, *Catal. Lett.*, Vol. 85, pp. 91–96, 2003.
146. (i) S. Biella, L. Prati, M. Rossi, J. Mol. Catal. A: Chem., 197, 2003, 187–207; [125](j) S. Biella, L. Prati, M. Rossi, *Inorg. Chem. Acta.*, Vol. 349, pp. 253–257, 2003.
147. (k) S. Biella, L. Prati, M. Rossi, *J. Catal.*, Vol. 206, pp. 242–247, 2002.
148. (l) S. Biella, G.L. Castiglioni, C. Fumagalli, L. Prati, M. Rossi, *Catal. Today*, Vol. 72, pp. 43–49, 2002.
149. (m) L. Prati, F. Porta, *Appl Catal. A*, Vol. 291, p. 199, 2005.
150. (n) F. Porta, L. Prati, *J. Catal.*, Vol. 224, pp. 397–403, 2004.

151. E.G. Rodrigues, M.F.R. Pereira, J.J. Delgado, X. Chen, J.J.M. Orfao, *Catal. Commun.*, Vol. 16, pp. 64–69, 2011.
152. I. Sobczak, K. Jagodzinska, M. Ziolek, *Catal. Today*, Vol. 158, pp. 121–129, 2010.
153. K. Musialska, E. Finocchio, I. Sobczak, G. Busca, R. Wojcieszak, E. Gaigneaux, M. Ziolek, *Appl. Catal. A: Gen.*, Vol. 384, pp. 70–77, 2010.
154. S. Carrettin, P. McMorn, P. Johnston, K. Griffin, C.J. Kiely, G.A. Attard, G.J. Hutchings, *Top. Catal.*, Vol. 27, pp. 131–136, 2004.
155. S. Carrettin, P. McMorn, P. Johnston, K. Griffin, C.J. Kiely, G.J. Hutchings, *Phys. Chem. Chem. Phys.*, Vol, 5, pp. 1329–1336, 2003.
156. P. Haider, A. Baiker, *J. Catal.*, Vol. 248, pp. 175–187, 2007.
157. A. Abad, P. Concepcion, A. Corma, H. Garcia, *Angew. Chem.*, Vol. 117, pp. 4129–4134, 2005.
158. A. Abad, P. Concepcion, A. Corma, H. Garcia, *Angew. Chem., Int. Ed.*, Vol. 44, pp. 4066–4069, 2005.
159. A. Corma, M.E. Domine, *Chem. Commun.*, 2005, 4042–4044.
160. D.I. Enache, J.K. Edwards, P. Landon, B. Solsona-Espriu, A.F. Carley, A.A. Herzing, M. Watanabe, C.J. Kiely, D.W. Knight, G.J. Hutchings, *Science*, Vol. 311, pp. 362–365, 2006.
161. S. Mandal, K.K. Bando, C. Santra, S. Maity, O.O. James, D. Mehta, B. Chowdhury, *Appl. Catal. A: Gen.*, Vol. 452, pp. 94–104, 2013.
162. S. Mandal, C. Santra, K.K. Bando, O.O. James, S. Maity, D. Mehta, B. Chowdhury, *J. Mol. Catal. A: Chem.*, Vol. 378, pp. 47–56, 2013.
163. C. Santra, S. Rahman, S. Bojja, O.O. James, D. Sen, S. Maity, A.K. Mohanty, S. Mazumder, B. Chowdhury, *Cat. Sci & Tech.*, Vol. 3, pp. 360–370, 2013.
164. V.V. Pushkarev, V.I. Kovalchuk, J.L. Ditri, *J. Phys. Chem. B*, Vol. 108, pp. 5341–5348, 2004.
165. D.A. Outka, R.J.J. Madix, *J. Am. Chem. Soc.*, Vol. 109, pp. 1708–1714, 1987.
166. X.Y. Liu, R.J. Madix, C.M. Friend, *Chem. Soc. Rev.*, Vol. 37, pp. 2243–2261, 2008.
167. D.A. Outka, R.J. Madix, *Surf. Sci.*, Vol. 179, pp. 361–370, 1987.
168. B.J. Xu, X.Y. Liu, J. Haubrich, R.J. Madix, C.M. Friend, *Angew. Chem., Int. Ed.*, Vol. 48, pp. 4206–4209, 2009.
169. M.A. Lazaga, D.T. Wickham, D.H. Parker, G.N. Kastanas, B.E. Koel, *ACS Symp. Ser.*, Vol. 523, pp. 90–109, 1993.
170. J. Gong, D.W. Flaherty, R.A. Ojifinni, J.M. White, C.B. Mullins, *J. Phys. Chem. C*, Vol. 112, pp. 5501–5509, 2008.
171. M.A. Lazaga, D.T. Wickham, D.H. Parker, G.N. Kastanas, B.E. Koel, *ACS Symp. Ser.*, Vol. 523, pp. 90–109, 1993.
172. J.L. Gong, C.B. Mullins, *J. Am. Chem. Soc.*, Vol. 130, pp. 16458–16459, 2008.
173. X.Y. Liu, B.J. Xu, J. Haubrich, R.J. Madix, C.M. Friend, *J. Am. Chem. Soc.*, Vol. 131, pp. 5757–5759, 2009.
174. M. Boronat, A. Leyva-Perez, A. Corma, *Accounts Chem. Res.*, DOI: 10.1021/ar400068w.

175. (a) A. Abad, A. Corma, H. Garcia, *Chem. Eur. J.*, Vol. 14, pp. 212–222, 2008.
176. (b) Y. Guan, E.J.M. Hensen, *Appl. Catal. A: Gen.*, Vol. 361, pp. 49–56, 2009.
177. (c) S. Kanaoka, N. Yagi, Y. Fukuyama, S. Aoshima, H. Tsunoyama, T. Tsukuda, H. Sakurai, *J. Am. Chem. Soc.*, Vol. 129, pp. 12060–12061, 2007.
178. M. Boronat, A. Corma, *J. Catal.*, Vol. 284, pp. 138–147, 2011.

9

Nanosized Metal Oxide-Based Adsorbents for Heavy Metal Removal: A Review

Deepak Pathania* **and Pardeep Singh**

School of Chemistry, Faculty of Basic Sciences, Shoolini University, Himachal Pradesh, India

Abstract

Nanosized metal oxides (NMOs) mediated adsorption has been the subject of enormous studies related to water purification. The NMOs, mainly ferric oxides, manganese oxide and cerium oxides, provide high surface area and specific affinity for heavy metal adsorption from aqueous phase. Recently, new NMOs-based adsorbents development, characterization and evaluation of adsorption ability under varying experimental conditions are hot topics for researchers. The heavy metal adsorption mechanism has been explored on the basis of modern analytical techniques (SEM, TEM, XRD, FTIR, XPS, NMR, etc.). This review mainly illustrates various methods of preparing NMOs, their physicochemical properties, adsorption properties and applicability for heavy metal removal from aqueous phase. Magnetic NMOs have also been included, owing to their unique separation. Also presented are some porous material supported NMOs and NMO-based bionanocomposites, which are of importance due to their practical utility. Despite the many absorbents explored, various issues such as nanoparticle aggregation, reusability, larger size and nanotoxicity need to be discussed and resolved for highly efficient adsorption processes.

Keywords: Heavy metals, nanosized metal oxides, water treatment, hybrid adsorbent

**Corresponding author*: dpathania74@gmail.com

Ashutosh Tiwari and Mikael Syväjärvi (eds.) Advanced Materials for Agriculture, Food, and Environmental Safety, (243–264) 2014 © Scrivener Publishing LLC

9.1 Introduction

In recent years, heavy metal pollution has become one of the most serious environmental problems. The presence of heavy metals even in trace amounts can be toxic and hazardous to both flora and fauna. With the rapid growth of industries such as mining, surface finishing, energy producing, metallurgy, steel and iron, electroplating, electrolysis, leather and photography, the wastes containing heavy metals are causing a serious threat to the environment. Some elements like Fe, Zn, Cu, Co, Cr, Mn and Ni are required in small quantities for human metabolism, but may be toxic at higher levels. Others like lead, mercury, cadmium, arsenic, etc., have no beneficial role and are positively toxic [1, 2]. The environmental pollution from heavy metals can arise from anthropogenic and natural sources. With the rapid increase in industrialization, anthropogenic sources of pollution have been increased (Table 9.1). Toxic chemicals enter the human body

Table 9.1 Anthropogenic source of heavy metals [8,9].

Metal	Source
Mercury (Hg)	Chlor-alkali plants, thermal power plants, fluorescent lamps, hospital waste, electrical appliances, etc.
Arsenic (As)	Geogenic/natural processes, smelting operations, thermal power plants, fuel burning
Lead (Pb)	lead acid batteries, paints, E-waste, Smelting operations, coal-based thermal power plants, ceramics, bangle industry
Copper (Cu)	Mining, electroplating, smelting operations
Cadmium (Cd)	Zinc smelting, waste batteries, e-waste, paint sludge, incinerations and fuel combustion
Nickel (Ni)	Smelting operations, thermal power plants, battery industry
Molybdenum (Mo)	Spent catalyst
Zinc (Zn)	Smelting, electroplating
Vanadium (V)	Spent catalyst, sulphuric acid plant
Chromium(Cr)	Mining, industrial coolants, chromium salts manufacturing, leather tanning

mostly through food and water, and to a lesser extent through inhalation of polluted air. Thus, how to effectively remove even deeply hidden metals from water systems is a very important and challenging issue for environmental engineers. Nowadays, numerous methods have been proposed for efficient heavy metal removal from waters, including chemical precipitation, ion exchange, adsorption, membrane filtration and electrochemical technologies [3–5]. Among these methods, adsorption offers flexibility in design and operation and in many cases generates high-quality treated effluent. In addition, due to the reversible nature of most adsorption processes, the adsorbents can be regenerated by suitable desorption processes for multiple use, and many desorption processes have low maintenance costs, high efficiency, and are easy to operate [6, 7]. Therefore, the adsorption process has come to the forefront as one of the major techniques for heavy metal removal from wastewater.

Nanosized metal oxides (NMOs), including nanosized ferric oxides, manganese oxides, aluminum oxides, titanium oxides, magnesium oxides and cerium oxides are classified as promising for heavy metals removal from wastewater [10, 11]. This is because of their large surface areas and high activities caused by the size-quantization effect [12]. During the last decade, the research on metal oxide-based adsorbents for heavy metal removal has been gaining importance (Figure 9.1). Recent studies have suggested that many NMOs exhibit very favorable sorption to heavy metals in terms of high capacity and selectivity, which would result in deep removal of toxic metals to meet increasingly strict regulations. However, as the size of metal oxides is reduced from micrometer to nanometer levels, the increased surface energy unavoidably leads to their poor stability. Consequently, NMOs are prone to agglomeration. To improve the

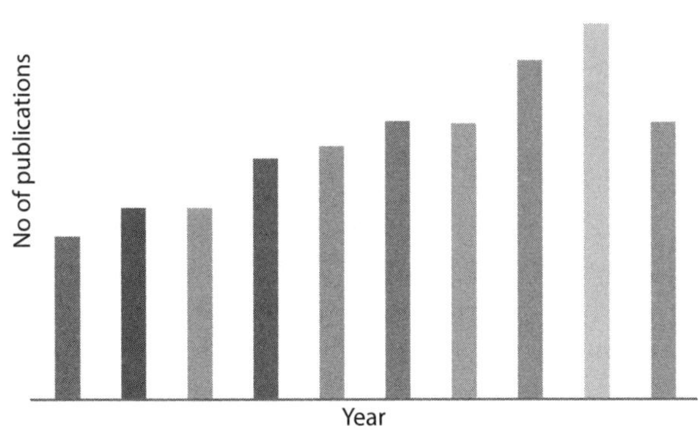

Figure 9.1 The number of publications produced during the last ten years on the topic of metal oxide-based adsorbents for heavy metal removal.

applicability of NMOs in real wastewater treatment, they were impregnated in porous supports of large size to obtain composite adsorbents [13]. The widely used porous supports include activated carbon, natural materials, synthetic polymeric hosts, etc.

This chapter presents a brief overview of several NMOs, including their synthesis and their sorption behavior towards heavy metals from aqueous systems under varying experimental conditions, the underlying mechanism responsible for the sorption, as well as their reusability.

9.2 Nanosized Metal Oxide

For adsorptional removal of heavy metals from aqueous systems, the most widely investigated NMOs include iron oxides, manganese oxides, aluminum oxides, and titanium oxides. The size and shape of NMOs are both important factors that affect their adsorption performance. During the last decades, facile synthetic methods have been widely applied to obtain shape-controlled, highly stable, and monodisperse metal oxide nanomaterials [14]. Generally, the synthesis methods can be classified into two categories: (1) physical methods, including inert gas condensation, severe plastic deformation, high-energy ball milling, ultrasound shot peening, and (2) chemical methods, including controlled chemical co-precipitation, chemical vapor condensation, pulse electrode position, liquid flame spray, liquid-phase reduction, etc. Among these synthesis protocols, co-precipitation, thermal decomposition and/or reduction, and hydrothermal synthesis techniques are used widely and are easily scalable with high yields [15–18]. The following section includes the synthesis, characterization, and application perspectives of NMOs.

9.2.1 Nano Ferric Oxides (NFeOs)

Iron is one of the most abundant elements found in earth. Availability of resources and environmental friendliness make iron a suitable adsorbent for heavy metal removal. The intensive NFeOs for heavy metal removal include hematite (α-Fe_2O_4) [19], goethite α(-FeOOH) [20], and amorphous hydrous Fe oxides [21] (Table 9.2). Hydrous ferric oxide (HFO) could be prepared by precipitation of ammonia with ferric chloride or nitrate solutions in carbonate-free environment by purging with N_2 [22]. Dzombak and Morel [23] prepared HFO with mean pore diameter of 3.8 nm and surface area of 600 m^2/g. The sorption of heavy metals to HFO seems poorly sensitive to the variation of ionic strength. For example, Swallow *et al.* [24]

Table 9.2 Nanosized metal oxides for heavy metal removal from water.

Adsorbent	Method	Morphology	Target metal	Isotherm	performance	Ref.
Hematite (α-Fe_2O_4)	Co-precipitation: $FeSO_4$+NaOH heated at 40°C for 2 days	Granular, crystal size 74 nm	Cu(II)	Langmuir	84.46 mg/g	[19]
Geothite (α-FeOOH)	Fe(NO3) precipitation	Needlelike length 200 nm, width < 50 nm	Cu(II)	–	100% removal	[20]
Hydrous amphorphous Fe oxide	Precipitation: $Fe(NO_3)_2$+NaOH	Particles; Diameter, 3.8 nm	Pb(II)	–	19.2 mg/g	[21]
	Sol-gel	Particles; Diameter, 10.8 nm	Cr(VI)	Freundlich	Cr(VI)	[27]
	Sol-gel	Particles; Diameter, 10.8 nm	Cr(VI), Cu(II), Ni(II)	Langmuir	17.2mg/g Cr(VI) 26.8 mg/g Cu(II), 23.6 mg/g Ni(II)	[28]
MnO_2	Precipitation:$Mn(NO_3)_2$ + NaMnO4 + NaOH	Particles; Diameter, 2.1 nm	Pb(II),		Surface diffusivities 1.7 × 10–16 cm2s-1	[34]
	Precipitation: $MnSO_4$+ $NaClO_4$	Particles	Pb(II), Cd(II)	Freundlich	-	[35]

(*Continued*)

Table 9.2 (*Cont.*)

Adsorbent	Method	Morphology	Target metal	Isotherm	performance	Ref.
TiO_2	Hydrolysis	Particles; Diameter, 2.1 nm	Zn(II), Cd(II),	Langmuir	Capacities:15.3(Zn), 7.9(Cd) mg/g	[42]
	Commercially	Particles; Diameter, 7.5 nm	Pb(II), Cd(II), Ni(II)	–	q_m= 401.4(Pb), 135.14(Cd), 114.4(Ni)	[43]
ZnO	Hydrothermal,	Nanosheets with square side of 1 μm and thickness in nanoscale	Pb(II)	Freundlich	6.7 mg/g	[49]
	Solvothermal	Nanoplates of 15 nm in thickness	Cu(II)	Langmuir	1600 mg/g	[50]
Hydrous Al oxides	Precipitation: NaOH+$Al(NO_3)$	Particles; Diameter 1.9 nm	Pb(II)	–	6.7 mg/g	[55]
γ-Al_2O_3	Precipitation:	Particles size 7.5	Ni(II)	–	176 mg/g	[56]

reported that Cu^{2+} and Pb^{2+} sorption to HFO was unaffected by different ionic strength from 0.005 to 0.5 M $NaClO_4$, or by change in the nature of the background electrolyte from $NaClO_4$ to a complex artificial seawater mixture.

Another important magnetic nanomaterial is nanosized magnetite. Chemical co-precipitation has been widely used to prepare magnetite nanoparticles by adding alkaline carbonate into solution containing Fe^{2+} and Fe^{3+} in a molar ratio of 1:2 [25, 26]. It was found that the particle size was reduced when a surfactant (such as oleic acid) was used during the preparation. Two methods were reported to prevent the change of the ratio caused by air oxidation. One is to conduct the reaction under an inert environment with nitrogen gas. Another is to set the initial Fe^{3+}: Fe^{2+}molar ratio less than 2:1 so that after the oxidation of Fe^{2+} to Fe^{3+}, the ratio approaches 2:1 [27, 28]. In addition, nano-Fe_3O_4 will be oxidized to nano-γ-Fe_2O_3. The size of the resultant Fe_3O_4 or γ-Fe_2O_3 hydrosols were considerably smaller than that reported in the literature, and no surfactants are needed to stabilize the sols. A decrease of specific saturation magnetization value was observed when the nanoparticles were coated with oleic acid. Scanning electron micrograph showed that the prepared Fe3O4 nanoparticle sol had an average diameter of 8.5 ± 1.3 nm, where needlelike γ-Fe_2O_3 nanoparticles with lengths of 20–50 nm and widths of 4–6 nm are visible [29]. Hu *et al.* investigted Cr(VI) removal by nano maghemite and found that the equilibrium period was independent of initial Cr(VI) concentration and the adsorptive capacity increased when pH decreased [30]. Nano maghemite has emerged as being highly selective for Cr(VI) from water. Negligible competition was observed from many coexisting ions. The adsorption capacity of nano maghemite for Cr(VI) (19.2 mg/g) is higher than that of diatomite (11.55 mg/g) [31], anatase (14.56 mg/g) [32], commercial activated carbon (15.47 mg/g) [33], and beech sawdust (16.13 mg/g) [34]. Based on X-ray diffraction (XRD), X-ray photoelectron spectroscopy (XPS), and Raman spectroscopic techniques, it could be deduced that no chemical redox reaction occurred during Cr(VI) retention, which also hints at the stability of nanoscale γ-Fe_2O_3. The adsorption mechanism of Cr(VI) onto γ-Fe_2O_3 is suggested to be electrostatic attraction, particularly at a relatively low pH.

9.2.2 Nano Manganese Oxides (NMnOs)

In the past decades, NMnOs have been exploited for sorption of cationic or anionic pollutants from natural waters, such as heavy metal ions, arsenate and phosphate, due to their polymorphic structures and higher specific surface area [34, 35]. The widely studied NMnOs for environmental applications include hydrous manganese oxide (HMO) and nanoporous/nanotunnel manganese oxides [36].

Parida *et al.* prepared HMO by adding $MnSO_4 \cdot H_2O$ into NaClO solution (containing active chlorine). The precipitate was washed with HCl to remove excessive alkali, followed by rinsing with double-deionized water. The BET surface area is around 100.5 m^2/g [37]. The BET surface of the resultant HMOs is 359 m^2/g. Heavy metal sorption onto HMOs, including Pb(II), Cd(II), and Zn(II), usually results in the inner-sphere complex formation, and it can be described by an ion-exchange process. Divalent metals on HMO consist of two similar steps as that of HFO: rapid adsorption of metal ions to the external surface followed by a slow intraparticle diffusion along the micropore walls [38]. The adsorption can be represented by the Freundlich model more reasonably than the Langmuir model, implying that the active sites of HMO surface are heterogeneous for metal sorption. The HMO prefers metal sorption in the order of $Pb^{2+} > Cd^{2+} > Zn^{2+}$, which might rest on the different softness of these metals [6]. Dyer *et al.* [39] explored the sorption behavior of radio nuclides on crystalline synthetic tunnel manganese oxides. Trace strontium (89 Sr) and cesium (137 Cs) ions were removed through ion-exchange mechanism. Selectivity coefficients were estimated as $K_{Cs/K} = 0.6$ and $K_{Sr/K} = 1.0$ for OMS-2 and $K_{Cs/Mg} = 7550$, $K_{Sr/Mg} = 50$ and $K_{Sr/Ca} = 10$ for OMS-1. It was concluded that OMS-2 was particularly effective for the separation of trace silver ions. The affinity trend for magnesium and calcium ion-extracted OMS-1 in HNO_3 was 137Cs > 59Fe > 51Cr ≈ 57Co ≈ 241Am > 54Mn > 63Ni > 65Zn > 236Pu > 89Sr. As reported by Pakarinen *et al.* [40], OMS materials exhibited selective adsorption of Cu^{2+}, Ni^{2+} and Cd^{2+} in the presence of Ca^{2+} and Mg^{2+}. The exchange rates were reasonably high due to the small particle dimensions. OMS materials are stable and their maximum Cu^{2+} uptake capacity was 0.9–1.3 mmol/g. Koivula *et al.* [41] found that hydrometallurgical wastewater (containing Al, Ca, Fe, Mg, Mn and Na) rich in manganese can be easily used as a manganese precursor for OMS synthesis. A synthetic raffinate solution, which contained 4500, 490, 300, 150, 200 and 3500 mg/L of Mn, Mg, Fe, Al, Ca and Na, respectively, could be used as the manganese source for preparing OMS-2.

9.2.3 Nano Titanium Oxides (NTOs)

Both bulk and nanoparticle TiO_2s show different chemical behavior, catalytic reactivity, and surface acidity based on their different structure [42, 43]. Engates *et al.* reported that specific surface area of the nanosized and bulk particles was 185.5 and 9.5 m^2/g, and the nominal particle sizes calculated from BET measurements were 8.3 and 329.8 nm, respectively [42]. The nanoparticles were able to simultaneously remove multiple metals (Zn, Cd, Pb, Ni, Cu) in the aqueous phase. The nanoparticles had more adsorption capacity than the bulk particles. Adsorption kinetics for heavy metals followed a modified first-order model, and the nanoparticles had a faster adsorption than the bulk ones. Langmuir isotherm was suitable to characterize metal adsorption onto TiO_2. By comparing the distribution coefficient (K_d), TiO_2 nanoparticles performed better than commercial activated carbon and other metal oxide nanoparticles. Liang *et al.* [44] concluded that nano-TiO2 (diameter = 10–50 nm, BET surface area = 208 m^2/g) showed adsorptive capacity to Zn and Cd as 15.3 and 7.9 mg/g, respectively, at pH = 9.0. The presence of common cations and anions (100–5000 mg/L) had no significant influence on the targeted metal (Zn^{2+} and Cd^{2+} ions of 1.0 mg/mL) adsorption under the given conditions.

9.2.4 Nano Zinc Oxides (NZnOs)

Nano zinc oxide is widely used in the catalyst industry, and for gas sensors, solar cells and so on [45, 46]. Besides some properties similar to TiO_2, ZnO nanoplates has many unique advantages, such as being simple, cheap, easy to prepare and convenient to tailor morphologically. As an adsorbent, ZnO was mostly applied to eliminate H_2S [47]. Lee *et al.* [48] prepared nanometer-size zinc oxide (ZnO) powder by solution-combustion method (SCM). The adsorption behavior of ZnO was compared with two titanium dioxide powders, P25 and ZnO, prepared by a homogeneous precipitation process at low temperature. The zinc oxide nanopowder showed higher removal rate of Cu^{2+} ions from the solution. The plate-like nanostructured ZnO with high specific surface area was fabricated by various methods such as hydrothermal [49], solvothermal [50], chemical vapor deposition [51], electrochemical deposition [52] and microwave methods [45]. The solvothermal-prepared ZnO nanoplates [35] are composed of two terminal non-polar planes with several microns in the planar dimensions and 10–15 nm in thickness. The nanoplates are porous with a pore diameter of 5–20 nm and a high specific surface area (147 m^2/g). These nanoplates have an adsorption capacity of >1600 mg/g for Cu(II) ions. The adsorption isotherm is subject to the Freundlich equation (K_F = 324.22 $(mg/g/mg)^{-n}$, n = 4.56), while the commercial ZnO nanopowders follow a Langmuir isotherm model. In addition, the metal sorbed nano-ZnO can be employed to fabricate other environmental materials. Ma *et al.* [53] reported a novel strategy to prepare ZnO/PbS heterostructured functional nanocomposite based on Pb^{2+} sorbed ZnO. In brief, ZnO nanosheets prepared via a hydrothermal approach were used to adsorb Pb^{2+} and then hydrothermally treated in aqueous solution containing sulfur source.

9.2.5 Nano Aluminum Oxides

Alumina (Al_2O_3) is a well-explored adsorbent for heavy metals. Chang *et al.* prepared [54] nanosized γ-Al_2O_3 by sol–gel method and employed it as solid phase extraction material for separation/preconcentration of trace metal ions. Sorption ability of γ-Al_2O_3 towards heavy metals can be enhanced by chemical or physical modification of γ-Al_2O_3 nanoparticles with certain functional groups containing some donor atoms such as oxygen, nitrogen, sulfur and phosphorus [55, 56]. The removal mechanism is changed on the immobilization of modifier at the surface of alumina. The target metals are not only removed by adsorption on the surface of the alumina but also by a surface attraction or chemical interaction due to modifier [56]. The fixing of γ-mercaptopropytrimethoxysilane (γ-MPTMS) on the surface of γ-Al_2O_3 would enhance the desorption capacity of γ-Al_2O_3 towards Cu, Hg, Au and Pd ions [57]. It was found that the γ- Al2O3 tends to be amorphous due to the formation of the chemical bond of Si–O–Al. The adsorption of metal ions onto γ-MPTMS-modified nano alumina was governed by three mechanisms: (1) metal ions adsorbed through the affinity of –SH, (2) the hydrolyzation of metal ions, and (3) electrostatic adsorption. The first mechanism plays a dominant role in acidic medium, while in basic solutions, the hydrolyzation and the electrostatic interactions are the significant forces of attraction. Sodium dodecyl sulfate-coated nano γ-Al_2O_3 was modified with 2, 4-dinitrophenylhydrazine (DNPH) as a solid-phase adsorbent for removal of trace Pb(II), Cr(III), Cd(II), Ni(II), Co(II) and Mn(II). The SEM image showed that naked alumina nanoparticles had a mean diameter of 53 nm, while that of the modified ones were in the range of 68–87 nm [58]. The BET surface area decreased from 42.62 m^2/g to 30.38 m^2/g after modification due to the bulk size of the organic ligand [59–61]. The sorption isotherms were better described by Freundlich model for Mn(II), Pb(II), Cr(III) and Cd(II) ions and by Langmuir model for Ni(II) and Co(II) ions. The maximum adsorption capacity values of the modified alumina nanoparticles (q_m, calculated from Langmuir equation) toward Cr(III), Cd(II) and Pb(II) ions from multiple-metal solution (mixture of six metal ions) were 100.0, 83.33 and 100.0 mg/g, respectively [62].

9.3 Hybrid Adsorbents

Nanosized metal oxides emerge as effective and adsorbent for heavy metal removal from aqueous phase. However, they are usually present as fine or ultrafine particles, which often lead to problems such as activity loss due to agglomeration and difficult separation [63]. An effective way to overcome these technical bottlenecks is to fabricate hybrid adsorbents by impregnating or coating NMOs particles into/onto porous materials of larger size [64]. The widely used supports include natural hosts such as bentonite [65], sand [66], montmorillonite [21] metallic oxide materials such as Al_2O_3 membrane [67] and porous manganese oxide complex [68], and synthetic polymer supports such as crosslinked ion-exchange resins [69, 70]. Some of the hybrid adsorbents are classified in Table 9.3.

9.3.1 Bentonite-Based Hybrid Nano-Metal Oxide Nanocomposites (B-NMOs)

Bentonite is a natural clay with a basic structural unit of two tetrahedrally coordinated sheets of silicon ions surrounding a sandwiched octahedrally coordinated sheet of aluminum ions with net negative surface charge [71]. Also, bentonite has a high exchange capacity, amphoteric pH-dependent surfaces and different modes of aggregation [72], which makes it a potential adsorbent for adsorption of heavy metals from aqueous solutions [73]. The iron oxide-coated bentonite (ICB) and magnesium oxide-coated bentonite (MCB) were prepared by precipitating the metal ions with sodium hydroxide on the surface of raw bentonite, followed by thermal treatment. The BET surface areas follow an order as ICB > RB > MCB. The ICB, RB and MCB were tested for the removal of Pb(II) from aqueous phase. The Langmuir monolayer adsorption capacities of RB, ICB and MCB toward Pb(II) from 0.1 M KNO3 solution were found to be 16.70, 22.20 and 31.86 mg/g, respectively. Both bentonite samples showed an increased uptake of Pb(II) with gradually increasing pH because H^+ can compete for the exchange site with Pb(II). Increasing the ionic strength from 0.01 to 0.1 M led to a significant decrease in Pb(II) adsorption. Eren *et al.* [74] also investigated magnesium oxide-coated bentonite for the removal of copper ions from aqueous solution. The adsorption of Cu(II) ions depends upon both the nature of the adsorbent surface as well as the Cu(II) species distribution solution, which are greatly affected by the pH of the system. The Langmuir monolayer adsorption capacity of MCB in 0.1 M KNO_3 solution was estimated to be 58.44 mg/g, whereas the adsorption capacity of RB was 42.41 mg/g, indicating that the treatment with magnesium oxide increased

Table 9.3 Some hybrid adsorbents with different support materials.

Adsorbent	Support materials	Target metal	Performance	Ref.
Hydrated ferric oxide	SiO_2	Cd(II)	0.2 mg/g at pH=8	[86]
	Polyacrilimide	Pb(II), Cd(II), Hg(II)	211.4 mg/g for Pb(II), 155.0 mg/g for Hg(II), 147.2 mg/g for Cd(II)	[87]
Fe_3O_4	Polymeric cation exchanger	Pb(II), Cd(II), Cu(II)	Ffrom 1ppm to <5ppb	[88]
	Cyclodexatrin	Cu(II)	47.2 mg/g	[89]
Fe_2O_4	Spoilite	Ni(II)	18.3 mg/g	[90]
MnO_2	Crushed brick	Pb(II)	0.030 mmol/g	[91]
	Sand	Ni(II), Pb(II), Mn(II)	3.33mg/g for Ni(II), 0.029mmol/g(Pb), 1.069mg/g for Mn(II)	[92, 91, 93]
	Silica	Mn(II)	0.396 mg/g	[94]
	Calcinated -starfish	Mn(II)	1.480 mg/g	[93]
	Zeolite	Cu(II), Pb(II)	0.116 mmol/g, 0.349mmol/g	[95]
	Bentonite	Cu(II),	105.38 mg/g	[96]
	Diaotomite	Cu(II), Pb(II)	99.00 mg/g (Pb), 27.86 mg/g(Cd)	[97]

Adsorbent	Support materials	Target metal	Performance	Ref.
MnO_2	Reduced graphene	Hg(II)	100 % removal	[98]
Hydrous MnO_2	Ploymeric cation eanger xcD-001,001 x7;D-113	Pb(II), Cd(II), Zn(II)	K_d inceased by 20-800 times as compared to host exchanger, sorption capacity incerasd by 50 – 300%	[99]
MgO_2	Bentonite	Cu(II)	58.44 mg/g	[100]
$Mg(OH)_2/MgO$	Al_2O_3 membrane	Ni(II)	147.2 mg/g($Mg(OH)_2$),	[101]
ZnO	Activated carbon	Pb(II)	100 % removal	[102]
Nanometer calicium titanate	Al_2O_3	Pb(II), Cd(II), Zn(II)	124 mg/g(Pb), 8.58mg/g (Cd) and 13.86 mg/g(Zn)	[103]
Geothite	Sand (quartz)	Cu(II), Pb(II)	0.259 mg/g (Cu/sand), 1.211 mg/g(Pb/Sand)	[104]
	clinoptilolite	Cu(II), Mn(II), Zn(II)	0.5 mmol/g(Cu), 0.09mmol/g(Mn), 0.2mmol/g(Zn)	[105]
	Sand	Cd(II), Pb(II)	Maximum adsorption capacity values:7.4 μg/g(Cd/sand), 702 μg/g(Pb/sand)	[106]
	Sand (quartz)	Cd(II),	0.2mg/g	[107]

the number of adsorption sites to a large extent, which may be attributed to an increase in surface charge due to the formation of magnesium oxide on the bentonite surface.

9.3.2 Polymer-Supported Nano-Metal Oxide Nanocomposites (P-NMOs)

Polymeric supports are attractive choices for nano-metal oxide support because of their controllable pore size and surface chemistry as well as their excellent mechanical strength for long-term use. Zhao *et al.* reviewed recent work on polymer-based nanocomposites for environmental application [75]. Su *et al.* prepared a new hybrid adsorbent HMO-001 by impregnating nanosized hydrous manganese dioxide (HMO) onto a porous polystyrene cation exchanger resin (D-001) [76]. The HMO-001 was tested for lead adsorption onto HMO-001 and the maximum capacity of HMO-001 towards lead ion was about 395 mg/g. As compared to a macroporous cation exchanger, D-001, HMO-001 exhibited highly selective lead retention from waters in the presence of competing $Ca^{2+,}$ Mg^{2+} and Na^{+} at high concentration levels. Fixed-bed column adsorption of a simulated water indicated that lead retention on HMO-001 resulted in a significant decrease of this toxic metal from 1 mg/L to below 0.01 mg/L (the drinking water standard recommended by WHO).

9.3.3 Zeolites-Supported Nano Metal Oxide Nanocomposites (P-NMOs)

Zeolites offer an attractive and inexpensive alternative for e-removal of organic and inorganic contaminants [77]. Natural zeolites are low cost and act as cation exchangers for metallic contaminants. The adsorption capacity of zeolite-based nanocomposites results from their high surface area and net negative charge on their channel structure, which causes an affinity for heavy metals [78]. The NaY zeolites serve as potential adsorbent for heavy metals with pore diameter of 0.78 nm, large surface area and high cation exchange capacity [79, 80]. Oliveira *et al.* [81] prepared NaY zeolite and magnetic iron oxides-based magnetic adsorbent. The NaY zeolite:iron oxide magnetic composites were prepared at a weight ratio of 3:1, which was chosen to keep a relatively high content of iron oxide and thereafter to avoid the decrease in adsorption capacity of the composites. They were prepared by precipitation of iron oxides or hydroxides onto the zeolite surface. Fe oxide in the composites had a smaller particle size of 25 nm for pure Fe oxide and 16 nm for the composite. The immobilized Fe oxides are mainly present as maghemite along with some goethite. Upon controlled H_2 treatment, the iron oxides were reduced to metallic iron with higher magnetization of composite. The adsorption of Cu^{2+}, Cr^{3+} and Zn^{2+} from aqueous solutions on the 3:1 zeolite:Fe oxide composite was investigated. The adsorption capacity was in the order of $Cr^{3+} > Cu^{2+} > Zn^{2+}$.

9.3.4 Metal Oxides-Based Nanocomposites

Research on metal oxides-based nanocomposites is an important strategy to overcome the drawbacks of binary metal oxide absorbents. Mahmoud *et al.* [82] prepared novel nanosorbents by surface impregnation of magnetic nano-iron oxide (Nano-Fe_3O_4) with nano-silicon oxide (Nano-SiO_2) for the formation of (Nano-Fe_3O_4–SiO_2) sorbent. The material was further functionalized with target nitrogen donor atoms via covalent surface binding and immobilization of triethylenetetramine (TETA) for the formation of (Nano-Fe_3O_4–SiO_2–TETA) sorbent. The average particle size was found in the range of 14–40 nm based on the SEM analysis. The nanosorbents were examined to evaluate their selectivity and efficiency in removal of Cu(II) and Pb(II) from water samples by the batch equilibrium and micro-column techniques. Nano-Fe_3O_4–SiO_2–TETA sorbent was identified by a high Cu(II) sorption capacity (480 μ mol g^{-1}) in pH 7.0, while Nano-Fe_3O_4 and Nano-Fe_3O_4–SiO_2 sorbents were characterized by high affinity to Pb(II).

Mahapatra *et al.* [83] prepared Fe_2O_4-Al_2O_3 nanocomposite fiber using electrospinning method metal ions from aqueous solution. Firstly, boehmite nanoparticle was synthesized by sol-gel method and was impregnated in PVP-Iron acetylacetonate solution in a ratio of 1:1:2. This boehmite-impregnated polymer solution was electrospun to form nanocomposite polymer fiber. Iron oxide alumina mixed nanocomposite fiber was characterized by UV-vis-DRS, IR, SEM-EDX and TEM. The removal percentage metal was found in the order of $Cu^{2+} < Pb^{2+} < Ni^{2+} < Hg^{2+}$. The maximum sorption capacities by applying the Langmuir equation were found to be 4.98 mg/g for Cu^{+2}, 32.36 mg/g for Ni^{+2}, 23.75 mg/g for Pb^{+2} and 63.69 mg/g for Hg^{+2} ions.

Lee *et al.* [84] reported hydrothermal preparation of flower-like TiO_2-graphene oxide (GO-TiO_2) hybrid by stirring a titanium oxide precursor in isopropyl alcohol with GO colloidal solution. The GO-TiO_2 hybrid was applied for the removal of heavy metal ions from water. The flower-like TiO_2 on GO structure significantly improved the removal efficiency of heavy metals. After 6 h and 12 h of hydrothermal treatment at 100°C, the GO-TiO_2 hybrid adsorption capacities of heavy metal ions were respectively 44.8 ± 3.4 and 88.9± 3.3 mg/g for removing Zn^{2+}, 65.1± 4.4 and 72.8 ± 1.6 mg/g for removing Cd^{2+} and 45.0 ± 3.8 and 65.6 ± 2.7 mg/g for removing Pb^{2+} at pH 5.6. In contrast, colloidal GO under identical conditions showed removal capacities of 30.1 ± 2.5 (Zn^{2+}), 14.9± 1.5 (Cd^{2+}) and 35.6 ± 1.3 mg/g (Pb^{2+}). The TiO_2 flowers were markedly formed upon GO as the hydrothermal treatment time at 100°C increased from 6 h to 12 h.

Yuan *et al.* [85] prepared novel multifunctional microsphere with a large pore size mesoporous silica shell (10.3 nm) and a magnetic core (Fe_3O_4) which has been successfully synthesized via a facile two-step sol–gel method. Microspheres were fabricated using cetyltrimethy lammonium bromide (CTAB) as pore-forming agents, tetraethyl orthosilicate (TEOS). Compared with small pore size mesoporous silica magnetic microspheres (2–4 nm), the large pore size one can graft 447 mg/g amino groups in order to adsorb more heavy metal ions (Pb^{2+}: 880.6 mg/g, Cu^{2+}: 628.3 mg/g, Cd^{2+}: 492.4 mg/g). The results suggest that the large pore size multifunctional microspheres are potentially useful materials which are highly effective for adsorbing and removing different heavy metal ions in aqueous solution.

9.4 Conclusion

Nowadays, NMOs are widely investigated as highly efficient adsorbents for heavy metal separation from wastewater. They exhibit various advantages such as fast kinetics, high capacity, and preferable sorption toward heavy metals in aqueous phase. Nevertheless, to further promote the practical application of NMOs in removal of heavy metal pollution, some technical issues still exist which need to be resolved. For example, when applied in aqueous solution, NMOs tend to aggregate into large-size particles and their capacity loss seems inevitable. The low-cost and efficient separation of heavy metals from wastewater still remains an attractive but challenging task. Fortunately, NMOs-based composite adsorbents seem to be an effective approach to respond to all the above technical problems. However, various issues need to be solved concerning the development of more facile processes to obtain the composite adsorbents, the answer to the interplay between the hosts and the supported NMOs, the long-term performance of the composite adsorbents, as well as their field application in heavy metal contaminated water treatment.

References

1. M. Jamil, M.S. Zia, and M. Qasim, *Journal of Chemical Society of Pakistan*, Vol. 32, p. 370, 2010.
2. S. Khan, Q. Cao, Y.M. Zheng, Y.Z. Huang, and Y.G. Zhu, *Environmental Pollution* Vol. 152, p. 686, 2008.

3. Y.H. Wang, S.H. Lin, and R.S. Juang, *Journal of Hazardous Material*, Vol. 102, p. 291, 2003.
4. D.W. O'Connell, C. Birkinshaw, T.F. O'Dwyer, *Bioresource Technology*, Vol. 99, p. 6709, 2008.
5. T.A. Kurniawan, G.Y.S. Chan, W.H. Lo, and S. Babel, *Chemical Engineering Journal*, Vol. 118, p. 83 2006.
6. Q. Su, B.C. Pan, S.L. Wan, W.M. Zhang, and L. Lv, *Journal of Colloid Interface Science*, Vol. 349, p. 607, 2010.
7. R. Koivula, J. Pakarinen, M. Sivenius, K. Sirola, R. Harjula, E. Paatero, *Separation and Purification Technology*, Vol. 70, p. 53, 2009.
8. D. Chakraborti, B. Das, and M.M. Rahman. *Molecular Nutrition and Food Research*, Vol. 53, p. 542, 2009.
9. http://insaindia.org/pdf/Hazardous_Metals.pdf.
10. J.F. Liu, Z.S. Zhao, and G.B. Jiang, *Environmental Science and Technology*, Vol. 42, p. 6949, 2008.
11. S.H. Huang, and D.H. Chen, *Journal of Hazardous Materials*, Vol. 163, p. 174, 2009.
12. X. Guan, Y. Qin, L. Wang, R. Yin, M. Lu, and Y. Chin, *Journal of Environmental Sciences*, Vol. 28, p. 436, 2007.
13. T. Pradeep, and Anshup, *Thin Solid Films*, Vol. 517, p. 6441, 2009.
14. B.L. Cushing, V.L. Kolesnichenko, and C.J. O'Connor, *Chemical Reviews*, Vol. 104, p. 3893, 2004.
15. J. Park, K.J. An, Y.S. Hwang, J.G. Park, H.J. Noh, J.Y. Kim, J.H. Park, N.M. Hwang, and T. Hyeon, *Nature Materials*, Vol. 3, p. 891, 2004.
16. X. Wang, J. Zhuang, Q. Peng, Y.D. Li, *Nature*, Vol. 437, p. 121, 2005.
17. Y. Ju-Nam, and J.R. Lead, *Science of the Total Environment*, Vol. 400, p. 396, 2008.
18. J. Wang, S. Zheng, Y. Shao, J. Liu, Z. Xu, and D. Zhu, *Journal of Colloid and Interface Science*, Vol. 349, p. 293, 2010.
19. P.R. Grossl, D.L. Sparks, and C.C. Ainsworth, *Environmental Science and Technology*, Vol. 28, p. 1422, 1994.
20. Y.H. Chen, and F.A. Li, *Journal of Colloid Interface Science*, Vol. 347, p. 277, 2010.
21. M. Fan, T. Boonfueng, Y. Xu, L. Axe, and T.A. Tyson, *Journal of Colloid Interface Science*, Vol. 281, p. 39, 2005.
22. J. Hu, G.H. Chen, and I.M.C. Lo, *Water Research*, Vol. 39, p. 4528, 2005.
23. D.A. Dzombak, and F.M.M. Morel, *Journal of Colloid Interface Science*, Vol. 112, p. 588, 1986.
24. K.C. Swallow, D.N. Hume, and F.M.M. Morel, *Environmental Science Technology*, Vol. 14, p. 1326, 1980
25. P. Tartaj, M.P. Morales, S. Veintemillas-Verdaguer, and T. Gonzalez-Carreno, *Journal of Physics D: Applied Physics*, Vol. 36, p. 182, 2003.
26. P. Tartaj, M.P. Morales, T. Gonzalez-Carreno, S. Veintemillas-Verdaguer, and C.J. Serna, *Journal of Magnetism and Magnetic Materials*, Vol. 290, p. 28, 2005.

27. S.E. Khalafalla, and G.W. Reimers, *Prep. IEEE Trans. Magn.*, Vol. 16, p. 178, 1980.
28. L. Fu, V.P. Dravid, and D.L. Johnson, *Applied Surface Science*, Vol. 181, p. 173, 2001
29. Y.S. Kang, S. Risbud, J.F. Rabolt, and P. Stroeve, *Chemistry Materials,* Vol. 8, p. 2209, 1996.
30. J. Hu, G.H. Chen, and I.M.C. Lo, *Water Research*, Vol. 39, p. 4528, 2005.
31. P. Trivedi, J.A. Dyer, and D.L. Sparks, *Environmental Sciences and Technology*, Vol. 37, p. 908, 2003.
32. T.N.D. Dantas, A.A.D. Neto, and M. Moura, *Water Research*, Vol. 35, p. 2219, 2001.
33. C.H. Weng, J.H. Wang, and C.P. Huang, *Water Science Technology*, Vol. 35, p. 55, 1997.
34. S. Babel, and T.A. Kurniawan, *Chemosphere*, Vol. 54, p. 951, 2004.
35. H.Q. Wang, G.F. Yang, Q.Y. Li, X.X. Zhong, F.P. Wang, Z.S. Li, and Y.H. Li, *New Journal of Chemistry*, Vol. 35, p. 469, 2011.
36. P. Larger-Trivedi, and L. Axe, *Journal of Colloid and Interface Science*, Vol. 218, p. 554, 1999.
37. K.M. Parida, S.B. Kanungo, and B.R. Sant, *Electrochemica Acta*, Vol. 26, p. 435, 1981.
38. M. Fan, T. Boonfueng, Y. Xu, L. Axe, and T.A. Tyson, *Journal of Colloid and Interface Science*, Vol. 281, p. 39, 2005.
39. A. Dyer, M. Pillinger, J. Newton, R. Harjula, T. Moller, and S. Amin, *Chemistry of Materials*, Vol. 12, p. 3798, 2000.
40. J. Pakarinen, R. Koivula, M. Laatikainen, K. Laatikainen, E. Paatero, and R. Harjula, *Journal of Hazardous Materials*, Vol. 180, p. 234, 2010.
41. R. Koivula, J. Pakarinen, M. Sivenius, K. Sirola, R. Harjula, and E. Paatero, *Separation and Purification Technology,* Vol. 70, p. 53, 2009.
42. K.E. Engates, and H.J. Shipley, *Environmental Science and Pollution Research*, Vol. 18, p. 386, 2011.
43. D.E. Giammar, C.J. Maus, and L.Y. Xie, *Environmental Engineering Science*, Vol. 24, p. 85, 2007.
44. P. Liang, T.Q. Shi, and J. Li, *International Journal of Environmental Analytical Chemistry*, Vol. 84, p. 315, 2004.
45. Z.H. Jing, and J.H. Zhan, *Advanced Materials*, Vol. 20, p. 4547, 2008.
46. T.P. Chou, Q.F. Zhang, G.E. Fryxell, and G.Z. Cao, *Advanced Materials*, Vol. 19, p. 2588, 2007.
47. X.F. Ma, Y.Q. Wang, M.J. Gao, H.Z. Xu, and G.A. Li, *Catalysis Today*, Vol. 158, p. 459, 2010.
48. H.S. Kim, S.-Y. Park, B.Y. Hur, and S.W. Lee, *Eco-Materials Processing and Design, VI*, Trans Tech Publications, Korea, pp. 510–513, 2005.
49. T.M. Shang, J.H. Sun, Q.F. Zhou, and M.Y. Guan, *Crystal Research and Technology*, Vol. 42, p. 1002, 2007.
50. T. Ghoshal, S. Kar, and S. Chaudhuri, *Crystal Growth and Design*, Vol.7, p. 136, 2007.

51. N. Zhang, R. Yi, R.R. Shi, G.H. Gao, G. Chen, and X.H. Liu, *Materials Letters*, Vol. 63, p. 496, 2009.
52. C. Xu, D. Kim, J. Chun, K. Rho, B. Chon, S. Hong, and T. Joo, *Journal of Physical Chemistry B*, Vol.110, p. 21741, 2006.
53. X.F. Ma, Y.Q. Wang, M.J. Gao, H.Z. Xu, and G.A. Li, *Catalysis Today*, Vol. 158, p. 459, 2010.
54. G. Chang, Z. Jiang, T. Peng, and B. Hu, *Acta Chimica Sinica*, Vol. 61, p. 84, 2003.
55. M. Hiraide, M.H. Sorouradin, and H. Kawaguchi, *Analytical Sciences*, Vol. 10, p. 125, 1994.
56. M. Ghaedi, K. Niknam, A. Shokrollahi, E. Niknam, H.R. Rajabi, and M. Soylak, *Journal of Hazardous Materials*, Vol. 155, p. 121, 2008.
57. D.J. Yang, B. Paul, W.J. Xu, Y. Yuan, E.M. Liu, X.B. Ke, R.M. Wellard, C. Guo, Y. Xu, Y.H. Sun, and H.Y. Zhu, *Water Research*, Vol. 44, p. 741, 2010.
58. D.J. Yang, B. Paul, W.J. Xu, Y. Yuan, E.M. Liu, X.B. Ke, R.M. Wellard, C. Guo, Y. Xu, Y.H. Sun, and H.Y. Zhu, *Water Research*, Vol. 44, p. 741, 2010.
59. R. Celis, M.C. Hermosin, and J. Cornejo, *Environmental Sciences and Technology*, Vol. 34, p. 4593, 2000.
60. Y. Mao, and B.M. Fung, *Chemistry of Materials*, Vol. 10, p. 509, 1998.
61. K.S. Walton, and R.Q. Snurr, *Journal of American Chemical Society*, Vol. 129, p. 8552, 2007.
62. Afkhami, M. Saber-Tehrani, and H. Bagheri, *Journal of Hazardous Materials*, Vol. 181, p. 836, 2010.
63. L. Cumbal, and A.K. Sengupta, *Environmental Science and Technology*, Vol. 39, p. 6508, 2005
64. M. Jang, W.F. Chen, and F.S. Cannon, *Environmental Science and Technology*, Vol. 42 p. 3369, 2008.
65. K.L. Chen, S.E. Mylon, and M. Elimelech, *Langmuir*, Vol. 23, p. 5920, 2007.
66. B.Ø. Hansen, P. Kwan, M.M. Benjamin, C.W. Li, and G.V. Korshin, *Environmental Science and Technology*, Vol. 35, p. 4905, 2001.
67. S. Zhang, F.Y. Cheng, Z.L. Tao, F. Gao, and J. Chen, *Journal of Alloys and Compounds*, Vol. 426, p. 281, 2006.
68. J. Otsu, and Y. Oshima, *Journal of Supercriticals Fluids*, Vol. 33, p. 61, 2005.
69. B.J. Pan, H. Qiu, B.C. Pan, G.Z. Nie, L.L. Xiao, L. Lv, W.M. Zhang, Q.X. Zhang, and S.R. Zheng, *Water Research*, Vol. 44, p. 815, 2010.
70. Q. Su, B.C. Pan, B.J. Pan, Q.R. Zhang, W.M. Zhang, L. Lv, X.S. Wang, J. Wu, and Q.X. Zhang, *Science of the Total Environmental*, Vol. 407, p. 5471, 2009.
71. P.F. Luckham, and S. Rossi, *Advances in Colloid* and *Interface Science*, Vol. 82, p. 43, 1999.
72. E. Eren, *Journal of Hazardous Materials*, Vol. 159, p. 235, 2008.
73. E. Eren, *Journal of Hazardous Materials*, Vol. 165, p. 63, 2009.
74. E. Eren, A. Tabak, and B. Eren, *Desalination*, Vol. 257, p. 163, 2010.
75. X. Zhao, L. Lv, B. Pan, W. Zhang, S. Zhang, and Q. Zhang, *Chemical Engineering Journal*, Vol. 178, p. 161, 2011

76. Q. Su, B.C. Pan, B.J. Pan, Q.R. Zhang, W.M. Zhang, L. Lv, X.S. Wang, J. Wu, and Q.X. Zhang, *Science of the Total Environmental*, Vol. 407 p. 5471, 2009.
77. H.H. Murray, *Applied Clay Science*, Vol. 17, p. 207, 2007.
78. F. Cadena, R. Rizvi, and R.W. Peter, "Feasibility studies for the removal of heavy metal from solution using tailored bentonite, hazardous and industrial wastes," in: *Proceedings 22nd Mid Atlantic Industrial Waste Conference*, Drexel University, p. 77, 1990.
79. K. Ryachi, and A. Bencheikh, *Annales de Chimie Science des Matériaux*, Vol. 23, p. 393, 1998.
80. M. Trgo, and J. Peric, *Journal of Colloid* and *Interface Science*, Vol. 260, p. 166, 2003.
81. L.C.A. Oliveir, D.I. Petkowicz, A. Smaniotto, and S.B.C. Pergher, *Water Research*, Vol. 38, p. 3699, 2004.
82. M.E. Mahmoud, M.S. Abdelwahab, and E.M. Fathallah, *Chemical Engineering Journal*, Vol. 223, p. 318, 2013.
83. http://dx.doi.org/doi:10.1016/j.jhazmat.2013.04.045 (accessed on 15/05/2013).
84. Y.C. Lee, and J.W. Yang, *Journal of Industrial and Engineering Chemistry*, Vol. 18, p. 1178, 2012.
85. Q. Yuan, N. Li, Y. Chi, W.C. Geng, W. Yan, Y. Zhao, X. Li, and B. Dong, *Journal of Hazardous Materials*, Vol. 254–255, p. 157, 2013.
86. C.H. Lai, C.Y. Chen, B.L. Wei, and S.H. Yeh, *Water Research*, Vol. 36, p. 4943, 2002.
87. G.N. Manju, K.A. Krishnan, V.P. Vinod, and T.S. Anirudhan, *Journal of Hazardous Materials*, Vol. 91, p. 221, 2002.
88. B.C. Pan, B.J. Pan, W.M. Zhang, Q.J. Zhang, Q.R. Zhang, P.J. Jiang, and Q.X. Zhang, Chinese Patent: CN 200710191355.3, 2007.
89. Z.M. Badruddoza, A.S.H. Tay, P.Y. Tan, K. Hidajat, and M.S. Uddin, *Journal of Hazardous Materials*, Vol. 185, p. 1177, 2011.
90. S. Lazarevic, I. Jankovic-Castvan, V. Djokic, Z. Radovanovic, D. Janackovic, and R. Petrovic, *Journal* of *Chemical and Engineering Data*, Vol. 55, p. 5681, 2010.
91. N. Boujelben, J. Bouzid, and Z. Elouear, *Adsorption Science and Technology*, Vol. 27, p. 177, 2009.
92. N. Boujelben, J. Bouzid, Z. Elouear, and M. Feki, *Environmental Technology*, Vol. 31, p. 1623, 2010.
93. S.M. Lee, W.G. Kim, J.K. Yang, and D. Tiwari, *Environmental. Technology*, Vol. 31, 445, 2010.
94. P.Y. Hu, Y.H. Hsieh, J.C. Chen, and C.Y. Chang, *Journal of Colloid* and *Interface Science*, Vol. 272, p. 308, 2004.
95. E.E. Turkey, *Journal of Hazardous Materials*, Vol. 159, p. 235, 2008.
96. M.A.M. Khraisheh, Y.S. Al-Degs, and W.A.M. McMinn, *Chemical Engineering Journal*, Vol. 99, p. 177, 2004.
97. T.S. Sreeprasad, S.M. Maliyekkal, K.P. Lisha, and T. Pradeep, *Journal of Hazardous Materials*, Vol. 186, p. 921, 2011.

98. B.C. Pan, Q. Su, W.M. Zhang, Q.X. Zhang, H.Q. Ren, Q.R. Zhang, and B.J. Pan, Chinese Patent: CN 101224408A, 2008.
99. Y. Kikuchi, Q.R. Qian, M. Machida, and H. Tatsumoto, *Carbon*, Vol. 44, p. 195, 2006.
100. P.Y. Hu, Y.H. Hsieh, J.C. Chen, and C.Y. Chang, *Journal of Colloid* and *Interface Science*, Vol. 272, p. 308, 2004.
101. S.M. Lee, W.G. Kim, C. Laldawngliana, and D. Tiwari, *Journal* of *Chemical and Engineering Data*, Vol. 55, p. 3089, 2010.
102. Y. Kikuchi, Q.R. Qian, M. Machida, and H. Tatsumoto, *Carbon*, Vol. 44, p. 195, 2006.
103. D. Zhang, *Advances in Materials Research*, Vol. 152, p. 670, 2010.
104. C.H. Lai, and C.Y. Chen, *Chemosphere*, Vol. 44, p. 1177, 2001.
105. M.K. Doula, *Water Research*, Vol. 43, p. 3659, 2009.
106. C.H. Lai, C.Y. Chen, B.L. Wei, and C.W. Lee, *Journal* of *Environmental Science* and *Health Part A*, Vol. 36, p. 747, 2001.
107. C.H. Lai, C.Y. Chen, B.L. Wei, and S.H. Yen, *Water Research*, Vol. 36, p. 4943, 2002.

10

Future Prospects of Phytosynthesized Transition Metal Nanoparticles as Novel Functional Agents for Textiles

Shahid-ul-Islam, Mohammad Shahid and Faqeer Mohammad*

Department of Chemistry, Jamia Millia Islamia (A Central University), New Delhi, India

Abstract

Synthesis of transition metal nanoparticles has become a promising field of research in recent years. Chemical approaches are the most popular methods for their production. However, some chemical methods cannot avoid the use of toxic chemicals in their synthesis protocol. Chemical methods are quite expensive and are potentially dangerous to the environment. Recently, the utilization of biological systems has emerged as a novel method for the green synthesis of nanoparticles. Among biological agents, the use of plants in the synthesis of transition metal nanoparticles is gaining importance due to their biocompatibility, low toxicity, green approach and environmentally-friendly nature. Owing to the biodiversity of plant biomasses, many bioconstituents like proteins, enzymes, antioxidants, amino acids, alkaloids, and other polyols have contributed to the synthetic processes. The intent of this chapter is to survey some of the important scientific research and developmental works pertaining to transition metal nanoparticle synthesis using plant materials. It also highlights how the phyto-inspired synthesized nanoparticles have been integrated into textile materials in order to improve their performances and extend their practical applications by developing new functionalities. Finally, a few recommendations have been proposed for further research efforts to advance this field of study.

Keywords: Phytosynthesis, bioconstituents, nanoparticles, functional textiles, antimicrobial

**Corresponding author*: faqeermohammad@rediffmail.com

Ashutosh Tiwari and Mikael Syväjärvi (eds.) Advanced Materials for Agriculture, Food, and Environmental Safety, (265–290) 2014 © Scrivener Publishing LLC

10.1 Introduction

Over the past few years, application of nanotechnology has been booming in many areas, including material science, mechanics, electronics, optics, medicine, plastics, energy, electronics and aerospace [1]. New techniques to manufacture nanoparticles with well-defined sizes, shapes and crystal structures are constantly being studied and developed [2]. Conventional methods of nanoparticle synthesis, physical (e.g., evaporation and laser ablation) [3, 4] and chemical (reduction of salts) [5–7] methods are the most popular methods for their production. Some chemical methods cannot avoid the use of toxic and/or inherently hazardous reagents, surfactants and solvents [2], generation of hazardous byproducts and high energy consumption [8]. Owing to these reasons, the development of efficient green chemistry methods for synthesis of metal nanoparticles has become a major focus of researchers [9]. Consequently, novel, environmentally-benign procedures for the synthesis of transition metal nanoparticles has led to the recent interest in using biological entities.

One of the most considered methods for the production of metal nanoparticles is by using organisms like microbes [10], plants and plant extracts [11] and enzymes [12]. Among these organisms plants seem to be the best candidates since they are suitable for large-scale biosynthesis of nanoparticles. Nanoparticles produced by plants are more stable and the rate of synthesis is faster than in the case of microorganisms. Moreover, nanoparticles are more various in shape and size in comparison with those produced by other organisms [9]. Although numerous pieces of research have been conducted on the subject of nanoparticle synthesis using biological agents, this chapter summarizes only some of the important scientific research and developmental works pertaining to transition metal nanoparticle synthesis using plant materials. It also highlights how the phyto-inspired synthesized nanoparticles have been integrated into textile materials to improve their performances and to extend their practical applications by developing new functionalities.

10.2 Synthesis of Transition Metal Nanoparticle Using Various Plant Parts

Researchers from many fields, such as biology, physics, chemistry, material and engineering, are showing much interest in the utilization of plants, mainly due to the elimination of the elaborate process of maintaining cell cultures compared to other environmentally-benign biological processes

[13, 14]. Up to now, several plant parts have been successfully utilized for efficient and rapid synthesis of transition metal nanoparticles. Recently, Iravani [9] reviewed the green synthesis of metal nanoparticles using plants. Mittal and coworkers shared their views about the synthesis and characterization of metallic nanoparticles using plant extracts in their recent review [15]. More recently, several plants have been studied in the synthesis of nanoparticles (Table 10.1).

10.2.1 Silver – Most Versatile Transition Metal Nanoparticle Synthesized by Using Plants

Silver nanoparticles have attracted intensive research interest because of their versatile applications in diverse areas of the biomedical, agricultural, electronic fields, etc. Biosynthesis of silver nanoparticles by a simple procedure using leaf extract of *Aloe vera* carried out by Chandran and colleagues [16] resulted in the production of spherical nanoparticles of 15.2 nm ± 4.2 nm size. Ramteke *et al.* [17] reported the synthesis of silver nanoparticles having an average particle size of about 18 nm using aqueous leaf extract of *Ocimum sanctum*. The Ag nanoparticles were stabilized by eugenols, terpenes and other aromatic compounds present in the leaves extract of tulsi plant. These nanoparticles were found to be highly active against *Staphylococcus aureus* and *E. coli*. Furthermore, the synthesis of quasi-spherical silver nanoparticles from silver nitrate using apiin, a bioconstituent isolated from henna leaves (*Lawsonia inermis*), has been reported. The formed silver nanoparticles had a mean size of 39 nm [85]. Zaheer and Rafiuddin [65] are also among the researchers who have studied the shape-directing role of cetyltrimethylammonium bromide (CTAB) on the green extracellular synthesis of Ag-nanoparticles using a *Ocimum sanctum* leaf extract. These silver nanoparticles ranged in size from about 18–35 nm and had diverse shapes (spherical, truncated triangular nanoplates) and were found to be highly polydispersed in the presence of CTAB. Shukla *et al.* [86] have discussed the production of silver nanoparticles (20–50 nm) using crude black pepper (*Piper nigrum*) extract at room temperature. Christensen *et al.* [87] produced silver nanoparticles using a leaf extract of *Murraya koenigii* and found particles of spherical shape with sizes ranging from 10–25 nm.

In a research experiment conducted by Logeswari *et al.* [88], a series of plant extracts (i.e., *Ocimum tenuiflorum*, *Solanum tricobatum*, *Syzygium cumini*, *Centella asiatica* and *Citrus sinensis)* were studied for their potential to act as reducing materials for the synthesis of silver nanoparticles. These silver nanoparticles were found to have an average size of 28 nm,

Table 10.1 Synthesis of transition metal nanoparticles by using different plant materials.

Plant name	**Reducing agents source**	**Type**	**Location**	**Size**	**Morphology**	**Reference**
Abelmoschus esculentus	Seeds	gold	Extracellular	45–75 nm	spherical	[18]
Acalypha indica	Leaves	silver	Extracellular	20–30 nm	spherical	[19]
Alfalfa sprouts	–	gold	Intracellular	2–20 nm	icosahedral	[20]
Aloe vera	Leaves	gold	Extracellular	50–350	nanotriangles	[16]
Annona squamosa	Peel	palladium	Extracellular	80 ± 5 nm	spherical	[21]
Arbutus unedo	Leaves	silver	Extracellular	<30 nm	spherical	[22]
Artemisia nilagirica	Leaves	silver	Extracellular	70–90 nm	nanotriangles, spherical	[23]
Avena sativa	Biomass	gold	Extracellular	5–85	tetrahedral, decahedral, hexagonal, icosahedral multitwinned, irregular shape, rod shape	[24]
Azadirachta indica	Leaves	silver and gold	Extracellular	50–100nm	spherical	[25]
Azadirachta indica	Leaves	silver	Extracellular	10–37	hexagonal	[26]
Bacopa monnieri	Leaves	gold	Extracellular	3–45 nm	spherical	[27]

Plant name	Reducing agents source	Type	Location	Size	Morphology	Reference
Brassica juncea	–	silver	Intracellular	2–35 nm	spherical	[28]
Brassica juncea	–	gold	Intracellular	2 nm – 2 μm	spherical, hexagonal, triangular, irregular	[29]
Brucea javanica	Rind	silver	Extracellular	38 ±14 nm	spherical	[30]
Boswellia ovalifoliolata	Stem bark	silver	Extracellular	30–40 nm	spherical	[31]
Calotropis procera	Flowers	silver	–	35nm	cubical	[32]
Camellia sinensis	Leaves	silver and gold	Extracellular	20 nm	spheroidal, nanoprisms, nano–rods	[33]
Catharanthus roseus	Dried leaves	silver	Extracellular	27±2 – 30±2 nm	spherical	[34]
Centella asiatica	Leaves	gold	Extracellular	–	fcc phase	[35]
Cinnamomum Camphora	Leaves	silver and gold	Extracellular	55–80 nm	triangular, spherical, quasi–spherical	[36]
Cinnamon zeylanicum	Bark	silver	Extracellular	31–100 nm	circular, ellipsoidal	[37]
Cissus quadrangularis	Leaves	silver	Extracellular	50–100nm	spherical	[38]

(*Continued*)

Table 10.1 (*Cont.*)

Plant name	Reducing agents source	Type	Location	Size	Morphology	Reference
Cissus quadrangularis	Leaves	gold	Extracellular	12.0 ± 3.2 nm	spherical	[39]
Citrus limon	Fruit	silver and gold		< 10 nm	spherical	[40]
Cochlospermum gossypium	Gum	silver, gold, platinum	Extracellular	2.4–5.5 nm	cubic	[41]
Coriandrum sativum	Leaves	gold	Extracellular	6.75–57.91 nm	decahedral	[42]
Crossandra infundibuliformis	Leaves	silver	Extracellular	38nm	fcc	[43]
Curcuma longa	Tuber	silver	Extracellular	71–100 nm	quasi–spherical, triangular, rod–shaped	[44]
Curcuma longa	Tuber	zirconia	Extracellular	41–45 nm	linear chains	[45]
Musa paradisiacal	Peel	palladium	Extracellular	50 nm	–	[46]
Diopyros kaki	Leaves	gold	Extracellular	5–300 nm	plate and spherical	[47]
Drypetes roxburghii	Fruit	silver	Extracellular	26.6 nm	quasi–spherical	[48]
Eclipta prostrate	Leaves	titanium dioxide	Extracellular	36 – 68 nm	spherical clusters	[49]

Plant name	Reducing agents source	Type	Location	Size	Morphology	Reference
Emblica officinalis	Fruit	silver, gold	Extracellular	10–20 nm (Ag), 15–15 nm (Au)	spherical	[50]
Euphorbia hirta	Leaves	gold	Extracellular	6–71nm	spherical	[51]
Eucalyptus chapmaniana	Leaves	silver	Extracellular	60 nm	fcc	[52]
Ficus benghalensis	Leaves	silver	Extracellular	16nm	spherical	[53]
Hibiscus cannabinus	Leaves	silver	Extracellular	9 nm	spherical	[54]
Iresine herbstii	Leaves	Silver	Extracellular	44–64 nm	cubic, fcc	[55]
Ixora coccinea	Leaves	silver	Extracellular	13–57 nm	spherical	[56]
Jatropha curcas	Latex	titanium dioxide	Extracellular	25–100 nm	spherical	[57]
Magnolia kobus	Leaves	gold	Extracellular	5 – 300 nm	plate, spherical	[47]
Magnolia kobus	Leaves	copper	Extracellular	37–110 nm	–	[58]

(Continued)

Table 10.1 (*Cont.*)

Plant name	Reducing agents source	Type	Location	Size	Morphology	Reference
Medicago sativa	–	gold	Intracellular	2 nm–1 μm	spherical, hexagonal, triangular, irregular	[29]
Medicago sativa	Seeds	silver	Extracellular	12 nm	spherical	[59]
Mentha piperita	Leaves	silver and gold	Extracellular	5–150 nm	spherical	[60]
Mentha spp.	Leaves	gold	–	3–16 nm	spherical	[61]
Moringa oleifera	Leaves	silver	Extracellular	57 nm	–	[62]
Origanum vulgare	Leaves	silver	Extracellular	136 ± 10.09 nm	spherical	
Ocimum sanctum	Leaves	silver	Extracellular	10–20 nm	spherical	[63]
Ocimum sanctum	dried stem and root	silver	Extracellular	5–10 nm	spherical	[64]
Ocimum sanctum	Leaves	gold	Extracellular	30 nm	crystalline, hexagonal, triangular	[63]
Ocimum sanctum	Leaves	silver	Extracellular	18 nm	–	[17]

Plant name	Reducing agents source	Type	Location	Size	Morphology	Reference
Ocimum sanctum	Leaves	silver	Extracellular	18–35 nm	spherical	[65]
Papaver somniferum	Seed	silver	Extracellular	3.2–7.6 nm	spherical	[66]
Pedilanthus tithymaloides	Leaves	silver	Extracellular	15–30 nm	spherical	[67]
Pelargonium graveolens	Leaves	gold	Extracellular	20–40 nm	rods, flat sheets, triangles	[68]
Pelargonium graveolens	Leaves	silver	Extracellular	16–40 nm	quasilinear	[69]
Piper betle	Leaf petiole	silver	Extracellular	4–30 nm	spherical, truncated triangular polyhedral plate, irregular	[70]
Piper Pedicellatum	Leaves	silver and gold	Extracellular	2–30 nm	spherical, triangular, hexagonal pentagonal	[71]
Piper longum	Leaves	silver	Extracellular	17.6–41 nm	spherical	[72]
Portulaca oleracea	Leaves	silver	Extracellular	<60 nm	–	[73]
Polylthia longifolia	Leaves	silver	Extracellular	57 nm	spherical	[74]
Rhizophora mucronata	Leaves	silver	Extracellular	60–95 nm	spherical	[75]

(Continued)

Table 10.1 (*Cont.*)

Plant name	Reducing agents source	Type	Location	Size	Morphology	Reference
Sesuvium portulacastrum	callus, Leaves	gold	Extracellular	5–20nm	spherical	[76]
Sorbus aucuparia	Leaves	silver and gold	Extracellular	16–18nm	spherical, triangular and hexagonal	[77]
Stigmaphyllon littorale	Leaves	silver	Extracellular	5– 25nm	spherical	[78]
Syzygium aromaticum	Flower	silver	Extracellular	20–149 nm	spherical	[79]
Terminalia chebula	Fruit	silver	Extracellular	<100 nm	pentagons, spherical and triangular	[80]
Terminalia chebula	Fruit	gold	Extracellular	6–60 nm	anisotropic	[81]
Terminalia chebula	Fruit	palladium and iron	Extracellular	<80 nm (Fe), < 100 nm (Pd)	amorphous(Fe) and cubic(Pd)	[82]
Trachyspermum ammi	Seed	silver	Extracellular	87–998 nm	triangular	[66]
Tribulus terrestris	Fruit	silver	Extracellular	16–28 nm	spherical	[83]
Zingiber officinale	Rhizome	gold	Extracellular	5–15nm	spherical	[84]

26.5 nm, 65 nm, 22.3 nm and 28.4 nm, respectively, and were highly active against *Staphylococcus aureus*, *Pseudomonas aeruginosa*, *Escherichia coli* and *Klebsiella pneumoniae*. Kaviya *et al.* [89] also reported the production of antimicrobial silver nanoparticles by using peel extract of *Citrus sinensis* as reducing and capping agent. Gopinath and coworkers studied the feasibility of using *Tribulus terrestris* L for the synthesis of silver nanoparticles [83]. The process was quite fast and spherical-shaped silver nanoparticles were observed in the size range of 16–28 nm. These silver nanoparticles were shown to possess antimicrobial activity against multi-drug-resistant bacteria such as *Streptococcus pyogens*, *Pseudomonas aeruginosa*, *Escherichia coli*, *Bacillus subtilis* and *Staphylococcus aureus*. Vijayakumar *et al.* [23] have investigated the rapid and extracellular formation of antibacterial silver nanoparticles (70–90 nm) by employing *Artemisia nilagirica*. Spherical nanoparticles with an average size of about 38.00 ± 14.00 nm have been synthesized by the rind extract of *Brucea javanica* L. [30]. Amaladhas *et al.* [90] reported for the first time the use of sunlight for rapid synthesis of silver nanoparticles using aqueous extract of *Achyranthes aspera* plant. They characterized the silver nanoparticles using UV–Vis, Fourier transform infrared spectroscopy (FTIR), transmission electron microscopy (TEM), and energy dispersive X-ray analysis (EDAX) techniques and found that the silver nanoparticles formed were monodispersed and spherical in shape with an average size of 12.82 nm. Biological activity of the silver nanoparticles was compared with standard antibacterial (Amikacin) and antifungal (Fluconazole) drugs and they were found to have considerable activity against bacterial and fungal pathogens.

Vijayaraghavan *et al.* [66] reported a study for the one-step green synthesis of silver nano/microparticles using extracts of *Trachyspermum ammi* and *Papaver somniferum*. For the formation of biocompatible silver nanoparticles, the main essential oil (constituents such as thymol, p-cymene and terpinene) in *T. ammi* was found to be a better reducing agent than the alkaloids (morphine and codeine) present in *P. somniferous*. Biosynthesis of silver nanoparticles by peel extract of *Citrus sinensis* has also been reported [43]. The synthesized silver nanoparticles were found to show antibacterial activity against *E. coli*, *P. aeruginosa* (Gram-negative), and *S. aureus* (Gram-positive).

Kotakadi *et al.* [34] produced stable nanoparticles using dried leaves of *Catharanthus roseus* L. The nanoparticles ranged in size from about 27 ± 2 and 30 ± 2 nm and had very good antimicrobial activity. Murugan and Dipankar [55] reported for the first time the formation of face-centered cubic silver nanoparticles with size ranging from 44 to 64 nm using *Iresine herbstii* leaf aqueous extracts. The synthesized nanoparticles showed strong antioxidant activity as well as cytotoxicity against HeLa cervical cell lines. In a series of experiments, Khan *et al.* [26, 70] reported the shape-directing role of cetyltrimethylammonium bromide in the green synthesis of silver nanoparticles using leaf extract of neem (*Azadirachta indica*) and leaf petiole extract of paan (*Piper betle*) as reducing agents. The extracts of *A. indica* produced silver nanoparticles (10–37 nm) of various shapes and the extract of *P. betle* resulted in spherical, truncated triangular polyhedral plate and some irregular-shaped silver nanoparticles (4–30 nm). Bankar *et al.* [91] used banana peel extract to synthesize silver nanoparticles and found that the peel extract results in the production of 100 nm-size nanoparticles with antifungal activity against the yeasts *Candida albicans* and *Candida lipolytica,* and antibacterial activity against *E. coli*, *Shigella sp.*, *Klebsiella sp.* and *Enterobacter aerogenes.* Ponarulselvam *et al.* [92] have discussed the production of silver nanoparticles with an average size of 35–55 nm using aqueous leaves extracts of *Catharanthus roseus* (*C. roseus*) Linn. Nabikhan and coworkers [76] studied the effect of extracts from tissue culture-derived callus and leaf of the saltmarsh plant, *Sesuvium portulacastrum* L., on the synthesis of antimicrobial silver nanoparticles using $AgNO_3$ as a substrate. In the FTIR spectra, peaks corresponding to amide I, II and III indicate the presence of protein, and peaks for aromatic rings, geminal methyls and ether linkages indicate the presence of flavones and terpenoids were observed. These compounds were supposed to be responsible for the stabilization of synthesized silver nanoparticles.

10.2.2 Synthesis of Gold Nanoparticles

Gold nanoparticles are among the important transition metal nanoparticles whose applications in diverse industries have soared in recent years. Recently, Nagaraj *et al.* [93] reported the extracellular biosynthesis of AuNPs using flower extracts of *Plumeria alba* as reducing agent. They found that the synthesized nanoparticles have significant antimicrobial activity against *A. niger* and *A. flavus.* Based on a research study administered by Bankar *et al.* [94], the synthesis of gold (Au) nanoparticles from aqueous solutions of chloroauric acid using banana peel extract (BPE) as a simple, nontoxic, ecofriendly green procedure have been elucidated. Fourier transform infrared spectroscopic studies were carried out to investigate the role of carboxyl, amine and hydroxyl groups in the reduction process. Likewise, Jayaseelan and coworkers [18] reported the synthesis of biologically active gold nanoparticles by using seed aqueous extract of *Abelmoschus esculentus.* Field emission scanning electron microscopy analysis of Au nanoparticles indicated the formation of spherical nanoparticles with a narrow size range of 45–75 nm. The synthesized nanoparticles were highly active against a range of fungal pathogens such as *Puccinia graminis tritci*, *Aspergillus flavus*, *Aspergillus niger* and *Candida albicans*. Basavegowda *et al.* [95] have used pineapple (*Ananas comosus* L.) fruit extracts to produce gold nanoparticles. The authors demonstrated very high inhibitory activity of the synthesized gold nanoparticles against some important human pathogens like *E. coli* and *Streptobacillus sp.*

At a low temperature, Song *et al.* [47] reported production of a mixture of plate (triangles, pentagons, and hexagons) and spherical nanoparticles (size, 5–300 nm) using leaf extracts of *Magnolia kobus* and *Diopyros kaki*. Castro *et al.* [96] synthesized gold nanowires by using pulp of sugar beet as reducing and capping agent. Khalil *et al.* [97] have elucidated that olive leaf extract synthesized gold nanoparticles (AuNPs) of various shapes (triangle, hexagonal, and spherical). Rastogi and Arunachalam [98] have produced gold nanoparticles via reduction of $HAuCl_4$ by using aqueous extract of garlic cloves as reducing/stabilizing agent. Similarly, Singh *et al.* [99] used a rapid, reproducible and a green biogenic approach for the biosynthesis of gold and silver nanoparticles by using biomolecules present in the leaf extract of *Dalbergia sissoo* as a reducing material.

In a recent research study, Babu and coworkers [27] reported a controlled biosynthesis of gold nanoparticles by using ethanolic leaf extract of *Bacopa monnieri* [27] and UV irradiation. The X-ray diffraction, UV-Visible spectra and TEM demonstrated the particles were crystalline in nature whose size ranged from 3–45 nm with a mean diameter of 11 nm.

10.2.3 Gold/Silver Bimetallic Nanoparticles

In recent research carried out by Mondal *et al.* [100], the role of aqueous extract of dried leaves of mahogany (*Swietenia mahogani* Ja.) in the rapid synthesis of stable monometallic Au and Ag nanoparticles and also Au-Ag bimetallic alloy nanoparticles were investigated. The reduction of aqueous Ag^+ and $AuCl_4^-$ ions was ascribed to the presence of various polyhydroxy limonoids in the extract. Extracellular synthesis of pure metallic silver, gold nanoparticles and bimetallic Au-Ag nanoparticles using neem (*Azadirachta indica*) leaf broth has been reported by Shankar and coworkers [25]. Synthesis of bimetallic Au core–Ag shell nanoparticles in solution takes place when $Au3^+$ and Ag^+ ions are present simultaneously in the solution, with silver nanoparticles being adsorbed onto the gold nanoparticles. In a related report, Sheny *et al.* [101] have investigated the biosynthesis of Au, Ag and Au–Ag bimetallic nanoparticles with the aqueous extract and dried leaf of *Anacardium occidentale*. They characterized the synthesized nanoparticles formed by leaf powder and leaf extract at different pH values of 4–8 and in the temperature range of 300–373 K. Using leaf powder was found to be advantageous over leaf extract due to the elimination of the seasonal effects and also because its concentration can be optimized; it was also concluded that Au nanoparticles were more stable at pH 6, while Ag nanoparticles were found to be stable at pH 8.

10.2.4 Palladium Nanoparticles

Currently, the applications of palladium nanoparticles in many fields have increased the interest of researchers in the development of green chemistry, efficient methods for its synthesis. Fabrication of palladium nanoparticles with mean size of nanoparticles, ranging from 3.2 to 6.0 nm, using broth of *Cinnamomum camphora* leaf without adding surfactant, capping agent, and/or template have been reported by Yang and coworkers [102]. *Kanchana et al. [103]* used leaf extract of *Solanum trilobatum* to produce palladium nanoparticles of different sizes. Sathishkumar and coworkers [104] also reported for the first time the production of nanocrystalline palladium nanoparticles by using *Curcuma longa* tuber extract as biomaterial. They proposed that the polyphenolic compounds present in the extract are responsible for the reduction of palladium nanoparticles. Likewise, in their research experiment Bankar and colleagues [46] synthesized palladium nanoparticles by using boiled, crushed, acetone precipitated, and air-dried peel extract of banana as a reducing material. Song *et al.* [47] investigated the reducing properties of the leaf extract of *Diopyros kaki* for the extracellular synthesis of platinum nanoparticles from an aqueous H_2PtCl_6 $6H_2O$ solution. They reported that a leaf broth concentration of >10% could result in the 90% conversion of platinum ions to nanoparticles with average particle size ranging from 2 to12 nm.

Sheny *et al.* [101] have studied the formation of spherical palladium nanoparticles by using the dried leaf powder of *Anacardium occidentale*. X-ray diffraction (XRD) results proved that the palladium nanoparticles had crystalline forms with fcc structure. The TEM analysis showed that the formed nanoparticles were about 2–5 nm in size. This synthesis proposed that the polyols found in the leaf extracts are responsible for the reduction of Pd(II) ions. Another biosynthesis of Pd nanoparticles (3–5 nm) in the presence of palladium chloride using antioxidants such as geniposide, chlorogenic acid, crocins and crocetin of *Gardenia jasminoides* Ellis as reducing and stabilizing agents has been reported by Jia *et al.* [105]. Nadagouda and Varma [106] have prepared silver and palladium nanoparticles by a simple green approach using coffee and tea extract at room temperature. The nanoparticles produced were found to be in the size range of 20–60 nm and crystallized in face-centered cubic symmetry. More recently, Mallikarjuna and coworkers [107] examined the one-pot synthesis of palladium (4 ± 1 nm) nanoparticles using a leaf extract of *Piper betle*. Fourier transform infrared (FTIR) spectroscopy revealed that the water soluble flavonoids of *P. betle* are responsible for reduction of palladium nanoparticles. The produced nanoparticles displayed high antifungal activity. In another study, Raut *et al.* [108] reported extracellular synthesis of platinum and palladium nanoparticles, of spherical shape, using root extract of *Asparagus racemosus* L. The palladium nanoparticles synthesized were characterized using UV-Visible spectroscopy, transmission electron microscopy (TEM) and cyclic voltammetry (CV) techniques. The authors found that both the nanoparticles are produced within 5 min, having crystalline form, and are in the size range of 1–6 nm.

10.2.5 Synthesis of Other Transition Metal Nanoparticles

Many other transition metal nanoparticles such as copper, zinc and cadmium oxide have attracted huge attention due to their applications in catalysis, solar cells, sensors and other photonic devices [109, 110]. The common routes for their production involve chemical reductions using synthetic agents [111]. The use of plants in their synthesis is not well documented in the literature; however Lee and coworkers [58] have reported the synthesis of stable copper nanoparticles (37–110 nm) by treating aqueous solution of $CuSO_4.5H_2O$ with the leaf extract of *Magnolia kobus* as a reducing material.

Sangeetha *et al.* [112] have produced polydispersed zinc oxide nanoparticles having an average size of 25–40 nm by using leaf extract of *Aloe barbadensis* Miller plant. Their results have proved that a leaf broth concentration of greater than 25% could result in more than 95% conservation of zinc salts to zinc nanoparticles. Likewise, spherical zinc oxide nanoparticles (50–200 nm) have been produced by Samat and colleagues [113] using extracts of *Citrus aurantifolia*.

10.3 Proposed Mechanisms

Several mechanisms for the synthesis of nanoparticles using plants have been proposed, however, an exact mechanism is still not known. An emerging area of scientific research being developed is the investigation of the role of biomolecules in the synthesis of transition metal nanoparticles. Scientists all around the world are of the opinion that the presence of

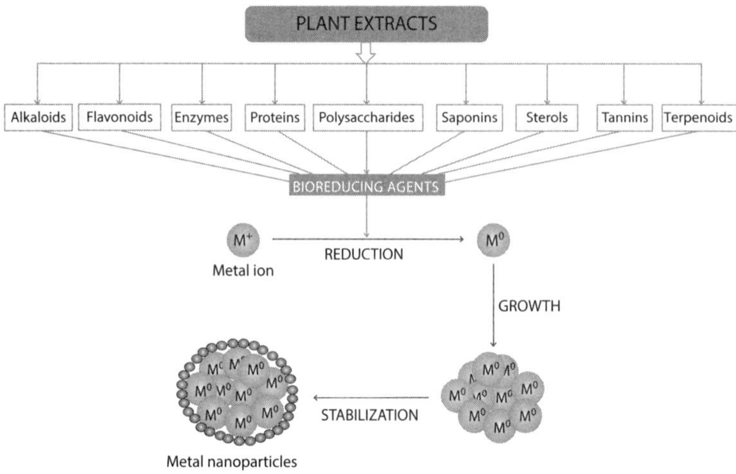

Figure 10.1 Synthesis of nanoparticles using bioconstituents from plants.

antioxidants, proteins/enzymes, amino acids, polysaccharides, alkaloids, alcoholic compounds, vitamins and other water-soluble heterocyclic components in plants are mainly responsible for the production and stabilization of the nanoparticles (Figure 10.1) [9, 13].

For example, Zhan and coworkers [114] verified by FTIR the presence of functional groups like C=O and C-O groups in the leaf extract of *Cacumen platycladi* and their role in the synthesis of Au–Pd bimetallic nanoparticles (~7 nm). Likewise, MubarakAli *et al.* [60] showed that the leaf extract of menthol is very good bioreductant for the synthesis of silver and gold nanoparticles.

Recently, palm oil mill effluent was used by Gan *et al.* [115] as a reducing agent for the synthesis of spherical gold nanoparticles with an average size of 18.75 ± 5.96 nm. The FTIR analysis of palm oil mill effluent before and after the gold nanoparticle formation showed the disappearance of N–H or C=O stretching vibrations bands at 2351cm^{-1}. This fact can be attributed to the presence of proteins or flavonoids that contain abundant C=O or N-H groups, which are presumed to act as reducing and capping agents for the gold nanoparticles.

Jagtap and Bapat [116] reported the use of *Artocarpus heterophyllus* seed powder extract (ASPE) as a reducing agent for the synthesis of spherical-shaped silver nanoparticles. Their study indicated the involvement of amides, carboxyl, amino groups and amino acid residues present in ASPE extract for the synthesis of silver nanoparticles. Similarly, Bankar and colleagues [46] proposed the role of carboxyl, amine and hydroxyl groups in the formation of palladium nanoparticles using peel extract of banana fruit.

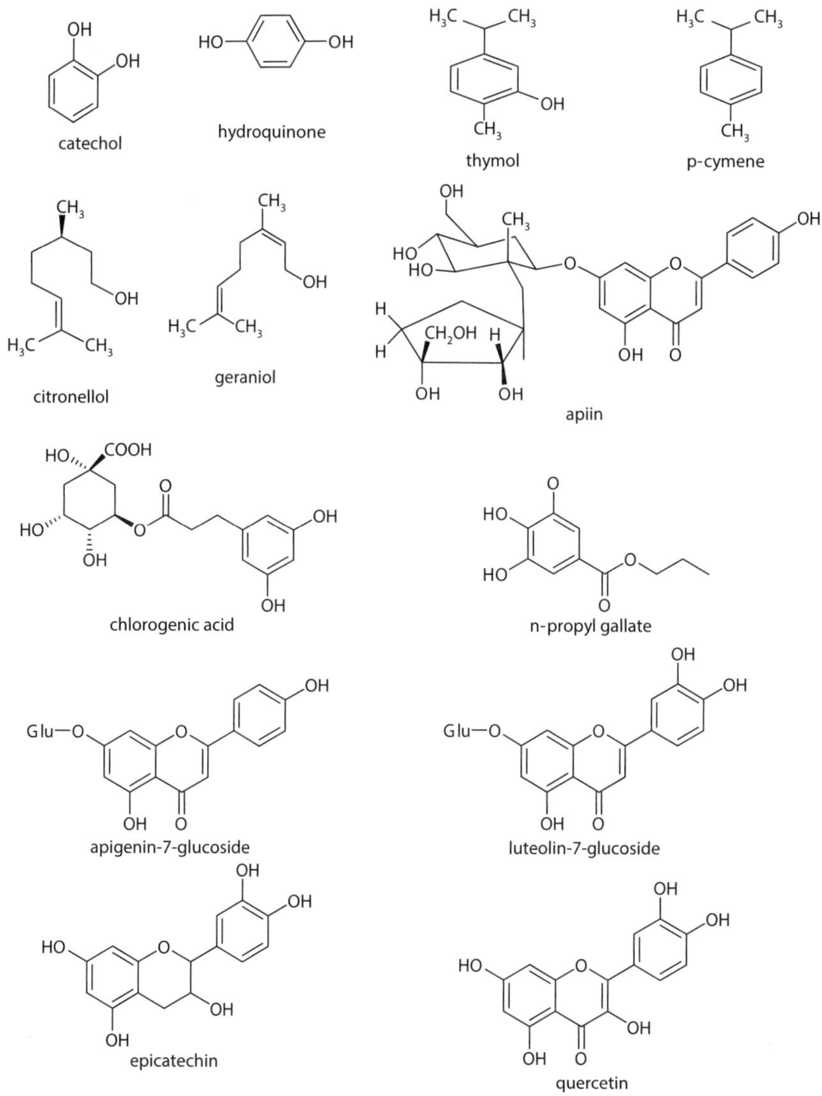

Figure 10.2 Representative structures of some bioconstituents used in the synthesis of nanoparticles.

In a very recent experiment, Suman and coworkers [117] used *Morinda citrifolia* root extract for the synthesis of silver nanoparticles and concluded that functional groups like flavonoids and terpenoids present in *M. citrifolia* root are responsible for capping and stabilization of silver nanoparticles. Structures of some of the phytoconstituents used in the synthesis of nanoparticles are shown in Figure 10.2.

10.4 Transition Metal Nanoparticles as Novel Antimicrobial Agents for Textile Modifications

Conventional fibers and polymers including both proteinous materials such as wool, silk and cellulosic materials such as cotton, jute, flax and other fibers are susceptible to microbial attack, as they provide large surface area and absorb moisture, which facilitates microbial growth. Contamination by microorganisms is of great concern in a variety of areas such as medical devices, health care and hygienic applications. They may contribute to transmission of diseases in hospitals. To overcome pathogenic infections caused by textile materials, the use of antimicrobial textiles in healthcare facilities is a new concept. Therefore, research involving antimicrobial modifications of textiles to improve performance or to create unprecedented functions is flourishing [118–120]. As of now, various kinds of antimicrobial agents have been tested such as quaternary ammonium compounds, N-halamines, halogenated phenols, triclosan and polybiguanides [121]. Owing to the possible harmful or toxic effects of many chemical antimicrobial agents, much of the recent research has been focused on using natural materials for textile modifications [120, 122]. In recent years, several attempts have also been made by researchers and industries to utilize transition metal nanoparticles and their nanocomposites for antimicrobial and multifunctional modifications of textiles. The use of nanoparticles for antimicrobial textile modifications have gained momentum due to their recently discovered water repellence [123], antimicrobial [124], UV-protection [125], soil resistant, anti-static, anti-infrared, flame retardant [126] and wrinkle resistant properties [127], in addition to their role in the coloration of different types of textile fabrics (Figure 10.3). Globally, researchers are actively engaged in nanotechnology research aimed at applications in the textile sector.

In view of the above benefits that nanoparticles have on textiles, in the past few years, several research groups worldwide have initiated work on the application of plant-synthesized nanoparticles for various textile modifications (Table 10.2).

Sathishkumar *et al.* [44] reported the synthesis of silver (Ag) nanoparticles from silver precursor using plant biomaterials such as *Curcuma longa* tuber powder and extract. The nanoparticles were found to be quasi-spherical, triangular and small rod-shaped. The application of these silver nanoparticles on cotton fabrics was shown to impart bactericidal activity to cotton, which was retained even after consecutive launderings. In another study, aqueous leaf extract of the lemon tree (*Citrus limon*) was examined to act as reducing material and encapsulating cage for the

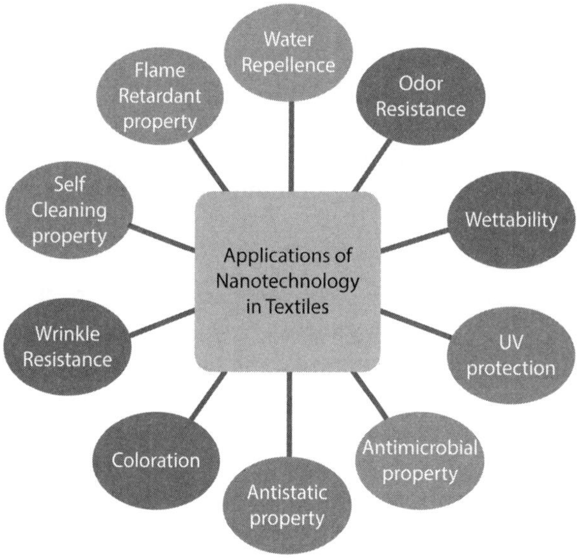

Figure 10.3 Applications of nanoparticles in textile modifications.

Table 10.2 Antimicrobial textile modifications using some green chemistry synthesized nanoparticles.

Plant	Part	Fibre/ Fabric	Tested micro-organism	Reference
Azadirachta indica	Leaves	nylon and cotton	*Klebsiella pneumoniae* and *Sarcina lutea*	[128]
Citrus limon	Leaves	cotton and silk	*Fusarium oxysporum* and *Alternaria brassicicola*	[129]
Curcuma longa	Tuber powder and extract	cotton	*Escherichia coli*	[44]
Eucalyptus citriodora	Leaves	cotton	*Escherichia coli*	[130]
Ficus bengalensis	Leaves	cotton	*Escherichia coli*	[130]

synthesis of silver nanoparticles. The synthesized nanoparticle particles displayed high antifungal activity after application on cotton and silk fabrics against *Fusarium oxysporum* and *Alternaria brassicicola* [129]. Likewise, leaf extracts of *Eucalyptus citriodora* and *Ficus bengalensis* plants were used to produce silver nanoparticles in the range of 20 nm [130].

Through the UV-Vis and electron microscopy data, it was verified that the inclusion of silver nanoparticles in cotton fibers improved their thermal stability and elongation properties. The developed silver nanoparticles-loaded cotton fibers were highly active against gram-negative *E. coli.* The author further suggested that such kinds of fibers have great potential for utilization in hospitals and healthcare applications. In an interesting experiment, Hebeish *et al.* [131] synthesized silver nanoparticles (6–8 nm) by employing the reducing and stabilizing power of hydroxypropyl starch (HPS). They reported that the deposition of nanosilver on the surface of cotton fabrics in the presence and absence of binder had been an effective way to prevent microbial contaminations. It was observed that through the use of binder, the antibacterial activity of the treated fabrics against *Escherichia coli* (gram -ve) and *Staphylococcus aureus* (gram +ve) bacteria was maintained even after 20 washings in home laundering. Shastri and colleagues [128] also reported the production of silver nanoparticles using the wet reduction method and biologically by using leaf extract of neem (*Azadirachta indica*) plant. Formation of silver nanoparticles of various shapes and sizes (5–50 nm) were revealed by UV-Visible spectroscopy and transmission electron microscopy. The authors of this study investigated the antibacterial efficacy of the nanosilver-finished fabrics (nylon and cotton sock fabrics) against a wide range of organisms such as *Sarcina lutea*, an odor-producing organism, *Klebsiella pnuemoniae*, *Pseudomonas aeruginosa*, methicillin-resistant *Staphylococcus aureus* and *Candida albicans*, causing foot infections, by employing a zone inhibition test, antibacterial test and wash fastness test.

10.5 Concluding Remarks and Future Aspects

Synthesis of green chemistry-based transition metal nanoparticles using inexpensive and efficient plant materials is a new and emerging area of research in the scientific world. The intent of this chapter was to provide a concise insight into the recent developmental works pertaining to transition metal nanoparticle synthesis by using plant material. The synthesis of nanoparticles is characterized by processes that result in the production of stable nanoparticles of more various shapes and sizes. Such kinds of nanoparticles have numerous applications and have created new interesting fields in all the sciences. Their applications have already led to the development of novel textile properties. Existing studies are confined mostly to their use in antimicrobial textiles and little information is available about their role in other important applications such as self-cleaning and UV protection. Further research efforts should be focused on understanding the real effect caused by the bioconstituents in the synthesis of green nanoparticles, followed by their practical applications in textiles using green techniques. By achieving this goal, there is a promising future for this field of study.

References

1. L. Qian, and J.P. Hinestroza, *J. Text. Apparel Technol. Manag.*, Vol. 4, p. 1, 2004.
2. J.A. Dahl, B.L. Maddux, and J.E. Hutchison, *Chem. Rev.*, Vol. 107, p. 2228, 2007.
3. Z. Wang, X. Li, M. Gao, and X. Zeng, *Powder Technol.*, Vol. 215–216, p. 147, 2012.
4. G. Cristoforetti, E. Pitzalis, R. Spiniello, R. Ishak, F. Giammanco, M. Muniz-Miranda, and S. Caporali, *Appl. Surf. Sci.*, Vol. 258, p. 3289, 2012.
5. K.D. Kim, D.N. Han, J.B. Lee, and H.T. Kim, *Scr. Mater.*, Vol. 54, p. 143, 2006.
6. Z. Khan, S.A. Al-Thabaiti, A.Y. Obaid, and A.O. Al-Youbi, *Colloid. Surf. B*, Vol. 82, p. 513, 2011.
7. A.P. Kumar, B.P. Kumar, A.B.V.K. Kumar, B.T. Huy, and Y.I. Lee, *Appl. Surf. Sci.*, Vol. 265, p. 500, 2013.
8. K.N. Thakkar, S.S. Mhatre, and R.Y. Parikh, *Nanomed: Nanotechnol.*, Vol. 6, p. 257, 2010.
9. S. Iravani, *Green Chem.*, Vol. 13, p. 2638, 2011.
10. D. Mandal, M.E. Bolander, D. Mukhopadhyay, G. Sarkar, and P. Mukherjee, *Appl. Microbiol. Biotechnol.*, Vol. 69, p. 485, 2006.
11. V. Kumar, and S.K. Yadav, *J. Chem. Technol. Biotechnol.*, Vol. 84, p. 151, 2009.
12. N. Durán, P.D. Marcato, M. Durán, A. Yadav, A. Gade, and M. Rai, *Appl. Microbiol. Biotechnol*, Vol. 90, p. 1609, 2011.
13. D. Hebbalalu, J. Lalley, M.N. Nadagouda, and R.S. Varma, *ACS Sust. Chem. Eng.*, Vol. 1, p. 703, 2013.
14. O.V. Kharissova, H.V.R. Dias, B.I. Kharisov, B.O. Pérez, and V.M.J. Pérez, *Trends Biotechnol.*, Vol. 31, p. 240, 2013.
15. A.K. Mittal, Y. Chisti, and U.C. Banerjee, *Biotechnol. Adv.*, Vol. 31, p. 346, 2013.
16. S.P. Chandran, M. Chaudhary, R. Pasricha, A. Ahmad, and M. Sastry, *Biotechnol. Prog.*, Vol. 22, p. 577, 2006.
17. C. Ramteke, T. Chakrabarti, B.K. Sarangi, and R.A. Pandey, *J. Chem.*, 2013. dx.doi.org/10.1155/2013/278925.
18. C. Jayaseelan, R. Ramkumar, A.A. Rahuman, and P. Perumal, *Ind. Crop. Prod.*, Vol. 45 p. 423, 2013.
19. C. Krishnaraj, E.G. Jagan, S. Rajasekar, P. Selvakumar, P.T. Kalaichelvan, and N. Mohan, *Colloid. Surf. B*, Vol. 76, p. 50, 2010.

20. J.L. Gardea-Torresdey, E. Gomez, J.R. Peralta-Videa, J.G. Parsons, H. Troiani, and M. Jose-Yacaman, *Langmuir*, Vol. 19, p. 1357, 2003.
21. S.M. Roopan, A. Bharathi, R. Kumar, V.G. Khanna, and A. Prabhakarn, *Colloid. Surf. B*, Vol. 92, p. 209, 2012.
22. P. Kouvaris, A. Delimitis, V. Zaspalis, D. Papadopoulos, S.A. Tsipas, and N. Michailidis, *Mater. Lett.*, Vol. 76, p. 18, 2012.
23. M. Vijayakumar, K. Priya, F.T. Nancy, A. Noorlidah, and A.B.A. Ahmed, *Ind. Crop. Prod.*, Vol. 41, p. 235, 2013.
24. V. Armendariz, I. Herrera, M. Jose-Yacaman, H. Troiani, P. Santiago, and J.L. Gardea-Torresdey, *J. Nanopart. Res.*, Vol. 6, p. 377, 2004.
25. S.S. Shankar, A. Rai, A. Ahmad, and M. Sastry, *J. Colloid Interface Sci.*, Vol. 275, p. 496, 2004.
26. Z. Khan, J.I. Hussain, and A.A. Hashmi, *Colloid. Surf. B*, Vol. 95, p. 229, 2012.
27. P.J. Babu, P. Sharma, S. Saranya, and U. Bora, *Mater. Lett.*, Vol. 93, p. 431, 2013.
28. R. Haverkamp, and A. Marshall, *J. Nanopart. Res.*, Vol. 11, p. 1453, 2009.
29. R. Bali, and A.T. Harris, *Ind. Eng. Chem. Res.*, Vol. 49, p. 12762, 2010.
30. S. Yudha S.D. Notriawan, E. Angasa, T. Eka Suharto, J. Hendri, and Y. Nishina, *Mater. Lett.*, Vol. 97, p. 181, 2013.
31. S. Ankanna, E. Elumalai, and N. Savithramma, *Dig. J. Nanomater. Bios.*, Vol. 5, p. 369, 2010.
32. S.A. Babu, and H.G. Prabu, *Mater. Lett.*, Vol. 65, p. 1675, 2011.
33. N.A. Begum, S. Mondal, S. Basu, R.A. Laskar, and D. Mandal, *Colloid. Surf. B*, Vol. 71, p. 113, 2009.
34. V.S. Kotakadi, Y.S. Rao, S.A. Gaddam, T.N.V.K.V. Prasad, A.V. Reddy, and D.V.R.S. Gopal, *Colloid. Surf. B*, Vol. 105, p. 194, 2013.
35. R.K. Das, B.B. Borthakur, and U. Bora, *Mater. Lett.*, Vol. 64, p. 1445, 2010.
36. J. Huang, Q. Li, D. Sun, Y. Lu, Y. Su, X. Yang, H. Wang, Y. Wang, W. Shao, and N. He, *Nanotechnology*, Vol. 18, p. 105104, 2007.
37. M. Sathishkumar, K. Sneha, S.W. Won, C.W. Cho, S. Kim, and Y.S. Yun, *Colloid. Surf. B*, Vol. 73, p. 332, 2009.
38. J.S. Valli, and B. Vaseeharan, *Mater. Lett.*, Vol. 82, p. 171, 2012.
39. S.R. Bhuvanasree, D. Harini, A. Rajaram, and R. Rajaram, *Spectrochim. Acta Part A*, Vol. 106, p. 190, 2013.
40. W. Liang, T.L. Church, and A.T. Harris, *Green Chem.*, Vol. 14, p. 968, 2012.
41. V.T.P. Vinod, P. Saravanan, B. Sreedhar, D.K. Devi, and R.B. Sashidhar, *Colloid. Surf. B*, Vol. 83, p. 291, 2011.
42. K.B. Narayanan, and N. Sakthivel, *Mater. Lett.*, Vol. 62, p. 4588, 2008.
43. S. Kaviya, J. Santhanalakshmi, and B. Viswanathan, *Mater. Lett.*, Vol. 67, p. 64, 2012.
44. M. Sathishkumar, K. Sneha, and Y.S. Yun, *Bioresour. Technol.*, Vol. 101, p. 7958, 2010.
45. M. Sathishkumar, K. Sneha, and Y.S. Yun, *Mater. Lett.*, Vol. 98, p. 242, 2013.

46. A. Bankar, B. Joshi, A.R. Kumar, and S. Zinjarde, *Mater. Lett.*, Vol. 64, p. 1951, 2010.
47. J.Y. Song, H.K. Jang, and B.S. Kim, *Process Biochem.*, Vol. 44, p. 1133, 2009.
48. K. Haldar, B. Haldar, and G. Chandra, *Parasitol. Res.*, Vol. 112, p. 1451, 2013.
49. G. Rajakumar, A.A. Rahuman, B. Priyamvada, V.G. Khanna, D.K. Kumar, and P.J. Sujin, *Mater. Lett.*, Vol. 68, p. 115, 2012.
50. B. Ankamwar, C. Damle, A. Ahmad, and M. Sastry, *J. Nanosci. Nanotechnol.*, Vol. 5, p. 1665, 2005.
51. A. Annamalai, V.L.P. Christina, D. Sudha, M. Kalpana, and P.T.V. Lakshmi, *Colloid. Surf. B.*, Vol. 108, p. 60, 2013.
52. G.M. Sulaiman, W.H. Mohammed, T.R. Marzoog, A.A.A. Al-Amiery, A.A.H. Kadhum, and A.B. Mohamad, *Asian Pac. J. Trop. Biomed.*, Vol. 3, p. 58, 2013.
53. A. Saxena, R.M. Tripathi, F. Zafar, and P. Singh, *Mater. Lett.*, Vol. 67, p. 91, 2012.
54. M.R. Bindhu, M. Umadevi, *Spectrochim. Acta Part A*, Vol. 101, p. 184, 2013.
55. C. Dipankar, and S. Murugan, *Colloid. Surf. B*, Vol. 98, p. 112, 2012.
56. M. Karuppiah, and R. Rajmohan, *Mater. Lett.*, Vol. 97, p. 141, 2013.
57. M. Hudlikar, S. Joglekar, M. Dhaygude, and K. Kodam, *Mater. Lett.*, Vol. 75, p. 196, 2012.
58. H.J. Lee, J.Y. Song, and B.S. Kim, *J. Chem. Technol. Biotechnol.*, Vol. 88, p. 1971, 2013.
59. A.I. Lukman, B. Gong, C.E. Marjo, U. Roessner, and A.T. Harris, *J. Colloid Interface Sci.*, Vol. 353, p. 433, 2011.
60. D. MubarakAli, N. Thajuddin, K. Jeganathan, and M. Gunasekaran, *Colloid. Surf. B*, Vol. 85, p. 360, 2011.
61. A.K. Jha, and K. Prasad, *Int. J. Green Nanotechnol.*, Vol. 4, p. 219, 2012.
62. T. Prasad, and E.K. Elumalai, *Asian Pac. J. Tropic. Biomed.*, Vol. 1, p. 439, 2011.
63. D. Philip, and C. Unni, *Physica E: Low Dimens. Syst. Nanostruct.*, Vol. 43, p. 1318, 2011.
64. N. Ahmad, S. Sharma, M.K. Alam, V.N. Singh, S.F. Shamsi, B.R. Mehta, and A. Fatma, *Colloid. Surf. B*, Vol. 81, p. 81, 2010.
65. Z. Zaheer, and Rafiuddin, *Colloid. Surf. B.*, Vol. 108, p. 90, 2013.
66. K. Vijayaraghavan, S.P.K. Nalini, N.U. Prakash, and D. Madhankumar, *Colloid. Surf. B*, Vol. 94, p. 114, 2012.
67. C. Sundaravadivelan, M.N. Padmanabhan, P. Sivaprasath, and L. Kishmu, *Parasitol. Res.*, Vol. 112, p. 303, 2013.
68. S.S. Shankar, A. Ahmad, R. Pasricha, and M. Sastry, *J. Mater. Chem.*, Vol. 13, p. 1822, 2003.
69. S.S. Shankar, A. Ahmad, and M. Sastry, *Biotechnol. Prog.*, Vol. 19, p. 1627, 2003.
70. Z. Khan, O. Bashir, J.I. Hussain, S. Kumar, and R. Ahmad, *Colloid. Surf. B*, Vol. 98, p. 85, 2012.
71. C. Tamuly, M. Hazarika, S.C. Borah, M.R. Das, and M.P. Boruah, *Colloid. Surf. B*, Vol. 102, p. 627, 2013.

72. S. Justin Packia Jacob, J.S. Finub, and A. Narayanan, *Colloid. Surf. B*, Vol. 91, p. 212, 2012.
73. M.J. Firdhouse, and P. Lalitha, *Asian J. Pharm. Clin. Res.*, Vol. 6, p. 92, 2012.
74. T.N.K.V. Prasad, E.K. Elumalai, and S. Khateeja, *Asian Pac. J. Trop. Disease*, Vol. 2, Suppl. 2, p. S663, 2012.
75. M. Gnanadesigan, M. Anand, S. Ravikumar, M. Maruthupandy, V. Vijayakumar, S. Selvam, M. Dhineshkumar, and A.K. Kumaraguru, *Asian Pac. J. Trop. Med.*, Vol. 4, p. 799, 2011.
76. A. Nabikhan, K. Kandasamy, A. Raj, and N.M. Alikunhi, *Colloid. Surf. B*, Vol. 79, p. 488, 2010.
77. S.P. Dubey, M. Lahtinen, H. Särkkä, and M. Sillanpää, *Colloid. Surf. B*, Vol. 80, p. 26, 2010.
78. K.R. Kudle, M.R. Donda, R. Merugu, Y. Prashanthi, and M.P. Rudra, *Int. J. Nanomater. Bios.* Vol. 3, p. 13, 2013.
79. K. Vijayaraghavan, S.P.K. Nalini, N.U. Prakash, and D. Madhankumar, *Mater. Lett.*, Vol. 75, p. 33, 2012.
80. K.M. Kumar, M. Sinha, B.K. Mandal, A.R. Ghosh, K.S. Kumar, and P.S. Reddy, *Spectrochim. Acta Part A*, Vol. 91, p. 228, 2012.
81. K.M. Kumar, B.K. Mandal, M. Sinha, and V. Krishnakumar, *Spectrochim Acta Part A*, Vol. 86, p. 490, 2012.
82. K.M. Kumar, B.K. Mandal, K.S. Kumar, P.S. Reddy, and B. Sreedhar, *Spectrochim. Acta Part A*, Vol. 102, p. 128, 2013.
83. V. Gopinath, D. MubarakAli, S. Priyadarshini, N.M. Priyadharsshini, N. Thajuddin, and P. Velusamy, *Colloid. Surf. B*, Vol. 96, p. 69, 2012.
84. K.P. Kumar, W. Paul, and C.P. Sharma, *Process Biochem.*, Vol. 46, p. 2007, 2011.
85. J. Kasthuri, S. Veerapandian, and N. Rajendiran, *Colloid. Surf. B*, Vol. 68, p. 55, 2009.
86. V.K. Shukla, R.P. Singh, and A.C. Pandey, *J. Alloy. Compd.*, Vol. 507, p. L13, 2010.
87. L. Christensen, S. Vivekanandhan, M. Misra, and A.K. Mohanty, *Adv. Mater. Lett.*, Vol. 2, p. 429, 2011.
88. P. Logeswari, S. Silambarasan, and J. Abraham, *J. Saudi Chem. Soc.*, 2012. dx.doi.org/10.1016/j.jscs.2012.04.007.
89. S. Kaviya, J. Santhanalakshmi, B. Viswanathan, J. Muthumary, and K. Srinivasan, *Spectrochim. Acta Part A*, Vol. 79, p. 594, 2011.
90. T.P. Amaladhas, N. Usha, and S. Naveen, *Adv. Mater. Lett.*, Vol. 4, p. 779, 2013.
91. A. Bankar, B. Joshi, A.R. Kumar, and S. Zinjarde, *Colloid. Surf. A*, Vol. 368, p. 58, 2010.
92. S. Ponarulselvam, C. Panneerselvam, K. Murugan, N. Aarthi, K. Kalimuthu, and S. Thangamani, *Asian Pac. J. Trop. Biomed.*, Vol. 2, p. 574, 2012.
93. B. Nagaraj, M. Barasa, T.K. Divya, N.B. Krishnamurthy, P. Liny, and R. Dinesh, *Int. J. Drug Dev. Res.*, Vol. 4, p. 144, 2012.

94. A. Bankar, B. Joshi, A. Ravi Kumar, S. Zinjarde, *Colloid. Surf. B*, Vol. 80, p. 45, 2010.
95. N. Basavegowda, A. Sobczak-Kupiec, D. Malina, H.S. Yathirajan, V.R. Keerthi, N. Chandrashekar, S. Dinkar, and P. Liny, *Adv. Mat. Lett.*, Vol. 4, p. 332, 2013.
96. L. Castro, M.L. Blázquez, F. González, J.A. Muñoz, and A. Ballester, *Chem. Eng. J.*, Vol. 164, p. 92, 2010.
97. M.M.H. Khalil, E.H. Ismail, and F. El-Magdoub, *Arabian J. Chem.*, Vol. 5, p. 431, 2012.
98. J.A.L. Rastogi, *Adv. Mater. Lett.*, Vol. 4, p.548, 2013.
99. R.K.B.C. Singh, P.K. Naik, and H. Singh, *Adv. Mater. Lett.*, Vol. 3, p. 279, 2012.
100. S. Mondal, N. Roy, R.A. Laskar, I. Sk, S. Basu, D. Mandal, and N.A. Begum, *Colloid. Surf. B*, Vol. 82, p. 497, 2011.
101. D.S. Sheny, D. Philip, and J. Mathew, *Spectrochim. Acta Part A.*, Vol. 91, p. 35, 2012.
102. X. Yang, Q. Li, H. Wang, J. Huang, L. Lin, W. Wang, D. Sun, Y. Su, J. Opiyo, L. Hong, Y. Wang, N. He, and L. Jia, *J. Nanopart. Res.*, Vol. 12, p. 1589, 2010.
103. A. Kanchana, S. Devarajan, and S.R. Ayyappan, *Nano-Micro Lett.*, Vol. 2, p. 169, 2010.
104. M. Sathishkumar, K. Sneha, and Y. Yun, *Int. J. Mater. Sci.*, Vol. 4, p. 11, 2009.
105. L. Jia, Q. Zhang, Q. Li, and H. Song, *Nanotechnology*, Vol. 20, p. 385601, 2009.
106. M.N. Nadagouda, and R.S. Varma, *Green Chem.*, Vol. 10, p. 859, 2008.
107. K. Mallikarjuna, N.J. Sushma, B.V.S. Reddy, G. Narasimha, and B.D.P. Raju, *Int. J. Chem. Analy. Sci.*, Vol. 4, p. 14, 2013.
108. R.W. Raut, A.S.M. Haroon, Y.S. Malghe, B.T. Nikam and S.B. Kashid, *Adv. Mater. Lett.*, Vol. 4, p. 650, 2013.
109. W.J.E. Beek, M.M. Wienk, and R.A.J. Janssen, *Adv. Mater*, Vol. 16, p. 1009, 2004.
110. X. Luo, A. Morrin, A.J. Killard, and M.R. Smyth, *Electroanalysis*, Vol. 18, p. 319, 2006.
111. K. Manickathai, S.K. Viswanathan, and M. Alagar, *Indian J. Pure Appl. Phys.*, Vol. 46, p. 561, 2008.
112. G. Sangeetha, S. Rajeshwari, and R. Venckatesh, *Mater. Res. Bull.*, Vol. 46, p. 2560, 2011.
113. N.A. Samat, R.M. Nor, *Ceramics Int.*, Vol. 39, Suppl. 1, p. S545, 2013.
114. G. Zhan, J. Huang, M. Du, I. Abdul-Rauf, Y. Ma, and Q. Li, *Mater. Lett.*, Vol. 65, p. 2989, 2011.
115. P.P. Gan, S.H. Ng, Y. Huang, and S.F.Y. Li, *Bioresour. Technol.*, Vol. 113, p. 132, 2012.
116. U.B. Jagtap, and V.A. Bapat, *Ind. Crop. Prod.*, Vol. 46, p. 132, 2013.
117. T.Y. Suman, S.R.R. Rajasree, A. Kanchana, and S.B. Elizabeth, *Colloid. Surf. B*, Vol. 106, p. 74, 2013.
118. Y. Gao, and R. Cranston, *Text. Res. J.*, Vol. 78, p. 60, 2008.

119. B. Simoncic, and B. Tomsic, *Text. Res. J.*, Vol. 80, p. 1721, 2010.
120. S. Islam, M. Shahid, and F. Mohammad, *Ind. Eng. Chem. Res.*, Vol. 52, p. 5245, 2013.
121. M.M.G. Fouda, E.S. Abdel-Halim, and S.S. Al-Deyab, *Carbohydr. Polym.*, Vol. 92, p. 943, 2013.
122. M. Shahid, S. Islam, and F. Mohammad, *J. Clean. Prod.*, Vol. 53, p. 310, 2013.
123. G.Y. Bae, B.G. Min, Y.G. Jeong, S.C. Lee, J.H. Jang, and G.H. Koo, *J. Colloid Interface Sci.*, Vol. 337, p. 170, 2009.
124. R. Dastjerdi, and M. Montazer, *Colloid. Surf. B*, Vol. 79, p. 5, 2010.
125. A. Becheri, M. Dürr, P.L. Nostro, and P. Baglioni, *J. Nanopart. Res.*, Vol. 10, p. 679, 2008.
126. A. Yadav, V. Prasad, A.A. Kathe, S. Raj, D. Yadav, C. Sundaramoorthy, and N. Vigneshwaran, *Bull. Mater. Sci.*, Vol. 29, p. 641, 2006.
127. C. Yuen, S. Ku, Y. Li, Y. Cheng, C. Kan, and P. Choi, *J. Text. I*, Vol. 100, p. 173, 2009.
128. J.P. Shastri, M.G. Rupani, and R.L. Jain, *J. Text. I*, Vol. 103, p. 1234, 2012.
129. P.S. Vankar, and D. Shukla, *Appl. Nanosci.*, Vol. 2, p. 163, 2012.
130. S. Ravindra, Y.M. Mohan, N.N. Reddy, and K.M. Raju, *Colloid. Surf. A*, Vol. 367, p. 31, 2010.
131. A. Hebeish, M.E. El-Naggar, M.M.G. Fouda, M.A. Ramadan, S.S. Al-Deyab, and M.H. El-Rafie, *Carbohydr. Polym.*, Vol. 86, p. 936, 2011.

11

Functionalized Magnetic Nanoparticles for Heavy Metal Removal from Aqueous Solutions: Kinetics and Equilibrium Modeling

Ravindra Kumar Gautam*, Amita Jaiswal and Mahesh Chandra Chattopadhyaya

Environmental Chemistry Research Laboratory, Department of Chemistry, University of Allahabad, Allahabad, India

Abstract

Continuous discharge of heavy metals into aquatic bodies poses toxic effects to biological systems. Functionalized magnetic nanoparticles have recently been examined as promising platforms for removal of heavy metals from aqueous solutions. Magnetic nanoparticles offer a new vista of separation and purification technology for heavy metals. The multifunctional magnetic nanoparticles have been successfully applied for the reduction of toxic metal ions up to ppb level in treated wastewater. This chapter highlights the potential application of magnetic nanoparticles for the removal of heavy metals from aqueous solutions.

Keywords: Magnetic nanoparticles, nanocomposites, adsorption, heavy metals, wastewater treatment, kinetics, modeling

11.1 Introduction

Pollution by heavy metals is a serious threat to aquatic ecosystems because some of these metals are potentially toxic, even at very low concentrations. Contamination of heavy metals in the surroundings has done great harm to human and other organisms due to their high toxicity and carcinogenicity

**Corresponding author*: ravindragautam1987@gmail.com

Ashutosh Tiwari and Mikael Syväjärvi (eds.) Advanced Materials for Agriculture, Food, and Environmental Safety, (291–332) 2014

[1, 2, 3, 4, 5, 6]. The United States Environmental Protection Agency (USEPA) has classified heavy metals as priority pollutants. Heavy metal pollution in aqueous solutions has become a serious threat today and are of great environmental concern as they are non-biodegradable, and thus persistent in nature. Lead (Pb), cadmium (Cd), chromium (Cr), nickel (Ni), cobalt (Co), zinc (Zn) and copper (Cu) are such metals, which are frequently present in polluted water from various industrial processes such as catalysts, mineral processing, electrical apparatus, painting and coating and agricultural materials [7, 8, 9]. Additionally, heavy metals are not biodegradable and tend to accumulate in living organisms, and can cause severe problems to both human health and wildlife. Natural processes and human activities have polluted and reduced the quality of water resources all over the world. Groundwater and superficial water have been contaminated in various ways; e.g., by mining wastes, by agricultural chemicals, by industrial and domestic sewage that is sometimes discharged without treatment into waterways and by natural trace elements (e.g., arsenic). Since it is important to eliminate or reduce the concentration of heavy metals in the aquatic ecosystems, various methods and technologies are commonly applied in the treatment of mining and refining industrial effluents before they are discharged to receiving water. These methods include precipitation, ion exchange, and membrane processes. However, the application of some of these methods may be impractical due to economic constraints or may be insufficient to meet strict regulatory requirements. Furthermore, they may generate hazardous products or products which are difficult to treat [10, 11]. The traditional coagulation and flocculation processes use inorganic coagulants such as aluminium hydroxides in drinking water treatment. Metal coagulants can be used to partially remove heavy metal from wastewater [12], but, the use of metal coagulants is not 100% effective for removing metal cations from water at pH 7. The high cost of adsorbents such as activated carbon and some ion-exchange resins used for the treatment of water and wastewater have led to new, more effective and cheaper adsorbents. Bailey *et al.* [13] have mentioned that natural materials that are available in large quantities or industrial waste products can also be used as adsorbents.

11.2 Sources of Heavy Metals in the Environment

Various industries that generate wastewater containing heavy metals include the tanning, battery, glassware, ceramics, electroplating, fertilizer, mining, paints, and photographic industries. These wastewaters contain heavy metals such as chromium, lead, cadmium, arsenic, copper, iron, manganese, nickel, mercury and cobalt, among others. The amount and the number of metals present in any wastewater are related directly to the operations carried out in an industry. For example, tanneries discharge chromium in wastewater; copper, chromium, zinc, and cadmium are widely generated from metal plating; the production of electrical equipment and mining, smelting, and fossil fuel combustion contribute to mercury pollution; and lead is generated from a number of industrial and mining sources. In most wastewaters, the concentration of heavy metals present is much larger than the safe permissible limits and, therefore, they need to be removed. Table 11.1 summarizes the anthropogenic sources of heavy metals in the environment.

Arsenic is ubiquitous and ranks 20th in natural abundance, comprising about 0.00005% of the Earth's crust, 14th in seawater, and 12th in the human body [14]. Its concentration in most rocks ranges from 0.5 to 2.5 mg/kg, though higher concentrations are found in finer-grained argillaceous sediments and phosphorites. Since its isolation in 1250 A.D. by Albertus Magnus [15], this element has been a continuous center of controversy. Arsenic is mobilized by natural weathering reactions, biological activity, geochemical reactions, volcanic emissions and other anthropogenic activities. Soil erosion and leaching contribute to 612×10^8 and 2380×10^8 g/year of arsenic, respectively, in dissolved and suspended forms in the oceans. Most environmental arsenic problems are the result of mobilization under natural conditions. However, mining activities, combustion of fossil fuels, use of arsenic pesticides, herbicides, and crop desiccants and use of arsenic additives to livestock feed create additional impacts [14, 16].

Mercury is a highly toxic element that is found both naturally and as an introduced contaminant in the environment. Although its potential for toxicity in highly contaminated areas such as Minamata Bay, Japan, in the 1950s and 1960s, and San Francisco Bay, USA, is well documented [17, 18, 19, 20, 21, 22, 23, 24, 25], research has shown that mercury can be a threat to the health of people and wildlife in many environments that are not obviously polluted. The risk is determined by the likelihood of exposure, the form of mercury present, and the geochemical and ecological factors that influence how mercury moves and changes form in the environment.

Cadmium is a naturally occurring metallic element, one of the components of the Earth's crust, and present everywhere in our environment. However, in certain forms and in high concentrations, cadmium may produce adverse human health and environmental effects. Cadmium emissions to the environment may arise from both natural and anthropogenic or man-made sources. Estimates of the proportion of total cadmium

Table 11.1 Significant anthropogenic sources of metal ions in the environment and their effect on human beings and animals.

Pollutants	Major Sources	Typical Effects	References
As	Arsenic containing fungicides, pesticides and herbicides, metal smelters, by-product of mining activities, chemical wastes	Inhalation, ingestion or absorption though skin can cause mild bronchitis, nasal irritation or dermatitis. Carcinogenic activity also is suspected. Attack SH group of enzymes, coagulate proteins	[14, 15, 16]
Cd	Cadmium producing industries, electroplating, welding. By-product of refining of Pb, Zn and Cu, fertilizer industry, pesticide manufacture, cadmium nickel batteries, nuclear fission plants, production of TEL used in additive in petrol	Inhalation of fumes and vapours causes kidney damage, bronchitis, gastric and intestinal disorders, cancer, disorder of heart, liver and brain. Chronic and acute poisoning may result renal dysfunction, anaemia, hypertension, bone-marrow disorder and cancer	[26]
Cr	Metallurgical and chemical industries, processes using chromate compounds, cement and asbestos units	Toxic to body tissues, can cause irritation, dermatitis, ulceration of skin, perforation of nasal septum. Carcinogenic action suspected	[27]
Pb	Automobile emissions, lead smelters, burning of coal and oil, lead arsenate pesticides, smoking mining and plumbing	Absorption through gastrointestinal and respiratory tract and deposition in mucous membranes, causes liver and kidney damages, gastrointestinal damages, mental retardation in children, abnormalities in fertility and pregnancy	[28]

Pollutants	Major Sources	Typical Effects	References
Zn	Zinc refineries, galvanizing processes, brass manufacture, metal plating, plumbing	Zinc fumes have corrosive effects on skin and can cause irritation and damage mucous membranes	[31]
Mn	Ferromanganese production, organo-manganese fuel additives, welding rods, incineration of manganese containing substances	Poisoning of central nervous system (CNS), absorption, inhalation, or skin contact may cause manganic pneumonia	[38]
Ni	Metallurgical industries using nickel, combustion of fuels containing nickel additives, burning of coal and oil, electroplating units using nickel salts, incineration of nickel containing substances	Respiratory disorders, dermatitis, cancer of lungs and sinus. Histological alterations in the epididymis and seminal vesicles; decreases in sperm concentration, motility, abnormalities, and in fertility. The observed effects include inflammatory changes in the lungs, atrophy of the nasal olfactory epithelium, hyperplasia in the bronchial and mediastinal lymph nodes, impaired immune function, and decreases in body weight gain, which are probably secondary to the lung damage	[32, 33, 34, 35, 36]
Se	Burning of fuels and residual oils, fumes and gases from refinery wastes, incineration of paper and other wastes	Irritation of gastro-intestinal and respiratory tracts, irritation of eyes, nose and throat, damages to lungs, liver and kidneys	[39]

(*Continued*)

Table 11.1 (*Cont.*)

Pollutants	Major Sources	Typical Effects	References
Hg	Mining and refining of mercury, organic mercurials used in pesticides, laboratories using mercury	Inhalation of mercury vapours may cause toxic effects and protoplasmic poisoning. Organo-mercurials are highly toxic and may cause irreversible damage to nervous system and brain	[17, 18, 19, 20, 21, 22, 23, 24, 25]
Cu	Iron and steel industry, fertilizer industry and burning of wood, discharge of mine tailings, disposal of fly ash and disposal of municipal and industrial wastes are the sources of copper in the atmosphere	Defects in pigmentation, bone formation, reproduction, myelination of the spinal cord, cardiac function and connective tissue formation, in addition to defects in growth and hematopoiesis were found to be the manifestations of copper deficiency	[37]
Co	The primary anthropogenic sources of cobalt in the environment are from the burning of fossil fuels, application of cobalt-containing sludge or phosphate fertilizers, mining and smelting of cobalt-containing ores, processing of cobalt-containing alloys, and industries that use or process cobalt compounds.	Inhalation exposure to cobalt-containing particles, the primary target of exposure is the respiratory tract. Occupational exposure of humans to cobalt metal or cobalt-containing hard metal have reported primarily respiratory effects, including decreased pulmonary function, asthma, interstitial lung disease, wheezing, and dyspnea. Many of the respiratory tract effects are believed to be the result of the generation of oxidants and free radicals by the cobaltion.	[29]

emissions due to natural sources have ranged from as low as 10% to as high as 50% [26]. Some of these natural emission sources include weathering and erosion of parent rocks, volcanic activity, and forest fires. Anthropogenic sources of cadmium releases include products and materials to which cadmium has deliberately been added to impart a specific chemical, mechanical or physical property and products or materials in which cadmium is naturally present as a residual or impurity element. Because of its unique chemical, mechanical and physical properties, cadmium is utilized deliberately in a number of important industrial and consumer applications. These include nickel-cadmium (NiCd) batteries for both industrial and consumer uses; cadmium corrosion-resistant coatings on iron, steel, aluminum and titanium; cadmium pigments in plastics, glasses, ceramics, enamels and artists' colors; weathering and ultraviolet light stabilizers for polyvinyl chloride (PVC); as an alloying element in soldering, brazing, electrical contact, high electrical conductivity and high thermal conductivity alloys; and in electronic compounds such as cadmium telluride and cadmium sulfide employed in solar cells, detectors, electronic gates, switches, sensors and relays.

Chromium is a naturally-occurring element found in rocks, animals, plants, and soil, where it exists in combination with other elements to form various compounds. The three main forms of chromium are Cr(0), Cr(III), and Cr(VI). Small amounts of Cr(III) are needed for human metabolism. Chromium is released into the environment from natural and anthropogenic sources, with the largest release occurring from industrial releases. Chromium is widely used in manufacturing processes to make various metal alloys such as stainless steel. Chromium can be found in air, soil, and water after release from industries that use chromium, such as industries involved in electroplating, leather tanning, textile production, and the manufacture of chromium-based products [27]. Chromium can also be released into the environment from the burning of natural gas, oil, or coal.

Lead is a heavy, low melting, bluish-gray metal that occurs naturally in the Earth's crust. However, lead rarely occurs in its elemental state, but rather in its +2 oxidation state in various ores throughout the earth. The most important lead-containing ores are galena (PbS), anglesite ($PbSO_4$) and cerussite ($PbCO_3$). It is usually found combined with two or more other elements to form lead compounds. Lead and lead alloys are commonly found in pipes, storage batteries, weights, shot and ammunition, cable covers, and sheets used to shield from radiation. The largest use for lead is in storage batteries in cars and other vehicles [28]. Lead compounds are used as a pigment in paints, dyes, and ceramic glazes. Tetraethyl lead may still be used in gasoline for off-road vehicles and airplanes. It is also still used in a number of developing countries. Lead used in ammunition, which is the largest non-battery end-use, has remained fairly constant in recent years. However, even the use of lead in bullets and shot as well as in fishing sinkers is being reduced because of its harm to the environment. Most lead used by industry comes from mined ores or from recycled scrap metal or batteries. However, most of the high levels found throughout the environment come from human activities. Environmental levels of lead have increased more than 1,000 fold over the past three centuries as a result of human activity. Lead can enter the environment through releases from mining lead and other metals, and from factories that make or use lead, lead alloys, or lead compounds.

Zinc is one of the most common elements in the Earth's crust. Zinc is found in the air, soil, and water and is present in all foods. Metallic zinc has many uses in industry. A common use for zinc is to coat steel and iron as well as other metals to prevent rust and corrosion. Metallic zinc is also used to make dry cell batteries. Zinc can also combine with other elements, such as chlorine, oxygen, and sulfur, to form zinc compounds. Zinc compounds that may be found at hazardous waste sites are zinc chloride, zinc oxide, zinc sulfate, and zinc sulfide. Most zinc ore found naturally in the environment is in the form of zinc sulfide. Zinc compounds are widely used in industry. Zinc sulfide and zinc oxide are used to make white paints, ceramics, and other products. Zinc enters the air, water, and soil as a result of both natural processes and human activities. Most zinc enters the environment as the result of mining, purifying of zinc, lead, and cadmium ores, steel production, coal burning, and burning of wastes. These activities can increase zinc levels in the atmosphere. Waste streams from zinc and other metal manufacturing and zinc chemical industries, domestic wastewater, and run-off from soil containing zinc can discharge zinc into waterways [31].

Cobalt is a naturally-occurring element that has properties similar to those of iron and nickel. It has an atomic number of 27. Small amounts of cobalt are naturally found in most rocks, soil, water, plants, and animals, typically in small amounts. Cobalt is also found in meteorites. Soils near ore deposits, phosphate rocks, or ore smelting facilities, and soils contaminated by airport traffic, highway traffic, or other industrial pollution may contain high concentrations of cobalt [29]. Small amounts of cobalt may be released into the atmosphere from coal-fired power plants and incinerators, vehicular exhaust, industrial activities relating to mining and processing of cobalt-containing ores, and the production and use of cobalt alloys and chemicals. Cobalt is even found in water in dissolved or ionic form, typically in small amounts. A biochemically important cobalt compound is vitamin B_{12} or cyanocobalamin. Vitamin B_{12} is essential for good health in animals and humans.

Nickel is a very hard metal that occurs naturally in soils and volcanic dust. Nickel is used in combination with other metals to form alloys used for coins, jewelry, and stainless steel. Nickel compounds are used for electroplating, to color ceramics, and in battery production [30]. Nickel is released into the atmosphere by windblown dust, volcanoes, combustion of fuel oil, municipal incineration, and industries involved in nickel refining, steel production, and other nickel alloy production. Background levels of nickel in soils vary widely depending on local geology and anthropogenic inputs, but concentrations typically range between 4 and 80 ppm. Concentrations of nickel in household dust can be high and therefore pose an increased risk to young children.

11.3 Toxicity to Human Health and Ecosystems

It becomes a matter of alarm when metals accumulate in a locality in concentrations approaching or exceeding those that can harm organisms, including human beings. Some heavy metals are essential for biological systems in small amounts, but if ingested in larger amounts can disrupt physiological processes. Heavy metals, such as Cd, Hg, Pb, and Sn, may serve no particular biological function and can be toxic at relatively low concentrations. Heavy metals tend to bioaccumulate in the food chain, and their toxicity to organisms ranges from chronic to acute and from sublethal to deadly.

At high exposure levels, lead causes encephalopathy, cognitive impairment, behavioral disturbances, kidney damage, anemia and toxicity to the reproductive system. In its hexavalent form, chromium is widely recognized to exert toxic effects on biological systems. Human exposure to Cr(VI) compounds is associated with a higher frequency of respiratory cancers [40]. Cadmium is associated with nephrotoxic effects, particularly at high exposure levels; long-term exposure may cause bone damage as well. High concentrations of mercury can lead to neurobehavioral disorders and developmental disabilities including dyslexia, attention deficit hyperactivity disorder, and intellectual retardation. Excessive copper concentrations can lead to weakness, lethargy and anorexia, as well as damage to the gastrointestinal tract.

Arsenic occurrence in the environment, its toxicity, health hazards, and the techniques used for speciation analysis are well known and have been reviewed [14, 41]. Long-term drinking water exposure causes skin, lung, bladder, and kidney cancer as well as pigmentation changes, skin thickening (hyperkeratosis) neurological disorders, muscular weakness, loss of appetite, and nausea. This differs from acute poisoning, which typically causes vomiting, esophageal and abdominal pain, and bloody "rice water" diarrhea [14]. Arsenic in natural waters is a worldwide problem. Arsenic pollution has been reported recently in the USA, China, Chile, Bangladesh, Taiwan, Mexico, Argentina, Poland, Canada, Hungary, New Zealand, Japan and India. The largest population at risk among the 21 countries with known groundwater arsenic contamination is in Bangladesh, followed by West Bengal in India.

The toxic effects of mercury depend on its chemical form and the route of exposure. Methylmercury (CH_3Hg) is the most toxic form. It affects the immune system, alters genetic and enzyme systems, and damages the nervous system, including coordination and the senses of touch, taste, and sight [42]. It is particularly damaging to developing embryos, which are five to ten times more sensitive than adults. Exposure to CH_3Hg is usually by ingestion, and it is absorbed more readily and excreted more slowly than other forms of mercury. Elemental mercury (Hg^0), the form released from broken thermometers, causes tremors, gingivitis, and excitability when vapors are inhaled over a long period of time. Although it is less toxic than CH_3Hg, Hg^0 may be found in higher concentrations in environments such as gold mine sites, where it has been used to extract gold. If Hg^0 is ingested, it is absorbed relatively slowly and may pass through the digestive system without causing damage. Ingestion of other common forms of mercury, such as the salt $HgCl_2$, which damages the gastrointestinal tract and causes kidney failure, is unlikely from environmental sources.

Cadmium is a relatively rare element that is a minor nutrient for aquatic organisms at low concentrations, but is toxic to aquatic biota at only slightly higher concentrations. In aquatic environments, organisms are exposed to cadmium in dissolved and particulate-bound forms. Cadmium can be taken up by bacteria, phytoplankton, zooplankton and fish directly or through the food chain as a potentially toxic metal, which can enter the organism's body via waterborne and dietary pathways, although regulatory assessments of metal toxicity to aquatic organisms assume that toxic effects are caused by dissolved metals. In recent years, concern has been expressed about the possible effects on aquatic organisms from exposure to cadmium.

The toxicity of a chromium compound depends on the valence state of the chromium atom and the nature of its ligands. There are several routes to human chromium exposure which include inhalation, ingestion, dermal contact and respiratory tract, etc. The ingestion of Cr(VI) compounds produces toxic effects which results in irritation and ulcer of the alimentary canal and can cause anemia; Cr(III) compounds are much less toxic and do not appear to cause these problems. The most common health problems in workers exposed to chromium involves the respiratory tract. These health effects include irritation of the lining of the nose, runny nose, and breathing problems (asthma, cough, shortness of breath, wheezing). Workers have also developed allergies to chromium compounds, which can cause breathing difficulties and skin rashes. Sperm damage and damage to the male reproductive system have also been seen in laboratory animals exposed to Cr(VI). The International Agency for Research on Cancer (IARC) has determined that Cr(VI) compounds are carcinogenic to humans. In workers, inhalation of Cr(VI) has been shown to cause lung cancer [43, 44]. Mixed results have been found in studies of populations living in areas with high levels of Cr(VI) in the drinking water [45]. In laboratory animals, Cr(VI) compounds have been shown to cause tumors in the stomach, intestinal tract, and lung.

The toxic effects of lead can be observed when it enters the body through breathing or swallowing. The main target for lead toxicity is the nervous system, both in adults and children. Long-term exposure of adults to lead at work has resulted in decreased performance in some tests that measure functions of the nervous system [46, 47]. Lead exposure may also cause weakness in fingers, wrists, or ankles. Lead exposure also causes small increases in blood pressure, particularly in middle-aged and older people [48, 49]. Lead exposure may also cause anemia. At high levels of exposure, lead can severely damage the brain and kidneys in adults or children and ultimately cause death. In pregnant women, high levels of exposure to lead may cause miscarriage. High-level exposure in men can damage the organs responsible for sperm production.

Inhaling large amounts of zinc (as zinc dust or fumes from smelting or welding) can cause a specific short-term disease called metal fume fever, which is generally reversible once exposure to zinc ceases. However, very little is known about the long-term effects of breathing zinc dust or fumes. Ingesting high levels of zinc for several months may cause anemia, damage to the pancreas, and decrease levels of high-density lipoprotein (HDL) cholesterol. Skin irritation from exposure to these chemicals would probably occur in humans. Following inhalation of zinc oxide, and to a lesser extent zinc metal and many other zinc compounds, the most commonly reported effect is the development of "metal fume fever." Metal fume fever is characterized by chest pain, cough, dyspnea, reduced lung volumes, nausea, chills, malaise, and leukocytosis.

Cobalt has both beneficial and harmful effects on human health. Cobalt is beneficial for humans because it is part of vitamin B_{12}, which is essential to maintain human health. However, if too much cobalt gets into our body, harmful health effects can occur. Serious effects on the lungs, including asthma, pneumonia, and wheezing have been found in people exposed to 0.005 mg cobalt/m^3. Absorption of nickel following oral exposure has been shown to vary (3–40%) depending on whether the nickel was in drinking water or food, with greater absorption occurring with drinking water. About 20–35% of the inhaled nickel that is retained in the lungs is absorbed into the blood. Most of the absorbed nickel is excreted in the urine, regardless of the route of exposure. Adverse respiratory effects have been reported in humans and animals exposed to nickel compounds at concentrations much higher than typically found in the environment. In nickel workers, exposure to nickel did not result in increases in the risk of death from nonmalignant respiratory system disease. The cellular bioavailability of nickel (Ni), ie, the ability of nickel to enter target cells and to release nickel ions, appears to be a major determinant of the carcinogenic and toxic effects of nickel compounds [50, 51, 52, 53].

Due to their detrimental effects on human beings and biodiversity, it has become necessary to remediate the wastewater containing heavy metals and industrial effluents. Various methods exist for the removal of heavy metal ions from wastewater which include chemical precipitation, coagulation, membrane technology, electrolytic reduction, ion exchange, and adsorption. Ion exchange is the most widely used method for heavy metal removal from aqueous streams. During removal, recovery, or processing of metals, ion exchange acts as a concentrator of metals. Coagulation–flocculation can also be employed to treat wastewater ladened with heavy metals, wherein the coagulation process destabilizes colloidal particles by adding a chemical agent and results in sedimentation. Membrane filtration has received considerable attention for the treatment of inorganic effluent, since it is capable of removing not only suspended solid and organic compounds, but also inorganic contaminants such as heavy metals. Amongst all the treatment processes mentioned, adsorption using sorbents is one of the most popular and effective processes for the removal of heavy metals from wastewater. The adsorption process offers flexibility in design and operation and in many cases produces treated effluent suitable for reuse, free of color and odor. In addition, because adsorption is sometimes reversible, the regeneration of the adsorbent with resultant economy of operation may be possible. Activated carbon adsorbents are used widely in the removal of organic contaminants and, to a lesser extent, in the removal of heavy metal contaminants in product purification and pollution control. In spite of its prolific use, activated carbon remains an expensive material and the higher the quality of activated carbon, the higher its cost. Also, in practice, activated carbon is employed more frequently for adsorption of organic compounds rather than heavy metal ions. Magnetic nanoparticles offer a green approach for the treatment of heavy metals containing wastewater for management and sustainable development.

11.4 Magnetic Nanoparticles

The special and superior properties of nanomaterials have attracted much attention in the past two decades. Particularly, magnetic nanoparticles (MNPs) with inherent magnetic properties and high surface-to-volume ratio have continued to draw considerable interest because of their diverse potential applications in biological, environmental and medical diagnostic fields. One of the rapidly emerging research subjects involving MNPs is their application in biological systems, including their application in magnetic resonance imaging (MRI), targeted drug delivery, rapid biological separation, biosensors, magnetic hyperthermia therapy [55, 56], and environmental applications. The most commonly used magnetic particles are magnetite (Fe_3O_4) and maghemite (γ-Fe_2O_3). Other types of magnetic particles are pure metal (Fe and Co) and spinel-type ferromagnets ($MeO.Fe_2O_3$, where M = Ni, Co, Mg, Zn, Mn). Due to their unique size, biocompatibility, low toxicity and superparamagnetic properties, magnetic nanoparticles are emerging as promising tools in various fields such as physics, medicine, biology, environmental and material science [54]. Several types of iron oxides have been investigated recently in the field of nanosized magnetic particles (mostly maghemite [γ-Fe_2O_3], or magnetite [Fe_3O_4], single domains of about 5–20 nm in diameter), among which Fe_3O_4 is a very promising and popular candidate since its biocompatibility has already been proven [54]. Over the last decades, numerous synthetic methods have been developed to synthesize iron oxide nanoparticles including co-precipitation [57], sol-gel synthesis [58], microemulsion synthesis [59], sonochemical reaction [60], hydrothermal reaction [61], thermal decomposition [62], laser pyrolysis [63], etc. Among these reported methods, chemical co-precipitation may be the most promising one because of its simplicity and high productivity.

11.4.1 Properties of Magnetic Nanoparticles

In contrast to bulk iron oxide, which is a multi-domain ferromagnetic material (exhibits a permanent magnetization in the absence of a magnetic field), iron oxide magnetic nanoparticles smaller than approximately

20–30 nm in size contain a single magnetic domain with a single magnetic moment and exhibit superparamagnetism [64]. A material in a paramagnetic phase is characterized by randomly oriented (or uncoupled) magnetic dipoles, which can be aligned only in the presence of an external magnetic field and along its direction. This type of material has no coercivity nor remanence, which means that when the external magnetic field is switched off, the internal magnetic dipoles randomize again; no extra energy is required to demagnetize the material and hence the initial zero net magnetic moment is spontaneously recovered. A nanoparticle with such magnetic behavior is superparamagnetic (SPM).

The size reduction of magnetic materials shows interesting advantages that make them more suitable for therapeutic, diagnostic and environmental techniques compared to their bulk counterparts. Magnetic parameters such as the coercivity of the nanoparticles can be finely tuned by decreasing their size. Moreover, a further reduction of the size below a certain value of the radius, the so-called superparamagnetic radius (rSP), induces a magnetic transition in particles where both ferro- and ferrimagnetic nanoparticles (FM) become superparamagnetic and, as previously stated, high magnetic moments are observed under the effect of a magnetic field, but no remanent magnetic moment will be present when the external magnetic field is removed. Superparamagnetism is a property strictly associated with nanostructured magnetic materials and arises when the thermal energy is sufficiently high to overcome the magnetic stabilization energy of the particle. With increasing size, the particles will become blocked as thermal energy becomes insufficient to allow the free rotation of spins. Superparamagnetic particles are usually ordered below a blocking temperature.

11.5 Synthesis of Magnetic Nanoparticles

During the past few years, a large number of the published articles about nanoparticles have described efficient routes to attain shape-controlled, highly stable, low-cost, ecofriendly, and narrow size distribution magnetic nanoparticles. Currently, several popular methods including co-precipitation, microemulsion, thermal decomposition, sonochemical, microwave assisted, chemical vapor deposition, combustion synthesis, carbon arc, laser pyrolysis synthesis, hydrothermal synthesis and sol-gel synthesis have been reported for synthesis of magnetic nanoparticles.

The synthesis of superparamagnetic nanoparticles is a complex process because of their colloidal nature. Over the last decades, much research has been devoted to the synthesis of magnetic nanoparticles. Especially during the last few years, many publications have described efficient synthetic routes to shape-controlled, highly stable, functionalized, and monodisperse magnetic nanoparticles. These methods have been used to prepare particles with homogeneous composition and narrow size distribution. However, the most common method for the production of magnetite nanoparticles is the chemical co-precipitation technique of iron salts.

11.5.1 Co-precipitation

Co-precipitation is a very facile and convenient way to synthesize iron oxide nanoparticles (either Fe_3O_4 or γ-Fe_2O_3) from aqueous Fe^{2+}/Fe^{3+} salt solutions by the addition of a base under inert atmosphere at room temperature or at elevated temperature. The chemical reaction of Fe_3O_4 formation may be written as [65]:

$$Fe^{2+} + 2Fe^{3+} + 8OH^{-} \rightarrow Fe_3O_4 + 4H_2O \tag{11.1}$$

According to the thermodynamics of this reaction, complete precipitation of Fe_3O_4 should be expected at a pH between 8 and 14, with a stoichiometric ratio of 2:1 (Fe^{3+}/Fe^{2+}) in a non-oxidizing oxygen environment. The size, shape, and composition of the magnetic nanoparticles very much depend on the type of salts used (e.g., chlorides, sulfates, nitrates), the Fe^{2+}/Fe^{3+} ratio, the reaction temperature, the pH value and ionic strength of the media. With this synthesis, once the synthetic conditions are fixed, the quality of the magnetite nanoparticles is fully reproducible. However, magnetite nanoparticles are not very stable under ambient conditions, and are easily oxidized to maghemite or dissolved in an acidic medium.

$$Fe_3O_4 + 2H^{+} \rightarrow \gamma - Fe_2O_4 + Fe^{2+} + H_2O \tag{11.2}$$

Oxidation of maghemite is the lesser problem, as it is a ferrimagnet. Therefore, magnetite particles can be subjected to deliberate oxidation to convert them into maghemite. This transformation is achieved by dispersing them in acidic medium, then adding iron(III) nitrate. The maghemite particles obtained are then chemically stable in alkaline and acidic medium [66]. The main advantage of the co-precipitation process is that a large amount of nanoparticles can be synthesized in a short time. However, the control of particle size distribution is limited, because only kinetic factors are controlling the growth of the crystal [65]. In the syntheses of magnetic nanoparticles by co-precipitation route, two stages are mainly involved, a short burst of nucleation occurs when the concentration of the species reaches critical supersaturation, and then, there is a slow growth of the nuclei by diffusion of the solutes to the surface of the crystal. To fabricate monodisperse iron oxide magnetic nanoparticles, these two key stages should be separated, i.e., nucleation should be avoided during the period of enlargement.

During the last few years, significant advances in preparing monodisperse magnetite nanoparticles of different sizes have been made by the use of organic additives as stabilization or functionalization agents. Recently, a novel magnetic nanoadsorbent for the removal of Cu^{2+} was first developed by the surface modification of chitosan-coated magnetic nanoparticles (CCMNPs) with α-ketoglutaric acid (α-KA), which is a natural, inexpensive, harmless and environmentally-friendly biological reagent containing active functional groups like carboxyl groups [67]. The TEM pattern of γ-Fe_2O_3 shows monodisperse rodlike and rhombohedral shape with the particle sizes of 110 and 10 nm, respectively, while the TEM picture of the CCMNPs shows discrete spherical shape. It was earlier reported that the shapes of the Fe_2O_3 products are rodlike to spherical, and the particle size of the sample had been clearly changed after modification by the surfactants. The TEM images also reveal that the CCMNPs synthesized in this study are multidispersed with a mean diameter of around 30 nm, and no marked aggregation occurs. It can be clearly seen that the chitosan coating process significantly altered the morphology of γ-Fe_2O_3 nanoparticles. It has been shown that the γ-Fe_2O_3 nanoparticles were coated by the chitosan, and the prepared CCMNPs exhibited good distribution.

Tai *et al.* [68] fabricated magnetic porous microspheres of Fe_3O_4@poly(methylmethacrylate-co-divinylbenzene) by a simple strategy of co-precipitation reaction. The magnetic microspheres materials, consisting of polymer-coated iron oxide nanoparticles, were synthesized by the modified suspension polymerization of methacrylate and divinylbenzene in the presence of a magnetic fluid. The large surface area and stabilized magnetic porous polymeric microspheres of Fe_3O_4@poly(methylmethacrylate-co-divinylbenzene) were used in the adsorptive removal of phenol from aqueous solution. It had been observed that the distribution of the oleic acid-coated Fe_3O_4 nanoparticles is comparatively uniform; in addition, the oleic acid-coated magnetite nanoparticles easily aggregate. Recent studies showed that oleic acid is an excellent candidate for the stabilization of magnetic Fe_3O_4 and γ-Fe_2O_3 nanoparticles [68, 69].

Hong *et al.* [70] demonstrated that precipitation of Fe_3O_4 nanoparticles with NH_4OH instead of NaOH gives a better crystallinity, higher saturation magnetization and smaller size. Nitrogen passes during the precipitation, and protects oxidation of the Fe_3O_4 nanomaterials but also reduces the particle size when compared to methods without oxygen removal. Magnetic nanoparticles, Fe_3O_4 or γ-Fe_2O_3, prepared under alkaline condition of NH_4OH, can be suitably stabilized with silica to form well-dispersed magnetic silica nanospheres. It gives several advantages over other stabilization materials. The change in the SiO_2/Fe_3O_4 ratio controls the particle size distribution.

11.5.2 Hydrothermal Synthesis

Several scientists have reported the hydrothermal synthesis of magnetic nanoparticles for a wide range of applications [71, 72, 73]. Hydrothermal reactions are performed in aqueous solutions at high temperature and pressure. The reaction conditions, such as solvent, temperature, and time, usually have important effects on the products synthesized by the hydrothermal method. There are mainly two routes for the synthesis of magnetic nanoparticles via hydrothermal conditions: hydrolysis and oxidation or neutralization of mixed metal hydroxides. The hydrothermal method offers high production yields with reduced production costs and is environmentally friendly as a result of the simple equipment, readily available source materials, absence of explosive or corrosive gases, and use of only deionized water as the reaction medium [72]. Wan *et al.* [73] have successfully synthesized Fe_3O_4 nanostructures with different morphologies such as Fe_3O_4 nanoparticles, nanorods and nanowire bundles via a developed hydrothermal route. The Fe_3O_4 nanoparticles with uniform size and relatively better monodispersity were synthesized at room temperature. They have reported that the synthesized product was mainly composed of uniform Fe_3O_4 nanoparticles with the average diameter of about 20 nm. The selected area electron diffraction (SAED) pattern indicated the good crystalline nature of the nanoparticles. It is worthwhile to note that the complexation between the surfactant molecules and ferrites leads to an improvement in the uniformity of crystallite size and morphology. As a result, the prepared Fe_3O_4 nanoparticles showed uniform size and relatively better monodispersity.

Cao *et al.* [74] reported a new reaction route to obtain rice- and cube-shaped single-crystalline α-Fe_2O_3 nanostructures via a reaction between $Fe(NO_3)_3 \cdot 9H_2O$ and $NH_3 \cdot H_2O$ in ethylene glycol at 200 and 280°C, respectively. The morphology of the products was studied using transmission electron microscopy and high-resolution TEM. The average size of a rice-shaped α-Fe_2O_3 nanocrystal along its major axis was approximately 84.9

nm and along its minor axis approximately 41.2 nm; the length-to-diameter ratio was about two. The single-crystalline nature of synthesized α-Fe_2O_3 was confirmed by the strong spots of the selected area electron diffraction (SAED) pattern. On the basis of the results, the authors have indicated that the reaction temperature plays a key role for varying the morphology of the synthesized α-Fe_2O_3. The sensitivity to the reaction temperature implies that the reaction route is under kinetic rather than thermodynamic control. The reaction between $Fe(NO_3)_3$ solution and $NH_3{\cdot}H_2O$ is shown in equation:

$$Fe_3(NO_3)_3 + 3NH_3 \cdot H_2O \rightarrow Fe(OH)_3 + 3NH_4NO_3 \quad (11.3)$$

and α-Fe_2O_3 particles will form through a two-step phase transformation, according to the following equation:

$$Fe(OH)_3 \rightarrow \beta - FeOOH \rightarrow \alpha - Fe_2O_3 \quad (11.4)$$

Ethylene glycol will have an effect on preventing the $Fe(OH)_3$ nanoparticles from agglomeration [74]. Therefore, uniform α-Fe_2O_3 nanocrystals were obtained after being heated at 200–280°C to complete the phase transformation.

11.5.3 Microemulsion

The microemulsion route is presented to obtain magnetic nanoparticles of different sizes, offering several advantages with respect to other similar methods [75, 76, 77, 78]. A microemulsion can be defined as a thermodynamically stable isotropic dispersion of two immiscible liquids consisting of nanosized domains of one liquid in the other, stabilized by an interfacial film of surface-active molecules. The surfactant molecules provide a confinement effect that limits particle nucleation, growth and agglomeration. In this method, co-precipitation occurs in tiny droplets of water ("water pools") embedded with surfactant molecules and homogenously distributed in an oil phase. The size of these "water pools," so called reverse micelles, which act as micro reactors for the synthesis of the nanoparticles, is thermodynamically defined by the water-to-surfactant molar ratio [79].

In this field the main challenge is to create the suitable surface of magnetic nanoparticles in order to functionalize and allow strong interactions with specific biological components. In an effort to make ultrasmall, coated, and uniform-size magnetic nanoparticles, Santra *et al.* performed water-in-oil microemulsion mediated sonochemical synthesis of superparamagnetic iron oxide nanoparticles in nonionic surfactants [80]. The magnetic core was coated by a thin layer of silica for stabilization of magnetic nanoparticles. The synthesized magnetic nanoparticles have diameters as small as 1–2 nm, and the thickness of the silica coating is as thin as 1–2 nm.

Vestal and Zhang have used normal and reverse micelle microemulsion methods to synthesis single-phase $CoCrFeO_4$ nanoparticles with a controlled size range of 6–16 nm [81]. Results obtained in this study showed that microemulsion methods provide a synthetic approach that allows for high-quality nanoparticles with a narrow size distribution. This method can be utilized for the synthesis of magnetic nanoparticles of various morphologies. However, this method needs large amounts of solvents and the yield is low.

11.5.4 Thermal Decomposition

Magnetic nanoparticles with smaller size and different morphology can be synthesized through the thermal decomposition of organometallic compounds in high-boiling organic solvents containing stabilizing surfactants [66, 82]. Iron oxide magnetic nanoparticles can also be synthesized by the thermal decomposition method. In thermal decomposition, two different approaches have been used for this purpose. The first approach is thermal decomposition of metal carbonyl precursors followed by an oxidation step using air, or oxidation by using an oxidant at elevated temperatures. The second is decomposition of precursors with a cationic metal center in the absence of reducing agents. The presence of reducing agents leads to metal nanoparticles even by the use of cationic precursors.

The thermal decomposition route appears to be the best synthesizing method for controlling size and morphology of magnetic nanoparticles. It gives scalable and high production yield. Although, one of the major disadvantages of this method is the production of organic soluble nanomaterials which limit the extent of application uses of them in biological and environmental fields, besides which surface treatment is needed after synthesis; also, thermal decomposition methods usually lead to complicated processes or require relatively high temperatures. However, success in precise particle size control of Fe_3O_4 nanoparticle has only been achieved through thermal decomposition using large quantities of toxic and expensive precursors and surfactants in organic solvent [83].

11.6 Magnetic Nanoparticles in Wastewater Treatment

Magnetic nanoparticles such as Fe_3O_4 and α-Fe_2O_3 have been investigated to resolve various environmental problems, such as removing toxic metal ions and radioactive elements, capturing microbial pathogens and organic dyes, accelerating the coagulation of sewage, and remediation of contaminated soils [84, 85, 86]. The magnetic nanoparticles possess high surface area, high chemical stability and optimal magnetic properties, which lead to high adsorption efficiency, highly efficient removal of contaminants, and easy and rapid separation of adsorbent from solution via external magnetic field.

Magnetism of the nanomaterials is a unique property that autonomously assists in water/wastewater purification by influencing the physical properties of contaminants in aqueous solutions. Thus, magnetic separations of pollutants have been used extensively in wastewater treatment and environmental cleanup. Magnetic iron oxide nanomaterials are promising candidates for larger scale wastewater treatment due to their easy functionalization, low cost, high adsorption capacity, strong physicochemical stability, and easy separations. Thus, these nanomaterials can be efficiently utilized for the treatment of wastewater containing hazardous dyes and toxic metal ions. In the following sections we have described the application of magnetic nanoparticles for rapid removal of heavy metals from aqueous environments.

11.6.1 Magnetic Nanoparticles as Nanosorbents for Heavy Metals

Heavy metal contamination of aqueous environments is an increasing worldwide problem. Heavy metals are well known to cause serious environmental and health problems due to their bioaccumulation through the food chain, and cause serious damage to the nervous and endocrine systems of human beings. Therefore, the development of selective and efficient methodologies for detecting and removing heavy metals from aqueous media is in great demand.

During the past few years, the majority of scientific research and large-scale field applications of materials for treatment of wastewater containing heavy metals have focused on magnetic nanoparticles, activated carbon, carbon nanotubes, biosorbents, zero-valent iron, and photocatalysts [87, 88, 89, 90, 91, 92, 93]. Among all these materials, magnetic nanoparticles possess the capability to treat wastewater on an industrial scale and are convenient for magnetic separation, are the most promising materials for heavy metal removal.

Recently, Feng *et al.* synthesized monodisperse carboxyl-functionalized superparamagnetic iron oxide nanoparticles, Fe_3O_4(np)-COOH [88]. Representative TEM images of the Fe_3O_4(np)-COOH nanoparticles deposited from a water dispersion and dried under ambient conditions revealed the presence of clearly defined features of an average size of ~8–17 nm with narrow size distribution. They examined their ability to adsorb Cd^{2+} from aqueous solutions. The adsorption efficiency of the particles for Cd^{2+} was on the order of 0.17–0.20 μmol Cd/mg Fe_3O_4(np)-COOH, which was among the highest reported in literature. As reported by the authors, the adsorption efficiency of the 8 nm particles was a bit higher than that for the 17 nm analogues. It had been clearly observed that an increase in the acidity of the media led to desorption of Cd^{2+}.

In order to provide long-term, high quality water, or to enable water recycling, Girginova *et al.* [94] recently synthesized silica-coated magnetite particles for magnetic removal of Hg^{2+} from water. Silica-coated magnetite particles derivatized with dithiocarbamate groups were evaluated as magnetic nanodrivers to remove trace levels of Hg^{2+} from aqueous solutions using a simple external NdFeB magnet. The authors reported that the uptake efficiency for Hg^{2+} was significantly higher for the particles bearing dithiocarbamate groups at the surface (74%), as compared to the non-derivatized silica-coated magnetite (24%). On the basis of this study, one can suggest that this efficiency of nanomaterials is related to the high stability of the chelates formed between the dithiocarbamate group and Hg^{2+}.

To further facilitate the adsorption affinity of guest molecules on nanomaterials, surface modification, including inorganic and organic coating, and covalent binding, has often been explored to enable specific metal complexation. Notably, the amino-functionalized materials demonstrated an outstanding ability to remove a wide variety of heavy metal ions such as Cu(II), Co(II), Ni(II), Zn(II), Pb(II), Cr(VI), and Cd(II) from aqueous solutions owing to the strong metal complexing capability of amino groups [95]. The SiO_2 is stable under acidic conditions and inert to redox reactions, as compared with the organic coating materials, and hence functions as an ideal shell composite to protect the inner magnetite core. Silica-coated core–shell magnetite nanoparticles (Fe_3O_4@SiO_2) have recently been investigated for potential environmental and biomedical applications [95]. The SiO_2 coating provides plenty of surface hydroxyl groups, which offers ease of succedent functionalization of magnetite nanoparticles.

Wang *et al.* [95] have developed a novel amino-functionalized Fe_3O_4@SiO_2 magnetic nanomaterial with a core–shell structure for the adsorptive removal of heavy metals from aqueous solutions. The average particle sizes of Fe_3O_4, Fe_3O_4@SiO_2, and Fe_3O_4@SiO_2–NH_2 were observed to be 13.4, 16.6 and 18.4 nm, respectively. High adsorption affinity for aqueous Cu(II), Pb(II), and Cd(II) was achieved through the complexation of metal ions by amino groups grafted on the silica surface of the nanomaterial, and the adsorption was not much impacted by the presence of cosolute of humic acid or alkali/earth metal ions. Thus, the result observed shows that core–shell structure of the nanoadsorbent formed by a magnetite wrapped with an inert silica layer provides ease of magnetic separation and protection from acid leaching in regeneration. These unique characteristics of synthesized Fe_3O_4@SiO_2–NH_2 nanoparticles give a high potential for effective and rapid removal of toxic metal ions in water treatment.

Very recently, Song *et al.* [96] prepared polyrhodanine-encapsulated magnetic nanoparticles via one-step chemical oxidation polymerization and investigated their potential for adsorptive removal of heavy metal ions from aqueous solutions. The TEM images showed that the polyrhodanine-encapsulated magnetic nanoparticles were synthesized with an average diameter of ca. 10 nm. Typically, 5.0 mg of the PR-MNPs was added into 10 mL of Hg(II) ion solution (80 mg/L) at a pH value of 4.0. Then, the prepared samples were shaken at 350 rpm using a mechanical shaker at 25°C in a batch mode. After a desired contact time, the PR-MNPs were removed from the solution using an external magnetic field. An inductively coupled plasma (ICP) analysis was observed for the residual Hg(II) ion concentration in the aqueous media. Similarly, the adsorption capacities of Cd(II), Mn(II), and Cr(III) ions were also determined. It was found that the adsorption capacity increases sharply with an increase in pH from 2.0 to 4.0, but increases slightly with a pH rise from 4.0 to 8.0. This means that the concentration of H^+ ion significantly affects the adsorption behavior of PR-MNPs. The negative surface charge enhances the affinity to positively charged metal ions. The increment rate of adsorption capacity diminished in the pH range of 4.0 to 8.0. The highest Hg(II) ion adsorptivity was 94.5% at the initial mercury ion concentration of around 1.3 mg/L. Parameters of two isotherm models, Langmuir and Freundlich, were tested for the adsorption of Hg(II) ion onto the PR-MNPs. Comparing the R^2 value and F-test result of both models, it was concluded that the experimental data were well fitted to the Freundlich model, which suggests the heterogeneous metal ion adsorption activity. This may result from the different adsorption sites (oxygen, nitrogen, and sulfur groups) of polyrhodanine-magnetic nanoparticles that have different metal-binding energies.

Hu *et al.* [97] have synthesized various types of magnetic nanoparticles that were prepared by chemical co-precipitation method and used for the removal of Cr(VI) from synthetic electroplating wastewater. The size of these magnetic nanoparticles was measured using transmission electron microscopy (TEM) and found to be about 20 nm. The adsorption of Cr(VI) was investigated in batch mode in acidic medium using 5g/L of different magnetic nanoparticles. The Cr(VI) removal performances were compared and the adsorption capacities followed the order: $MnFe_2O_4$ > $MgFe_2O_4$ > $ZnFe_2O_4$ > $CuFe_2O_4$ > $NiFe_2O_4$ > $CoFe_2O_4$. The points of zero charge pHpzc for various magnetic nanoparticles were measured to be around 7.0–8.0. The $MnFe_2O_4$ showed a very high BET surface area in comparison to other synthesized ferrite nanoparticles. The equilibrium time for Cr uptake by $MnFe_2O_4$, $MgFe_2O_4$, $ZnFe_2O_4$, $CuFe_2O_4$, $NiFe_2O_4$ and $CoFe_2O_4$ nanoparticles is 5, 45, 30, 20, 15 and 60 min, respectively. It was clear that Cr adsorption onto $MnFe_2O_4$ particles reached equilibrium in the shortest time compared to the other ferrites, which was due to the rapid redox reaction that occurred between the Cr species and the external adsorbent surface. The adsorption of Cr(VI) onto MeFe2O4 nanoparticles is highly pH-dependent; hence the desorption of Cr(VI) can be accomplished by increasing the solution pH. An 0.01M of NaOH revealed the highest desorption efficiency compared to other concentrations of NaOH or the same concentration of other eluents such as $NaHCO_3$, Na_2CO_3 and Na_3PO_4. Among all the ferrite nanoparticles, $CoFe_2O_4$ showed higher desorption efficiency (98.1%).

Water-soluble hollow spherical Fe_3O_4 nanocages with high saturation magnetization were prepared in a one-pot reaction by sol-gel method [98], and subsequent annealing to synthesize the maghemite (γ-Fe_2O_3) nanocages with similar nanostructures was also performed. Figure 11.1a,b displays the TEM images of the Fe_3O_4 and γ- Fe_2O_3 nanocages. It was found that the Fe_3O_4 nanocages had a hollow structure and the overall diameter of the nanocages is around 100 nm, which indicated an oriented aggregation of small Fe_3O_4 nanoparticles. It has been observed that the shape and size of the γ-Fe_2O_3 nanocages are similar to those of Fe_3O_4 nanocages. However, the size of the central hole of the nanocages becomes smaller after annealing, owing to the thermal diffusion of the small nanoparticles. The SAED pattern in the insets of Figure 11.1a,b reveals the polycrystal-like feature of the samples, and their pattern agrees well with the structure planes of iron oxide nanocages. When $FeSO_4$ and KOH are mixed together, the solution generates $Fe(OH)_2$ gels; subsequently, upon addition of KNO_3 to this mixture, many small magnetite nanoparticles (Figure 11.1c) were formed through homogeneous nucleation. In contrast, if the Gla was not added, these small magnetite nanocages neither demonstrate obvious growth nor do they aggregate, due to the gel

Figure 11.1 (a) Fe_3O_4 nanocages with glutamic acid; (b) Fe_3O_4 nanocages with glutamic acid annealed at 500°C for 2 h in oxygen atmosphere; (c) Fe_3O_4 nanocages without glutamic acid. Adapted from [98].

network structure in the gel solution; when several nucleations take place, the gels begin to flocculate during the aging period and form large Fe_3O_4 nanoparticles in a size range from 50 to 100 nm. The HRTEM images were obtained near the center region of the hollow nanocages for observation of morphologies and structures of the iron oxide nanocages (Figure 11.2). Figure 11.2a shows that the Fe_3O_4 nanocages include three single crystalline

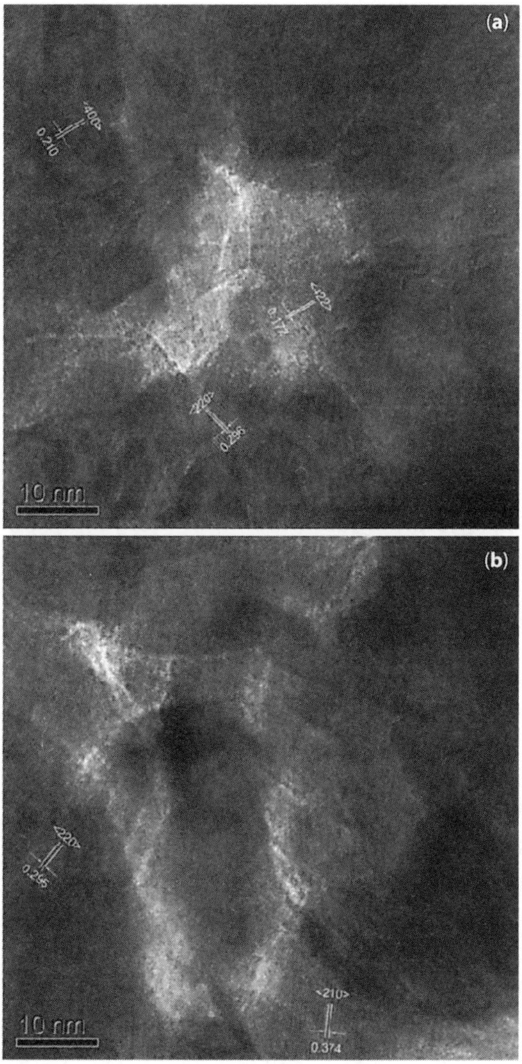

Figure 11.2 HRTEM images of the center region of hollow (a) Fe_3O_4 nanocages and (b) γ- Fe_2O_3 nanocages. Adapted from [98].

with an interplanar spacing of 0.296 nm for the {220} plane, 0.210 nm for the {400} plane, and 0.172 nm for the {422} plane, respectively. Figure 11.2b shows that the γ-Fe_2O_3 nanocages included two single crystalline obviously with an interplanar spacing of 0.374 nm for the {210} plane and 0.295 nm for the {220} plane, respectively. The studies by the authors have confirmed that the nanocages consist of randomly small iron oxide nanocrystals, and the nanocages present polycrystalline feature on the whole.

11.7 Modeling of Adsorption: Kinetic and Isotherm Models

Assessment of a solid–liquid sorption system is usually based on two types of investigations: equilibrium batch sorption tests and dynamic continuous-flow sorption studies. The kinetic study of adsorption processes provides useful data regarding the efficiency of the adsorption and the feasibility for scale-up operations. The kinetic data of adsorption can be evaluated using different types of mathematical models, of which the one most widely used is Lagergren's rate equation [99]. Equilibrium isotherm model equations such as Langmuir, Freundlich, Temkin, and Redlich-Peterson are used to describe experimental adsorption data in batch mode [100, 101, 102, 103]. It is important to find best-fit isotherm to evaluate the efficacy of the prepared adsorbent to develop suitable industrial adsorption system designs.

11.7.1 Kinetic Studies in Adsorption of Heavy Metals

Adsorption equilibria studies are important to conclude the efficacy of adsorption. In spite of this, it is also necessary to identify the adsorption mechanism type in a given system. For the purpose of investigating the mechanism of adsorption and its potential rate-controlling steps that include mass transport and chemical reaction processes, kinetic models have been exploited to test the experimental data. In addition, information on the kinetics of metal uptake is required to select the optimum condition for full-scale batch metal removal processes.

Several adsorption kinetic models have been established to understand the adsorption kinetics and rate-limiting step. These include pseudo-first-order and pseudo-second-order rate model, Weber and Morris sorption kinetic model, Adam–Bohart–Thomas relation, first-order reversible reaction model, external mass transfer model, first-order equation of Bhattacharya and Venkobachar, Elovich's model and Ritchie's equation.

The pseudo-first-order and pseudo-second-order kinetic models are the most well-liked models to study the adsorption kinetics of heavy metals and quantify the extent of uptake in adsorption kinetics.

11.7.1.1 Pseudo-First-Order Kinetic Model

Lagergren's first-order rate expression based on solid capacity is generally expressed as follows [99]:

$$\frac{dq}{dt} = k_1(q_e - q) \tag{11.5}$$

where q and q_e are amounts of adsorbate adsorbed (mg/g) at time, t (min) and at equilibrium, respectively, k_1 is the rate constant of adsorption (l/min). Integration of the above equation with the boundary conditions: t=0, q=0, and t=t, q=q, gives

$$\text{In}\,(q_e - q) = \text{In}\,q_e - k_1 t \tag{11.6}$$

This equation can be written in the linear form:

$$q = q_e(1 - exp(-k_1 t)) \tag{11.7}$$

Kinetic adsorptions by numerous magnetic nanomaterials have been studied using the pseudo-first-order kinetic model.

11.7.1.2 Pseudo-Second-Order Kinetic Model

Predicting the rate of adsorption for a given system is among the most important factors in adsorption system design, as the system's kinetics determine adsorbate residence time and the reactor dimensions [104]. As previously noted, various factors govern the adsorption capacity, i.e., initial heavy metals concentration, temperature, pH of solution, nanoparticle size, heavy metals nature; a kinetic model is only concerned with the effect of observable parameters on the overall rate. The pseudo-second-order model is derived on the basis of the sorption capacity of the solid phase, expressed as [105, 106, 107, 108]:

$$\frac{dq}{dt} = k_2(q_e - q)^2 \tag{11.8}$$

Integration of the above equation with the boundary conditions $t = 0$, $q = 0$, and at $t = t$, $q = q$, results in:

$$\frac{1}{q_e - q} = \frac{1}{q_e} + k_2 t \tag{11.9}$$

This equation can be stated in the linear form as:

$$\frac{t}{q} = \frac{t}{q_e} + \frac{1}{k_2 q_e^2} \tag{11.10}$$

where k_2 is the equilibrium rate constant of pseudo-second-order adsorption (g/mg/min). The pseudo-second-order rate constants can be determined experimentally by plotting t/q against t. As such, in comparison to the pseudo-first-order kinetic model, this model is considered more appropriate to represent the kinetic data in adsorption systems [104].

Pseudo-first-order and pseudo-second-order rate expressions have been and still are in wide use for studying the adsorption of heavy metals from aqueous solutions. In the chemisorption process, the pseudo-second-order is superior to the pseudo-first-order model, as it takes into account the interaction of adsorbent-adsorbate through their valency forces [104].

11.7.1.3 Weber and Morris Sorption Kinetic Model

The Weber and Morris sorption kinetic model [109] was initially employed by Pasavant *et al.* [110] to describe their biosorption experimental data. This model has the following form:

$$q = K_{WM} \sqrt{t} \tag{11.11}$$

In their investigation, the sorption process by *C. lentillifera* biomass for Cu(II), Cd(II), Pb(II), and Zn(II) was regulated by two main mechanisms: intraparticle diffusion and external mass transfer. Intraparticle diffusion can be estimated with:

$$D = \frac{\pi}{8640} \left(\frac{d_p K_{WM}}{q_e} \right)^2 \tag{11.12}$$

The external mass transfer process was determined by:

$$\frac{dq}{dt} = K_L' A\left(C - C_s^i\right) \tag{11.13}$$

They observed that the external mass transfer coefficients can be ordered from high to low values as Cu(II) > Pb(II) > Zn(II) > Cd(II), while the intraparticle diffusion coefficients were as follow: Cd(II) > Zn(II) > Cu(II) > Pb(II).

11.7.2 Equilibrium Modeling of Adsorption

Equilibrium isotherm models are usually classified into the empirical equations and the mechanistic models. The mechanistic models are based on mechanism of metal ion adsorption, which are able to not only represent, but also to explain and predict the experimental behavior. Some empirical models for single solute systems are listed in Table 11.2. The Langmuir model (based on monolayer adsorption of solute) and the Freundlich model (developed for heterogeneous surfaces) are the most widely accepted and used in literature [111, 112, 113, 114, 115]. The Brunauer-Emmett-Teller (BET) model [116] describes the multi-layer adsorption at the nanoparticle surfaces and assumes that the Langmuir isotherm applies to each layer. These models can provide information about metal uptake capacity and difference in metal uptake between various species.

11.7.2.1 Freundlich Isotherm

The Freundlich isotherm is an empirical equation. This equation is among one of the most widely used isotherms for the explanation of adsorption equilibrium. The Freundlich isotherm is capable of describing the adsorption of heavy metals on a wide variety of adsorbents. This equation has the following form [111]:

$$q_e = K_F C_e^{1/n} \tag{11.14}$$

It can also be expressed in the linearized logarithmic form [117]:

$$\log q_e = \log K_F + \frac{1}{n}\log C_e \tag{11.15}$$

Table 11.2 Summary of widely used isotherms for adsorption systems with their advantages and disadvantages.

Isotherm	Functional form	Linear form	Plot	Advantages	Disadvantages	References
Langmuir	$q_e = \frac{q_m K_L C_e}{1+K_L C_e}$	$\frac{C_e}{q_e} = \frac{1}{K_L q_m} + \frac{1}{q_m} C_e$	$\frac{C_e}{q_e} VS. C_e$	Has Henry law and finite saturation limit so valid over a wide range of concentration	Based on monolayer assumption	[100]
Freundlich	$q_e = K_F C_e^{1/n}$	$\ln q_e = \ln K_F + \frac{1}{n} \ln C_e$	$\ln q_e \; VS. \ln C_e$	Simple expression and has parameter for surface heterogeneity	Does not have Henry law and no saturation limit, not structured, not applicable over wide range of concentration	[101]
Temkin	$q_e = \frac{RT}{b} \ln (K_T C_e)$	$q_e = \frac{RT}{b} \ln K_T + \frac{RT}{b} \ln C_e$	$q_e \; VS. \ln C_e$	Simple expression	Same as Freundlich. It does not have correct Henry law limit and finite saturation limit, not applicable over wide range of concentration	[102]
Redlich–Peterson (R–P)	$q_e = \frac{K_{RP} C_e}{1+a_{RP} C_e^{\beta}}$	$\ln \left[\left(\frac{K_{RP} C_e}{q_e} \right) - 1 \right] = \ln a_{RP} + \beta \ln C_e$	$\ln \left[\left(\frac{K_{RP} C_e}{q_e} \right) - 1 \right] VS. \ln C_e$	Approaches Freundlich at high concentration	No special advantages	[103]

where C_e and q_e are equilibrium concentration of metal (mg/L) and amount of solute adsorbed at equilibrium conditions (mg/g), respectively; K_F and n are the characteristic constant related to the adsorption capacity (L/mg) and adsorption intensity or degree of favorability of adsorption, respectively.

The plot of log q_e versus log C_e has a slope with the value of $1/n$ and an intercept magnitude of log K_F. Log K_F is equivalent to log q_e when C_e equals unity. Though, in other case when $1/n \neq 1$, the K_F value depends on the units upon which q_e and C_e are expressed. On average, a favorable adsorption tends to have Freundlich constant n between 1 and 10. A larger value of n (smaller value of $1/n$) implies stronger interaction between adsorbent and heavy metal, while $1/n$ equal to 1 indicates linear adsorption leading to identical adsorption energies for all sites [104].

11.7.2.2 Langmuir Isotherm

The Langmuir isotherm is a well-known model used for describing heavy metal sorption onto magnetic nanoparticles. The Langmuir equation relates the coverage of molecules on a solid surface to concentration of a medium above the solid surface at a fixed temperature. This isotherm is based on three assumptions, namely, adsorption is limited to monolayer coverage, all surface sites are alike and only can accommodate one adsorbed atom, and the ability of a molecule to be adsorbed on a given site is independent of its neighboring sites occupancy [104, 118]. By applying these assumptions and a kinetic principle (rate of adsorption and desorption from the surface is equal), the Langmuir equation can be written in the following form [119]:

$$q_e = q_{max}\frac{K_L C_e}{1+K_L C_e} \tag{11.16}$$

The linear form of this equation is often written as [117]:

$$\frac{C_e}{q_e} = \frac{1}{q_{max}}C_e + \frac{1}{K_L q_{max}} \tag{11.17}$$

where C_e is equilibrium concentration of metal (mg/L) and q_e is the amount of metal adsorbed at equilibrium conditions (mg/g), q_{max} is maximum monolayer adsorption capacity (mg/g), and K_L is sorption equilibrium constant (L/mg), which is related to the energy of adsorption. Within the

Langmuir model, the saturation capacity q_{max} is supposed to coincide with the saturation of a fixed number of identical surface sites, and as such, it should logically be independent of temperature.

Sometimes, these empirical models do not reflect any mechanisms of sorbate uptake and hardly have a meaningful physical interpretation for adsorption. Some authors have pointed out that the results from the empirical models cannot be extrapolated, and no predictive conclusions can be drawn for systems operating under different conditions [120]. Also, since both are simple, basic models (Langmuir and Freundlich models) they do not incorporate the effects of any external variable environmental factors, although they are capable of describing many biosorption isotherms in most cases. The mechanistic conclusions from the good fit of the models alone should be avoided. Moreover, adsorption isotherms may exhibit an irregular pattern due to the complex nature of both the adsorbents and their varied multiple active sites, as well as the complex solution chemistry of some metallic compounds.

11.7.2.3 Two Additional Parameter Isotherms

A particular model might be invalid in a particular circumstance, and in most cases, more than one model can explain the adsorption mechanism. Several models initially developed for gas-phase adsorption can be implemented to correlate heavy metals adsorption processes. Some of these equations contain two fitting parameters (Temkin isotherm, Flory–Huggins, and Dubinin–Raduskevich equations), whereas others can have more than two parameters (Redlich–Paterson and Sips isotherms) [104].

Superior adsorption performance and magnetic properties of nanomaterials tend to be a promising approach to deal with a variety of environmental pollutants in aqueous media. Advances in magnetic nanoparticles could provide opportunities for developing new generation adsorbents with high surface area, new functionalities, high capacity, easy separation, and recyclability. The novel physicochemical and magnetic properties of nanoparticles can facilitate many advanced applications in the development of adsorptive technologies, and thus generate more efficient and cost-effective remediation approaches as compared with conventional technologies [121].

11.8 Thermodynamic Analysis

The Langmuir isotherm constant can be used in biosorption to estimate the thermodynamic parameters of Gibbs free energy (ΔG), change in enthalpy

(ΔH), and change in entropy (ΔS). Negative ΔG indicates the spontaneity of the adsorption process. The value of ΔH is used to identify the nature of adsorption [122]. A positive value of ΔH indicates the reaction is endothermic, and the negative value of ΔH shows that the reaction is exothermic [122, 123]. A positive value of ΔS indicates increased randomness of adsorbate molecules on the solid surface than in the solution [124].

The free energy of adsorption (ΔG) can be related with the Langmuir equilibrium constant by the following expression [125]:

$$\Delta G = -RT\ ln\ K_L \tag{11.18}$$

Enthalpy and entropy changes are also related to the Langmuir equilibrium constant by the following expression:

$$\text{In}\ K_{\text{L}} = \frac{\Delta S}{R} - \frac{\Delta H}{RT} \tag{11.19}$$

Thus, a plot of ln K_L versus 1/T should be a straight line. The ΔH and ΔS values could be obtained from the slope and intercept of this plot.

11.9 Metal Recovery and Regeneration of Magnetic Nanoparticles

Desorption is very much necessary when the material synthesis is costly. Regenerability of loaded magnetic nanoparticles is a key factor for improving the economy of the adsorption process. The economic feasibility of using adsorbent based on magnetic nanoparticles to remove heavy metals from aqueous solution relies on its regeneration ability during multiple adsorption/desorption cycles. It is possible to decrease the process cost and also the dependency of the process on a continuous supply of the nanoparticles through desorption. A successful desorption process requires the proper selection of elutants, which strongly depends on the type of material and the mechanism of adsorption [126]. Elutant must be (i) non-damaging to the magnetic nanoparticles, (ii) less costly, (iii) eco-friendly, and (iv) effective. Some workers have conducted exhaustive experiments to identify appropriate elutants for the recovery of precious metals.

Recently, Ge *et al.* synthesized novel Fe_3O_4 magnetic nanoparticles (MNPs) modified with 3-aminopropyltriethoxysilane (APS) and copolymers of acrylic acid (AA) and crotonic acid (CA). The authors explored the ability of the MNPs for removing heavy metal ions (Cd^{2+}, Zn^{2+}, Pb^{2+} and Cu^{2+}) from aqueous solution [127]. Furthermore, they explored desorption and reuse of MNPs. The Fe_3O_4@APS@AA-co-CA MNPs were excellent for removal of heavy metal ions such as Cd^{2+}, Zn^{2+}, Pb^{2+} and Cu^{2+} from aqueous solution. The Fe_3O_4@APS@AA-co-CA MNPs did not significantly adsorb metal ions at $pH < 2$, which suggested that the adsorbed metal ions could possibly be desorbed in a solution with such pH values. The authors investigated different concentrations of H^+ to determine the best conditions for reusing MNPs. The concentration of metal ions increased with increasing H^+ in water but there was little change with 0.1 mol L^{-1} H^+. Meanwhile, the adsorbent was corroded by the H^+, which could be seen from the content of the Fe^{3+}. The adsorbent completely disappeared at 2 mol L^{-1} H^+ after 3 h. Considering the desorption efficiency, and the reusability cycles of the adsorbents, the initial concentration of 0.1 M H^+ was better than others. The metal ion adsorption capacity of Fe_3O_4@APS@AA-co-CA MNPs remained almost constant for the 4 cycles, which indicates no irreversible sites on the surface of Fe_3O_4@APS@AA-co-CA MNPs for desorption with 0.1 mol L^{-1} H^+, and the reusability of the adsorbents was satisfactory.

11.10 Conclusions

Magnetic nanoparticles offer a tremendous opportunity to treat wastewater containing toxic metal ions because of their high surface area, high physicochemical stability, multifunctionality, ease in coating, and low-cost of synthesis. Several routes for the synthesis of magnetic nanoparticles have been developed and practiced *viz.*, co-precipitation, microemulsion, hydrothermal and thermal decomposition for rapid and highly efficient removal of guest molecules from aqueous solutions. However, the co-precipitation method is one of the most celebrated synthesis routes for magnetic nanoparticles due to its low-cost, easy operation, ease in design and high yield.

Magnetic nanoparticles offer a new vista of separation and purification technology for heavy metals. These multifunctional magnetic nanoparticles have been successfully applied for the reduction of toxic metal ions up to ppb level in treated wastewater. Magnetic nanoparticle loaded with toxic metal ions in water can be effectively recovered and regenerated with the help of an external magnetic field. The regenerated nanoadsorbent can be successfully and effectively applied for the recycling and treatment of wastewater. As reported in some literature, extreme acidic or basic conditions of solution can corrode the nanoparticles, causing loss of function groups. Thus, the stability of the nanoparticles in the treatment of heavy metals strongly depends on the pH of the solution.

Nanotechnology offers lots of advantageous potential for today's society but similar with other techniques it also has some drawbacks. Increasing use of nanoproducts and nanoadsorbents will increase the risk of nanopollutants in the environment. Hence, there is an essential need for research to solve the problems and overcome the challenges that have been discussed in this chapter. In addition, there is a further need for the study of nanoparticles behavior, fate, and effects in the different spheres of the ecosystem on a timely basis. Recent trends and widespread use of nanomaterials in the environmental area are extremely new, and further research is needed before there are on-site applications because of the nanotoxicity of the materials.

Magnetic nanoparticles have tremendous potential for treating wastewater containing toxic metal ions because of their high surface area, high physicochemical stability, multifunctionality, ease in coating, and low-cost of synthesis. Overall these nanomaterials can be successfully and effectively applied for the recycling, treatment and sustainable management of water.

Acknowledgements

R.K. Gautam is thankful to the University Grants Commission, New Delhi, for the award of Junior Research Fellowship (JRF). M.C. Chattopadhyaya acknowledges the support and necessary facilities provided by the University of Allahabad, Allahabad, India. We are thankful to Prof. V.S. Tripathi, Department of Chemistry, University of Allahabad, Allahabad, India, for his support and encouragement during the study.

References

1. R.K. Gautam, *Environmental Magnetism: Fundamentals and Applications*, Saarbrucken, LAP Lambert Academic Publishing, 2012.
2. R.K. Gautam, M.C. Chattopadhyaya, and S.K. Sharma, "Biosorption of heavy metals: Recent trends and challenges," in: S.K. Sharma and R. Sanghi, eds., *Wastewater Reuse and Management*, Springer, pp. 305–322, 2013.
3. Z. Lin, Y. Zhang, Y. Chen, and H. Qian, *Chemical Engineering Journal*, Vol. 200, p. 104, 2012.

4. S.R. Smith, *Environment International*, Vol. 35, p. 142, 2009.
5. N.K. Srivastava, and C.B. Majumder, *Journal of Hazardous Materials*, Vol. 151, p. 1, 2008.
6. A. Santos, and S. Judd, *Journal of Environmental Monitoring*, Vol. 12, p. 110, 2010.
7. M.A. Keane, *Colloids and Surfaces A: Physicochemical and Engineering Aspects*, Vol. 138, p. 11, 1998.
8. M. Monier, D.M. Ayad, Y. Wei, and A.A. Sarhan, *Journal of Hazardous Materials*, Vol. 177, p. 962, 2010.
9. K. Swayampakula, V.M. Boddu, S.K. Nadavala, and K. Abburi, *Journal of Hazardous Materials*, Vol. 170, p. 680, 2009.
10. M. Gavrilescu, *Engineering in Life Sciences*, Vol. 4, p. 219, 2004.
11. Z. Reddad, C. Gerente, Y. Andres, and P. Le Cloirec, *Environmental Science and Technology*, Vol. 36, p. 2067, 2002.
12. W.J. Eilbeck, and G. Mattock, *Chemical Process in Wastewater Treatment*, Ellis Harwood, Chichester, 1987.
13. S.E. Bailey, T.J. Olin, R.M. Bricka, and D.D. Adrian, *Water Research*, Vol. 33, p. 2469, 1999.
14. D. Mohan, and C.U. Pittman Jr, *Journal of Hazardous Materials*, Vol. 142, p. 1, 2007.
15. B.K. Mandal, and K.T. Suzuki, *Talanta*, Vol. 58, p. 201, 2002.
16. E.T. Mackenzie, R.J. Lamtzy, and V. Petorson, *Journal of the International Association for Mathematical Geology*, Vol. 6, p. 99, 1979.
17. J.T. Ackerman, C.A. Eagles-Smith, J.Y. Takekawa, S.A. Demers, T.L. Adelsbach, J.D. Bluso, A.K. Miles, N. Warnock, T.H. Suchanek, and S.E. Schwarzbach, *Science of the Total Environment*, Vol. 384, p. 452, 2007.
18. J.T. Ackerman, C.A. Eagles-Smith, J.Y. Takekawa, J.D. Bluso, and T.L. Adelsbach, *Environmental Toxicology and Chemistry*, Vol. 27, p. 897, 2008.
19. J.T. Ackerman, J.Y. Takekawa, C.A. Eagles-Smith, and S.A. Iverson, *Ecotoxicology*, Vol. 17, p. 103, 2008.
20. J.T. Ackerman, and C.A. Eagles-Smith, *Environmental Science and Technology*, Vol. 43, p. 2166, 2009.
21. J.T. Ackerman, and C.A. Eagles-Smith, *Environmental Science and Technology*, Vol. 44, p. 1451, 2010.
22. C.A. Eagles-Smith, J.T. Ackerman, T.L. Adelsbach, J.Y. Takekawa, A.K. Miles, and R.A. Keister, *Environmental Toxicology and Chemistry*, Vol. 27, p. 2136, 2008.
23. C.A. Eagles-Smith, J.T. Ackerman, S.W. De La Cruz, and J.Y. Takekawa, *Environmental Pollution*, Vol. 157, p. 1993, 2009.
24. C.A. Eagles-Smith, and J.T. Ackerman, *Environmental Science and Technology*, Vol. 43, p. 8658, 2010.
25. M. Harada, *Critical Reviews in Toxicology*, Vol. 25, p. 1, 1995.
26. M. Cook, and H. Morrow, Workshop Proceedings, National Workshop on Cadmium Transport into Plants, Canadian Network of Toxicology Centres, Ottawa: 1995.

27. J. Barnhart, *Journal of Soil Contamination*, Vol. 6, p. 561, 1997.
28. D.C. Bellinger, *Pediatrics*, Vol. 113, p. 1016, 2004.
29. Agency for Toxic Substances and Disease Registry, Toxicological profile for cobalt, U.S. Department of Health and Human Services, Atlanta, 2004.
30. H. Lalhruaitluanga, M.N.V. Prasad, and K. Radha, *Desalination*, Vol. 271, p. 301, 2011.
31. M.R. Broadley, P.J. White, J.P. Hammond, I. Zelko, and A. Lux, *New Phytologist*, Vol. 173, p. 677, 2007.
32. J.M. Benson, I.-Y. Chang, and Y.S. Cheng, *Fundamental and Applied Toxicology*, Vol. 28, p. 232, 1995.
33. J.E. Evans, M.L. Miller, A. Andringa, and L. Hastings, *Toxicology and Applied Pharmacology*, Vol. 130, p. 209, 1995.
34. B. Adkins Jr, J.H. Richards, and D.E. Gardner, *Environmental Research*, Vol. 20, p. 33, 1979.
35. J.A. Graham, F.J. Miller, M.J. Daniels, E.A. Payne, and D.E. Gardner, *Environmental Research*, Vol. 16, p. 77, 1978.
36. A.R. Oller, M. Costa, and G. Oberdorster, *Toxicology and Applied Pharmacology*, Vol. 143, p. 152, 1997.
37. E.J.V. Genderen, A.C. Ryan, J.R. Tomasso, and S.J. Klaine, *Environmental Toxicology and Chemistry*, Vol. 24, p. 408, 2005.
38. A.B. Santamaria, *Indian Journal of Medical Research,* Vol. 128, p. 484, 2008.
39. Literature review of chemistry and toxicity of mercury, cadmium, selenium and antimony in metal mining effluents, A TIME Network Project, Canada, 2002.
40. D. Mohan, and C.U. Pittman Jr, *Journal of Hazardous Materials*, Vol. B137, p. 762, 2006.
41. C.K. Jain, and I. Ali, *Water Research*, Vol. 34, p. 4304, 2000.
42. S.A. Counter, and L.H. Buchanan, *Toxicology and Applied Pharmacology*, Vol. 198, p. 209, 2004.
43. D. Benova, V. Hadjidekova, R. Hristova, *et al.*, *Mutation Research*, Vol. 514, p. 29, 2002.
44. H.J. Gibb, P.S.J. Lees, P.F. Pinsky, *et al.*, *American Journal of Industrial Medicine*, Vol. 38, p. 115, 2000.
45. J. Blasiak, and J. Kowalik, *Mutation Research*, Vol. 469, p. 135, 2000.
46. L.D. De Gennaro, *Growth, Development, and Aging,* Vol. 66, p. 43, 2002.
47. E.L. Baker, R.A. Goyer, B.A. Fowler, S. Can, and B. Cengiz, *American Journal of Industrial Medicine*, Vol. 1, p. 138, 1980.
48. C. Bagci, A.I. Bozkurt, E.A. Cakmak, *et al.*, *International Journal of Clinical Practice*, Vol. 58, p. 568, 2004.
49. J. Schwartz, *Archives of Environmental Health*, Vol. 50, p. 31, 1995.
50. J. Bonura, and R.M. Caprioli, *Cancer Research*, Vol. 41, p. 2868, 1981.
51. F.W. Sunderman Jr, S.M. Hopfer, J.A. Knight, *et al.*, *Carcinogenesis*, Vol. 8, p. 305, 1987.

52. F.W. Sunderman Jr, *Scandinavian Journal of Work, Environment & Health*, Vol. 15, p. 1, 1989.
53. A.R. Oller, *Environmental Health Perspectives*, Vol. 110, p. 841, 2002.
54. A.K. Gupta, and M. Gupta, *Biomaterials,* Vol. 26, p. 3995, 2005.
55. I.J. Bruce, and T. Sen, *Langmuir*, Vol. 21, p. 7029, 2005.
56. E.K.U. Larsen, T. Nielsen, T. Wittenborn, H. Birkedal, T. Vorup-Jensen, M.H. Jakobsen, L. Ostergaard, M.R. Horsman, F. Besenbacher, K.A. Howard, and J. Kjems, *ACS Nano,* Vol. 3, p. 1947, 2009.
57. W. Jiang, H.C. Yang, S.Y. Yang, H.E. Horng, J.C. Hung, Y.C. Chen, and C.Y. Hong, *Journal of Magnetism and Magnetic Materials*, Vol. 283, p. 210, 2004.
58. G.M. da Costa, E.D. Grave, P.M.A. de Bakker, and R.E. Vandenberghe, *Journal of Solid State Chemistry*, Vol. 113, p. 405, 1994.
59. Y. Deng, L. Wang, W. Yang, S. Fu, and A. Elassari, *Journal of Magnetism and Magnetic Materials*, Vol. 257, p. 69, 2003.
60. R. Abu Mukh-Qasem, A. Gedanken, *Journal of Colloid and Interface Science*, Vol. 284, p. 489, 2005.
61. D. Chen, and R. Xu, *Materials Research Bulletin*, Vol. 33, p. 1015, 1998.
62. S. Sun, and H. Zeng, *Journal of American Chemical Society*, Vol. 124, p. 8204, 2002.
63. S. Veintemillas-Verdaguer, M. Morales, and C. Serna, *Materials Letters*, Vol. 35, p. 227, 1998.
64. D.L. Huber, *Small*, Vol. 1, p. 482, 2005.
65. S. Laurent, D. Forge, M. Port, A. Roch, C. Robic, L.V. Elst, and R.N. Muller, *Chemical Reviews*, Vol. 108, p. 2064, 2008.
66. A.H. Lu, E.L. Salabas, and F. Schuth, *Angewandte Chemie International Edition*, Vol. 46, p. 1222, 2007.
67. Y.T. Zhou, H.L. Nie, C. Branford-White, Z.Y. He, and L.M. Zhu, *Journal of Colloid and Interface Science*, Vol. 330, p. 29, 2009.
68. Y. Tai, L. Wang, J. Gao, W.A. Amer, W. Ding, and H. Yu, *Journal of Colloid and Interface Science*, Vol. 360, p. 731, 2011.
69. A.L. Willis, N.J. Turro, S. O'Brien, *Chemistry of Materials*, Vol. 17, p. 5970, 2005.
70. R.Y. Hong, J.H. Li, H.Z. Li, J. Ding, Y. Zheng, and D.G. Wei, *Journal of Magnetism and Magnetic Materials*, Vol. 320, p. 1605, 2008.
71. Z. Feng, S. Zhu, D.R. Martins de Godoi, A.C.S. Samia, and D. Scherson, *Analytical Chemistry*, Vol. 84, p. 3764, 2012.
72. S.J. Lee, J.H. Cho, C. Lee, J. Cho, Y.R. Kim, and J.K. Park, *Nanotechnology*, Vol. 22, p. 375, 2011.
73. J. Wan, Y. Yao, and G. Tang, *Applied Physics A*, Vol. 89, p. 529, 2007.
74. H. Cao, G. Wang, L. Zhang, Y. Liang, S. Zhang, and X. Zhang, *ChemPhysChem*, Vol. 7, p. 1897, 2006.
75. J.A. López Pérez, M.A. López Quintela, J. Mira, J. Rivas, *IEEE Transactions on Magnetics*, Vol. 33, p. 4359, 1997.

76. J.A. López Pérez, M.A. López Quintela, J. Mira, J. Rivas, S.W. Charles, *Journal of Physical Chemistry B*, Vol. 101, p. 8045, 1997.
77. C. Okoli, M. Sanchez-Dominguez, M. Boutonnet, S. Järås, C. Civera, C. Solans, and G.R. Kuttuva, *Langmuir*, Vol. 28, p. 8479, 2012.
78. J. Vidal-Vidal, J. Rivas, and M. López-Quintela, *Colloids and Surfaces A: Physicochemical and Engineering Aspects*, Vol. 288, p. 44, 2006.
79. D. Makovec, and A. Košak, *Journal of Magnetism and Magnetic Materials*, Vol. 289, p. 32, 2005.
80. S. Santra, R. Tapec, N. Theodoropoulou, J. Dobson, A. Hebard, and W. Tan, *Langmuir*, Vol. 17, p. 2900, 2001.
81. C.R. Vestal, and Z.J. Zhang, *Chemistry of Materials*, Vol. 14, p. 3817, 2002.
82. E. Fantechi, G. Campo, D. Carta, A. Corrias, C. de Julián Fernández, D. Gatteschi, C. Innocenti, F. Pineider, F. Rugi, and C. Sangregorio, *The Journal of Physical Chemistry C*, Vol. 116, p. 8261, 2012.
83. S.F. Chin, S.C. Pang, and C.H. Tan, *Journal of Materials and Environmental Science*, Vol. 2, p. 299, 2011.
84. Y.F. Lin, H.W. Chen, P.S. Chien, C.S. Chiou, and C.C. Liu, *Journal of Hazardous Materials*, Vol. 185, p. 1124, 2011.
85. Z. Yong-Gang, S. Hao-Yu, P. Sheng-Dong, and H. Mei-Qin, *Journal of Hazardous Materials*, Vol. 182, p. 295, 2010.
86. H. Yong-Mei, C. Man, and H. Zhong-Bo, *Journal of Hazardous Materials*, Vol. 184, p. 392, 2010.
87. R.A. Crane, and T.B. Scott, *Journal of Hazardous Materials*, Vol. 211–212, p. 112, 2012.
88. Z. Feng, S. Zhu, D.R. Martins de Godoi, A.C.S. Samia, and D. Scherson, *Analytical Chemistry*, Vol. 84, p. 3764, 2012.
89. Z. Geng, Y. Lin, X. Yu, Q. Shen, L. Ma, Z. Li, N. Pan, and X. Wang, *Journal of Materials Chemistry*, Vol. 22, p. 3527, 2012.
90. Y.H. Chen, and Y.D. Chen, *Journal of Hazardous Materials*, Vol. 185, p, 168, 2011.
91. R. Chand, T. Watari, K. Inoue, T. Torikai, and M. Yada, *Separation and Purification Technology*, Vol. 65, p. 331, 2009.
92. S.G. Wang, W.X. Gong, X.W. Liu, Y.W. Yao, B.Y. Gao, and Q.Y. Yue, *Separation and Purification Technology*, Vol. 58, p. 17, 2007.
93. A. Ahmadpour, and D.D. Do, *Carbon*, 1995, Vol. 33, p. 1393.
94. P.I. Girginova, A.L. Daniel-da-Silva, C.B. Lopes, P. Figueira, M. Otero, V.S. Amaral, E. Pereira, and T. Trindade, *Journal of Colloid and Interface Science*, Vol. 345, p. 234, 2010.
95. J. Wang, S. Zheng, Y. Shao, J. Liu, Z. Xu, and D. Zhu, *Journal of Colloid and Interface Science*, Vol. 349, p. 293, 2010.
96. J. Song, H. Kong, and J. Jang, *Journal of Colloid and Interface Science*, Vol. 359, p. 505, 2011.
97. J. Hu, I.M.C. Lo, and G. Chen, *Separation and Purification Technology*, Vol. 56, p. 249, 2007.

98. W. Wu, X. Xiao, S. Zhang, H. Li, X. Zhou, and C. Jiang, *Nanoscale Research Letter*, Vol. 4, p. 926, 2009.
99. S. Lagergren, *Kongliga Svenska Vetenskaps Academiens nya Handlingar*, Vol. 24, p. 1, 1898.
100. I. Langmuir, *Journal of American Chemical Society*, Vol. 40, p. 1361, 1918.
101. H.M.F. Freundlich, *Journal of Physical Chemistry*, Vol. 57, p. 385, 1906.
102. M.J. Temkin, and V. Pyzhev, *Acta Physico-Chimica USSR*, Vol. 12, p. 217, 1940.
103. O. Redlich, and D.L. Peterson, *Journal of Physical Chemistry*, Vol. 63, p. 1024, 1959.
104. J. Febrianto, A.N. Kosasih, J. Sunarso, Y.H. Ju, N. Indraswati, and S. Ismadji, *Journal of Hazardous Materials*, Vol. 162, p. 616, 2009.
105. Y.S. Ho, *Water Research*, Vol. 40, p. 119, 2006.
106. Y.S. Ho, G. Mckay, D.J. Wase, and C.F. Foster, *Adsorption Science and Technology*, Vol. 18, p. 639, 2000.
107. Y.S. Ho, and G. McKay, *Process Biochemistry*, Vol. 34, p. 451, 1999.
108. Y.S. Ho, and G. McKay, *Chemical Engineering Journal*, Vol. 11, p. 5, 1998.
109. W.J. Weber, and J.C. Morris, *Journal of the Sanitary Engineering Division: American Society of Civil Engineers*, Vol. 89, p. 31, 1963.
110. P. Pasavant, R. Apiratikul, V. Sungkhum, P. Suthiparinyanont, S. Wattanachira, and T.F. Marhaba, *Bioresource Technology*, Vol. 97, p. 2321, 2006.
111. A. Afkhami, and R. Moosavi, *Journal of Hazardous Materials*, Vol. 174, p. 398, 2010.
112. A. Afkhami, T. Madrakian, and A. Amini, *Desalination*, Vol. 243, p. 258, 2009.
113. S. Chatterjee, D.S. Lee, M.W. Lee, and S.H. Woo, *Bioresource Technology*, Vol. 100, p. 2803, 2009.
114. H. Parham, B. Zargar, and R. Shiralipour, *Journal of Hazardous Materials*, Vol. 205–206, p. 94, 2012.
115. Y. Tian, M. Wu, X. Lin, P. Huang, and Y. Huang, *Journal of Hazardous Materials*, Vol. 193, p. 10, 2011.
116. S. Brunauer, P.H. Emmett, and E. Teller, *Journal of the American Chemical Society*, Vol. 60, p. 309, 1938.
117. L. Fan, C. Luo, M. Sun, X. Li, F. Lu, and H. Qiu, *Bioresource Technology*, Vol. 114, p. 703, 2012.
118. M. Ghaedi, A. Najibi, H. Hossainian, A. Shokrollahi, and M. Soylak, *Toxicological and Environmental Chemistry*, Vol. 94, Iss. 1, p. 40, 2012.
119. N. Gupta, A.K. Kushwaha, and M.C. Chattopadhyaya, *Journal of the Taiwan Institute of Chemical Engineers*, Vol. 43, p. 604, 2012.
120. J. Wang, and C. Chen, *Biotechnology Advances*, Vol. 27, p. 195, 2009.
121. P. Xu, G.M. Zeng, D.L. Huang, C.L. Feng, S. Hu, M.H. Zhao, C. Lai, Z. Wei, C. Huang, G.X. Xie, and Z.F. Li, *Science of the Total Environment*, Vol. 424, p. 1, 2012.
122. A. Bhatnagar, A.K. Minocha, M. Sillanpaa, *Biochemical Engineering Journal*, Vol. 48, p. 181, 2010.

123. N. Tewari, P. Vasudevan, and B.K. Guha, *Biochemical Engineering Journal*, Vol. 23, p. 185, 2005.
124. R. Ahmad, and R. Kumar, *Journal of Environmental Management*, Vol. 91, p. 1032, 2010.
125. S. Nethaji, A. Sivasamy, G. Thennarasu, and S. Saravanan, *Journal of Hazardous Materials*, Vol. 181, p. 271, 2010.
126. N. Das, *Hydrometallurgy*, Vol. 103, p. 180, 2010.
127. F. Ge, M.M. Li, H. Ye, B.X. Zhao, *Journal of Hazardous Materials*, Vol. 211–212, p. 366, 2012.

12

Potential Application of Nanoparticles as Antipathogens

Pratima Chauhan*[1], Mini Mishra[2] and Deepika Gupta[2]

[1]Department of Physics, University of Allahabad, Allahabad, India
[2]Department of Environmental Studies, University of Allahabad, India

Abstract
Pathogens are responsible for millions of diseases, disabilities and deaths every year. The ability to control and destroy pathogens is therefore of enormous importance to healthcare, food, drink, water treatment and agriculture, as well as the military. Nanomaterials have potential of wide -ranging industrial, biomedical and electronic applications. Recently, many researchers have been trying to develop nanostructures which can find wide applications for use as Antipathogens. Nanomaterials lay mid way between atomic scale and bulk, having properties different from that of bulk. Biomaterials with nanoscale organizations can be used as controlled-release reservoirs for drug delivery as a promising tool to support cell therapy. Metal-based nanoparticles constitute an effective antimicrobial agent against common pathogenic microorganisms. Some nanoparticles such as silver, gold, titanium dioxide and zinc oxide are receiving considerable attention as antimicrobials. Different nanosystems control, kill or disable pathogens differently.

Keywords: Nanomaterials, microorganisms, Antipathogens

12.1 Introduction

The first research in nanoscience and technology can be traced back to over five decades and was described in a lecture entitled "There's Plenty of Room at the Bottom," by Richard P. Feynman in 1959. The word

**Corresponding author*: mangu167@yahoo.co.in

Ashutosh Tiwari and Mikael Syväjärvi (eds.) Advanced Materials for Agriculture, Food, and Environmental Safety, (333–368) 2014 © Scrivener Publishing LLC

nanoparticle means a particle with atleast one dimension in the range of 1nm to 100nm. A particle whose diameter is 1nm has approximately 1000 atoms, as atomic size is of the order of an angstrom [1]. The substance in nano-regime exhibits a large number of new properties. The properties of substances in nanosize are entirely different from the corresponding bulk or atomic and molecular counterparts. Nanostructures have large surface-to-volume ratio and show quantum confinement [2–5]. The band gap of a substance in nanorange is higher than in the bulk form. Due to their small size, nanoparticles can penetrate the cell interior of bacteria, fungi, viruses, plant and animal cells, etc. [6–7]. A pathogen is a microorganism in the widest sense, such as a virus, bacterium, prion, or fungus that causes disease in its host. They are responsible for various diseases in plants and animals. They are responsible for various diseases in plants and animals and are responsible for millions of diseases, disabilities and deaths every year. The ability to control or destroy pathogens is therefore of enormous importance in areas such as healthcare, food, drink, water treatment, agriculture and the military. Antipathogens are substances which are used to kill, deactivate and control the pathogens. Conventionally, Antipathogens are drugs, antibiotics, other microorganisms, etc. Due to increasing resistance in pathogens, the search is on for new materials and composites. There are drawbacks in current therapies, including drug-induced disease and increasing prevalence of multiple drug resistant microorganisms. Nanoparticles have immense potential in biomedicine. They can be used for diagnosis, drug delivery system, prevention and as Antipathogens [8–14].

Pathogens may affect an animal (including humans), a plant, or even another microorganism. There are several substrates including pathways whereby pathogens can invade a host. The principal pathways have different episodic time frames, but soil contamination has the longest or most persistent potential for harboring a pathogen. Diseases caused by organisms in humans are known as pathogenic diseases. Some of the diseases that a pathogen can cause are smallpox, influenza, mumps, measles, chickenpox, tuberculosis and rubella. Not all pathogens are necessarily undesirable to humans.

12.1.1 Types of Pathogens

Viral: Pathogenic viral diseases are mainly caused by the families of Adenoviridae, Picornaviridae, Herpesviridae, Hepadnaviridae, Flaviviridae, Retroviridae, Orthomyxoviridae, Paramyxoviridae, Papovaviridae,

Polyomavirus, Rhabdoviridae and Togaviridae. Viruses typically range between 20–300 nanometers in length.

Bacterial: Although the vast majority of bacteria are harmless or beneficial, a few pathogenic bacteria can cause infectious diseases. Bacteria can often be killed by antibiotics because the outside of the cell wall is destroyed, expelling the DNA out of the body of the pathogen; therefore the pathogen is incapable of producing proteins and dies. They typically range between 1 and 5 micrometers in length.

Fungal: Fungi comprises eukaryotic kingdom of microbes that are usually saprophytes (consume dead organisms) but can cause diseases in humans, animals and plants. Fungi are the most common cause of diseases in crops and other plants. The typical fungal spore size is 1–40 micrometer in length.

Prionic: According to the prion theory, prions are infectious pathogens that do not contain nucleic acids. These abnormally folded proteins are characteristically found in some diseases such as scrapie, bovine spongiform encephalopathy (mad cow disease) and Creutzfeldt–Jakob disease. Although prions fail to meet the requirements laid out by Koch's postulates, the hypothesis of prions as a new class of pathogen led Stanley B. Prusiner to receive the Nobel Prize in Physiology or Medicine in 1997.

Animal pathogens: Animal pathogens are disease-causing agents of wild and domestic animal species, at times including humans.

12.1.2 Virulence

Virulence (the tendency of a pathogen to cause damage to a host's fitness) evolves when that pathogen can spread from a diseased host, despite that host being much debilitated. Horizontal transmission occurs between hosts of the same species, in contrast to vertical transmission, which tends to evolve symbiosis (after a period of high morbidity and mortality in the population) by linking the pathogen's evolutionary success to the evolutionary success of the host organism. Evolutionary medicine has found that under horizontal transmission, the host population might never develop tolerance to the pathogen.

12.1.3 Transmission

Transmission of pathogens occurs through many different routes, including airborne, direct or indirect contact, sexual contact, through blood, breast milk, or other body fluids, and through the fecal-oral route.

12.2 Applications of Nanoparticles

Since nanoparticles are very small particles which measure not more than three or four atoms across, they are also versatile enough, and therefore they can be used in many types of technological applications, from delicate electronics to revolutionary medical procedures [15–21].

Discussed below are the various biological roles of nanoparticles and their potential applications in different fields of the biological arena.

12.2.1 Nanoparticles in Drug Delivery

There is increasing optimism that nanotechnology, as applied to medicine, will bring significant advances in the diagnosis and treatment of disease [22], Anticipated applications in medicine include drug delivery in both *in vitro* and *in vivo* diagnostics, pharmaceuticals and production of improved biocompatible materials [23].

The primary goal of nanobiotechnolgy in drug delivery includes:

- More specific drug targeting and delivery;
- Reduction in toxicity while maintaining therapeutic effects;
- Greater safety and biocompatibility; and
- Faster development of new safe medicine [24].

Some chemicals used for drug delivery are albumin, cetylalcohol, chitosan, gold, poly (DL-lactic-Co-glycolic acid), PLGA, etc. [25].

The aim of nanoparticles entrapment of drugs is either enhanced delivery or uptake by target cells and/or a reduction in the toxicity of the free drugs to the nontargeted organ [26].

Nanoparticles have specific particles as tools to enable local drug delivery or drug targeting. These include benefits such as their small size, which allows penetration of cell membrane, binding and stabilization of protein and lysosomal escape after endocytosis [27–28].

So far, nanotechnology-based therapeutic products have been approved for clinical use, with total sales exceeding $5.4 billion [29]. Among these products, liposomal drugs and polymer-drug conjugate are two dominant classes, accounting for more than 80% of the total amount [27–28]. In 1995, Doxil was the first liposomal drug formulation approved by the US Food and Drug Administration (FDA) for the treatment of AIDS associated with Kaposi's sarcoma.

Another extensively used nanoparticle drug delivery platform currently in clinical practice is polymer–drug conjugates [30]. The conjugation of small molecule drugs to polymeric nanocarriers can improve undesirable adverse effects. Polyethylene glycol (PEG) was first introduced in clinical use in the early 1990s. It can enhance the plasma stability and solubility of the drug while reducing its immunogenicity. Besides PEG, other linear polymers such as polyglutamicacid, polysaccharide, and poly (allylamine hydrochloride) have also been harnessed as polymeric drug delivery carriers [31–32].

12.2.2 Role of Nanoparticles and Their Potential Application in Food Packaging

In the food industry, several novel applications of nanotechnologies aimed at ensuring food safety have become known, including the use of novel particle such as micelles, liposomes, nanoemulsions, biopolymeric nanoparticle, as well as the development of nanosensors.

Bio nanocomposites are hybrid nanostructured materials with improved mechanical, thermal and gas barrier properties. The use of bio nanocomposites for food packaging not only protects food and increases its shelf life but is also considered to be a more environmentally-friendly solution because it reduces the requirement of using plastic as packaging material. [33–34].

The nanoparticles are dispersed throughout the plastic and are able to block oxygen, carbon dioxide and moisture from reaching fresh meats or other food. Nanomaterials are being extensively researched for specific functions such as microbial growth inhibitor and carrier of antibiotics killing agents [35]. For example, zinc peroxide (ZnO2) nanoparticles exhibit antibacterial activity that increases with a decrease in particle size. Zinc peroxide is nontoxic, and because of its strong antibacterial activity in food packaging, antibacterial nanopackaging can create a modified atmosphere in packaging with controlled gaseous exchange so that the shelf life of vegetables may be increased weeks. The surface of ordinary packaging material such as plastic or paper can be adapted to make it suitable for food by coating it with one or more sharply defined layers of tens of nanometers thickness.

Nanocoatings for glass bottles are used for better preserving the quality of fruit and vegetable products by shielding them from light waves, thereby improving their shelf life.

12.2.3 Nanoparticles Used in Agriculture

Nanotechnology takes agriculture from the battlefield of genetically modified organisms to the brave new world of atomically modified organisms. Efforts have not only been initiated in the health sector but also in agriculture sectors. A nanotech research initiative in Thailand aims to atomically modify the characteristics of local rice varieties with the color of leaves and stems.

Nanosilica has a great impact on agro-based products; and the use of amorphous Nanosilica as a biopesticide is very effective [36].

Nanoparticles are increasingly utilized in the field of plant pathology, which results in the management of various plant diseases which cause trouble for farmers and indirectly affect all of nature.

Pepper anthracnose, a disease caused by *Colletotrichum* sp., is one of the most important limiting factors for pepper production in Korea. Silver has been evaluated to be a very effective nanoparticle against pepper anthracnose under different culture conditions instead of commercial fungicide. Silver nanoparticles (WA-PR-WB13R) were applied at various concentrations to determine antifungal activity *in vitro* as well as in the field.

Similarly, powdery mildew can also be cured by an application of 100ppm silver nanoparticles, which show the highest inhibition rate both before and after the outbreak of disease on cucumbers and pumpkins [37].

Nanoencapsulation of thiamine dilauryl sulphate (TDS), a vitamin B1 derivative, was proven to effectively inhibit the spore germination of *fusariumoxysporum*. Nanoparticles especially played a more efficient role in limiting the spore germination, due to their easy penetration into hard cell membrane and long resident time on the surface of spore cell wall.

Moreover, we can say that nanotechnology permits broad advances in agricultural research, such as reproductive science and technology, and by the conversion of agricultural and food wastes to energy and other useful byproducts through enzymatic nanobioprocessing, disease prevention and treatment of plants using various nanocides. Nanoscale devices with novel properties make the agricultural systems "smart." Such devices are capable of responding to different situations by themselves, thus taking appropriate remedial action. "Smart delivery systems" in agriculture should possess combinations of time controlled, specifically targeted, highly controlled, remotely regulated/preprogrammed/self-regulated and multifunctional characteristics to avoid biological barriers for successful targeting. Agriculture and food system security, novel delivery systems for disease treatment, new tools for molecular and cellular biology and new materials for plant pathogen detection are some of the nanotechnological links to agriculture.

12.2.4 Nanotechnology for the Health Sector

Nanotechnology has been increasingly implemented in the area of health therapeutics in nanomedicine for the preservation and improvement of human health, using molecular tools and molecular knowledge of the human body. Today there is the utmost need for fabricating nanodevices for early detection and location of cancer at a molecular level, delivering anticancer drugs to the specific site and simultaneously determining their efficacy in reducing cancer load [38–40].

Nanotechnology will ring in the era of dental nanorobots, which find application in treating hypersensitive teeth; and also in the identification and destruction of pathogenic bacteria residing in plaque, nanocomposites with nanofillers for enhanced durability and aesthetics, and impressions material with nanofillers for precise tissue detail.

12.2.5 Nanoparticles Applicable in the Area of Textile Fibers

Recently, an awareness of general sanitation, contact disease transmission, and personal protection has led to the development of antimicrobial textiles. In view of the textile industry's innovative history, it is no wonder that nanotechnology has found its way into this sector so quickly. Nanotechnology is forecasted as the second industrial evolution of the world. Its novel properties and low material consumption amount have attracted global interest across disciplines and industries. With the increase in health awareness, many people have focused their attention on educating and protecting themselves against harmful pathogens. It soon became more important for antimicrobial-finished textiles to protect the wearer from bacteria than it was to simply protect the garment from fiber degradation [41]. It is evident that the ZnO nanoparticles-treated fabrics showed higher antibacterial activity when compared with ZnO bulk-treated fabrics, whereas the untreated fabrics showed no antibacterial activity. In general, the antibacterial activity was higher against *S. aureus* than *E. coli* in both ZnO nanoparticles- and bulk-treated fabrics. It has already been proven that both nanosized and micron-sized ZnO suspensions are active in inhibiting bacterial growth; the nanosized ZnO suspension clearly has a much higher activity than the micron-sized ZnO suspension [42].

The application of silver NPs in the antibacterial finishing of textiles is favorable due to the fact that silver atoms on the surface of the NPs are exposed to fabric [1]. Therefore, due to NPs, considerable surrounding medium provides significant bactericidal efficiency. Recent studies have indicated that pretreatment of textiles by low-pressure plasmas can improve loading of NPs from colloids [43]. The laundering durability was studied by using silver colloids of higher concentrations.

12.2.6 Nanoparticles Used in Water Treatment

Today, many of us are facing a critical drinking water problem; the whole world is facing formidable challenges to meet the rising demand for clean and germ-free water. Therefore, in the area of water purification, nanotechnology provides an efficient way to remove germs and contaminants from wastewater, making it useful for drinking and other purposes.

Nanotechnology, due to its size scale of less than 100nm, holds the promise of creating new materials and devices which take advantage of phenomenon realized at that length scale, because of their high reactivity due to the large surface-to-volume ratio. Many nano-effects can contribute to better water quality by killing bacteria, removing toxic metals, decomposing organic contaminants, or even removing oil. In the field of groundwater remediation, the use of zero-valent iron nanoparticles has already become quite popular [44].

12.3 Nanoparticles in Biology

The nanomaterials level is the most advanced at present, both in scientific knowledge and in commercial application. A decade ago, nanoparticles were studied because of their size-dependent physical and chemical properties [45]. Now they have entered a period of commercial exploration.

Living organisms are built of cells that are typically 10μm across. However, the cells parts are much smaller and are in the sub-micron size. Even smaller are the proteins with the dimensions of the smallest man-made nanoparticles. This simple size comparison creates the idea of using nanoparticles as very small probes that would allow us to spy on cellular machinery without introducing too much interference. The understanding of biological processes on the nanoscale level is a strong driving force behind the development of nanotechnology.

Nanomedicine is the medical application of nanotechnology. Nanomedicine ranges from medical applications of nanomaterials, to nanoelectronic biosensors, and possibly even future applications of molecular nanotechnology. Current problems for nanomedicine involve understanding the issues related to the toxicity and environmental impact of nanoscale materials. One nanometer is one-millionth of a millimeter.

The biological and medical research communities have exploited the unique properties of nanomaterials for various applications (e.g., contrast agents for cell imaging and therapeutics for treating cancer). Terms such as biomedical nanotechnology, nanobiotechnolgy, and nanomedicine are used to describe this hybrid field. Functionalities can be added to nanomaterials by interfacing them with biological molecules or structures. The size of nanomaterials is similar to that of most biological molecules and structures; therefore, nanomaterials can be useful for both *in vivo* and *in vitro* biomedical research and applications [46–48]. Thus far, the integration of nanomaterials with biology has led to the development of diagnostic devices, contrast agents, analytical tools, physical therapy applications, and drug delivery vehicles.

12.4 Uses and Advantages of Nanoparticles in Medicine

Some of the uses of nanoparticles in biology and medicine include:

- Creating fluorescent biological labels for important biological markers and molecules for research and diagnosis of diseases;
- for drug delivery systems [49].
- for gene delivery systems in gene therapy;
- For biological detection of disease-causing organisms and diagnosis [22], [50].
- for detection of proteins;
- For the isolation and purification of biological molecules and cells in research [51].
- for probing of DNA structure;
- for genetic and tissue engineering;
- for the destruction of tumors with drugs or heat;
- in MRI studies; and
- In pharmacokinetic studies.

Nanoparticles are being increasingly used in drug delivery systems. Nanomedical approaches to drug delivery center on developing nanoscale particles or molecules to improve drug bioavailability. Bioavailability refers to the presence of drug molecules where they are needed in the body and where they will do the most good. Drug delivery focuses on maximizing bioavailability both at specific places in the body and over a period of time. This can potentially be achieved by molecular targeting by nanoengineered devices. It is all about targeting the molecules and delivering drugs with cell precision. More than $65 billion are wasted each year due to poor bioavailability. The advantages of using nanoparticles as a drug delivery system include:

- The size and surface characteristics of nanoparticles can be easily manipulated. This could be used for both passive and active drug targeting.
- Nanoparticles can be made to control and sustain drug release during the transportation as well as at the location of the release. Since distribution and subsequent clearance of the drug from the body can be altered, an increase in drug therapeutic efficacy and reduction in side effects can be achieved. Choosing an appropriate matrix also helps increase drug efficiency and reduces side effects [52].
- Targeted drugs may be developed.
- Various routes of administration including oral, nasal, injection, intraocular (within the eyes), etc., can be used.

Nanoparticles can be used in combination therapy for decreasing antibiotic resistance [39]. It has been shown that zinc oxide nanoparticles can decrease antibiotic resistance and enhance the antibacterial activity of Ciprofloxacin against microorganisms. Nanoparticles can interfere with the different proteins which are interacting in the antibiotic resistance or pharmacologic mechanisms of drugs.

12.5 Antibacterial Properties of Nanomaterials

Nanosized inorganic particles, of either simple or composite nature, display unique physical and chemical properties and represent an increasingly important material in the development of novel nanodevices which can be used in numerous physical, biological, biomedical, and pharmaceutical applications.

The preparation of uniform nanosized drug particles with specific requirements in terms of size, shape, and physical and chemical properties is of great interest in the formulation of new pharmaceutical products [45]. Resistance of bacteria to bactericides and antibiotics has increased in recent years due to the development of resistant strains. Some antimicrobial agents are extremely irritating and toxic and there is much interest in finding ways to formulate new types of safe and cost-effective biocidal materials.

Previous studies have shown that antimicrobial formulations in the form of nanoparticles could be used as effective bactericidal materials [53–54]. Recently, Klabunde and coworkers demonstrated that highly reactive metal oxide nanoparticles exhibit excellent biocidal action against Gram-positive and Gram-negative bacteria. Thus, the preparation, characterization, surface modification, and functionalisation of nanosized inorganic particles open the possibility of formulation of a new generation of bactericidal materials.

12.5.1 Gold Nanoparticles

The major clinical use of gold NPs are in the treatment of rheumatic diseases including psorosis, juvenile arthritis, palindromic rheumatism and discoid lupus erythematosus. The efficacy of antibacterial activity of gold nanoparticles can be increased by adding antibiotics. The antimicrobial activity of vancomycin was enhanced on coating with gold nanoparticles against vancomycin resistant enterococci.

Further, the gold nanoparticles generate holes in the cell wall, resulting in the leakage of cell content and cell death. It is also possible that gold nanoparticles bind to the DNA of bacteria and inhibit the uncoiling and transpiration of DNA.

12.5.2 Magnesium Oxide Nanoparticles

The magnesium oxide (AP-MgO) nanoparticles exhibited biocidal activity against certain vegetative gram-positive bacteria, gram-negative bacteria and the spores..Their extremely small size allows many particles to cover the bacteria cells to a high extent and bring halogen in an active form in high concentrations in proximity to the cell. Standard bacteriological tests have shown excellent activity against *E.coli* and *Bacillus megaterium* and a good activity against spores of *Bacillus subtilis*.

12.5.3 Copper Oxide Nanoparticles

Copper oxide (CuO) nanoparticles were effective in killing a range of bacterial pathogens involved in hospital-acquired infections [55]. However, a high concentration of copper nanoparticles is required to achieve a bactericidal effect [56]. These are very active against *B. subtilis*. This may be attributed to the greater abundance of amines and carboxyl groups on the cell surface of *B.subtilis* and greater affinity of copper towards these groups.

12.5.4 Titanium Dioxide Nanoparticles

Titanium dioxide (TiO_2) nanoparticles catalyze the killing of bacteria on illumination by near-UV light. The generation of active free hydroxyl radicals (OH) by photo excited TiO_2 particles is probably responsible for the antimicrobial activity. The use of TiO_2 photocatalysts as an alternate means of self-disinfecting contaminated surfaces on further development may provide potent disinfecting solution for prevention of biofilm formation. In recent years, visible-light absorbing photocatalysts with Ag/AgBr/TiO_2 have proven to be successful at killing *S. aureus* and *E. coli* [33–34], [57].

12.5.5 Zinc Oxide Nanoparticles

The antimicrobial activity of zinc oxide nanoparticles has been studied against the food-related bacteria *Bacillus subtilis, E. coli* and *Pseudomonas fluorescence.* The ZnO NPs could potentially be used as an effective antibacterial agent to safely protect agriculture and food from food borne pathogens, especially *E.coli* [58–60]. There are also other studies which confirm that antimicrobial activity of ZnO nanoparticles could be used against the food borne bacteria *Salmonella typhimurium* and *Staphylococcus aureus.* The antibacterial effect of zinc oxide (ZnO) nanoparticles on *Campylobacter jejuni* was investigated for inhibition and inactivation of cell growth. The results showed that *C. jejuni* was extremely sensitive to treatment with ZnO nanoparticles. The MIC of ZnO nanoparticles for *C. jejuni* was determined to be 0.05 to 0.025 mg/ml, which is 8- to 16-fold lower than that for *Salmonella enterica, Serovar enteritidis* and *Escherichia coli* O157:H7 (0.4 mg/ml). The action of ZnO nanoparticles against *C. jejuni* was determined to be bactericidal, not bacteriostatic. Scanning electron microscopy examination revealed that the majority of the cells transformed from spiral shapes into coccoid forms after exposure to 0.5 mg/ml of ZnO nanoparticles for 16 h, which is consistent with the morphological changes of *C. jejuni* under other stress conditions. These coccoid cells were found by ethidium monoazide-quantitative PCR (EMA-qPCR) to have a certain level of membrane leakage. To address the molecular basis of ZnO nanoparticle action, a large set of genes involved in cell stress response, motility, pathogenesis, and toxin production were selected for a gene expression study. Reverse transcription-quantitative PCR (RT-qPCR) showed that in response to treatment with ZnO nanoparticles, the expression levels of two oxidative stress genes (*katA* and *ahpC*) and a general stress response gene (*dna K*) were increased 52-, 7-, and 17-fold, respectively. These results suggest that the antibacterial mechanism of ZnO nanoparticles is most likely due to disruption of the cell membrane and oxidative stress in *Campylobacter* [42].

The antibacterial potential of zinc oxide (ZnO) nanoparticles (NPs), compared with conventional ZnO powder, against nine bacterial strains, mostly food borne including pathogens, was evaluated using qualitative and quantitative assays. Gram-positive bacteria were found to be more sensitive to ZnO NPs; their disruptive effect on bacteria completely exploded or lysed the cells. Promising results of ZnO NP antibacterial activity suggested its usage in food systems as preservative agent.

12.6 Antiviral properties of Nanoparticles

Metal nanoparticles have been proposed as antiviral systems taking advantage of the core material and/or the ligands shell. Silver nanoparticles have proven to exert antiviral activity against HIV-1 at noncytotoxic concentrations. Silver nanoparticles prevent CD4-dependent virion binding, fusion and infectivity, acting as an effective Virucidal agent for both cell-free virus and cell-associated virus [61].

Gold nanoparticles have also been explored for their anti-HIV activity. The gold nanoparticles coated with multiple copies of an amphiphilic sulphate-ended ligand, bind the HIV envelop glycoprotein gp120 and inhibit the HIV infection of T-cells at nanomolar concentration *in vitro*.

12.6.1 Silver

Silver in ionic or nanoparticle forms has a high antimicrobial activity and is therefore widely used for various sterilization purposes including materials of medical devices and water sterilization. It is well known that silver ions and silver-based compounds are highly toxic to microorganisms showing strong biocidal effects on as many as 16 species of bacteria, including *E. coli* [62]. Thus, silver ions, as an antibacterial component, have been used in the formulation of dental resin composites and ion-exchange fibers and in coatings of medical devices. Recently, Tiller and coworkers showed that hybrids of silver nanoparticles with amphiphilic hyper branched macromolecules exhibit effective antimicrobial surface coatings [63].

12.6.2 Selenium Nanoclusters

Surface characterized by selenium nanoparticles or selenium nanoclusters are herein referred to as being “antipathogenic” to the extent that the surface presents a toxic environment for pathogens such that they will be killed or rendered nonvirulent when present on the surface; therefore resulting in reduced proliferation and amount of pathogens and reduced risk of infection or illness due to the presence of pathogens. Hence, medical devices which may or may not be incorporated into the body, such as needles or implants and the like, that have been modified to include a surface characterized by selenium nanoparticles or selenium nanoclusters, have antipathogenic properties.

12.6.3 Metal Oxides

Nanoparticulate metal oxides show antimicrobial activity against peri-implantitis pathogens. Dental plaque accumulation may result in peri-implantitis, an inflammatory process causing loss of supporting bone that may lead to dental implant failure. The antimicrobial activities of six metal and metal oxide nanoparticles and two of their composites against bacterial pathogens associated with peri-implantitis were examined under anaerobic conditions. The activities of nanoparticles of silver (Ag), cuprous oxide (Cu_2O), cupric oxide (CuO), zinc oxide (ZnO), titanium dioxide (TiO_2), tungsten oxide (WO_3), Ag+CuO composite and Ag+ZnO composite were assessed by minimum inhibitory (bacteriostatic) concentration (MIC) and minimum bactericidal concentration (MBC) determination against *Prevotella intermedia*, *Porphyromonas gingivalis*, *Fusobacterium nucleatum* and *Aggregatibacter actinomycetemcomitans*. Time-kill assays were carried out to examine the dynamics of the antimicrobial activity with ZnO nanoparticles. The MIC and MBC values were in the range of <100 μg/mL to 2500 μg/mL and <100 μg/mL to >2500 μg/mL, respectively. The activity of the nanoparticles tested in descending order was $Ag > Ag+CuO > Cu_2O > CuO > Ag+ZnO > ZnO > TiO_2 > WO_3$. Time-kill assays with ZnO demonstrated a significant decrease in growth of all species tested within 4h, reaching 100% within 2h for *P. gingivalis* and within 3h for *F. nucleatum* and *P. intermedia*. Coating titanium surfaces of dental and orthopedic implants with antimicrobial nanoparticles should lead to an increased rate of implant success.

12.6.4 N-phenyl- and N-benzoylthiourea Derivatives

N-phenyl-and N-benzoylthiourea derivatives display selective antimicrobial activities against *Cladosporium cladosporioides*, *B. subtilis* and *Micrococcus luteus*

12.6.5 FeO_4/C_{12} Nanostructures and 2-((4-ethyl-phenoxy) methyl)-N-(substituted-phenyl carbamothioyl)-benzamides

Antipathogenic surface coating nanofluid which was obtained by combining FeO_4/C_{12} nanostructures and 2-((4-ethylphenoxy)metyl)-N-(substituted-phenylcarbamothioyl)-benzamides, showed an improved resistance to *Staphylococcus aureus* ATCC 25923 and *Pseudomonas aeruginosa* ATCC 27853 *in vitro* biofilm development. This was demonstrated by the viable cell counts of biofilm-embedded bacterial cells and by scanning electron microscopy examination of the colonized surface. The nanofluid proved to be noncytotoxic and did not influence the eukaryotic cell cycle [64].

12.6.6 Graphene Nanosheets

The impact of graphene-based materials on health and the environment needs to be thoroughly evaluated before their potential applications. Graphene has strong cytotoxicity toward bacteria. To better understand its antimicrobial mechanism, many researchers compared the antibacterial activity of four types of graphene-based materials (graphite Gt), graphite oxide (GtO), graphene oxide (GO), and reduced graphene oxide (rGO)) toward a bacterial model—*Escherichia coli.*

Under similar concentration and incubation conditions, GO dispersion shows the highest antibacterial activity, sequentially followed by rGO, Gt and GtO. Scanning electron microscope (SEM) and dynamic light scattering analyses show that GO aggregates have the smallest average size among the four types of materials. Scanning electron microscope images show that direct contact with graphene nanosheets disrupts cell membrane. No superoxide anion ($O_2^{\bullet -}$)-induced reactive oxygen species (ROS) production is detected. However, the four types of materials can oxidize glutathione, which serves as redox state mediator in bacteria. Conductive rGO and Gt have higher oxidation capacities than insulating GO and GtO. Results suggest that antimicrobial actions are contributed by both membrane and oxidation stress. This study proposes that a three-step antimicrobial mechanism, previously used for carbon nanotubes, is applicable to graphene-based materials. It includes initial cell deposition on graphene-based materials, membrane stress caused by direct contact with sharp nanosheets, and the ensuing superoxide anion-independent oxidation. It envisions that physicochemical properties of graphene-based materials, such as density of functional groups, size, and conductivity, can be precisely tailored to either reducing their health and environmental risks or increasing their application potentials [65].

12.6.7 Photoactivated Carbon Nanotube–Porphyrin Conjugates

The design of antimicrobial nanocomposite films based on conjugates of multiwalled carbon nanotubes (MWNT) and protoporphyrin IX (PPIX) are highly effective against *Staphylococcus aureus* (*S. aureus*) upon irradiation with visible light. *Staphylococcus aureus* infections can lead to life-threatening situations, especially when caused by antibiotic-resistant strains. While the light-activated antimicrobial activity of porphyrins against such pathogens is well known, a facile way to incorporate porphyrins into coatings may lead to their more effective use. [66]. To that end, synthesize and characterize MWNT–PPIX conjugates combine the biocidal capacity of porphyrins with the mechanical strength of MWNTs [67–70]. The conjugates could effectively deactivate *S. aureus* cells in solution upon irradiation with visible light [71].

12.7 Antifungal Activity

12.7.1 Silver

One application in which silver can be utilized is in the management of plant disease. Since silver displays multiple modes of inhibitory action against microorganisms, it may be used for controlling various plant pathogens in a relatively safer way compared to synthetic fungicide. Various forms of silver ions and nanoparticles were tested in the current study to examine the antifungal activity on two plant-pathogenic fungi, *Bipolaris sorokiniana* and *Magnoporthe grisea. In vitro* petridish assay indicated that silver ion and nanoparticle had a significant effect on the colony formation of these two pathogens. Effective concentrations of the silver compounds inhibiting colony formation by 50% (EC50) were higher for *B. sorokiniana* than for *M. grisea*. The inhibitory effect on colony formation significantly diminished after silver cations were neutralized with chloride ions. Growth chamber inoculation assays further confirmed that both ionic and nanoparticle silver significantly reduced these two fungal diseases on perennial rye grass (*Lolium perenne*) [72]. Particularly, silver ions and nanoparticles effectively reduced disease severity with an application at 3 h before spore inoculation, but their efficacy significantly diminished when applied at 24 h after inoculation. The *in vitro* evaluations of silver indicated that both silver ions and nanoparticles influence colony formation of spores and disease progress of plant-pathogenic fungi. In plants, efficacy of silver ions and nanoparticles is much greater with preventative application, which may promote the direct contact of silver with spores and germ tubes, and inhibit their viability [73].

Silver nanoparticles have an inhibitory effect against powdery mildew on cucumbers and pumpkins. Powdery mildew is one of the most devastating diseases in cucurbits. Crop yield can decline as the disease severity increases. A study evaluated the effect of silver nanoparticles against powdery mildew under different cultivation conditions *in vitro* and *in vivo*. Silver nanoparticles (WA-CV-WA13B) at various concentrations were applied before and after disease outbreak in plants to determine antifungal activities. In the field tests, the application of 100 ppm silver nanoparticles showed the highest inhibition rate for both before and after the outbreak of disease on cucumbers and pumpkins. Also, the application of 100 ppm silver nanoparticles showed maximum inhibition for the growth of fungal hyphae and conidial germination in *in vivo* tests. Scanning electron microscope results indicated that the silver nanoparticles caused detrimental effects on both mycelial growth and conidial germination.

12.8 Mechanism of Action of Nanoparticle inside the Body

The fact that nanoparticles exist in the same size domain as proteins makes nanomaterials suitable for biotagging or labeling. However, size is just one of many characteristics of nanoparticles that in itself are rarely sufficient if one is to use nanoparticles as biological tags. In order to interact with a biological target, a biological or molecular coating or layer acting as a bioinorganic interface should be attached to the nanoparticle. Examples of biological coatings may include antibodies, biopolymers like collagen, or monolayers of small molecules that make the nanoparticles biocompatible [47].

Nanoparticle usually forms the core of nanobiomaterial. It can be used as a convenient surface for molecular assembly, and may be composed of inorganic or polymeric materials. It can also be in the form of nanovesicle surrounded by a membrane or a layer. The shape is more often spherical but cylindrical, plate-like and other shapes are possible. The size and size distribution might be important in some cases, for example, if penetration through a pore structure of a cellular membrane is required. The size and size distribution become extremely critical when quantum-sized effects are used to control material properties. A tight control of the average particle size and a narrow distribution of sizes allow creating very efficient fluorescent probes that emit narrow light in a very wide range of wavelengths. This helps with creating biomarkers with many well-distinguished colors. The core itself might have several layers and may be multifunctional. For example, by combining magnetic and luminescent layers one can both detect and manipulate the particles.

The core particle is often protected by several monolayers of inert material, for example, silica. Organic molecules that are adsorbed or chemisorbed on the surface of the particle are also used for this purpose. The same layer might act as a biocompatible material. However, more often an additional layer of linker molecules is required to proceed with further functionalisation. This linear linker molecule has reactive groups at both ends. One group is aimed at attaching the linker to the nanoparticle surface and the other is used to bind various moieties like biocompatibles (dextran), antibodies, fluorophores, etc., depending on the function required by the application.

12.9 Detecting the Antipathogenicity of Nanoparticles on Microorganisms *in Vitro*

For clarifying and proving the antipathogenic effect of nanoparticles against various microbes, the most suited technique employed in the laboratory is zone inhibition technique, which is performed in petridishes containing different culture medium for the growth of microorganism, such as potato-dextrose medium, zepeck medium, and many others according to the specificity of the particular microbe [74].

This method is also known as zone of inhibition technique. In this method, in a petridish autoclave culture medium (e.g., agar media) is uniformly poured. One selects any microbe, fungi, bacteria, etc., to be tested for the antimicrobial activity (e.g. *penicillium*). The desired microorganism is inoculated in the petridish. Next, any of the nanoparticle or nanocomposite material of appropriate concentration is poured in the media and incubated for 24 hours. Then the efficacy of nanoparticle treatment can be evaluated at intervals of 2, 4, 6, 8 days by measuring the diameter of zone of inhibition. It was shown that different concentrations of ZnO NPs treatment inhibit the growth of *penicilium* differently.

12.10 Types of Nanoparticles

Nanoparticles exist in several forms, shapes and morphology such as:

- **Nanocrystals:** They are any crystalline nanomaterial with at least one dimension ≤ 100nm. Nanocrystals can provide efficient light emission even while bulk material does not [50, 75–76].
- **Quantum dots:** Semiconductor nanocrystals having all the three dimensions in nano-range are often referred to as quantum dots. These are zero-dimensional structures [77–78].
- **Nanowires:** Nanowires are the structures that have a thickness or diameter constrained to tens of nanometers or less and an unconstrained length. Nanowires can be used to link tiny components into extremely small circuits. These are 1-dimensional structures.
- **Nanoribbons:** These are strips of graphene or silicate with ultrathin width (<50nm).They can be produced by unzipping or cutting open nanotubes.

12.11 Synthesis of Nanoparticles by Conventional Methods

For the synthesis of nanoparticles, the first priority is to achieve large-scale production in an economical manner [79]. There are two main methods for the preparation of nanoparticles:

1. Top-down approach via milling
2. Bottom-up approach via chemical processes

12.11.1 Top-down approach

This method is solely based on physical and microlithographic philosophy, in contrast to the other approach, where atomic or molecular units are used to assemble molecular structure, ranging from atomic dimensions up to supramolecular structure in nanometer range. It mainly involves mechanical/thermal cycle and yields:

- Broad size distribution (10–1000nm)
- Varied particle shape or geometry
- Impurities

This method is useful for preparation of nanocomposites and nanograined bulk material.

12.11.2 Bottom-up approach

This approach is solely based upon chemical principles, it involves:

- Pyrolysis
- Inert gas condensation
- Solvothermal reaction
- Sol-gel fabrication
- Structured media

Bottom-up Synthesis

I. **Gas Phase Synthesis**

- Pyrolysis
- Inert gas condensation

II. **Liquid Phase Synthesis**

- Solvothermal process
- Sol-gel process
- Micellar structure media

III. **Vapor Phase Synthesis**

Nanoparticles of wide range can be prepared by a variety of methods. Hahn presented a useful overview of gas-phase synthesis of nanocrystalline material. In vapor-phase synthesis, conditions are created where the vapor-phase mixture is thermodynamically unstable relative to formation of solid material to be prepared in nanoparticulate form [80].

- **Pyrolysis:** In pyrolysis, vaporous precursor is forced through a hole or opening at high pressure and burned. The resulting solid in air is classified to recover oxide particles from byproduct gases.
- **Inert Gas Condensation:** In this process solid material is heated to evaporate it into a background gas, and then the vapor is mixed with cold gas to reduce the temperature. This method is well suited for production of metal nanoparticles. Maisels *et al.* prepared nanocomposites of

PbS with Ag by separate evaporation/condensation of two materials followed by coagulation of oppositely charged PbS and Ag particles.

IV. **Liquid Phase Synthesis**

Liquid-phase fabrication entails a wet-chemical route and involves the following processes:

- Solvothermal method
- Sol-gel method
- Synthesis in structured media (microemulsion)

i. **Solvothermal Synthesis**

In this method, precursors are dissolved in hot solvent (e.g., n-butyl alcohol). Solvents other than water can provide milder and friendlier reaction condition.

ii. **Sol-Gel Process**

The sol-gel process is a wet-chemical technique widely used in the field of material science and ceramic engineering; it involves the following steps:

- Formation of stable sol solution
- Gelation via polycondensation or polyesterification
- Gel aging into solid mass
- Drying of gel to remove liquid phases
- Dehydration at 8000°C to remove M-OH group for stabilizing gel
- Densification and decomposition of gel at high temperature (T> 8000°C)

The precursor sol can be either deposited on a substrate to form a film, cast into a suitable container with the desired shape or used to synthesize powders (e.g., nanospheres).

iii. **Synthesis in Structured Medium**

This is the mixing of two microemulsion carrying metal salt and reducing agent.

12.12 Biological Synthesis of Nanoparticles

Nanoparticles have been produced physically and chemically for a long time, but only recently has the critical role of microorganisms and other biological entities been explored for the synthesis of metal nanoparticles. The use of highly-structured physical and biosynthetic activities of microbial cells for the synthesis of nanosized materials has recently emerged as a novel approach for the synthesis of metal nanoparticles, e.g., synthesis of nanoscale, semiconducting CdS crystals in the yeast *Schizosaccha Schizosaccharomyces pombe* and production of silver nanoparticles with periplasmic space of *Pseudomonas stutzeri* [81].

Biosynthesis of metal nanoparticle is an eco-friendly method which does not use harsh, toxic and expensive chemicals. Biosynthesis of metallic nanoparticles using microorganisms is an emerging fabulous eco-friendly science of well-defined sizes, shapes and controlled monodispersity [82–83]. The green synthesis of metallic nanoparticles includes the use of biological agents such as bacteria, fungi, actinomycetes, yeast and plants [84–87]. In green nanotechnology, different microorganisms produce inorganic materials, either intracellularly or extracellularly, with properties similar to chemically synthesized material.

Literatures have reported green synthesis of nanoparticles using different microbes for example silver nanoparticles was synthesized by *Penicillium fellutanum* [88] and copper oxide by *Streptomyces Sp.* for the development of antimicrobial textiles which can be useful in hospitals to prevent or to minimize infection from pathogenic bacteria [89] and copper oxide was also synthesized by using *Escherichia coli* with a variable size and shapes [90]. Meanwhile, the use of fungi is potentially exciting. Compared to bacteria, fungi hold great promise for large-scale nanoparticles production because they secrete a large amount of enzymes which are involved in MNP biosynthesis and are simpler to grow both in the laboratory as well as on an indrustrial scale. Nanoparticles can also be synthesized using different fungal species for example *Fusarium oxysporium* was used for the synthesis of silver nanoparticles [91].

The organisms used in the synthesis of nanoparticles vary from simple prokaryotic bacterial cells to complex eukaryotes. Important steps are:

- Selection of the best organism which shows enzyme activities and biochemical pathways.
- Optimal condition for all growth and enzyme activity, e.g., nutrient, inoculums size, light, temperature and pH should be optimized.
- Optimal reaction condition.

The biological synthesis of metal nanoparticles by using plant extract has received more attention as a suitable alternative to chemical procedures and physical methods.

Synthesis of nanoparticles using plant extract is very cost effective, and therefore can be used as an economic and valuable alternative for large-scale production of metal nanoparticles. Extract from plants can act both as reducing and capping agent in nanoparticle synthesis. Shastry *et al.* have studied bioreduction of gold and silver ions by leaf broth of *Pelargonium graveolens* and *Azadirachta indica* [92].

Silver nanoparticles were successfully synthesized from $AgNo_3$ through a simple green and natural route using the latex of 6 different plant taxa belonging to 6 different families [93]. Some specific plant parts or whole plant, especially angiospermic plants, are used for great synthesis of nanoparticles. Alfalfa roots are capable of absorbing Ag (0) from agar medium and transferring them to the shoot of the plant in the same oxidation state. In the shoot these Ag atoms arrange themselves to form nanoparticles by joining themselves to form larger arrangements. With the use of *Emblica officinalis* fruit extract as reducing agent, the extracellular synthesis of highly stable Ag and AuNPs has been achieved [94].

Achlypha indica leaf extracts have produced silver nanoparticles (20–30nm) in 30 min. These nanoparticles had excellent antimicrobial activity against waterborne pathogens like *E.coli* and *V. cholera* [95].

12.12.1 Extraction of Nanoparticles

After synthesizing nanoparticles by using plant extract, methods such as freezing, osmotic shock and heating are used in order to extract them from cells.

In addition to this, in the case of extracellular production of nanoparticles, centrifugation could be used for extraction and purification of nanoparticles.

12.13 Characterizations of Nanoparticles

There are several techniques to analyze nanoparticles such as atomic absorbtion spectroscopy (AAS), TEM, surface characterization through X-ray photoelectron spectroscopy, dynamic light scattering, etc., some of which are described below.

- **X-ray differentiation:** XRD is a noncontact, nondestructive technique, which can be used to determine the crystalline phases presented in materials.

- **Transmission electron microscopy:** The morphology of the nanoparticles can be examined by TEM; ultrathin specimen with a thickness of about 50nm can be observed.
- **Atomic force microscopy (AFM):** The dried film of NPs can be scanned by AFM in tapping mode. From AFM image, the size of Cu NPs was measured using Neeco'sSPN lab analysis software.

12.14 Biocompatibility of Nanoparticles

- Biocompatibility refers to the ability of biomaterial to perform its desired function with respect to a medical therapy, without any harmful effects in the recipient or beneficiary of that therapy [42], [47].
- Nanoparticles are somehow permeable to membrane cells and spread along nerve cells, blood vessels and lymphatic vasculature in the body. The NPs selectively accumulate in different cells and certain cellular structure [48].
- For biomedical application, the NPs enter the body and contact with tissues and cells directly; therefore it is necessary for exploring their biocompatibility.
- Gold NPs can be readily functionalized with probe molecules such as antibodies, enzymes and nucleotides. These are important factors to enhance the immune response. Nanoparticles often become entangled or coated with mucous produced by epithelial cells or extracellular matrix structure on the apical surface of cells adjacent to the largest cells, reducing the number of nanoparticles that could be localized at the desired site.
- The TiO_2 and FeO_3nanostructured materials are used for simultaneously improving the antimicrobial properties of PMMA resin.

12.15 Toxic Effects of Nanoparticles

Humans have been exposed to nanoparticles throughout their evolutionary phases; however, this exposure has been increased to a great extent in the past century because of the industrial revolution [96]. Nanoparticles constitute a part of particulate matter (PM). Epidemiological studies have shown that urban pollution with airborne PM deriving from combustion sources such as motor vehicle and industrial emissions contributes to respiratory and cardiovascular morbidity and mortality. The respiratory risks associated with air pollution have been known as the London Fog Episode of 1952.

The growing use of nanotechnology in high-tech industries is likely to become another way for humans to be exposed to intentionally generate engineered nanoparticles. Nanotechnology is also being applied in medical sciences in an attempt to achieve a personalized medicine. However, the same properties (small size, chemical composition, structure, large surface area and shape) which make nanoparticles so attractive in medicine, may contribute to the toxicological profile of nanoparticles in biological systems [97]. In fact, the smaller particles are, the more surface area they have per unit mass; and this property makes nanoparticles very reactive in the cellular environment [98]. Therefore, any intrinsic toxicity of the particle surface will be enhanced. The respiratory system, blood, central nervous system (CNS), gastrointestinal (GI) tract and skin have been shown to be targeted by nanoparticles [97–99].

12.15.1 Respiratory System

One of the most important gateways of entry and organ target for nanoparticles is the respiratory system. It is well known that lungs are easily exposed to atmospheric pollutants [100]. In this regard, combustion-derived nanoparticles have been largely studied as a possible etiologic factor for several adverse health effects, including exacerbations of airways disease as well as deaths and hospitalization from cardiovascular disease. One of the main mechanisms of lung injury caused by combustion-derived nanoparticles is via oxidative stress leading to activation of different transcription factors with the regulation of inflammatory protein synthesis. As nanotechnology is being applied in aerospace and computing, the release of high amounts of nanoparticles in an enclosed environment may be of great concern for airline crews and hardware engineers [101]. It has been found that nanoparticles can induce increased lung toxicity compared to larger particles with the same chemical composition at equivalent mass concentration. In addition, it has also been shown that nanoparticles of different diameters can induce inflammatory reactions in the lungs of experimental animals [102. In fact, a significant correlation between the surface area of nanoparticles and the induced inflammation was observed via increased oxidative stress.

12.15.2 Translocation of nanoparticle to the Blood Stream and Central Nervous System

Nanoparticles could avoid normal phagocytic defenses in the respiratory system and gain access to the systemic circulation or even to the CNS. Once inhaled and deposited, nanoparticles can translocate to extra pulmonary sites and reach other target organs by different mechanisms [103]. The first mechanism involves passing of nanoparticles across the epithelia of the respiratory tract into the interstitial and access to the blood stream directly or via lymphatic pathways, resulting in systemic distribution of nanoparticles. Nanoparticles can be rapidly observed in rat platelets after intra tracheal instillation of particles of colloidal gold (30 nm). Interestingly, human studies have shown that elevated levels of Mn could be associated with increased rate of Parkinson's disease [104–105], Recently, it has been found that exposure of PC-12 neuroendocrine cell line to nanosized Mn induced an increase in reactive oxygen species and dopamine depletion [106–107].

12.15.3 Gastrointestinal Tract and Skin

Nanoparticles can also enter the body through the GI tract and skin. Nanoparticles can be ingested into the gut in many ways. For example, nanoparticles can be directly ingested from food, water, drugs and cosmetics, but inhaled nanoparticles can also be ingested by the GI tract once they are cleared by the respiratory tract. It is known that the kinetics of particle uptake in the GI tract depends on diffusion and accessibility through mucus, initial contact with enterocytes, cellular trafficking and post-translocation events [108]. The smaller the particle diameter is the faster it could diffuse through GI secretion to reach the colonic enterocytes. Following uptake by the GI tract, nanoparticles can translocate to the blood stream and distribute all over the body. Recently, it has been shown that Cu nanoparticles can induce adverse effects and heavy injuries in the kidney, liver and spleen of experimental mice compared to micro-Cu particles [109].

As with lungs, the GI tract is easily exposed to stimuli that can induce an inflammatory response. Inflammatory bowel disease (IBD), including both ulcerative colitis and Crohn's disease (CD), is an inflammatory chronic condition whose etiology still remains unclear [108].

Nanoparticles can also be taken up by lymphatic nodes at skin level, translocating to the blood stream via lymphatic pathways. It has been found that SWCNT can induce oxidative stress and pro-inflammatory responses in human keratinocytes cells *in vitro*.

Today the propensity of NPs to cross cell barriers, enter cells and interact with subcellular structure is well established, yet there is increasing concern that environmental exposure to engineered nanoparticles may result in significant adverse effects [110]. Hence proper characterization of the nanomaterials as well as an understanding of the interaction of nanoparticles with humans when in contact with them, is crucial for understanding possible toxic effects.

12.16 Conclusion

Today, metal nanoparticles have influenced many scientists due to their extensive applications in new technologies in different fields like medicine, optics, electronics, chemistry and biotechnology. Nanoparticles are very much different and improved than their corresponding bulk material, actually due to their very small size their surface area-to-volume ratio increases due to which their physical and chemical properties are quite different and enhanced. We have seen that besides many physical and chemical methods which have been developed for preparing metal nanoparticles, nanobiotechnology also serves as an important method in the development of nontoxic and eco-friendly processes for synthesizing nanoparticles. Hence, biosynthesis of MNPs present a promising route for MNPs production. It has been shown that metal nanoparticles can be synthesized by using microorganisms, and microbial nanoparticles have been observed in numerous fungal and bacterial species. The most recent researchhas shownthat nanoparticles can also be beautifully synthesized by using plant extract. Nanoparticles have really proven their potential applications in each and every field, such as in agriculture,where nanoparticles have shown the potential to replace insecticides and pesticides and are becoming a boon for curing a number of plant diseases; similarly they are used as food packaging material in many food industries to reduce microbial contamination in food products, andare showing a revolutionary response in the field of biomedical applications in drug delivery, cancer therapy, dentistry and many more.

Last but not the least, even though nanoparticles are of utmost importance in today's technological world, every coin have two sides; in the same way, although nanoparticles are very useful in all fields, they are also harmful to humans and other living beings in other ways. Therefore, while using them, we must examine the toxicity of particular MNPs towards the target on which they are applied, and for this a complete knowledge of its biocompatibility and toxicity is required.

References

1. R. Czajka, "Development of medical textiles", Fibers Text. East. Eur., Vol. 13, No. 1, 13–15.
2. P. Webster, "World nanotechnology market Frost and Sullivan", *Nanomedicine*, Vol. 1, No. 2, pp. 140–42, 2005.
3. P. Holister, J.W. Weener, C.V. Román, T. Harper, "Nanoparticles,*Technology*" *White Papers, Nr. 3*, Cientifica,pp. 1–11, Oct. 2003.
4. V. Wagner, "Nanobiotechnologie II: Anwendungen in der Medizin und Pharmazie". Dusseldorf, Germany: Zukunftigetechnologien, Nr 50, 2004:184. ISSN 1436–5928. (Evaluation techniques).
5. G.M. Whitesides, "Nanoscience, nanotechnology, and chemistry", *Small*, Vol. 1, pp. 172–179, 2005.
6. Q.L. Feng, J. Wu, G.Q. Chen, F.Z. Cui, T.N. Kim, J.O. Kim, "A mechanistic study of the antibacterial effect of silver ions on *Escherichia coli* and *Staphylococcus aureus*", *J. Biomed. Mater. Res.*, Vol. 2, p. 662, 2000.
7. S. Prabhu, and E.K. Poulose, "Silver nanoparticles: Mechanism of antimicrobial action, synthesis, medical applications, and toxicity effect", *International Nanoletters*, Vol. 2, p. 32, 2012.
8. R.A. FreitasJr., "Nanotechnology, nanomedicine and nanosurgery", *International Journal of Surgery*, Vol. 3, Iss. 4, pp. 243–247, 2005.
9. P. Ghosh, G. Han, M. De, K.C. Kim, M.V. Rotello, "Gold nanoparticles in delivery applications", *Advanced Drug Delivery Rev.*, Vol. 60, pp. 1307–1315, 2008.
10. S.K. Gogoi, P. Gopinath, A. Paul, A. Ramesh, S.S. Ghosh, A. Chattopadhyay, "Green fluorescent protein-expressing *Escherichia coli*as a model system for investigating the antimicrobial activities of silver nanoparticles", *Langmuir*, Vol. 22, pp. 9322–9328, 2006.
11. P. Hoet, I. Bruske-Hohlfeld, O. Salata, "Nanoparticles – known and unknown health risks", *J. Nanobiotechnol.*, Vol. 2, p. 12, 2004.
12. J.S. Kim, E. Kuk, K.N. Yu, J.-H. Kim, S.J. Park, H.J. Lee, S.H. Kim, Y.K. Park, Y.H. Park, C.-Y. Hwang, Y-K. Kim, Y-S. Lee, D.H. Jeong, and M.-H. Cho, "Antimicrobial effects of silver nanoparticles", *Nanomedicine: Nanotechnol. Biol. Medic.*, Vol. 3, pp. 95–101, 2007.
13. E.E. Jredget, H.A. Shankowsky, A. Groeneveld, and R. Rurrell, "Antibacterial nature of Ag used in infectious wounds", *J. Burn Care Rehabil.*, Vol. 19, pp. 531–537, 1998.
14. W.K. Jung, H.C. Koo, K.W. Kim, S. Shin, S.H. Kim, Y.H. Park, "Antibacterial activity and mechanism of action of the silver ion in *Staphylococcus aureus* and *Escherichia coli*", *Appl. Environ. Microbiol.*, Vol.74, pp. 2171–2178, 2008.
15. J.F. Kukowska-Latallo, *et al.*, "Nanoparticle targeting of anticancer drug improves therapeutic response in animal model of human epithelial cancer", *Cancer Res.*, Vol. 65, pp. 5317–5324, 2005.

16. S.R. Kumar, and R. Vijaylakshmi, "Nanotechnology in dentistry", *Indian J. Dent. Res.*,Vol. 17, No. 2, pp. 62–65, 2006. http://dx.doi.org/10.4103/0970-9290.29890.
17. S.V. Kyriacou, W.J. Brownlow, and X.-H.N. Xu, "Using nanoparticle optics assay for direct observation of the function of antimicrobial agents in single live bacterial cells", *Biochemistry*, Vol. 43, pp. 140–7, 2004.
18. R. Langer, "Biomaterials in drug delivery and tissue engineering: One laboratory's experience", *Acc. Chem. Res.*,Vol. 33, pp. 94–101, 2000.
19. P. Li, J. Li, C. Wu, Q. Wu, J. Li, "Synergistic antibacterial effects of β-lactam antibiotic combined with silver nanoparticles", *Nanotechnology*, Vol. 16, pp. 1912–1917, 2005.
20. Y. Liu, H. Miyoshi, and M. Nakamura, "Nanomedicine for drug delivery and imaging: A promising avenue for cancer therapy and diagnosis using targeted functional nanoparticles", *Int. J. Cancer*, Vol. 120, pp. 2527–2537, 2007.
21. M.K. Rai, S.D. Deshmukh, A.P. Ingle, and A.K. Gade, "Silver nanoparticles: The powerful nanoweapon against multidrug-resistant bacteria", *J. Appl. Microbiol.*, ISSN: 1364–5072.
22. J.P. Mathuria, "Nanoparticles in tuberculosis diagnosis,treatment and prevention: A hopefor the future", *Digest Journal Nanomaterials Biostructure*, Vol.4, No.2, pp. 309–312, June 2009.
23. V. Wagner, A. Dullaart, A.K. Bock, A. Zweck, "The emerging nanomedicine landscape", *Nat. Biotechnol.*, Vol. 24, No. 10, pp. 1211–1217, 2006. doi:10.1038/nbt1006-1211. PMID 17033654.
24. R.H. Muller, and B.H.L. Bohm, "Nanosuspensions," in: R.H. Muller, S. Benita, B.H.L. Bohm, eds., "Emulsions and Nanosuspensions for the Formulation of Poorly Soluble Drugs", Stuttgart: Medpharm, pp. 149–174, 1998.
25. D.F. Emerich, and C.G. Thanos, "Targeted nanoparticle-based drug delivery and diagnosis", *J. Drug Target*, Vol. 15, pp. 163–183, 2007.
26. R.O. Esenaliev, "Radiation and nanoparticles for enhancement of drug delivery in solid tumors", PCT Int Appl 2000, WO 2000002590.
27. A.J. Thote, R.B. Gupta, "Formation of nanoparticles of a hydrophilic drug using supercritical carbon dioxide and microencapsulation for sustained release", *Nanomedicine: Nanotech. Biology Medicine*, Vol. 1, pp. 85–90, 2005.
28. V.P. Torchilin, "Recent advances with liposomes as pharmaceutical carriers", *Nat. Rev. Drug Discov.*, Vol. 4, pp. 145–159, 2005.
29. O.C. Farokhzad, and R. Langer, "Nanomedicine: developing smarter therapeutic", *Adv. Drug Deliv. Rev.*, Vol. 58, pp. 1456–1459, 2006.
30. R. Duncan, "Polymer conjugates as anticancer nanomedicines", *Nat. Rev. Cancer*, Vol. 6, pp. 688–701, 2006.
31. R. Jurgons, *et al.*, "Drug loaded magnetic nanoparticles for cancer therapy", *J. Phys. Condens. Matter*,Vol. 18, pp. S2893-S2902, 2006.
32. J. Kreuter, ed., "Nanoparticles, in: *Colloidal Drug Delivery Systems*", Marcel Dekker: New York, pp 219–342, 1994.

33. B. Kim, D. Kim, D. Cho, S. Cho, "Bactericidal effect of TiO2 photocatalyst on selected food-bornepathogenic bacteria", *Chemosphere,*Vol. 52, pp. 277–281, 2003.
34. C. Chawengkijwanich, Y. Hayata, "Development of TiO2 powder-coated food packaging film and its ability to inactivate *Escherichia coli* in vitro and in actual tests", *Int. J. Food Microbiol.*, Vol. 123, pp. 288–292, 2008.
35. Nesli Sozer and Jozef L. Kokini, "Nanotechnology and its applications in the food sector", *Trends in Biotechnology* , 27(2):82–9, 2009.
36. H. J. Park, S.H. Kim, H.J. Kim, and S.-H. Choi, "A newcomposition of nanosized silica-silver for control of various plant diseases", *Plant Pathol. J.*, Vol. 22, pp. 295–302, 2006.
37. K. Lamsal, S.-W. Kim, J.H. Jung, Y.S. Kim, K.S. Kim, and Y.S. Lee, "Inhibition effects of silver nanoparticles against powdery mildews on cucumber and pumpkin", Mycobiology, Vol. 39, No. 1, pp. 26–32, 2011.
38. D.S. Thakur, P. Kumar, P. Kumar, "Current concepts and newer developments in cancer", *International Journal of Pharmaceutical Sciences Review and Research*, Vol. 3, No. 2, pp. 85–89, 2010.
39. M. Ferari, "Cancer nanotechnologies: Opportunities and challenges", *Nat. Rev. Cancer*, Vol. 5, pp. 161–71, 2005.
40. I. Roy, *et al.*, "Ceramic-based nanoparticles entrapping water-insoluble photosensitizing anticancer drugs: A novel drug-carrier system for photodynamic therapy", *J. Am. Chem. Soc.*, Vol. 125, Issue 26, pp. 7860–7865, 2003.
41. Y.Gao andR. Cranston, "Recent advances in antimicrobial treatments of textiles", *Textile Res. J.* Vol. 78, pp. 60–72, 2008.
42. Y. Zheng, and R. Li, "In vitro and in vivo biocompatibility studies of ZnO nanoparticles", *International Journal of Modern Physics: B*, Vol. 23, Nos. 6&7, pp. 1566–1571, 2009.
43. S. Liu and G. Sun, "Durable and regenerable biocidal polymers: Acyclic n-halamine cotton cellulose", *Ind. Eng. Chem. Res.*, Vol. 45, pp. 6477–6482, 2006.
44. D.K. Tiwari, J. Behari, and Prasenjit Sen, "Application of nanoparticles in waste water treatment", *World Applied Sciences Journal* 3 (3): 417–433, 2008 ISSN 1818-4952.
45. S. Pal, Y.K. Tak, J.M. Song, "Does the antibacterial activity of silver nanoparticles depend on the shape of the nanoparticle? A study of the gram-negative bacterium *Escherichia coli*", *App. Environ. Microbiol.*, Vol. 73, No. 6, pp. 1712–1720, 2007.
46. G. Fontana, G. Pitarresi, V. Tomarchio, B. Carlisi, P.L. San Biagio, "Preparation, characterization and in vitro antimicrobial activity of ampicillin-loaded polyethylcyanoacrylate nanoparticles", *Biomaterials*, Vol. 19, No. 11, pp. 1009–1017, 1998.
47. V.A. Sinani, D.S. Koktysh, B.G. Yun, R.L. Matts, T.C. Pappas, M. Motamedi, S.N. Thomas, N.A. Kotov, "Collagen coating promotes biocompatibility of semiconductor nanoparticles in stratified LBL films", *Nano Letters*, Vol. 3, pp. 1177–1182, 2003.

48. 7- E. Tasciotti, A. Parodi, N. Quattrocchi, A. van de Ven, C. Chiappini, M. Evangelpoulos, J. Martinex, B. Brown, *et al.*, "Synthetic nanoparticles functionalized with biomimetic leukocyte membranes possess cell-like functions", *Nature Nanotechnology*, Vol. 8, pp. 61–68, 2012. doi:10.1038/nnano.2012.212.
49. F.L. Yen, T. H. Wu, L.T. Lin, T. M. Cham, C.C. Lin, "Nanoparticles formulation of *Cuscuta chinensis* prevents acetaminophen- induced hepato toxicity in rats", Food Chem. Toxicol., Vol. 46, No. 5, pp 1771–1777, 2008.
50. R. Langer, "Biomaterials in drug delivery and tissue engineering: One laboratory's experience", *Acc. Chem. Res.*,Vol. 33, pp. 94–101, 2000.
51. Q, Shen, Y. Wang,Y. Zhang, "Improvement of colchicine oral bioavailability by incorporating eugenol in the nanoemulsion as an oil excipient and enhancer", *International Journal of Nanomedicine*, Vol. 6, pp. 1237–1243, 2011.
52. A. Radovic-Moreno, T. Lu, V. Puscasu, C. Yoon, R. Langer, O. Farokhzad, Omid, "Surface charge-switching polymeric nanoparticles for bacterial cell wall-targeted delivery of antibiotics", *ACS Nano*, Vol.6, Issue5, pp. 4279–4287, 2012. doi:10.1021/nn3008383. Retrieved May 24, 2012.
53. J.S. Kim, E. Kuk, K.N. Yu, J.H. Kim, S.J. Park, H.J. Lee, S.H. Kim, Y.K. Park, Y.H. Park, C.Y. Hwang, Y.K. Kim, Y.S. Lee, D.H. Jeong, M.H. Cho, "Antimicrobial effects of silver nanoparticles", *Nanomedicine*,Vol. 3, pp. 95–101, 2007.
54. M. Kolar, K. Urbanrk, T. Latal, "Antibiotic selective pressure and development of bacterial resistence", *Int. J.Antimicrob.Ag.*, Vol.17, pp. 357–363, 2001.
55. C.O. Dimpka, J.E. McLean, D.E. Latta, E. Manangon, D.W. Britt, W.P.Johnson, M.I. Boyanov, A.J. Anderson, "CuO and ZnO nanoparticles: Phytotoxicity, metal speciation, and induction of oxidative stress in sand-grown wheat", *J. Nanopart. Res.*, Vol. 14, p. 1125, 2012, DOI 10.1007/s11051-012-1125-9,
56. E.K. Athanassiou, R.N. Grass, and W.J. Stark, W J 2006 "Large-scale production of carbon-coated copper nanoparticles for sensor applications", *Nanotechnology*, Vol. 17 pp. 1668–73, 2006.
57. Archana Maurya, Pratima Chauhan, Amita Mishra, Abhay K. Pandey, "Surface functionalisation of TiO_2 with plant extracts and their combined antimicrobial activities against E. coli", *Journal of Research Updates in Polymer Science*, Vol.1, pp. 43–51, 2012.
58. T. Jin, D. Sun, Y. Su, H. Zhang, H.J. Sue, "Antimicrobial efficacy of zinc oxide quantum dots against *Listeria monocytogenes*, *Salmonella enteritidis* and *Escherichia coli*" O157:H7,*J. Food Sci.*, Vol. 74, pp. 46–52, 2009.
59. A.T. Ahmed, F.T. Wael, M.A. Shaaban, F.S. Mohammed, "Antibacterial action of zinc oxide nanoparticle agents, foodborne pathogens", *Journal of Food Safety*, Vol. 31, Iss. 2, pp. 211–218, May 2011.
60. D. Sharma, J. Rajput, B.S. Kaith, M. Kaur, S. Sharma, "Synthesis of ZnO nanoparticles and study of their antibacterial and antifungal properties", *Thin Solid Films*, Vol. 519, p. 1224, 2010.
61. Stefania Galdiero, Annarita Falanga, Mariateresa Vitiello, Marco Cantisani, Veronica Marra, and Massimiliano Galdiero, "Silver Nanoparticles as

Potential Antiviral Agents", *Molecules*, *16*, 8894–8918; doi:10.3390/molecules16108894, **2011.**

62. M. Rai, A. Yadav, A. Gade, "Silver nanoparticles as a new generation of antimicrobials", *Biotechnol. Adv.*, Vol. 27, pp. 76–83, 2009.
63. R.V. Rai, and J.A. Bai, "Nanoparticles and their potential application as antimicrobials", in: *Communicating Current Research and Technological Advances (Microbiology Book Series - No. 3)*, Formatex Research Centre, pp. 197–209, 2011.
64. Ion Anghel, Carmen Limban, Alexandru Mihai Grumezescu, Alina Georgiana Anghel, Coralia Bleotu, and Mariana Carmen Chifiriuc, " *In vitro* evaluation of anti-pathogenic surface coating nanofluid, obtained by combining Fe_3O_4/C_{12} nanostructures and 2-((4-ethylphenoxy)methyl)-*N*-(substituted-phenylcarbamothioyl)-benzamides", Nanoscale Res Lett. 2012; 7(1): 513.
65. Wenbing Hu , Cheng Peng , Weijie Luo , Min Lv , Xiaoming Li , Di Li , Qing Huang and Chunhai Fan, "Graphene-Based Antibacterial Paper", *ACS Nano*, 2010, 4 (7), pp 4317–4323
66. I. Banerjee, D. Mondal, J. Martin, R.S. Kane, "Photoactivated antimicrobial activity of carbon nanotube-porphyrin conjugates", *Langmuir*, Vol. 26, No. 22, pp. 17369–17374, 2010
67. R.J. Chen, *et al.*, "Noncovalent functionalization of carbon nanotubes for highly specific electronic biosensors", *Proc. Natl. Acad. Sci. USA*, Vol. 100, pp. 4984–4989, 2003.
68. D. Panatarotto, C.D. Prtidos, J. Hoebeke, F. Brown, E. Kramer, J.P. Briand, S. Muller, M. Prato, A. Bianco, "Immunization with peptide-functionalized carbon nanotubes enhances virus-specific neutralizing antibody responses", *Chemistry & Biology*, Vol. 10, pp. 961–966, 2003.
69. A.A. Shvedova, V. Castranova, E.R. Kisin, D. Schwegler-Berry, A.R. Murray, V.Z. Gandelsman, *et al.*, "Exposure to carbon nanotube material: Assessment of nanotube cytotoxicity using human keratinocyte cells", *J. Toxicol. Environ. Health: A*,Vol. 66, pp. 1909–1926, 2003.
70. V.N. Popov, "Carbon nanotubes: Properties and application", *Materials Science and Engineering: R*, Vol. 43, pp. 61–102, 2004.
71. Q. Huo, "A perspective on bioconjugated nanoparticles and quantum dots", *Colloids Surfaces B:* Biointerfaces, Vol. 59, pp. 1–10, 2007.
72. Liyan Yin, Yingwen Cheng, Benjamin Espinasse, Benjamin P. Colman, Melanie Auffan, Mark Wiesner, Jerome Rose, Jie Liu, and Emily S. Bernhardt, "More than the Ions: The Effects of Silver Nanoparticles on Lolium multiflorum", *Environ. Sci. Technol.*, 45 (6), pp 2360–2367, 2011.
73. Y.-K. Jo, B.H. Kim, and G. Jung, "Antifungal activity of silver ions and nanoparticles on phytopathogenic fungi", *Plant Dis.*, Vol. 93, pp. 1037–1043, 2009.
74. G. Zhao, J. Stevens, "Multiple parameters for the comprehensive evaluation of the susceptibility of Escherichia coli to the silver ion", *Biometals*, Vol. 11, No. 1, pp. 27–32, 1998.

75. S. Alexander, K.J. Klabunde, M.R. George, M.S. Christopher, "Biocidal activity of nanocrystalline silver powders and particles", *Langmuir*, Vol. 24, pp. 7457–7464, 2008.
76. A.P. Alivisatos, *et al.*, "Organization of nanocrystal molecules using DNA", *Nature*, Vol. 382, pp. 609–611, 1996.
77. D.R. Larson, *et al.*, "Water-soluble quantum dots for multiphot on fluorescence imaging in vivo", *Science*, Vol. 300, pp. 1434–1436, 2000.
78. W.J. Parak, R. Boudreau, M.L. Gros, D. Gerion, D. Zanchet, C.M. Micheel, S.C. Williams, A.P. Alivisatos, C.A. Larabell, "Cell motility and metastatic potential studies based on quantum dot imaging of phagokinetic tracks", *Adv. Mater.*, Vol. 14, pp. 882–885, 2002.
79. R.W. Siegel, "Nanomaterials: Synthesis, properties and applications", *Ann. Rev. Mater. Sci.*, Vol. 21, pp. 559–578, 1991.
80. M.T. Swihart, "Vapour-phase synthesis of nanoparticles", *Curr. Opin. Colloid. In.*, Vol. 8, pp. 127–133, 2003.
81. Meenal Kowshik, Neelima Deshmukh, W. Vogel, J. Urban, S. K. Kulkarni and K. M. Paknikar, "Microbial synthesis of semiconductor CdS nanoparticles, their characterization, and their use in the fabrication of an ideal diode", Article *first published online: 17 APR 2002*
82. F. Ramezani, M. Ramezani, S. Talebi, "Mechanisitc aspect of biosynthesis of nanoparticles by several microbes", Olomouc, Czech Republic, EU, 2010.
83. Xiangqian Li, Huizhong Xu, Zhe-Sheng Chen, and Guofang Chen, " Biosynthesis Of Nanoparticles By Microorganisms And Their Applications", *Journal of Nanomaterials, Volume, Article ID 270974, 16 pages, 2011.*
84. Virender K. Sharma, Ria A. Yngard, Yekaterina LinVirender K. Sharma , Ria A. Yngard, Yekaterina Lin, "Silver nanoparticles: Green synthesis and their antimicrobial activities", Advances in Colloid and Interface Science 145 83–96(2009).
85. Kavitha K.S., Syed Baker, Rakshith D., Kavitha H.U., Yashwantha Rao H.C., Harini B.P. and Satish S. "Plants as Green Source towards Synthesis of Nanoparticles", *International Research Journal of Biological Sciences,* ISSN 2278–3202, Vol. 2(6), 66–76, June (2013).
86. S. Iravani, "Green synthesis of metal nanoparticles using plants", *Green Chem.*,Vol. 13, pp. 2638–2650, 2011.
87. S. Honary, H. Barabadi, E. Gharaei-Fathabad, F. Naghibi, "Green synthesis of copper oxide nanoparticles using *Penicillium aurantiogriseum, Penicillium citrinum* and *Penicillium waksmanii*," *Digest Journal of Nanomaterials and Biostructures*, Vol. 7, No.3, pp. 999–1005, 2012.
88. K. Kathiresan, S. Manivannan, M.A. Nabeel, B. Dhivya, "Studies on silver nanoparticles synthesized by a marine fungus, *Penicillium fellutanum* isolated from coastal mangrove sediment", *Digest Journal of Nanomaterials and Biostructures* Vol. 7, No. 3, July - September 2012, p. 999 – 1005
89. R. Usha, E. Prabu, M. Palaniswamy, C.K. Venil and R. Rajendran, "Synthesis of Metal Oxide Nano Particles by *Streptomyces* Sp. for Development of Antimicrobial Textiles", Global Journal of Biotechnology & Biochemistry 5 (3): 153–160, 2010

90. V. Singh, Ajay; Patil, R.; Anand, Atul; Milani, Paolo; Gade, W.N., "Biological Synthesis of Copper Oxide Nano Particles Using Escherichia coli", *Current Nanoscience,* Volume 6, Number 4, August 2010 , pp. 365–369(5).
91. Absar Ahmad, Priyabrata Mukherjee , Satyajyoti Senapati , Deendayal Mandal , M. Islam Khan, Rajiv Kumar, Murali Sastry, "Extracellular biosynthesis of silver nanoparticles using the fungus Fusarium oxysporium", *Colloids and Surfaces B: Biointerfaces* 28 (2003) 313_/318
92. S.Shivshankar, Akhilesh Rai, Absar Ahmad, Murali Shastry, "Rapid synthesis of Au, Ag and bimetallic Au core-Ag shell nanoparticle using Neem (*Azadirachta indica*) leaf broth", Journal of colloid and interface science, Vol. 275,pp. 496–80, July 2004.
93. V. Parashar, R. Prashar, B. Sharma, A.C. Pandey, "Parthenium leaf extract mediated synthesis of silver nanoparticles: A novel approach towards weed utilization", *Digest Journal of Nanomaterials and Biostructures,* Vol. 4, No. 1, pp. 45–50, 2009.
94. Ankamwar, Balaprasad, Damle, Chinmay, Ahmad, Absar, Shastry, Murali, "Biosynthesis of gold and silver nanoparticle using *Emblica officinalis* fruit extract, Their phase transfer and transmetallation in an organic solution", Journal of Nanoscience and Nanotechnology, Vol. 5, No. 10, pp. 1665–1671, oct.2005.
95. C.Krishnaraj, E.G. Jagan, S. Rajashekhar, P.Selvakumar, P.T. Kalaichelnan, N. Mohan, "Synthesis of silver nanoparticles using *Achlypha indica* leaf extracts and its antibacterial activity against water borne pathogens", Colloids and Surfaces. B: Biointerfaces 76, pp.50–56, 2010.
96. S.J. Stohs, D. Bagchi, "Oxidative mechanisms in the toxicity of metal ions", *Free Radic. Biol. Med.*, Vol. 18, pp. 321–36, 1995.
97. A.M. Schrand, M.F. Rahman, S.M. Hussain, J.J. Schlager, D.A. Smith, A.F. Syed, "Metal-based nanoparticles and their toxicityassessment", *Nanomed. Nanobiotechnol.*, Vol. 2, pp. 544–568, 2010.
98. G. Oberdörster, E. Oberdörster, J. Oberdörster, "Nanotoxicology: An emerging discipline evolving from studies of ultrafine particles", *Environ. Health Perspect*, Vol. 113, pp. 823–839, 2005b.
99. C.A. Pope, "Particulate air pollution, C-reactive protein, and cardiac risk", *Eur. Heart J.*,Vol. 22, pp 1149–1150, 2001.
100. P.D. Bragg, D.J. Rainnie, "The effect of silver ions on the respiratory chain of *Escherichia coli*", *Can. J. Microbiol.*, Vol. 228, pp. 883–889, 1974.
101. S.S. Salvi, C. Nordenhall, A. Blomberg, B. Rudell, J. Pourazar, F.J. Kelly, *et al.*, "Acute exposure to diesel exhaust increases IL-8 and GRO-alpha production in healthy human airways", *Am. J. Respir. Crit. Care Med.*,Vol. 161, pp. 550–557, 2000.
102. A.L. Semeykina, and V.P. Skulachev, "Submicromolar Ag^+ increases passive Na^+ permeability and inhibits the respiration supported formation of Na^+ gradient in *Bacillus* FTU vesicles", *FEBS Lett.*, Vol. 269, pp. 69–72, 1990.

103. R.P.F. Schins, A. McAlinden, W. MacNee, L.A. Jimenez, J.A. Ross, K. Guy, *et al.*, "Persistent depletion of I kappa B alpha and interleukin-8 expression in human pulmonary epithelial cells exposed to quartz particles", *Toxicol. Appl. Pharmacol.*, Vol. 167, pp. 107–117, 2000.
104. A. Peters, C.A. Pope III, "Cardiopulmonary mortality and air pollution", *Lancet*, Vol. 360, pp. 1184–1185, 2002.
105. A. Radomski, P. Jurasz, D. Alonso-Escolano, M. Drews, M. Morandi, T. Malinski, *et al.*, "Nanoparticle-induced platelet aggregation and vascular thrombosis", *Br. J. Pharmacol.*,Vol. 146, pp. 882–893, 2005.
106. A. Radomski, P. Jurasz, D. Alonso-Escolano, M. Drews, M. Morandi, T. Malinski, *et al.*, "Nanoparticle-induced platelet aggregation and vascular thrombosis", *Br. J. Pharmacol.*,Vol. 146, pp. 882–893, 2005.
107. D. Panatarotto, C.D. Prtidos, J. Hoebeke, F. Brown, E. Kramer, J.P. Briand, S. Muller, M. Prato, A. Bianco, "Immunization with peptide-functionalized carbon nanotubes enhances virus-specific neutralizing antibody responses", *Chemistry & Biology*, Vol. 10, pp. 961–966, 2003.
108. C.A. Pope, "Particulate air pollution, C-reactive protein, and cardiac risk", *Eur. Heart J.*,Vol. 22, pp 1149–1150, 2001.
109. P. Jani, G.W. Halbert, J. Langridge, A.T. Florence, "Nanoparticle uptake by the rat gastrointestinal mucosa: Quantitation and particle size dependency", *J. Pharm. Pharmacol.*,Vol. 42, pp. 821–826, 1990.
110. A.A. Shvedova, V. Castranova, E.R. Kisin, D. Schwegler-Berry, A.R. Murray, V.Z. Gandelsman, *et al.*, "Exposure to carbon nanotube material: Assessment of nanotube cytotoxicity using human keratinocyte cells", *J. Toxicol. Environ. Health: A*,Vol. 66, pp. 1909–1926, 2003.

13

Gas Barrier Properties of Biopolymer-based Nanocomposites: Application in Food Packaging

Sarat Kumar Swain

Department of Chemistry, Veer Surendra Sai University of Technology, Burla, Sambalpur, Odisha, India

Abstract

A series of biopolymer- and protein-based nanobiocomposites have been prepared with reinforcement of different nanoparticles such as silicon carbide (SiC), boron nitride (BN), carbon nanotubes and nanoclays. The bionanocomposites were prepared by solution technique at various wt% of nonomaterials and Fourier transform infrared spectroscopy (FTIR) and field emission scanning electron microscopy (FESEM) were used for characterizing them. The mechanical, thermal, rheological, biodegradable, chemical resistance and oxygen barrier properties of biopolymer-based nanocomposites were studied. In this chapter, gas barrier properties of bionanocomposites were investigated for use in food packaging applications. A substantial reduction in oxygen permeability was achieved due to uniform distrubution of nanoparticles with biopolymer matrix. Since nanoparticles have a higher aspect ratio than their microscale counterparts, gas barrier properties were improved. The well-dispersed nanomaterials within the biopolymer- and protein-based matrixes created huddles for gas penetration. The theories of substantial reduction in gas permeability of nanocomposites due to insertion of nanofillers are studied. Also, antimicrobial properties and detection of gas produced by food spoilage are discussed.

Keywords: Proteins, nanocomposites, aspect ratio, permeability, antimicrobial

**Corresponding author*: swainsk2@yahoo.co.in

Ashutosh Tiwari and Mikael Syväjärvi (eds.) Advanced Materials for Agriculture, Food, and Environmental Safety, (369–384) 2014 © Scrivener Publishing LLC

13.1 Introduction

A nanocomposite is a multiphase material derived from the combination of two or more components, including a matrix (continuous phase) and a discontinuous nanodimensional phase with at least one nanosized dimension (i.e., with less than 100 nm). The nanodimensional phase can be divided into three categories according to the number of nanosized dimensions. Nanospheres or nanoparticles have the three dimensions on the nanoscale. Both nanowhiskers (nanorods) and nanotubes have two nanometric dimensions, with the difference that nanotubes are hollow, while nanowhiskers are solid. Finally, nanosheets or nanoplatelets have only one nanosized dimension. Most nanosized phases have a structural role, acting as reinforcements to improve the mechanical properties of the matrix (usually a polymer), since the matrix transfers the tension to the nanoreinforcement through the interface. Nanoreinforcements are especially useful for biopolymers, because of their usually poor performance when compared to conventional petroleum-based polymers. The incorporation of nanosized reinforcements into biopolymers may open new possibilities for improving not only their properties but also their cost-price efficiency.

Nowadays, most materials used for packaging are practically undegradable, representing a serious global environmental problem. New bio-based materials have been exploited to develop edible and biodegradable films in a big effort to extend shelf life and improve quality of food while reducing packaging waste [1]. However, the use of edible and biodegradable polymers has been limited because of problems related to performance, processing and cost. Starch, for example, has received considerable attention as a biodegradable thermoplastic polymer. However, it has a poor performance by itself because of its water sensitivity and limited mechanical properties with high brittleness, which is related to the anarchical growth of amylase crystals with time [2]. The application of nanotechnology to these polymers may open new possibilities for improving not only the properties but also the cost-price efficiency. The main advantage of plastics compared with other packaging materials is that they are lightweight, can be processed quickly and easily and offer design and printing flexibility. However, petroleum resources are finite and fuel prices have recently been escalating. In addition, the end-of-life management of plastics has also depleted the world's rights for sustainability. Hence, the emergence of biopolymers as an alternative material option has been promising.

The majority of materials currently used for packaging are nondegradable, creating environmental problems. Several biopolymers and proteins have been exploited to develop materials for eco-friendly packaging materials. However, the use of biopolymers/proteins have been limited because of their usually poor mechanical and barrier properties, which may be enhanced by adding discontinuous phased reinforcing compounds, forming composites. Most reinforced materials have poor matrix–filler interactions, which tend to improve depending on filler dimensions. When fillers of nanodimension are used with bulk materials then the hybrid materials are known as nanocomposites.

Several composites [3] have been developed by adding reinforcing compounds to biopolymers/proteins to enhance their thermal, mechanical and barrier properties. Most of these reinforced materials present have poor interactions at the interface of both components. Macroscopic reinforcing components usually contain defects, which become less important as the particles of the reinforcing component are smaller. Biopolymers are polymers that are produced by living organisms and represent a special class of polymers that are biodegradable in nature [4]. They are produced from plants from the agricultural non-food crops directly and by means of fermentation, hence rendering them renewable. In another aspect, use of biopolymers promotes organic recycling either through composting, anaerobic digestion or by biomass conversion. These benefits have contributed to the rising significance of biopolymers. Some examples of biopolymers include polylactide acid (PLA), polyhydroxy alkanoate (PHA), cellulose and starch-based polymers or thermoplastic starch (TPS). Despite the biodegradability of biopolymers, there exist several factors that limit their applications as packaging materials. First, their mechanical properties are relatively poor compared to many petroleum-based plastics due to their inherent lower stiffness and strength, as well as brittleness. Second, many are relatively sensitive to water, with some materials dissolving rapidly, or have a substantial decrease in mechanical performances when they absorb water, especially in moist environments.

The introduction of nanofillers should be explored for sustainable packaging as an improvement in properties could take place with the use of less raw materials, thus conserving resources [5]. This, in conjunction with degradability of biopolymer, could impart ease in waste management as well. Hence, several research groups started the preparation and characterization of various kinds of biodegradable polymers/nanofiller systems. To date, the work carried out on biopolymeric nanocomposites has been focused mainly on the designing and making of biopolymeric nanocomposites by selecting an appropriate synthetic method and adjusting their structures and compositions [6].

13.2 Experimental

A series of bionanocomposites were prepared with variable wt% of nanofillers by using the solution technique. The nanoparticles were dispersed in double-distilled water by continuous stirring at 60°C for 30 minutes and then the solution was treated with ultrasound (120W/180KHz) for 30 minutes. Then the biopolymer solution prepared in distilled water was added to different wt% filler solutions. The viscous product obtained was filtered and washed with double-distilled water. The bionanocomposites obtained were dried in an oven for 24 hours. at a temperature of 50°C.

The functionalization of MWCNT was done as per our earlier publication [22]. The MWCNTs were treated with concentrated H_2SO_4 and HNO_3 in the volume ratio of 3:1 and sonicated by using ultrasonic cleaner (120W/60KHz) at 40°C for 24 hours in a flask. The solution was diluted by distilled water and filtered. The remaining residue was washed by distilled water. Then the open ended tubes were polished with hydrogen peroxide and H_2SO_4 in volume ratio of 1:4 with stirring at 70°C for thirty minutes. The resulting solution was diluted by distilled water and centrifuged to get f-MWCNT.

Starch/MWCNT nanocomposites film was prepared by a convenient solution casting and evaporation method. The f-MWCNTs were dispersed in double-distilled water with stirring for ten minutes taking in a flask. Calculated amount of glycerol was added and homogenized in an ultrasonic bath for 30 minutes. The required quantity of starch was added into the flask and the solution was charged by constant stirring with heating at 95°C for 30 minutes to form the plasticized starch. Oxygen permeability of the bionanocomposite was measured with STM F 316-86 by using oxygen permeation analyzer (PMI instrument, model GP-201-A, Texas, NY, USA). For testing oxygen permeability the synthesized powdered bionanocomposites were converted into films of 5 mm thickness with the help of a polymer press at a pressure of 9 tons. The results were recorded as average of the values obtained from five same samples.

13.3 Objective

Even though research activities have intensified over the last decade, there is still a limited amount of literature on the applications of biopolymeric nanocomposites-based primary, secondary and tertiary packaging. This indicates that an understanding on such bio-based nanocomposites as sustainable packaging materials is very much still in its infancy and much is needed to assess their performance and potential. In the present study, the sustainability of biopolymer-based nanocomposites for packaging applications has been discussed along with their theories and gas permeability properties. The main objective of the present review is to compare different kinds of nanofillers suitable for various biopolymers in order to enable the materials to be used for food packaging materials. The oxygen permeability of nanocomposites of different fillers was compared with the virgin polymer/biopolymers.

13.4 Background of Food Packaging

13.4.1 Oxygen Penetration

Oxygen (O_2) is responsible for the deterioration of many foods either directly or indirectly. Food deterioration by indirect action of O_2 includes food spoilage by aerobic microorganisms [7]. The incorporation of O_2 scavengers into food packaging can maintain very low O_2 levels, which is useful for several applications. In particular, attention has been focused on the photocatalytic activity of nanocrystalline titania (TiO_2) under ultraviolet radiation. Industries estimate the food expiration date, taking into consideration the distribution and storage conditions to which the food product is predicted to be exposed. However, it is known that such conditions are not always the real ones, and foods are frequently exposed to temperature abuse; this is especially worrying for products which require a cold chain. Moreover, micropores or sealing defects in packaging systems can lead to an unexpected high exposure of oxygen in food products, which can result in undesirable changes. When integrated into food packaging, nanosensors can detect certain chemical compounds, pathogens and toxins in food [8].

Food spoilage is also caused by microorganisms, whose metabolism produces gases which can be detected by conducting polymer nanocomposites (CPC) or metal oxides. This can be used for quantification and/or identification of microorganisms based on their gas emissions. Sensors based on CPC consist of conducting particles embedded into an insulating polymer matrix; the resistance changes of the sensors produce a pattern that corresponds to the gas under investigation. Oxygen allows aerobic microorganism to grow during food storage [9]. There has been increasing interest in developing nontoxic and irreversible oxygen sensors to assure oxygen absence in oxygen-free food packaging systems, such as packaging under vacuum or nitrogen [10].

The use of biopolymers by the food industry has faced feasibility problems related mainly to their relatively high cost and poor overall performance when compared to those of synthetic polymers [11]. However, since industries are concerned with sustainable development, the production cost of biopolymers has decreased, allowing biopolymer-based materials to be increasingly developed. More importantly, nanocomposites promise to expand the use of edible and biodegradable films, since the addition of nanoreinforcements has been related to improvements in overall performance of biopolymers, enhancing their mechanical, thermal and barrier properties, usually even at very low contents. Thus, nanoparticles have an important role in improving feasibility of the use of biopolymers for several applications, including food packaging [12].

However, there are many safety concerns about nanomaterials, as their size may allow them to penetrate into cells and eventually remain in the system. There is no consensus about categorizing nanomaterials as new materials. On one hand, the properties and safety of the materials in its bulk form are usually well known, but the nanosized counterparts frequently exhibit different properties from those found at the macroscale. There is limited scientific data about migration of most types of nanoparticles (NPs) from the packaging material into food, as well as their eventual toxicological effects. It is reasonable to assume that migration may occur, hence the need for accurate information on the effects of NPs to human health following chronic exposure is imperative [13].

13.4.2 Antimicrobial Systems

Antimicrobial food packaging systems have received considerable attention since they help control the growth of pathogenic and spoilage microorganisms on food surfaces, where microbial growth predominates. Antimicrobial nanocomposite systems are particularly interesting, since materials in the nanoscale range have a higher surface-to-volume ratio when compared with their microscale counterparts. Nanomaterials are thus more efficient, since they are able to attach more copies of microbial molecules and cells. Nanoscale materials have been investigated for antimicrobial activity as growth inhibitors or killing agents [14].

Silver is well known for its strong toxicity to a wide range of microorganisms, besides some processing advantages such as high temperature stability and low volatility. Silver nanoparticles have been shown to be effective antimicrobials, even more effective than larger silver particles, thanks to their larger surface area available for interaction with microbial cells. In fact, the most common nanocomposites used as antimicrobial films for food packaging are based on silver nanoparticles, whose antimicrobial activity has been ascribed to different mechanisms, namely: (a) adhesion to the cell surface, degradation of lipopolysaccharides and formation of "pits" in the membranes, largely increasing permeability; (b) penetration inside bacterial cell, damaging DNA, and; (c) releasing antimicrobial Ag^+ ions by dissolution of silver nanoparticles. Besides their antimicrobial activity, silver nanoparticles have been reported to absorb and decompose ethylene, which may contribute to their effects on extending shelf life of fruits and vegetables [15].

Titanium dioxide (TiO_2) is widely used as a photocatalytic disinfecting material for surface coatings. Titanium dioxide photocatalysis, which promotes peroxidation of the phospholipids present in microbial cell membranes, has been used to inactivate food-related pathogens. A TiO_2 powder-coated packaging film able to reduce *E. coli* contamination on food surfaces has been developed; and the efficacy of TiO_2-coated films exposed to sunlight to inactivate fecal coliforms in water has been demonstrated. Metal doping improves visible light absorbance of TiO_2, and increases its photocatalytic activity under UV irradiation. It has been demonstrated that doping TiO_2 with silver greatly improved photocatalytic bacterial inactivation. This combination was explored by Hulleman *et al.*, who have obtained effective antibacterial activity from a polyvinyl chloride nanocomposite with TiO_2/Ag+ nanoparticles [16].

Carbon nanotubes have also been reported to have antibacterial properties. Direct contact with aggregates of carbon nanotubes have been demonstrated to kill *E. coli*, possibly because the long and thin nanotubes puncture microbial cells, causing irreversible damage and leakage of intracellular material. On the other hand, there are studies suggesting that carbon nanotubes may also be cytotoxic to human cells, at least when in contact to skin, which would affect people manipulating the nanotubes in processing stages rather than consumers [17]. Anyway, once present in the food packaging material, the nanotubes might eventually migrate into food. Then, it is mandatory to know any eventual health effects of ingested carbon nanotubes.

13.4.3 Detection of Gases Produced by Food Spoilage

Food spoilage is caused by microorganisms whose metabolism produces gases, which may be detected by several types of gas sensors which have been developed to translate chemical interactions between particles on a surface into response signals. Nanosensors to detect gases are usually based on metal oxides or, more recently, conducting polymer nanocomposites, which are able to quantify and/or identify microorganisms based on their gas emissions. Sensors based on conducting polymers (or electroactive conjugated polymers) consist of conducting particles embedded into an insulating polymer matrix [18]. The resistance changes of the sensors produce a pattern corresponding to the gas under investigation. Conducting polymers are very important because of their electrical, electronic, magnetic and optical properties, which are related to their conjugated π electron backbones. Polyene and polyaromatic conducting polymers, such as polyaniline, polyacetylene and polypyrrole, have been widely studied. Electrochemically polymerized conducting polymers have a remarkable ability to switch between conducting oxidized (doped) and insulating reduced (undoped) states, which is the basis for several applications. Nanosensors containing carbon black and polyaniline have been developed which have been demonstrated to be able to detect and identify three foodborne pathogens by producing a specific response pattern for each microorganism.

13.4.4 Different Fillers for Nanocomposites

13.4.4.1 Nanoclay as Fillers

Nanoclays have been the most studied nanofillers due to their high availability, low cost, good performance and good processability. The clays for nanocomposites usually are bidimensional platelets with very tiny thicknesses (frequently around 1 nm) and several micrometers in length. In contrast with the typical tactoid structure of microcomposites (conventional composites), in which the polymer and the clay tactoids remain immiscible, the interaction between layered silicates and polymers may produce two types of nanoscale composites, namely: intercalated nanocomposites, which result from penetration of polymer chains into the interlayer region of the clay, producing an ordered multilayer structure with alternating polymer/inorganic layers, and exfoliated nanocomposites, which involve extensive polymer penetration, with the clay layers delaminated and randomly dispersed in the polymer matrix. Exfoliated nanocomposites have been reported to exhibit the best properties due to their optimal clay-polymer interactions.

The hydrophilicity of the surface of most clays make their dispersion in organic matrices difficult. Organoclays, produced by interactions of clays and organic compounds, have found an important application in polymer nanocomposites. An adequate organophilization is essencial for successful exfoliation of clays in most polymeric matrices, since organophilization

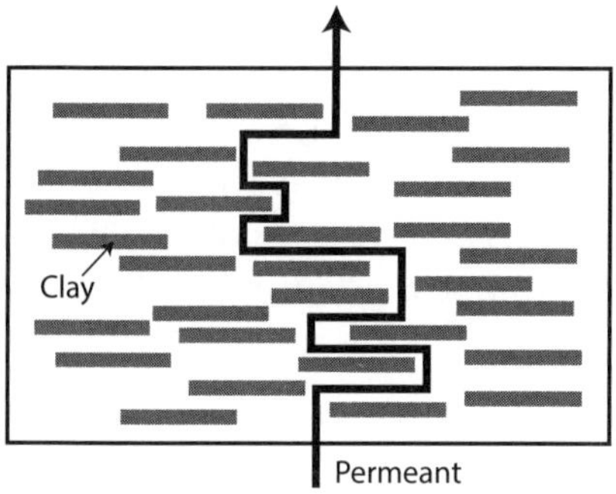

Figure 13.1 Tortuous path of a permeant in a clay-based nanocomposite.

reduces the energy of clays and improves their compatibility with organic polymers. Organo-montmorillonite (MMT) has been produced, for example, by exchanging inorganic cations of MMT with organic ammonium ions, improving compatibility of MMT with organic polymers, leading to a more regular organization of the layers, and decreasing the water uptake by the resulting nanocomposite.

The most widely known theories to explain the improved barrier properties of polymer–clay nanocomposites are based on a theory developed by Nielsen (Figure 13.1), which focuses on a tortuous path around the clay plates, forcing the gas permeation to travel a longer path to diffuse through the film. The increase in path length is a function of the high aspect ratio of the clay filler and the vol% of the filler in the composite. Nielsen's model predicts permeability of systems at clay loading rates of less than 1%, but experimental data deviate significantly from predicted values at higher loading rates and more extensively in certain polymers.

The improved barrier properties of polymer-clay nanocomposites seem to be due to an increased tortuosity of the diffusive path for permeants, forcing them to travel a longer path to diffuse through the film. The increase in path length is a function of the aspect ratio of the clay and the volume fraction of the filler in the composite. Nielsen's model has been used effectively to predict permeability of systems at clay loadings of less than 1%, but some experimental data have reported much lower permeabilities than predicted at higher loadings, and a new proposed model to predict permeability of nanocomposites focused on the polymer-clay interface as an additional governing factor to the tortuous path, thus providing a correction factor to Nielsen's model.

The oxygen permeability of polyacrylonitrile (PAN)/clay nanocomposites was measured with variation of clay concentration along with different frequencies and powers of ultrasound waves [19]. It was found that oxygen permeability of nanocomposites decreased substantially with an increase in clay loading (Figure 13.2). It was marked that the oxygen permeability was further reduced with an increase in power and frequency of ultrasound. This may be due to the removal of microvoids by the application of ultrasound.

The oxygen permeability of the virgin polymethylmethacrylate (PMMA) and PMMA/clay nanocomposites can be studied from Figure 13.3. The

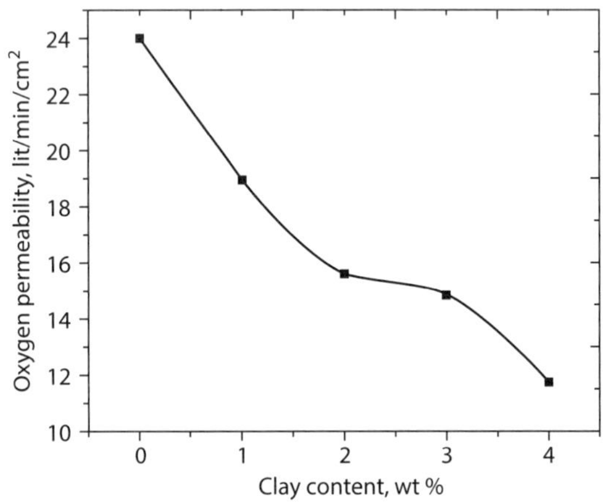

Figure 13.2 Oxygen permeability of PAN/clay nanocomposites as a function of clay content at 5 psi pressure [19].

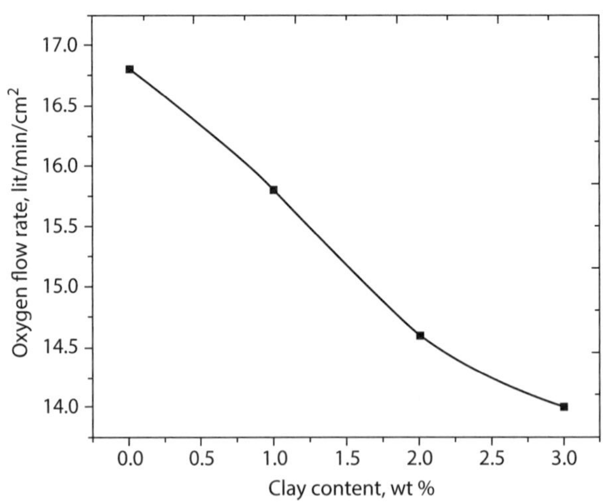

Figure 13.3 Oxygen flow rate of PMMA and PMMA/clay nanocomposites with variation of clay content at a constant pressure (5 psi) [20].

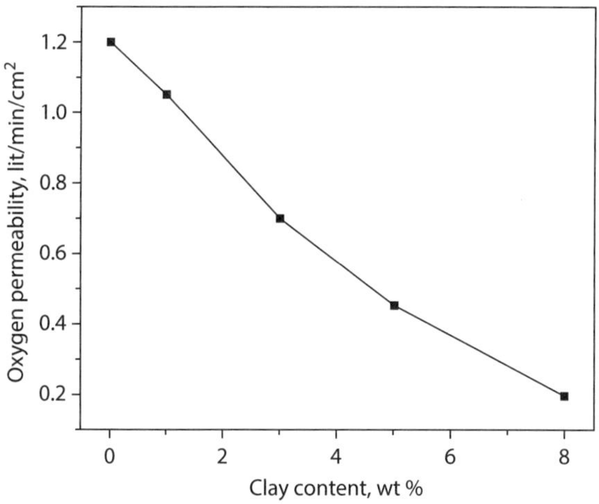

Figure 13.4 Oxygen permeability values of the soy/clay bionanocomposites as a function of clay weight percent at a pressure of 5 psi [21].

oxygen flow rate through all the nanocomposites was observed to be less in comparison to the virgin PMMA at different pressures up to 5 psi; the flow rate was found to decrease with an increase in percentage clay loading. The flow rate of PMMA/clay nanocomposites was reduced by 17% as compared to virgin PMMA at 3% clay concentration. This is due to the tortuous path created as a result of exfoliation of clay during sonication [20].

The oxygen flow rate of the soy/clay bionanocomposites was found to be decreased in proportion to clay loading of 2% (Figure 13.4). At 8% clay concentration, the oxygen permeability was reduced by 6 times as compared to the virgin protein. This is because the clay nanoparticles act as physical obstacle, retarding the movement of the gas [21].

13.4.4.2 Carbon Nanotube as Fillers

Carbon nanotubes may consist of a one-atom thick single-wall nanotube, or a number of concentric tubes called multiwalled nanotubes, having extraordinarily high aspect ratios and elastic modulus. Several polymers have been found to have their tensile strength/modulus improved by addition of carbon nanotubes, such as polyethylene naphtalate, polyvinyl alcohol, polypropylene and a polyamide. The polylactic acid not only had its tensile properties improved by carbon nanotubes, but also had its water vapor transmission rate decreased by 200%.

The oxygen permeability of virgin PMMA and PMMA/functionalized multiwalled carbon nanotube (f-MWCNT) nanocomposites has been

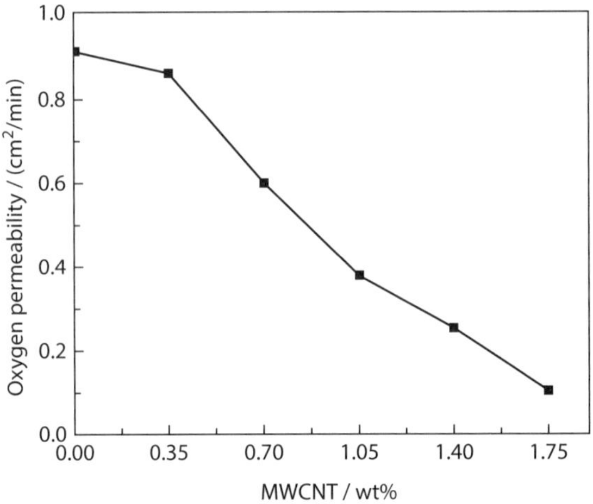

Figure. 13.5 Oxygen permeability of PMMA/f-MWCNT nanocomposites at constant pressure of (0.5 psi) [22].

studied [22], as shown in Figure 13.5. The oxygen permeability of PMMA/f-MWCNT nanocomposites with 1.75 wt% of MWCNTs loading is about eight times less than that of virgin PMMA. The reduction of permeability arises from the longer diffusive path of the penetration of the oxygen in the presence of MWCNTs. The incorporation of MWCNT in PMMA matrix is particularly excellent at maximizing the path length due to the high aspect ratio. Furthermore, the presence of MWCNT introduces a torturous path for which the oxygen travels longer diffusive path. Hence oxygen permeability of PMMA/MWCNT nanocomposites is less than that of PMMA matrix. The tremendous decrease in oxygen permeability with increasing MWCNT (wt%) is due to good dispersion of MWCNTs in polymer matrix of PMMA/f-MWCNT composites.

13.4.4.3 Other Nanomaterials

Silica nanoparticles ($nSiO_2$) have been reported to improve tensile properties of polypropylene, starch, starch/polyvinyl alcohol, besides decreasing water absorption by starch and improving oxygen barrier of polypropylene prepared nanocomposites of polyvinyl alcohol with $nSiO_2$ by radical copolymerization of vinyl silica nanoparticles and vinyl acetate. The nanocomposites had improved thermal and mechanical properties when compared to the pure polyvinyl alcohol, due to strong interactions between $nSiO_2$ and the polymer matrix via covalent bonding.

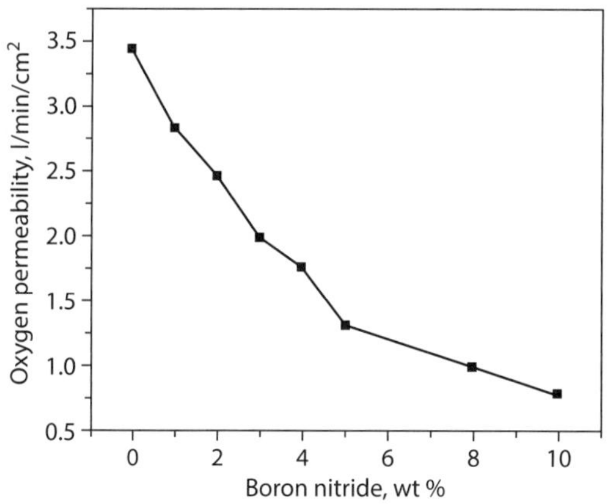

Figure 13.6 Oxygen permeability of the composites as a function of boron nitride content at a pressure of 0.010 MPa [23].

Polymer-based composites which have been shown to provide barrier to oxygen are being studied for packaging applications. The oxygen permeability of virgin chitosan and chitosan/BN composites can be seen in Figure 13.6. As chitosan is a porous polymer, the dispersion of boron nitride in chitosan matrix may provide the huddles for oxygen entrance, whereas virgin chitosan may have voids for oxygen permeation [23]. Oxygen permeability of chitosan/BN composites was conducted to measure the effect of boron nitride concentrations on the chitosan matrix. The oxygen flow rate through all the composites was observed to be less in comparison with the virgin chitosan at pressure 0.010 MPa (Figure 13.6). It was found that the flow rate decreased with an increase in percentages of boron nitride loading. The substantial reduction in oxygen permeability may be due to the dispersion of boron nitride within the chitosan matrix. The oxygen flow rates through all composites were also observed to be less in comparison with the virgin chitosan at different pressures up to 0.013 MPa.

Oxygen permeability of the synthesized cellulose/BN nanobiocomposites was conducted to measure the effect of boron nitride concentrations on oxygen barrier properties of the cellulose matrix [24]. The oxygen flow rate through all the nanobiocomposites was observed to be less in comparison to the virgin cellulose at constant pressures of 2 psi (Figure 13.7a). It was found that the flow rate was decreased with an increase in percentage of boron nitride loading. The remarkable reduction in oxygen permeability was due to nanostructure dispersion of boron nitride within the cellulose matrix at difference pressure up to 2 psi (Figure 13.7b).

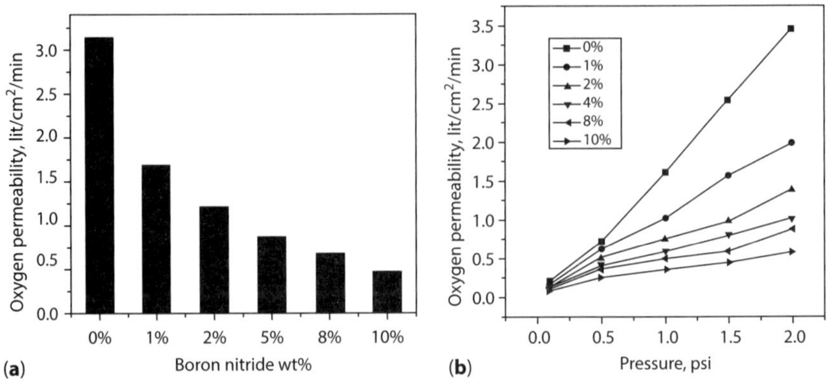

Figure 13.7 (a) The oxygen permeability of cellulose/BN nanobiocomposites with various loading of BN at constant pressure of 2 psi, and (b) the oxygen permeability of cellulose/BN nanobiocomposites with various loading of BN at difference pressure [24].

13.5 Conclusion

Bionanocomposites were prepared by solution technique with dispersion of reinforced nanomaterials with biopolymer matrix. This technology was developed to improve barrier performance to gases such as oxygen. It also enhances the barrier performance to ultraviolet rays, as well as adding strength, stiffness, dimensional stability, and heat resistance. New plastics created with this technology demonstrate an increased shelf life and are less likely to shatter. Once perfected, these plastics will offer these improved characteristics at competitive prices. It will also make them attractive for use in food and beverage packaging and pharmaceutical packaging applications.

References

1. A. Samir, F. Alloin, J.Y. Sanchez, A. Dufresne, Cellulose nanocrystals reinforced poly(oxyethylene), *Polymer*, Vol. 45, pp. 4149–4157, 2004.
2. K.M. Dean, M.D. Do, E. Petinakis, L. Yu, Key interactions in biodegradable thermoplastic starch/poly(vinyl alcohol)/montmorillonite micro- and nanocomposite, *Comp. Sci. Tech.*, Vol. 68, No. 6, pp. 1453–1462, 2008.
3. J.K. Pandey, A.P. Kumar, M. Misra, A.K. Mohanty, L.T. Drzal, R.P. Singh, Recent advances in biodegradable nanocomposites, *J. Nanosci. Nanotech.*, Vol. 5, No. 4, pp. 497–525, 2005.

4. R.X. Zhao, P. Torley, P.J. Halley, Emerging biodegradable materials: Starch- and protein-based bio-nanocomposites, *J. Mat. Sci.*, Vol. 43, No. 9, pp. 3058–3071, 2008.
5. A. Sorrentino, G. Gorrasi, M. Tortora, V. Vittoria, U. Costantino, F. Marmottini, F. Padella, Incorporation of Mg-Al hydrotalcite into a biodegradable Poly(ε-caprolactone) by high energy ball milling, *Polymer*, Vol. 46, Iss. 5, pp. 1601–1608, 2005.
6. S.S. Ray, M. Bousmina, Biodegradable polymers and their layered silicate nanocomposites: In greening the 21st century materials world, *Prog. Mats. Sci.*, Vol. 50, Iss. 8, pp. 962–1079, 2005.
7. L. Gram, L. Ravil, M. Rasch, J.B. Bruhn, A.B. Christense, and M. Givskov, Food spoilage– Interactions between food spoilage bacteria, *International Journal of Food Microbiology*, Vol. 78, Iss. 1–2, pp. 79–97, 2002.
8. A. Sorrentino, G. Gorrasi, V. Vittoria, Potential perspectives of bio-nanocomposites for food packaging applications, *Trends Food Sci. Tech.*, Vol. 18, Iss. 2, pp. 84–95, 2007.
9. J.W. Rhim, S.I. Hong, H.M. Park, P.K.W. Ng, Preparation and characterization of chitosan-based nanocomposite films with antimicrobial activity, *J. Agric. Food Chem.*, Vol. 54, No. 16, pp. 5814–5822, 2006.
10. J.W. Rhim, P.K.W. Ng, Natural biopolymer-based nanocomposite films for packaging applications, *Crit. Rev. Food Sci. Nutr.*, Vol. 47, Iss. 4, pp. 411–433, 2007.
11. H.J. Sue, K.T. Gam, N. Bestaoui, N. Spurr, A. Clearfield, Epoxy nanocomposites based on the synthetic α-zirconium phosphate layer structure, *Chem. Mater.*, Vol. 16, Iss. 2, pp. 242–249, 2004.
12. H.F. Zobel, Molecules to granules: A comprehensive starch review, *Starch*, Vol. 40, Iss. 2, pp. 44–50, 1988.
13. G.T. Oostergetel, and E.F.J. van Bruggen, On the origin of a low angle spacing in starch, *Starch*, Vol. 41, Iss. 9, pp. 331–335, 1989.
14. A. Imberty, and S. Perez, A revisit to the three-dimensional structure of B-type starch, *Biopolymers*, Vol. 27, Iss. 8, pp. 1205–1221, 1988.
15. H.F. Zobel, Starch crystal transformations and their industrial importance, *Starch*, Vol. 40, Iss. 1, pp. 1–7, 1988.
16. S.H.D. Hulleman, F.H.P. Janssen, H. Feil, The role of water during plasticization of native starches, *Polymer*, Vol. 39, Iss. 10, pp. 2043–2048, 1998.
17. J.J.G. van Soest, N. Knooren, Influence of glycerol and water content on the structure and properties of extruded starch plastic sheets during aging, *J. Appl. Polym. Sci.*, Vol. 64, No. 7, pp. 1411–1422, 1997.
18. H.G. Bader, D. Göritz, Investigations on high amylose corn starch films. Part 3: Stress strain behaviour, *Starch*, Vol. 46, Iss. 11, pp. 435–439, 1994.
19. S.K. Swain, S.K. Patra, Ultrasonic and viscometric study of synthesized PAN/clay nanocomposites, *Int. J. Polym Mater.*, Vol. 60, Iss. 12, pp. 559–566, 2011.
20. S.K. Patra, G. Prusty, and S.K. Swain, Ultrasound assisted synthesis of PMMA/clay nanocomposites: Study of oxygen permeation and flame retardant properties, *Bull. Mater. Sci.*, Vol. 35, pp. 27–32, 2012.

21. S.K. Swain, S.K. Patra, and P. Priyadarshini, Soyprotein/clay nanobiocomposites for ideal packaging materials, *Polymer-Plastic Technology and Engineering*, Vol. 51, Iss. 12, pp. 128–1287, 2012.
22. A.K. Pradhan, S.K. Swain, Oxygen barrier multiwalled carbon nanotube/polymethyl methacrylate nanocomposites prepared by in situ method, *J. Mater Sci. Tech.*, Vol. 28, No. 5, pp. 391–395, 2012.
23. S.K. Kisku, S.K. Swain, Synthesis and characterization of chitosan/bn composites, *Journal of the American Ceramic Society*, Vol. 95, Iss. 9, pp. 2753–2757, 2012.
24. S.K. Swain, S. Dash, C. Behera, S.K. Kisku, and L. Behera, Effect of nano BN on the thermal properties of cellulose, *Carbohydrate Polymers*, Vol. 95, pp. 728–732, 2013.

14

Application of Zero-valent Iron Nanoparticles for Environmental Clean Up

Ritu Singh[*,1,2] **and Virendra Misra**[2]

[1]*Department of Environmental Science, School of Earth Sciences, Central University of Rajasthan, Ajmer, Rajasthan, India*
[2]*Environmental Toxicology Division, CSIR-Indian Institute of Toxicology Research (IITR), Lucknow, Uttar Pradesh, India*

Abstract

The contamination of soil and groundwater through a variety of toxic/hazardous compounds has become a global environmental problem and is a great challenge before the scientific community of the world. In response to the continuously increasing need to address the problems of environmental contamination, several remediation technologies/methods have been investigated. In the advancement of the search for potential remediation technologies, nanotechnology has introduced a new dimension to the area of environmental clean up. Nanoparticles mediated remediation could be considered as an effective alternative to the current practices of site remediation. Their extremely small size and high surface area to volume ratio impart them properties which can be exploited for degradation/reduction of hazardous/toxic wastes. Among several nanoparticles evaluated for the purpose of contaminant degradation, zero-valent iron nanoparticles (nFe^0) have received considerable attention over the last decade as an excellent electron donor with the potential ability for *in situ* remediation of large contaminated sites. In recent years, nFe^0 has been successfully applied for degradation of a wide array of contaminants including polyhalogenated compounds, chlorinated solvents, dyes, inorganic anions, heavy metals, etc. Although nFe^0 is effective against a wide array of pollutants, still there are specific issues related to the reactivity and longevity of nFe^0, their transport and fate in the environment, etc., which are of concern. The aim of this chapter is to provide an overall perspective of the use of nFe^0 to

**Corresponding author*: ritu_ens@curaj.ac.in; ritu2735@gmail.com

Ashutosh Tiwari and Mikael Syväjärvi (eds.) Advanced Materials for Agriculture, Food, and Environmental Safety, (385–420) 2014 © Scrivener Publishing LLC

decipher potential issues related to the treatment of contaminated environmental matrices (ground water, surface water, aquifers and soil).

Keywords: Zero-valent iron nanoparticles, stabilized iron nanoparticles, bimetallic iron nanoparticles, contaminants, remediation/degradation

14.1 Introduction

Nanotechnology is an area which touches almost every aspect of the modern world ranging from research application, medical field, and information technology to consumer goods. In the past two decades, nanotechnology has advanced on all fronts: pharmaceutics, engineering, electronics, optics, etc. Here the question arises; why is nanotechnology gaining so much attention? The answer to this question resides in the unique features of nanoparticles, which are associated with their extremely small size, high surface area to volume ratio and high intrinsic energies. These properties significantly increase the proportion of atoms located at the surface of nanoparticles in comparison to the bulk, which in turn is responsible for the enhanced tendency to interact with atoms, ions and molecules or the complexes present in their ambience. The magnificence of nanoparticles is that the novel phenomenon and properties (optical, magnetic, electric, physical, etc.) which arise due to their nanometer length scale could be exploited for a wide array of applications in domestic as well as industrial processes. For instance, Copper (Cu) nanoparticles smaller than 50 nm, are considered as super hard materials that exhibit malleability and ductility completely different from bulk copper. Another example is Silver (Ag), which is considered chemically inert at macroscale, whereas nanosized Ag is used as antimicrobial agent. Likewise, Gold (Au) nanoparticles display remarkably different properties (color, melting point, etc.) when compared with bulk material of Au. Similar to these, numerous examples can be found in literature which show the utility of controlling and/or manipulating the material at nanometer-length scale.

With the continuous advancement of industrialization and urbanization, the level of pollution has increased many folds in the past few decades, contaminating almost every compartment of the environment. This is affecting the health of millions of people worldwide and incessantly degrading the environmental quality. In response to the continuously increasing need to address the problem of environmental contamination, several remediation technologies/methods have been proposed and investigated in the past. In the advancement of the search for potential remediation technologies, nanotechnology has introduced a new dimension to the area of environmental clean up. In the past two decades, a nanoparticles-based remediation approach has gained significant achievements in the field of environmental clean up, suggesting its potential as a good alternative to the current practices of site remediation. As compared to conventional treatment technologies such as pump and treat, soil flushing, incineration, etc., nanoparticles offer an effective, economic and time-efficient technology. Their extremely small size and excellently high surface area make them an efficient remediation tool for both *in situ* and *ex situ* application. Among several of the nanoparticles, zero-valent metal nanoparticles show good efficiency for reducing organic and inorganic contaminants. For instance, Zn^0, Cu^0, Fe^0, Mg^0, Pd^0, etc., have been reported for the removal of various contaminants from soil, sediments and groundwater [1–3]. Being inexpensive and environmentally benign, zero-valent iron (Fe^0) is the most widely studied nanoparticle for the treatment of environmental contaminants. The reducing ability of Fe^0 was first demonstrated in 1994 by Gillham and O'Hannesin [4] while studying the reduction of chlorinated compounds. Since then, a huge effort has been made to test the potential of Fe^0 for reducing toxic/hazardous substances such as chlorinated hydrocarbons, chlorinated solvents, pesticides, dyes, heavy metals, etc., as indicated by the number of related studies published after 1994 [5–10].

With the introduction of nanotechnology in the area of environmental clean up, attempts have been made to explore the potential of nanoscale zero-valent iron (hereafter referred to as nFe^0) for treatment of environmental toxicants. The combination of nanosize effect and excellent reducing capabilities of Fe^0 has resulted in the emergence of a very powerful remediation tool for reducing the burden of toxic/hazardous wastes from the environment. Zero-valent iron has been successfully applied as reductant to degrade chlorinated aliphatic and alicyclic compounds, chlorinated solvents, nitroaromatic compounds, textile dyes, chlorinated pesticides, heavy metals, etc. [11–15]. Two potential advantage of nFe^0 over their micro or macro counterparts is that nFe^0 can be injected deep into groundwater and aquifers to treat contaminated plume and that the reaction kinetics of contaminant degradation is much higher. The appealing traits of this technology have led the scientific and engineering community to rapidly adopt it as a new alternative tool for remediation. The performance of nFe^0 has been displayed by several investigators in batch-scale studies. Numerous field-scale demonstrations have also been tested and completed in the past few years.

Apart from their fruitful applications of nFe^0, there is also a wide debate among the scientific community, government bodies and public, regarding its toxicity, fate and behavior in the environment and its potential impact on the ecosystem. This chapter gives an overview of existing applications of nFe^0 in facilitating waste management and reducing the environmental burden of hazardous/toxic substances, factors governing nFe^0 reactivity, practices adopted for improving the performance of nFe^0, recent progress made in laboratory studies and field-scale studies. In addition, various concerns related with the application of nFe^0 have also been discussed at the end.

14.2 Zero-Valent Iron Nanoparticles: A Versatile Tool for Environmental Clean Up

14.2.1 Iron Chemistry

Iron is the fourth most abundant element in the earth's crust which primarily exists in two valence states—one is water-soluble ferrous iron (Fe^{2+}) and the other is water-insoluble ferric iron (Fe^{3+}). The zero-valent or elemental state of iron is a highly reactive species and is rarely formed on the earth's surface. In the environment, Fe^0 undergoes rapid oxidation/corrosion owing to its high unstability. This occurs through an electrochemical process, whereby Fe^0 dissolves at anode and reduction of redox amenable species take place at cathode. Under aerobic conditions, usually dissolved oxygen is the primary electron acceptor (Eq. 14.1), whereas water predominately accepts electron under anaerobic conditions (Eq. 14.2).

$$2Fe^0_{(s)} + 4H^+_{(aq)} + O_{2(aq)} \rightarrow 2Fe^{2+} + 2H_2O_{(l)} \tag{14.1}$$

$$2Fe^0_{(s)} + 2H_2O_{(l)} \rightarrow 2Fe^{2+} + H_{2(g)} + 2OH^-_{(aq)} \tag{14.2}$$

The ferrous ions formed during corrosion undergo further oxidation to form ferric ions (Eqs. 14.3 and 14.4);

$$2Fe^{2+}_{(s)} + 2H^+_{(aq)} + 1/2O_{2(aq)} \rightarrow 2Fe^{3+} + H_2O_{(l)} \tag{14.3}$$

$$2Fe^{2+}_{(s)} + 2H_2O_{(l)} \rightarrow 2Fe^{3+} + H_{2(g)} + 2OH^-_{(aq)} \tag{14.4}$$

As evident from the above results, nFe^0 mediated redox reactions produce a significant increase in solution pH and a concomitant decline in solution potential (Eh). In other words, introduction of nFe^0 leads to development of highly reducing conditions in the system. Sun *et al.* [16] reported an increase from pH ~6 to 8–9 in nFe^0 containing distilled water.

14.2.2 Synthesis

In general, the synthesis of nanoparticles can be grouped in two categories: one is the top-down approach and the other is the bottom-up approach. The top-down approach starts with large-size (i.e., granular or microscale) materials with the generation of nanoparticles by mechanical and/or chemical steps including milling, etching, and/or machining. This approach is usually not very well suited to prepare uniform nanoparticles; especially, problems are encountered in cases where required nanoparticles dimensions are very small. The bottom-up approach entails the "growth" of nanostructures atom-by-atom or molecule-by-molecule via chemical synthesis, self-assembling, and positional assembling. This approach illustrates the possibility of creating exact materials that are designed to have exactly the desired properties.

Numerous methods based on these two approaches are available in literature for the synthesis of nFe^0 such as vaccum sputtering, sonochemical method, spray pyrolysis, laser ablation, electrochemical method, gas-phase reduction, liquid-phase reduction, etc. [12, 17–20]. Among these methods, gas-phase reduction and liquid-phase reduction are the most commonly used methods. Several workers have synthesized nFe^0 by reducing goethite and hematite particles with hydrogen gas at elevated temperatures (200–600°C), by decomposition of iron pentacarbonyl [$Fe(CO_5)$] in organic solvents or in argon, by electrodeposition of ferrous salts, etc. [19, 21–23].

The generation of nFe^0 by the "bottom-up" reduction of ferric (Fe^{3+}) or ferrous (Fe^{2+}) salts with sodium borohydride has gained much popularity because of its relative simplicity with the need of only two common reagents and no need for any special equipment/instrument. In a typical preparation, 1:1 volume ratio of $NaBH_4$ and $FeCl_3$ or $FeSO_4$ was vigorously mixed in the flask reactor (Eqs. 14.5 and 14.6).

$$4Fe^{+3} + 3BH_4^- + 9H_2O \rightarrow 4Fe^0 + 3H_2BO_3^- + 12\ H^+ + 6H_2 \quad (14.5)$$

$$2Fe^{+2} + BH_4^- + 3H_2O \rightarrow 2Fe^0 + H_2BO_3^- + 4\ H^+ + 2H_2 \quad (14.6)$$

To ensure efficient use of the reducing agent BH_4^{-1}, the reactor system was operated under inert conditions. After about 15–30 min, when gas evolution ceased, nFe^0 was harvested with vacuum filtration. Maintaining a thin layer of ethanol on the top of nFe^0 can help to preserve the nanoparticles from oxidation. Experimental parameters such as pH, reactant concentrations, stirring speed, titration rate, reaction time and external temperature influence the composition and surface properties of produced nanoparticles, and hence the need to be maintained constant in the experiments to produce consistent samples [16, 24–26].

14.2.3 Structure

Zero-valent iron nanoparticle (nFe^0) has a core shell structure, wherein core is made of metallic iron and shell consists of iron oxides/hydroxides/oxyhydroxides. The metallic iron present at the center holds reducing power, whereas the oxides hydroxides coating exhibits sorption characteristics. Thus core shell structure bestows nFe^0 with dual characteristics of sorption and reduction, which could be of significant importance in the separation and transformation of contaminants. The constituent of shell, i.e., iron oxides, may have either metal-like or ligand-like coordination properties, depending on the solution chemistry. At acidic pH, iron oxides acquire positive charge and show affinity towards anionic ligands, in contrast, at higher pH, they have negative charge and form surface complexes with cationic species [16]. On one hand the shell endows nFe^0 with the properties of sorption, on the other hand it also acts as a passive film, which provides a physical barrier between the metal and dissolved contaminants [27]. As the nFe^0 mediated reactions are driven by oxidation of Fe^0 core, sustained reduction of contaminants requires localized defects in the passive film, which could facilitate charge and mass transport through it. These transport processes play a significant role in contaminant reduction kinetics. The overall reactivity of nFe^0 is a function of the surface concentration of these reactive sites/localized defects [28].

It is always difficult to measure the exact thickness of the shell due to the high reactivity of iron, which is responsible for continuously increasing shell thickness, as the oxidation reaction proceeds. However, Martin *et al.* [29] attempted to determine the thickness of the shell using three methods, i.e., high resolution transmission electron microscopy (HR-TEM), high resolution photoelectron microscopy (HR-XPS) and complete oxidation reaction of nFe^0. The HR-TEM images revealed that the shell thickness predominately lies in the range of 2–4 nm. The HR-XPS analysis also provides approximately similar thickness range with an average in the range of 2.3–2.8 nm. The complete oxidation of nFe^0 by Cu(II) indicated a shell thickness of 3.4 nm, which is consistent with results of HR-TEM and HR-XPS.

14.2.4 Environmental Application

14.2.4.1 Zero-Valent Iron Nanoparticle

Zero-valent iron nanoparticle (nFe^0) is widely used to detoxify and degrade various classes of environmental contaminants. It has diminutive size, high density of surface reactive sites and greater intrinsic reactivity; all these properties make it a potential remediation tool for organic as well as inorganic contaminants. Owing to its size and surface characteristics, nFe^0 can be injected or pumped straight into subsurface aquifers and groundwater to treat contaminated plumes. Not only does nFe^0 reduce the total content of contaminants, but it also lowers the amount of mobile and bioavailable fractions of contaminant. In addition, it also provides enormous flexibility for both *in situ* and *ex situ* applications.

14.2.4.1.1 Organic Contaminants

Zero-valent iron nanoparticle can reduce a number of halogenated hydrocarbons to benign products such as hydrocarbons, chloride and water. Wang and Zhang [26] reported complete dechlorination of 20 mg/L trichloroethylene (TCE) to hydrocarbons including ethene, ethane, propene, propane, butene, butane and pentane in the presence of 1.0 g nFe^0. Similarly, Zhang [30] studied the dechlorination reaction of trichloroethane, trichloroethene and tetrachloroethene by nFe^0 and found a 99% reduction within 24 hours with ethane as the major product. Other chlorinated compounds which are reported to undergo partial or complete dechlorination include tetrachloromethane, trichloromethane [31], polychlorinated biphenyls [32, 33], etc. Regarding the dechlorination pathway of chlorinated ethanes in a nFe^0 system, Song and Caraway [34] proposed that reductive β-elimination will be the major pathway for chlorinated ethanes possessing α,β-pairs of chlorine atoms whereas reductive α-elimination and hydrogenolysis will be concurrently followed by compounds having chlorine substitution on one carbon only. Zero-valent iron nanoparticle has also been proven as an efficient reductant for chlorinated pesticides such as Lindane and Atrazine [35–40]. Organophosphate compounds such as chlorpyrifos, tributyl phosphate have also been reported to degrade in the presence of nFe^0 [41, 42].

As a strong reductant, nFe^0 degrades organochlorine compounds following Eq. 7. The Fe^{2+} ion formed during iron oxidation reaction is a good electron donor and contributes in reducing the chlorinated compounds (Eq. 14.8).

$$Fe^0 + R\text{-}Cl + H^+ \rightarrow Fe^{2+} + R\text{-}H + Cl^- \qquad (14.7)$$

$$Fe^{2+} + R\text{-}Cl + H_2O \rightarrow 2Fe^{3+} + R\text{-}H + OH^- + Cl^- \qquad (14.8)$$

Other organic contaminants effectively degraded by nFe^0 include: nitroamines [43], nitroaromatic compounds [44], azo dyes [45, 46], polybrominated diphenyl ethers [47], polycyclic aromatic hydrocarbon [48], etc.

14.2.4.1.2 Inorganic Contaminants

Not only can nFe^0 reduce organic contaminants, but also the inorganic contaminants such as perchlorate, chromate, nitrate, arsenic [49, 50–53], etc. The ability of nFe^0 to reduce redox-sensitive elements has been demonstrated at both bench-scale and field-scale tests [11, 14, 15]. The degradation mechanism is based on the transformation of the toxic contaminant to nontoxic or less toxic form. For example, nFe^0 transforms highly soluble, highly mobile, extremely toxic Cr(VI) into relatively less soluble, less mobile and less toxic Cr(VI) [20, 54]. Arsenic is another example, whose +3 and +5 oxidation states are both reported to be effectively removed from groundwater by nFe^0 [53, 55]. Also, nFe^0 has been proven effective for degradation of alkaline earth metals such as barium, transition metals – copper, silver, lead, etc., and radioactive elements such as uranium and technetium [56–59]. Table 14.1 lists the organic and inorganic contaminants which are reported to be successfully degraded or transformed into less toxic or nontoxic entities by nFe^0.

As significant variations exist in the contaminant chemistry, there are numerous possible pathways for contaminant removal in nFe^0 mediated reactions such as sorption, complexation, precipitation/co-precipitation and surface-mediated chemical reduction. Generally the contaminant is removed via a combination of two or more processes. For example, the removal mechanism of Cr(VI) involve instantaneous adsorption of Cr(VI) on nFe^0 surface where electron transfer takes place, and Cr(VI) is reduced to Cr(III) with oxidation of Fe^0 to Fe(III). Subsequently, Cr(III) precipitates as Cr(III) hydroxides and/or mixed Fe(III) /Cr(III)hydroxides/ oxyhydroxides as per the following Eqs. (14.9–14.11).

$$3Fe^0 + Cr_2O_7^- + 7H_2O \rightarrow 3Fe^{2+} + 2Cr(OH)_3 + 8OH^- \quad (14.9)$$

$$_{(1-x)}Fe^{3+}{}_{(aq)} + {}_{(x)}Cr^{3+}{}_{(aq)} + 3H_2O \rightarrow Cr_x\,Fe_{1-x}(OH)_{3\,(s)} + 3H^+{}_{(aq)} \quad 14.10)$$

$$_{(1-x)}Fe^{3+}{}_{(aq)} + {}_{(x)}Cr^{3+}{}_{(aq)} + 2H_2O \rightarrow Cr_x\,Fe_{1-x}(OOH)_{(s)} + 3H^+{}_{(aq)} \quad (14.11)$$

where x varies from 0 to 1. Mixed hydroxides of Cr(III) and Fe(III) get incorporated into the iron oxy hydroxide shell of nFe^0 forming $(Cr_xFe_{1-x})(OH)_3$ or $Cr_xFe_{1-x}OOH$ at the surface; in this way Cr(III) gets stabilized/ immobilized on nanoparticle surface [20, 60]. Similarly, the removal mechanism of As(V) and As(III) involves spontaneous adsorption and co-precipitation with iron(II) and iron(III) oxides and hydroxides [55, 61]. Yan *et al.* [62] studied the removal mechanism of Hg(II), Zn(II) and hydrogen sulphide and reported that Hg(II) sequestrates via chemical reduction to

Table 14.1 List of organic and inorganic contaminants degraded by nFe^0.

Contaminant category	Contaminant removed	Reaction mechanism	Remarks	References
Chlorinated aliphatic, alicyclic and aromatic compounds	Hexachloroethane (HCA), Pentachloroethane (PCA), 1,1,2,2 Tetrachloroethane (1,1,2,2-TeCA), 1,1,1,2 Tetrachloroethane (1,1,1,2-TeCA), 1,1,2-Trichloroethane (1,1,2 TCA), 1,1,1-Trichloroethane (1,1,1 TCA), 1,2-Dichloroethane (1,2-DCA), 1,1-Dichloroethane (1,1-DCA).	β-elimination, Reductive α-elimination and hydrogenolysis	All chlorinated ethanes except 1,2-DCA transformed to less chlorinated ethanes or ethenes	[135]
	Polychlorinated Biphenyls (PCB)	Dechlorination to biphenyl	Catalyst, high temperature and aerobic conditions favours PCB destruction	[32]
	Biosolids	Sorption, Precipitation, Dechlorination, Immobilization	Forms stable, non volatile surface complexes with hydrogen sulphide, methyl sulphide, degrade pesticides, sequester toxic metals.	[136]
	Hexachlorocyclohaxanes (α, β, γ and δ isomers)	Dechlorination	Removal follows $\gamma = \alpha > \beta > \delta$ trend	[39]
	Atrazine	Reductive dechlorination to 2-ethyl-amino-4-isopropylamino1,3,3-triazine.	Dechlorination rate of nFe^0 7 times higher than that of microscale Fe^0. Fe or Al sulfate salts, catalyst and pH influences destruction rate.	[40]
	AB24 Dye	pH dependent adsorption and reduction	Rate constant increases linearly with increasing nFe^0 concentration	[45]

(*Continued*)

Table 14.1 (*Cont.*)

Contaminant category	Contaminant removed	Reaction mechanism	Remarks	References
Chlorinated aliphatic, alicyclic and aromatic compounds	Lindane	Dihaloelimination via TeCCH to benzene	Rate constant not affected by Fe^0 dose and type.	[137]
	Lindane	Dechlorination to benzene (major) and chlorobenzene (minor product)	Dechlorination rate increases with temperature and nFe^0 dosage; decreases with solution pH	[38]
	Pyrene	Reduction	Removal dominated by iron dosage, follows pseudo first order kinetics	[48]
	Pentachlorophenol (PCP)	Complete pyrolysis yielding CO_2, HCl, Cl^-	nFe^0 removal rate faster than that of activated carbon in the presence of microwave irradiation	[138]
	Alachlor	Complete dechlorination	Rate constant for nFe^0 is comparatively higher than micro Fe^0.	[139]
	Azo dye Congo Red	Reduction	Decolorization follows pseudo first order reaction. Increase in nFe^0 dosage enhance decolorization rate.	[140]
Inorganic contaminants	Cr(VI)	Reductive precipitation and immobilization	Surface area normalized rate constant 25 times greater than that of iron powder.	[60]
	As(V)	Removal by adsorption followed by reduction to As(III)	nFe^0-As(V) forms an inner-sphere surface complexation. HCO_3^-, $H_4SiO_4^0$, $H_2PO_4^{2-}$ are potential interfering agents in As(V) adsorption reaction.	[61]

Contaminant category	Contaminant removed	Reaction mechanism	Remarks	References
Inorganic contaminants	NO_3^-	Removal by reduction	Integrated nFe^0-cell system in removes nitrate more effectively than nFe^0 or cell alone	[141]
	Cr(VI)	Reduction followed by precipitation	Cr-Fe hydroxide shell: stable and sink for Cr(VI)	[142]
	NO_2^- , NO_3^-	Reduction to N_2	Removal efficiency: 65-83% for NO_2^- and 51-68% for NO_3^-	[143]
	As(V)	Removal by adsorption and coprecipitation	Humic acid decreased As(V) removal rate, UV light promotes removal efficiency	[53]
	As(III)	Adsorption	Anions (chloride, carbonate, nitrate, phosphate, sulfate and borate), manganese, and organic matter inhibit As(III) removal.	[144]
	Cd(II)	Chemisorption	Sorption- endothermic & spontaneous; follows pseudo second-order kinetic model	[145]
	U(VI), Cu(II), Cr(VI), Mo(VI)	Adsorption followed by reduction	U(VI) removal rate comparatively less in multielement system.	[146]
	PO_4^{3-}	Adsorption	Removal and recovery rate increases with increasing pH.	[147]

elemental mercury, Zn(II) undergoes sorption to iron oxide shell followed by zinc hydroxide precipitation and hydrogen sulfide gets immobilized on nFe^0 surface as disulfide (S_2^{2-}) and monosulfide (S^{2-}) species. The removal mechanism of metal ions is mainly dependent on the electrode potential of the metal [57]. The detailed explanation is given in Section 14.3.

14.2.4.1.3 Factors Affecting nFe^0 Reactivity

When nFe^0 is introduced into the contaminated sites, it gets exposed to different environmental factors. These factors all together determine the overall efficacy of nFe^0 for the contaminant in the environment. The following section deals with the factors which play a crucial role in determining the potential efficacy of nFe^0 for remediation of contaminants present in different environmental matrices.

- ***Size:*** The smaller particle size is accountable for greater density of reactive surface sites or surface sites of higher intrinsic reactivity. As the size of particle decreases, particle dimensions approach the size of certain physical length scales, such as the electron mean-free path and electron wavelength, which in turn results in quantum effects. These effects cause changes in the Fermi level and the band gap, which ultimately leads to an increase in reactivity [63].
- ***Specific surface area:*** It is an important factor which controls the physicochemical properties of nanoparticles. The surface area of a nanoparticle can be determined through BET gas adsorption isotherms. Alternatively, it can be calculated using Eq. 14.12.

$$r = 3[\rho * S]^{-1} \tag{14.12}$$

  where r is the radius of the nanoparticle, ρ is the density of Fe (7,870 kg m^{-3}) and S is the specific surface area. Specific surface area shares a direct relationship with the reactivity of nFe^0. Increasing the specific surface area results in an increase in the fraction of iron atom present on the particle surface, thereby creating a greater reductive capacity per unit of nanoparticles [64]. Several investigators have demonstrated the supremacy of nFe^0 over microscale Fe^0 in terms of reactivity and reaction rate constants [35, 40, 60, 64].
- ***Aggregation and oxidation:*** Generally, because of high reactivity of nFe^0, it reacts rapidly with the surrounding

media such as dissolved oxygen or water, resulting in formation of passive layers over the surface of nFe^0. This passive layer hinders the access of target contaminants to nFe^0, resulting in the loss of reactivity and degradation efficiency. Furthermore, due to size effects and high surface energy, nanoparticles tend to aggregate rapidly, resulting in formation of much larger clusters which may or may not reflect the properties of their nanoconstituents. In addition, aggregation also decreases the surface area, which in turn impedes the overall reactivity of the nanoparticle. To address these issues, several strategies such as use of stabilizing agent, supports, immobilization/encapsulation, etc., have been developed, which not only prevent aggregation but also resist oxidation, prolonging the life of reactive nFe^0 [24, 25, 35, 65].

- ***pH:*** The pH plays a detrimental role in corrosion of iron and thus indirectly controls the degradation of contaminants. The lower pH is responsible for (a) acceleration of corrosion of iron, and (b) an increase in the aqueous solubility of ferrous (Fe^{2+}) and ferric (Fe^{3+}) ions, yielding less iron hydroxide precipitates on the surface of nFe^0 and more exposed reactive sites [35]. However, an extremely low pH does not favor the contaminant reduction reaction. This can be explained by intensive corrosion of iron at extremely acidic pH, which results in an abrupt production of hydrogen which may produce a blanket of gas bubbles around the surface of nFe^0, inhibiting the contact of contaminant species to nanoparticles; hence the overall degradation rate decreases [66]. At higher pH, Fe^{2+} and Fe^{3+} from iron surface and OH ions in the alkaline solution react to precipitate iron oxides/hydroxides on the surface of iron occupying the reactive sites, thus hindering the access of contaminant species to the reactive sites, reducing the reaction rate [67, 68].
- ***Dosage/Loading:*** The reactivity of nanoparticles is a function of concentration of reactive sites. With the increase in nFe^0 dosage, there is a corresponding increase in the number of reactive sites. The greater the number of reactive sites, the greater the number of target molecules accessing the site, hence the rate of reaction is greater [20, 38, 69]. Alternatively, if the concentration of the target contaminant exceeds the number of reactive sites available for reaction, there will be a decrease in the rate of reaction.

- ***Reaction temperature:*** As per literature, the rise in temperature has an accelerating effect on the reaction process. This can be attributed to an increase in the mobility of target molecules with the increase in temperature, which in turn enhances the rate of reaction [35]
- ***Natural Organic Matter:*** The natural organic matter such as humic acid plays an important role in contaminant reduction by virtue of its functional groups such as quinones [70]. Humic acids have high binding affinity towards Fe^{2+} and Fe^{3+} [71] and also have a strong tendency for adsorption on iron oxides surfaces [20, 72]. The adsorption of humic acid inhibits iron corrosion, thereby prolonging the lifetime of nano Fe^0. On the other hand, adsorbed humic acid also transfers electron from inner Fe^0 to Fe^{3+} to facilitate reduction reaction. In addition, humic acid complexation with dissolved iron released from corrosion can regenerate reactive Fe^{2+} to reduce redox amenable species [73]. In contrast to the enhancing effect of humic acid, Liu *et al.* [74] reported that humic acid competes with the contaminants for the reactive sites on nFe^0 surface and alters the reduction potential of neighboring surface sites, thus decelerating the contaminant reduction rate.

These are some of the factors which have been widely investigated in nFe^0 mediated remediation. Besides these, other factors which determine the reactivity of nFe^0 include dissolved oxygen, hardness, oxidation-reduction potential, ionic strength of groundwater, aquifer's hydraulic properties, etc. [75].

14.2.4.2 *Bimetallic Iron Nanoparticles: Improving the Reactivity of nFe⁰*

The urge for enhancement of reaction kinetics and improvement in metal usage has lead to the synthesis of bimetallic nanoparticles. Iron-based bimetallic nanoparticles consist of a base metal, i.e., Fe^0, as the reductant, and a second metal such as Pd, Cu, Ni or Pt as the catalyst. These bimetallic nanoparticles have reaction rates that are orders of magnitude higher than the corresponding monometallic nanoparticles [26, 35, 64, 76]. Catalytic hydrodechlorination is considered as the main pathway of contaminant degradation in the case of bimetallic iron nanoparticles. Studies have shown that this pathway leads to formation of a lesser amount of toxic chlorinated intermediates during reaction [31, 77]. Table 14.2 shows the

Table 14.2 Common contaminants degraded by bimetallic iron nanoparticles and stabilized or supported nFe^0.

Contaminant category	Nanoparticle Type	Contaminant removed	Reaction mechanism	Remarks	References
Chlorinated aliphatic, alicyclic and aromatic compounds	nFe^0/Pd	Trichloroethene (TCE) and Poly-chlorinated biphenyl (PCB)	TCE decchlorinates to to hydrocarbons (Ethene, Ethane, Propene, Propane, Butene, Butane, Pentane) and PCB to biphenyl	Surface area normalized rate constants were 10-100 times higher than those of commercially available iron particles	[26]
	nFe^0/Ni	CCl_4 and $CHCl_3$	Reductive dechlorination	Reaction rate constant 2-8 times higher for nFe^0/Ni as compared to nFe^0	[148]
	nFe^0/Pd	Pentachlorophenol	Complete dechlorination to phenol via tetra-, tri-, di- & mono-chlorophenol	Ultrasonication improves dechlorination efficiency	[149]
	nFe^0/Pd	2,4-dichlorophenol (2,4-DCP)	Catalytic dechlorination to *o*-CP or *p*-CP initially, then to phenol finally.	Humic acid inhibits dechlorination	[150]
	nFe^0/Pd	2,2,'4,5,5'-penta-chloro-biphenyl	Catalytic hydrodechlorination	High Pd loading, high nFe^0/Pd dosage and slightly acidic conditions favours dechlorination	[151]
	nFe^0/Pd	Trichloroethene (TCE)	Dechlorination to *cis*-1,2-DCE and hydrocarbons (ethyne, ethene, ethane)	nFe^0 deactivation reaction follows first order kinetics, Clay sediments effect nanoparticles viability	[152]
	nFe^0/Pd	Lindane	Complete dechlorination to cyclohexane	Negligible dechlorination observed by commercial microscale Fe^0.	[153]

(*Continued*)

Table 14.2 (*Cont.*)

Contaminant category	Nanoparticle Type	Contaminant removed	Reaction mechanism	Remarks	References
Chlorinated aliphatic, alicyclic and aromatic compounds	nFe^0/Ag	Tetrabromo-bisphenol A (TBBPA)	Reductive debromination to tri-BBPA, di-BBPA, mono-BBPA and BPA.	Ultrasonication assists degradation	[154]
	nFe^0/Ni	Poly-brominated diphenyl ethers (PBDEs)	Catalytic debromination	Debromination rate increases with increasing amount of nFe^0/Ni, Ni/Fe ratio, and decreasing initial concentration of PBDEs.	[155]
	nFe^0/Ni	Aroclor 1242	Catalytic dechlorination to biphenyl	High nFe^0/Ni dosage and high Ni content favours dechlorination reaction.	[156]
	Cellulose acetate supported nFe^0	Trichloro-ethylene	Dechlorination	Membrane supported nFe^0 showed synergistic effect on dechlorination	[65]
	CMC stabilized nFe^0/Pd	Lindane	Dihaloelimination and dehydrohalogenation under anaerobic conditions, Oxidative degradation by dechlorination /dehydrohalogenation under aerobic conditions	Stabilized bimetallic nanoparticles more effective under anaerobic conditions	[157]
		Atrazine	Dechlorination followed by dealkylation under anaerobic conditions, dealkylation under aerobic conditions		

Contaminant category	Nanoparticle Type	Contaminant removed	Reaction mechanism	Remarks	References
Chlorinated aliphatic, alicyclic and aromatic compounds	CMC stabilized nFe^0/Pd	Trichloroethylene	Complete dechlorination without formation of toxic intermediates	Dechlorination rate of CMC-nFe^0/Pd 17 times faster that of nFe^0/Pd.	[24]
	CMC-Cu/nFe^0	1, 2, 4-trichlorobenzene	Sequential dechlorination and catalytic hydrogenation	Rate-determining step is the electrophilic H^* addition to the double bond of benzene ring followed by the C–Cl scission.	[158]
	CMC-Pd/nFe^0	*p*-nitrochlorobenzene (*p*-NCB)	Dechlorination to aniline with trace amount of *p*-chloroaniline as intermediate	*p*-NCB toxicity greatly reduces and biodegradability improves	[86]
	CMC-Pd/nFe^0	2,4-dichlorophenoxyacetic acid	Adsorption followed by reduction to 2-chlorophenoxyacetic acid and finally to phenoxyacetic acid	Reaction pH and CMC/nFe^0 ratio significantly affects the reduction process	[159]
	nFe^0/Cu with activated carbon support	γ-HCH	Simultaneous adsorption and dechlorination	Tetrachlorocyclohexene and chlorobenzene identified as major intermediate and final product.	[89]
	nFe^0/Pd-alginate	Trichloro-ethylene (TCE)	Complete dechlorination to ethane and butane	> 99% TCE reduction within 4 hours. Less than 3% Fe release from support.	[96]
	Bentonite supported nFe^0	Methyl orange (MO)	Adsorption followed by reductive cleavage of azo bonds	Degradation significantly affected by pH, nFe^0 dosage, initial conc. of MO, and temperature.	[90]

(*Continued*)

Table 14.2 (*Cont.*)

Contaminant category	Nanoparticle Type	Contaminant removed	Reaction mechanism	Remarks	References
Inorganic contaminants	nFe^0/Ni	NO_3^-	Reduction	Near neutral pH favourable for reduction.	[160]
	nFe^0/Ag	P	Reduction	Also exhibits antimicrobial and antifungal activities	[161]
	Resin supported nFe^0 (Ferragel)	Cr(VI) and Pb(II)	Reduction of Cr(VI) to Cr(III), Pb(II) to Pb(0)	Ferragel reduction rate 30 times higher than that of Fe^0 powder and fillings	[64]
	CMC stabilized nFe^0	Cr(VI)	Reductive immobilization	nFe^0 reduces Cr(VI) leachability	[50]
	Calcium alginate entrapped nFe^0	NO_3^-	Reduction	nFe^0 can be effectively entrapped in beads without significant decrease in reactivity towards NO_3^-	[92]
	Calcium alginate entrapped nFe^0	Cr(VI)	Reduction	Reduction follows pseudo first order kinetics.	[93]
	Bentonite supported nFe^0	Cr(VI)	Reduction	Also able to remove Pb(II), Cu(II) and Zn(II) from electroplating wastewater. Exhibit reusability.	[91]
	Pillared bentonite (Al-bent) supported nFe^0	Cr(VI)	Reduction	Reduction efficiency much higher than nFe^0 alone. nFe^0/Al-bent exhibit good stability and reusability	[162]

list of bimetallic iron nanoparticles which have been reported to exhibit a high efficacy for transformation of various chlorinated compounds, along with their predominant degradation mechanism. Generally, two methods are used for preparation of bimetallic nanoparticles. One method includes consecutive reduction of the second metal ions and subsequent deposition onto first metal particle, i.e., iron. This method leads to formation of nanoparticles with core-shell structure. Another method involves simultaneous reduction of two metal ions, resulting in formation of alloy structure nanoparticles [78, 79].

The reason behind the enhanced reaction kinetics of bimetallic iron nanoparticles may be attributed to the catalytic effect of the second metal, which not only acts as catalyst but also enhances the surface area of nanoparticles [80]. The second metal also collects hydrogen gas produced during the corrosion of iron and dissociates it into atomic hydrogen, which is considered as a strong reductant for dehalogenation reactions [77]. Furthermore, the deposition of second metal creates many galvanic cells on the surface of iron, which accelerates corrosion of iron and enhances the kinetics of the redox reactions [81].

In bimetallic nanoparticles, the loading of second metal plays a crucial role in determining the rate of reaction. He *et al.* [82] found that the dechlorination efficiency of nFe^0/Pd for polychlorinated biphenyls increases as the loading of second metal (Pd in this case) increases. Singh *et al.* [35] also noticed a similar effect while studying dechlorination of γ-HCH in the presence of stabilized nFe^0/Pd. The most likely explanation for this enhanced rate is: (a) concomitant increase in the amount of hydrogen adsorbed on Pd surface with the increases in Pd content, which in turn promotes the rate and extent of dechlorination, and (b) higher catalyst loading increase in the total number of galvanic cells, increasing the rate of iron corrosion, thus increasing degradation rate. However, beyond certain limits, the increase in Pd loading shows a declining effect on dechlorination efficiency. Wang *et al.* [38] suggested that the excessive amount of Pd on nFe^0 hinders the formation of hydrogen by nFe^0 corrosion, thereby reducing the dechlorination rate.

An important concern associated with the use of bimetallic nanoparticles is the dislodgement of secondary metal with the corrosion of iron. As the secondary metals are the reactive sites, this issue is a major setback in the use of iron-based bimetallic nanoparticles for targeting contaminants. In view of this, two regenerative approaches were investigated by Zhu and Lim [83] for the recycling and reutilization of Pd/nFe^0 particles. One approach utilized HCl and the other employed $NaBH_4$ for regeneration of Pd. Pretreatment of aged nanoparticles with HCl served the purpose of removing surface oxide layers, thereby re-exposing the covered Pd islets, whereas $NaBH_4$ was rejuvenated via reducing the iron oxide layer formed on Pd/nFe^0 surface to Fe^0. Of these two methods, regeneration via HCl was found relatively more efficient.

14.2.4.3 Stabilized and Supported Iron Nanoparticles: Improving the Mobility and Stability of nFe^0

The activity of nFe^0 is very high, thus stabilizing or supporting the particles is important to preserve their chemical nature until they can be contacted with the target contaminant. In general, two approaches are applied for stabilization of nanoparticles, i.e., pre-synthesis and post-synthesis. Between these two techniques, the pre-synthesis approach was reported to be more efficient in producing nanoparticles with narrow size distribution and low sedimentation rate/long suspension time. The most widely used stabilizers are carboxymethyl cellulose (CMC), polyvinylpyrrolidone (PVP) and guar gum. They differ in their stabilization mechanism and thus their stabilizing ability. The stabilization potential mainly depends on the functional group, molecular structure and molecular weight of the stabilizing agent. For instance, carboxylate group present in CMC forms strong complex with Fe^{2+} ions present in the solution, resulting in dispersion of the Fe^{2+} ions throughout the CMC network. These Fe^{2+} ions then undergo rapid nucleation on addition of reducing agent, which is followed by growth of nuclei to a critical size. At this point, steric and electrostatic hindrance produced by negatively charged CMC molecule limits further growth of nuclei [81]. An increase in molecular weight of CMC leads to formation of well-dispersed nanoparticles with narrow size distribution. In the case of PVP, carbonyl group forms weak bonds with Fe^{2+}, slowing the rate of nucleation, resulting in formation of relatively larger-sized nanoparticles. As PVP is a neutral molecule, growth of nanoparticles is predominately counter-checked by steric hindrance. Similar to PVP, guar gum is also neutral, but hydroxyl group of guar gum forms more stable complexes with Fe^{2+} as compared to carbonyl-Fe^{2+} complex, resulting in the synthesis of nanoparticles with relatively smaller-sized nanoparticles [84]. Among these three stabilizers, CMC produces the smallest nanoparticle. As far as stability is concerned, guar gum exhibits maximum stability followed by CMC and PVP. Moreover, high ionic strength (0.5 M NaCl and 3 mM $CaCl_2$) does not affect the stability of guar gum-stabilized nFe^0 [85]. An increase in molecular weight of PVP increases the nanoparticles suspension stability, but it is not effective in the case of CMC.

He *et al.* [24] reported that the degradation efficiency of CMC-stabilized Fe-Pd nanoparticles for trichloroethylene is ~17 times faster than their non-stabilized counterparts in terms of rate constant. In one more study conducted with para-nitrochlorobenzene, CMC-stabilized Fe-Pd nanoparticles were reported four times more efficient than non-stabilized nFe^0 [86]. Zhou *et al.* [87] successfully demonstrated the potential of CMC-Fe^0/Pd nanoparticles for dechlorination of 2,4-dichlorophenoxyacetic acid to phenoxyacetic acid.

The long-term stability of nFe^0 can also be enhanced by immobilizing it in a support. The immobilization of nFe^0 with some support could simultaneously provide three advantages. The first one lies with the ability of supporting material to control the growth of nanoparticle as well as aggregation; the second one is related to the protection provided to nanoparticles against oxidation and hydrolysis in water; the third one is the preconcentration of the target contaminant around nFe^0 via adsorption on the surface of supporting material, which in turn enhances the overall reactivity of nFe^0. In addition, supported nanoparticles are more convenient in terms of their real application [64, 65]. Park *et al.* [88] successfully immobilized nFe^0 on an ion-exchange resin sphere, which not only combats the problem of agglomeration but also reduces the amount of ammonia produced during nitrate reduction, which is otherwise a major limitation for nFe^0. Zhang *et al.* [89] proposed kaolin as an ideal support for nFe^0 and demonstrated its use in removing Pb(II) from aqueous solution. Cellulose acetate, activated carbon, bentonite, resin, etc., have been successfully used as supporting agent for nFe^0 [64, 65, 90–92].

The entrapment of iron in calcium alginate beads [93, 94] or chitosan beads [95] serves the purpose of preventing oxidation and agglomeration without compromising the reactivity of nFe^0. Krajangpan *et al.* [96] revealed that alginate gel cluster acts as a bridge that binds the nFe^0 particles together. Kim *et al.* [97] immobilized nFe^0 in alginate bead to investigate the degradation of trichloroethylene (TCE). The study reported that the iron released from alginate bead is <3% of the loaded iron. Liu *et al.* [98] utilized epichlorohydrin (ECH) as crosslinker to improve the mechanical strength of chitosan-nFe^0 (CS-nFe^0) beads. In comparison to CS-nFe^0 beads, ECH-CS-nFe^0 beads were found to possess higher thermal stability. Liu *et al.* [98] further demonstrated that ECH-CS-nFe^0 beads could be regenerated and reused by washing with dilute HCl. The contaminants which are successfully degraded by stabilized or supported nFe^0 are displayed in Table 14.2.

14.3 Reduction Mechanisms and Pathways

In the case of reductive dechlorination of organochlorine compounds by nFe^0, there are three possible mechanisms which are as follows:

i. Direct reduction at the metal surface by electron transfer from nFe^0 to adsorbed contaminant molecules.
ii. Indirect reduction by ferrous ions (Fe^{2+}).
iii. Hydrogen produced as a result of iron corrosion is also capable of reductive dechlorination in the presence of a catalyst.

The direct reduction pathway mainly includes five steps: mass transfer of target molecule from solution to the surface of nFe^0, adsorption of target molecule on the surface of nFe^0 via formation of organic chemisorptions complex, reaction between the adsorbed target molecule and nFe^0, desorption of products and diffusion of products into solution. The main reactions of direct reduction pathway include reductive α/β elimination (dihalo-elimination), hydrogenolysis (hydrogen substitution of halogen), or hydrogenation (multiple bonds cleavage). Indirect reduction pathway includes reduction by ferrous ions and/or hydrogen. Although Fe^{2+} is a good reducing agent, its reaction is generally quite slow and is probably not of great consequence. Hydrogen gas is an excellent reductant for dechlorination reactions, but is insufficient for reduction of a chlorinated organic compound in absence of a catalytic surface.

In bimetallic systems, the second metal (i.e., Pd, Cu, Ni, etc.) acts as a catalyst to generate active atomic hydrogen while Fe^0 acts as an electron donor. The coexistence of both e^- donor and catalyst leads to higher dechlorination efficiency. The catalytic reductive dechlorination of an organochlorine compound by a bimetallic iron nanoparticle follows three steps (second metal is denoted as M):

i. The corrosion of iron leads to production of hydrogen.
ii. M and hydrogen are combined to form a transitional compound $M.H_2$ with H_2 embedded in the crystal lattice.
iii. Transitional compound dechlorinates organochlorine compound.

All these steps are depicted in Eqs. 14.13–14.18. Figure 14.1 presents a schematic diagram of the dechlorination mechanism occurring in nFe^0 and

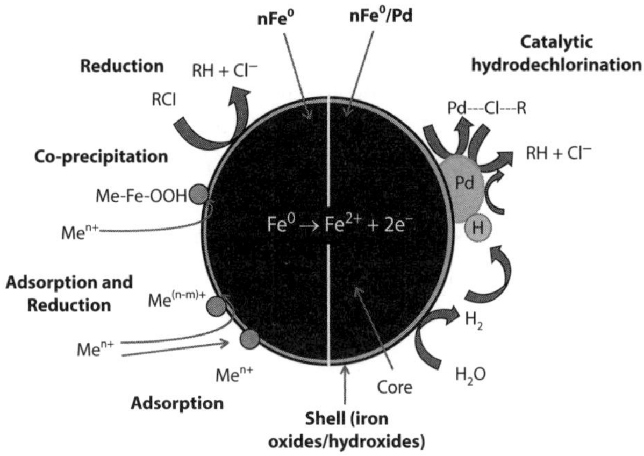

Figure 14.1 Schematic diagram of mechanisms involved in contaminant degradation in presence of nFe^0 (left portion) and nFe^0/Pd (right portion).

bimetallic nFe^0. The left and right portion represent direct reduction of an organochlorine by nFe^0 and catalytic reductive dechlorination respectively.

$$Fe^0 + 2H^+ \rightarrow Fe^{2+} + H_2 \text{ (in acidic solution)} \tag{14.13}$$

$$Fe^0 + 2H^+ \rightarrow Fe^{2+} + H_2 + 2OH^- \text{ (in alkaline solution)} \tag{14.14}$$

$$Fe^0 + RCl + H^+ \rightarrow Fe^{2+} + RH + 2Cl^- \text{ (direct reduction)} \tag{14.15}$$

$$M + H_2 \rightarrow M.H_2 \tag{14.16}$$

$$M + RCl \rightarrow M\ldots Cl\ldots R \tag{14.17}$$

$$M.H_2 + M\ldots Cl\ldots R \rightarrow RH + H^+ + Cl^- + M \tag{14.18}$$

Reduction and precipitation of metal ions by nFe^0 depend on transport of the dissolved metal ions to the surface and electron transfer (ET) to the metal ion. Potential ET pathways from the surface to the sorbed ions/molecules may include:

i. Direct electron transfer (DET) from nFe^0 through defects such as pits or pinholes, where the oxide layer is interpreted as a simple physical barrier.
ii. Indirect electron transfer (IET) from nFe^0 through the oxide layer via the oxide conduction band, impurity bands or localized bands.
iii. Electron transfer from sorbed or lattice Fe^{2+} surface site.

Li and Zhang [57] explained the reduction of metal cations on the basis of the standard oxidation-reduction potential of the metal ions. For metal with E_0 very close to or more negative than that of iron (-0.41 V), such as Zn(II) and Cd(II), the removal mechanism is sorption/surface complex formation. For metals with E_0 greatly more positive than iron, for instance Cu(II) and Hg(II), the removal mechanism is predominantly reduction. Metals with E_0 slightly more positive than that of iron, as in case of Ni(II) and Pb(II), can be immobilized at the nanoparticle surface by both sorption and reduction.

14.4 Pilot- and Field-Scale Studies

As the nFe^0-based remediation approach has considerable benefits over conventional treatment technologies, a rapid transfer of nFe^0 technology from laboratory to field-scale application has been done in the last decade. Several *in situ* field demonstrations of nFe^0 technology have been undertaken and are being developed to remediate contaminated soil and groundwater. In general, there are two ways to employ nFe^0 in groundwater and soil remediation: (i) through permeable reactive barriers (PRB), which contain the reactive iron, and (ii) through injection well, wherein nanoparticles establish a plume of reactive iron in subsurface environment.

Puls *et al.* [99] used Fe^0 in PRB for *in situ* remediation of groundwater contaminated from an old chrome-plating facility located on an US Coast Guard Air Station near Elizabeth City, North Carolina, and observed chromate reduced to less than 0.01 mg L^{-1}. Wilkin *et al.* [100] also reported Cr(VI) reduction in groundwater from 1500 μg L^{-1} to <1 μg L^{-1} in PRB installed at US Coast Guard Support Center located near Elizabeth City, North Carolina. Similar results were observed by Flury *et al.* [101] in PRB installed at Willisau, Switzerland, for *in situ* treatment of groundwater contaminated from wood preserving industries.

Elliott and Zhang [102] studied field-scale dechlorination of trichloroethene-contaminated groundwater through nFe^0/Pd at an industrial and manufacturing facility situated in Trenton, New Jersey, USA. As per their report, ~ 890 L of nFe^0/Pd suspension at concentration of 1.5 g/L on the first day and 450 L of 0.75 g/L nFe^0/Pd on subsequent days, was gravity fed into trichloroethene-contaminated groundwater. They observed a reduction efficiency of about 96% over a 4 week period following injection. Another field test was carried out at an industrial facility located in Research Triangle Park, North Carolina, wherein 1600 gallons of nFe^0 slurry (1.9 g/L) were injected into the groundwater contaminated with volatile organic carbon (VOCs), tetra-, tri- and di-chloroethene. After six weeks of nanoparticle injection, over a 90% reduction was observed in concentration of VOCs and the concentration of terta-, tri-, di-chloroethene were found below groundwater quality standards [103].

Quinn *et al.* [104] and O'Hara *et al.* [105] reported high dechlorination efficiencies for DNALP contaminant trichloroethylene (TCE) during a field-scale demonstration of emulsified nFe^0 at Cape Canaveral Air Force Station, Florida, USA. A pilot-scale demonstration of bimetallic nanoparticles has also been carried out at Jacksonville and Lakehurst (USA) for reduction of TCE. In comparison to emulsified nFe^0, bimetallic nanoparticles were found less efficient, possibly due to early passivation of nanoparticle or use of an insufficiently high ratio of iron and soil to generate reducing environment in the aquifer [14]. In another pilot-scale study, 150 gallons of 0.2 g/L Pd/nFe^0 were injected into 50-ft deep, aquifer contaminated with chlorinated ethenes (perchloroethylene (PCE) and trichloroethylene (TCE) and polychlorinated biphenyls (PCBs)). One month later, another 150 gallons 1.0 g/L Fe-Pd (CMC = 0.6 wt%, Pd/Fe = 0.1 wt%) was injected into the same aquifer at an injection pressure <5 psi. As per results, rapid degradation of PCE and TCE was observed in the first week of injection, however the reducing power of nFe^0 was exhausted after 2 weeks. To counteract this limitation CMC-stabilized nFe^0 was used, which not only enhanced the dechlorination, but also appeared to boost *in situ* biological dechlorination reactions. This study proposed that CMC-stabilized nFe^0 facilitated the early stage abiotic degradation and at the same time acted as a source of carbon for biotic processes, thereby sustaining the biological degradation of chlorinated ethenes in subsurface [106]. Singh *et al.* [107] studied remediation of Cr(VI) with nFe^0 in a tannery waste-contaminated site located in Rania, Kanpur, India. The site has been used for many years as a dumping ground for wastes generated from leather tanneries and basic chrome sulphate industries. They investigated the remediation process in constructed soil windrows and reported ~99% reduction over reaction period of 50 days.

Field-scale commercial application of nFe^0 has become common in the United States, but there have been only a few field-scale projects carried out in Europe so far. Among them, Bornheim, Germany, was the first contaminated site where nFe^0 had been applied for full-scale remediation. The site was contaminated with several tons of tetrachloroethene (PCE) from an industrial laundry/dry cleaner. The contaminant was spread over several square kilometers of area down to a depth of 20 m. Injection of one ton of stabilized nFe^0 along with 2 tons of microsized nFe^0, effectively reduced 90% of the PCE concentration initially present.

Other full-scale applications include remediation of polychlorinated ethenes (PCE, TCE, DCE) contaminated sites located in Hořice and Písečná, Czech Republic, using nFe^0 alone [108]. Yan *et al.* [109] have compiled several other field and pilot studies which have already been done or are still in progress.

14.5 Transport of nFe^0 in Environment

Over the past years, an increasing concern regarding the transport of nFe^0 in diverse matrices has been noticed. The reason could be attributed to the poor colloidal stability of nFe^0. As soon as the nanoparticles come in contact with the matrix, they undergo collisions with soil grains/sediments/aquifer materials and are subjected to adsorption, dispersion and retardation, thereby limiting their transport through matrix [110]. In addition, their tendency towards aggregation has caused a major setback for *in situ* remediation applications. To achieve satisfactory remediation in subsurface environment, it is important to improve the transport properties of nFe^0. The mobility of iron nanoparticles can be enhanced via meeting two requirements: (i) by preventing the aggregation of nanoparticles, and (ii) by reducing the attachment of nanoparticles to matrix. In view of this, several research groups have tested various types of surface modifiers such as surfactants, polymers, etc., to enhance the transport of nFe^0 in contaminated site. For instance, poly(acrylic acid) and anionic hydrophilic carbon form reduce aggregation among nanoparticles via electrostatic repulsion and enhance transport of nFe^0 through soil- and sand-packed columns [111]. Similarly, poly(vinyl alcohol-co-vinyl acetate-co-itaconic acid) (PV3A) and triblock copolymer also increase the stability and subsurface mobility potential of nFe^0 [112, 113].

He *et al.* [24] investigated the mobility of CMC-stabilized nFe^0 and non-stabilized nFe^0 in glass column packed with 2.7 mL loamy soil sand bead through column breakthrough and elution tests. In contrast to non-stabilized nFe^0, it was found that stabilized nFe^0 readily passes through the column, and is completely dispersed in the soil. Further, ~98% of stabilized nFe^0 eluted with three bed volumes of deionized water indicated that retention of stabilized nanoparticles in soil is irreversible. Vechhia *et al.* [114] investigated the transport of micro- and nanoscale Fe^0 through 0.46m-long sand-packed column in the presence of xanthan gum and obtained high elution concentration (>0.94 for micro- and >0.88 for nano-Fe^0 respectively) and recoveries (95% for micro- and 92% for nano-Fe^0 respectively) for both. Cameselle *et al.* [115] carried out a comparison study with eight different dispersants (aluminum lactate, sodium lactate, ethyl lactate, aspartic acid, polyacrylic acid, 2-hydroxypropyl-β-cyclodextrin, β-cyclodextrin and methyl-cyclodextrin) and found improved stability and transport for the all over bare nFe^0, with aluminum lactate as the best stabilizer demonstrating highest elution from soil column. Besides these polymers, several researchers have also used microemulsions, starch, guar gum, silica, etc. [25, 85, 116], as delivery vehicles for transport of nFe^0 in porous media.

Another interesting approach towards improving the mobility of nanoparticles in soil is the application of electric field, i.e., Electrokinetics. Yang *et al.* [117] reported that under the influence of an electric magnetic field, polyacrylic acid-modified nFe^0 could be effectively transported horizontally through packed loamy sand soil (18–25 m). Pamukcu *et al.* [118] also showed effective movement of polymer-coated nFe^0 from anode to cathode through clay soil. They further reported in addition to enhanced transport, electric field also activates nFe^0. Reddy *et al.* [119] investigated the electrokinetic delivery of lactate modified nanoscale iron particles (LM-NIP) into low permeability kaolin soil, spiked with dinitrotoluene (DNT). The application of constant voltage gradient (1VDC/cm) produced electro-osmotic flow in the soil, which in combination with electromigration enhanced the extent of transport of LM-NIP in the soil. Besides increasing the transport, enhanced degradation of the contaminant via electrokinetic remediation is an additional benefit of this approach [120]. Some recent studies also successfully demonstrated the effectiveness of electrokinetics in enhancing nFe^0 transport and contaminant degradation in several nFe^0 permeable reactive barriers [121, 122].

14.6 Integrated Approach

The integrated approach exploits the reductive capabilities of nFe^0-mediated abiotic degradation and biotic degradation/microbial degradation either sequentially or concurrently. The combination of nFe^0 aided remediation and microbial degradation processes, could provide an effective method for degradation of most recalcitrant compounds. Generally, highly chlorinated compounds with five or more chlorine atoms such as polychlorinated biphenyls remain biorefactory to aerobic bacteria, whereas low chlorinated compounds degrade easily in the environment. The basic concept of an integrated system is to transform highly chlorinated non-biodegradable compounds abiotically through nFe^0 to less chlorinated compounds, which can then be easily subjected to degradation by biotic communities under aerobic/anaerobic conditions.

Several studies in literature have demonstrated the applicability of integrated systems to degrade different chlorinated compounds. He *et al.* [69] successfully transformed 2,2,′4,5,5′-pentachlorobiphenyl to 2,2′,4-trichlorobiphenyl using nFe^0/Pd and then treated the end product with aerobic bacteria to further degrade the compound. Similarly, Triclosan (2,4,4′-trichloro-2′-hydroxydiphenyl ether, TCS) was also reported to get completely transformed into nontoxic products in a sequential treatment system comprised of nFe^0/Pd and laccase in the presence of syringaldehyde [123]. Polybrominated diphephenyl ether was also reported to undergo degradation in a sequential treatment system involving nFe^0 and *Sphingomonas sp.* PH-07 [124]. *Sphingomonas sp. PH-07* was also reported to degrade dechlorinated product of Triclosan in a nano-bio hybrid system [125]. The combined use of nFe^0 and autotrophic hydrogen-bacteria was found to have a 3.2-fold higher dechlorination rate for trichloroethylene (TCE) in comparison to their individual effect showing synergistic effect, of combined method on overall dechlorination reaction [126]. A similar synergistic effect was also observed by Singh *et al.* [127] while studying γ-HCH dechlorination with stabilized nFe^0/Pd and *Sphingomonas sp.* strain NM05 in soil system.

14.7 Challenges Ahead

14.7.1 Toxicity

Along with nanoremediation, nanoecotoxicology has also emerged as a discipline concerned with the potential risks associated with the release of nanoparticles in the environment. In recent years several articles have been devoted to the toxic effect of nFe^0 on bacterial cells, mammalian nerve cells, bronchial epithelial cells, fish embryo, etc. [128–132]. In general, nanoparticles exert toxic effect on microbes by disrupting cell membranes, increasing membrane permeability, interrupting energy transduction, producing reactive oxygen species, lipid peroxidation, DNA damage, etc. [128, 129]. The toxicity of iron is based on its ability to catalyze the formation of hydroxyl radicals (OH^-) from superoxide (O_2^-) and hydrogen peroxide (H_2O_2). These reactive species show strong biochemical activity, affecting antioxidant enzymatic activities, peroxidation of membrane lipids, modification of nucleic acids, dysfunction of cellular components, hypoxia and eventually cause cell death and tissue injury [129, 132]. Some researchers reported that partial oxidation/aging and surface modification of nFe^0, could decrease its toxic properties [130, 133, 134].

The increasing concern over safe use of nanoparticles demands: (i) development of rapid analytical tools and methods to detect their presence in different environmental compartments, (ii) monitoring their production, usage, release and disposal, and (iii) adequate models for toxicity testing and risk analysis.

14.7.2 Fate and Behavior in Environment

Nanoparticles hold enormous potential to change many aspects of our existence, some for the better and some for the worse. In other words, nanoparticles have a Janus face. The very properties which are exploited for numerous beneficial applications could also produce unpredictable ecotoxicological effects. For example, the toxic property of nanoparticles on one hand could be utilized for the purpose of water disinfection, on the other hand the same property could be harmful to the microbial community when nanoparticles enter environment. Similarly, their high catalytic property is of great significance for effective degradation of pollutants, but the same property could induce toxic responses when taken up by a cell [13]. When nFe^0 is introduced in a system (soil, sediment, groundwater, aquifer) with an aim of degrading a contaminant, it not only interacts with the contaminant but also with the biotic component of that system. As the safety of nFe^0 is uncertain till date, it is important to consider its aftermath in order to decide its fate. A proper understanding of mobility, bioavailability, toxicity and persistence of nFe^0 in the environment is needed for assessing its risks. Additionally, several factors such as oxidation and aging, interaction with common cations, ligands and complexes, surface coating with humic acids, etc., play an important role in controlling the fate and behavior of nFe^0 in the environment, and thus need to be studied in detail.

14.8 Concluding Remarks

This chapter presents an overall perspective of nFe^0 to decipher potential issues related to its application for remediation of sites contaminated with toxic/hazardous pollutants. As evident from numerous studies, nFe^0 has the potential ability to degrade or transform a wide array of contaminants to nontoxic or less toxic form through reduction, adsorption and precipitation/co-precipitation. Although nFe^0 has gained significant achievements in the area of environmental clean up at both bench and field scales, there are still a few grey areas, economic hurdles and unforeseen ecological threats which could limit their acceptance among the scientific community, regulating agencies, private and public sector. The potential benefits and environmental risks associated with the use of nFe^0 should be carefully weighed before the implementation of this technology in order to minimize any unintended impact on the ecosystem. Future studies should address issues concerning the mobility, reactivity, long-term fate and ecological impact of nFe^0. A fundamental understanding of behavior, interactions and overall impact of nFe^0 on environment could lead to universal acceptance of this technology.

References

1. C.C. Huang, S.L. Lo, and H.L. Lien, *Chem. Eng. J.*, Vol. 219, p. 311, 2013.
2. M.R. Song, M. Chen, and Z.J. Zhang, *Material Characterization*, Vol. 59, p. 514, 2008.
3. Y.H. Kim, and E.R. Canaway, *Environ. Technol.*, Vol. 24, p. 1455, 2003.
4. R.W. Gillham, and S.F. O'Hannesin, *Ground Water*, Vol. 32, p. 958, 1994.
5. M.S.H. Mak, P. Rao, and I.M.C. Lo, *Environ. Pollut.*, Vol. 159, p. 377, 2011.
6. S.C. Kim, J.E. Yang, Y.S. Ok, J. Skousen, D.G. Kim, and J.H. Joo, *Bull. Environ. Contam. Toxicol.*, Vol. 84, p. 459, 2010.
7. D.C.W. Tsang, N.J.D. Grahm, and I.M.C. Lo, *Chemosphere*, Vol. 575, p. 1338, 2009.
8. T. Liu, D.C.W. Tsang, and I.M.C. Lo, *Environ. Sci. Technol.*, Vol. 42, p. 2092, 2008.
9. R.A. Doong, and Y.L. Lai, *Chemosphere*, Vol. 64, p. 371, 2006.
10. I.M.C. Lo, C.S.C. Lam, and K.C.K. Lai, *Water Res.*, Vol. 40, p. 595, 2006.
11. X.Q. Li, D.W. Elliott, and W.X. Zhang, *Crit. Rev. Solid State Mater. Sci.*, Vol. 31, p. 111, 2006.
12. L. Li, M. Fan, R.C. Brown, and J.H.V. Leeuwen, *Crit. Rev. Environ. Sci. Technol.*, Vol. 36, p. 405, 2006.
13. B. Nowack, *Environ. Aspect*, Vol. 2, p. 1, 2008.
14. A.B. Cundy, L. Hopkinson, and R.L.D. Whitby, *Sci. Tot. Environ.*, Vol. 400, p. 42, 2008.
15. R.A. Crane, and T.B. Scott, *J. Hazard. Mater.*, Vol. 211–212, p. 112, 2012.
16. Y.P. Sun, J. Cao, W.X. Zhang, and H.P. Wang, *Adv. Colloid Interf. Sci.*, Vol. 37, p. 47, 2006.
17. L.T. Kuhn, A. Bojesen, L. Timmermann, M.M. Nielsen, and S. Morup, *J. Phys. Condens. Matter*, Vol. 14, p. 13551, 2002.
18. G.N. Glavee, K.J. Klabunde, C.M. Sorensen, and G.C. Hadjipanayis, *Inorg. Chem.*, Vol. 34, p. 28, 1995.
19. C.J. Choi, X.L. Dong, and B.K. Kim, *Mater. Trans.*, Vol. 42, p. 2046, 2001.
20. R. Singh, V. Misra, and R.P. Singh, *J. Nanoparticle Res.*, Vol. 13, p. 4063, 2011.
21. M.N.A. Karlsson, K. Deppert, B.A. Wacaser, L.S. Karlsson, and J.O. Malm, *Appl. Phys. A: Mater. Sci. Process.*, Vol. 80, p. 1579, 2005.

22. M. Uegami, J. Kawano, T. Okita, Y. Fujii, K. Okinaka, K. Kakuya, S. Yatagai, Toda Kogyo Corp. US Patent Application, 2003.
23. K. Elihn, F. Otten, M. Boman, F.E. Kruis, H. Fissan, J.O. Carlsson, *J. Nanostruct. Mater.*, Vol. 12, p. 79, 1999.
24. F. He, D. Zhao, J. Liu, and C.B. Robert, *Ind. Eng. Chem. Res.*, Vol. 46, p. 29, 2007.
25. F. He, and D. Zhao, *Environ. Sci. Technol.*, Vol. 39, p. 3314, 2005.
26. C.B. Wang, and W.X. Zhang, *Environ. Sci. Technol.*, Vol. 31, p. 2154, 1997.
27. M.M. Scherer, B.A. Balko, and P.G. Tratnyek, "The role of oxides in reduction reactions at the metal-water interface," in: *Mineral-Water Interfacial Reactions*, Vol. 715, Ch 15, p. 301, 1999.
28. T.L. Johnson, M.M. Scherer, and P.G. Tratnyek, *Environ. Sci. Technol.*, Vol. 30, p. 2634, 1996.
29. J.E. Martin, A.A. Herzing, W. Yan, X.Q. Li, B.E. Koel, C.J. Kieley, and W.X. Zhang, *Langmuir*, Vol. 24, p. 4329, 2008.
30. W.X. Zhang, *J. Nanopart. Res.*, Vol. 5, p. 323, 2003.
31. H.L. Lien, and W.X. Zhang, *J. Environ. Eng.*, Vol. 125, p. 1042, 1999.
32. P. Varanasi, A. Fullana, and S. Sidhu, *Chemosphere*, Vol. 66, p. 1031, 2007.
33. G.V. Lowry, and K.M. Johnson, *Environ. Sci. Technol.*, Vol. 38, p. 5208, 2004.
34. H. Song, and E.R. Carraway, *Environ. Sci. Technol.*, Vol. 39, p. 6237, 2005.
35. R. Singh, V. Misra, M.K.R. Mudiam, L.K.S. Chauhan, and R.P. Singh, *J. Hazard. Mater.*, Vol. 237–238, p. 355, 2012.
36. R. Singh, V. Misra, and R.P. Singh, *J. Bionanosci.*, Vol. 5, p. 82, 2011.
37. R. Singh, A. Singh, V. Misra, and R.P. Singh, *J. Biomed. Nanotechnol.*, Vol. 7, p. 175, 2011.
38. Z. Wang, P. Peng, and W. Huang, *J. Hazard. Mater.*, Vol. 166, p. 992, 2009.
39. D.W. Elliott, H.L. Lien, and W.X. Zhang, *J. Environ. Qual.*, Vol. 37, p. 2192, 2008.
40. T. Satapanajaru, P. Ankurakpongsator, P. Pengthamkeerati, and H. Boparai, *Water Air Soil Pollut.*, Vol. 192, pp. 349–359, 2008.
41. R.D. Ambashta, E. Repo, and M. Sillanpää, *Ind. Eng. Chem. Res.*, Vol. 50, p. 11771, 2011.
42. A.V.B. Reddy, V. Madhavi, K.G. Reddy, and G. Madhavi, *J. Chem.*, doi.org/10.1155/2013/521045, 2013.
43. G. Naja, A. Halasz, S. Thiboutot, G. Ampleman, and J. Hawari, *Environ. Sci. Technol.*, Vol. 42, p. 4364, 2008.
44. X. Zhang, Y.M. Lin, X.Q. Shan, and Z.L. Chen, *Chem. Eng. J.*, Vol. 158, p. 566, 2010.
45. Y. Lin, C. Weng, and F. Chen, *Sep. Purif. Technol.*, Vol. 64, p. 26, 2008.
46. J. Fan, Y. Guo, J. Wang, and M. Fan, *J. Hazard. Mater.*, Vol. 166, p. 904, 2009.
47. Y.H. Shih, and Y.T Tai, *Chemosphere*, Vol. 78, p. 1200, 2010.
48. M.C. Chang, and H.Y. Kang, *J. Environ. Sci. Health, A: Tox. Hazard. Subst. Environ. Eng.*, Vol. 44, p. 576, 2009.
49. Z. Xiong, D. Zhao, and G. Pan, *Water Res.*, Vol. 41, p. 3497, 2007.
50. Y. Xu, and D. Zhao D, *Water Res.*, Vol. 41, p. 2101, 2007.

51. D.V. Franco, L.M.D. Silva, and W.F. Jardim, *Water Air Soil Pollut.*, Vol. 197, p. 49, 2009.
52. Y.H. Liou, S.L. Lo, C.J. Lin, W.H. Kuan, and S.C. Weng, *J. Hazard. Mater.*, Vol. 127, p. 102, 2005.
53. M.E. Morgada, I.K. Levy, V. Salomone, S.S. Farias, G. Lopez, and M.I. Litter, *Catal. Today*, Vol. 143, p. 261, 2009.
54. R. Singh, V. Misra, and R.P. Singh, *Environ. Monit. Assess.*, Vol. 184, p. 3643, 2011.
55. S.R. Kanel, B. Manning, L. Charlet, and H. Choi, *Environ. Sci. Technol.*, Vol. 39, p. 1291, 2005.
56. U.C.O Çelebi, T. Shahwan, and H.N. Erten, *J. Hazard. Mater.*, 148, p. 761, 2007.
57. X.Q. Li, and W.X. Zhang, *J. Phys. Chem.*, Vol. 111, p. 6939, 2007.
58. T.B. Dickinson, and T.B. Scott, *J. Hazard. Mater.*, Vol. 178, p. 171, 2010.
59. J.G. Darab, A.B. Amonette, D.S.D. Burke, and R.D. Orr, *Chem. Mater.*, Vol. 19, p. 5703, 2007.
60. J. Cao, and W.X. Zhang, *J. Hazard. Mater.*, Vol. B132, p. 213, 2006.
61. S.R. Kanel, J.M. Greneche, H. Choi, *Envrion. Sci. Technol.*, Vol. 40, p. 2045, 2006.
62. W. Yan, A.A. Herzing, C.J. Kiely, and W.X. Zhang, *J. Contam. Hydrol.*, Vol. 118, p. 96, 2010.
63. J.T. Nurmi, P.G. Tratnyek, V. Sarathy, D.R. Baer, J.E. Amonette, K. Pecher, C. Wang, J.C. Linehan, D.W. Matson, R.L. Penn, and M.D. Driessen, *Environ. Sci. Technol.*, Vol. 39, p. 1221, 2005.
64. S.M. Ponder, J.G. Darab, and T.E. Mallouk, *Environ. Sci. Technol.*, Vol. 34, p. 2564, 2000.
65. L. Wu, M. Shamsuzzoha, and S.M.C. Ritchie, *J. Nanopart. Res.*, Vol. 7, p. 469, 2005.
66. T. Zhou, Y. Li, T.T. Lim, *Sep. Purif. Technol.*, Vol. 76, p. 206, 2010.
67. M.J. Alowitz, and M. Scherer, *Environ. Sci. Technol.*, Vol. 36, p. 299, 2002.
68. T. Satapanajaru, P.J. Shea, S.D. Comfort, and Y. Roh, *Environ. Sci. Technol.*, Vol. 37, p. 5219, 2003.
69. N. He, P. Li, Y. Zhou, S. Fan, and W. Ren, *Chemosphere*, Vol. 76, p. 1491, 2009.
70. P.G. Tratnyek, M.M. Scherer, B. Deng, and S. Hu, *Water Res.*, Vol. 35, p. 4435, 2001.
71. E. Tipping, *Cation Binding by Humic Substances*, Cambridge University Press, Cambridge, UK, 2002.
72. B. Gu, J. Schmiu, Z. Chen, L. Liang, and J.F. MaCarthy, *Environ. Sci. Technol.*, Vol. 28, p. 38, 1994.
73. D.C.W. Tsang, N.J.D. Graham, I.M.C. Lo, *Chemosphere*, Vol. 75, p. 1338, 2009.
74. T. Liu, D.C.W. Tsang, and I.M.C. Lo, *Environ. Sci. Technol.*, Vol. 42, p. 2092, 2008.

75. E. Keane, Fate, transport, and toxicity of nanoscale zero-valent iron (nZVI) used during superfund remediation, USEPA, Wahington D.C., 2009.
76. Y.H. Tee, L. Bachas, and D. Bhattacharyya, *J. Phys. Chem. C*, Vol. 113, p. 9454, 2009.
77. H.L. Lien, and W.X. Zhang, *Colloid. Surf.*, Vol. 191, p. 97, 2001.
78. M. Nutt, J.B. Hughes, and M. Wong, *Environ. Sci. Technol.*, Vol. 39, p. 1343, 2005.
79. J. Yang, J.Y. Lee, L.X. Chen, and H.P. Too, *J. Phys. Chem. B*, Vol. 109, p. 5468, 2005.
80. S.M. Hosseini, B.A. Ashtiani, and M. Kholghi, *Desalination*, Vol. 276, p. 214, 2011.
81. F. He, D. Zhao, *Environ. Sci. Technol.*, Vol. 41, p. 6216, 2007.
82. N. He, P. Li, Y. Zhou, S. Fan, and W. Ren, *Chemosphere*, Vol. 76, p. 1491, 2009.
83. B.W. Zhu, T. Lim, and J. Feng, *Chemosphere*, Vol. 65, p. 1137, 2006.
84. N. Sakulchaicharoen, D.M. O'Carroll, and J.E. Herrera, *J. Contam. Hydrol.*, Vol. 118, p. 117, 2010.
85. A. Tiraferri, K.L. Chen, R. Sethi, and M. Elimelech, *J. Colloid Interf. Sci.*, Vol. 324, p. 71, 2008.
86. T. Dong, H. Luo, Y. Wang, B. Hu, and C. Hua, *Desalination*, Vol. 271, p. 11, 2011.
87. H. Zhou, J. Han, S.A. Baig, and X. Xu, *J. Hazard. Mater.*, Vol. 198, p. 7, 2011.
88. H. Park, Y.M. Park, K.M. Yoo, and S.H. Lee, *Water Sci. Technol.*, Vol. 59, p. 2153, 2009.
89. X. Zhang, S. Lin, Z. Chen, M. Megharaj, and R. Naidu, *Water Res.*, Vol. 45, p. 3481, 2011.
90. C. Chang, F. Lian, and L. Zhu, *Environ. Pollut.*, Vol. 159, p. 2507, 2011.
91. J. Chen, Z. Xiu, G.V. Lowry, and P.J. Alvarez, *Water Res.*, Vol. 45, p. 1995, 2011.
92. L.N. Shi, X. Zhang, and Z.L. Chen, *Water Res.*, Vol. 45, p. 886, 2011.
93. A.N. Bezbaruah, S. Krajangpan, B.J. Chisholm, E. Khan, and J.J. Bermudez, *J. Hazard. Mater.*, Vol. 166, p. 1339, 2009.
94. R. Singh, V. Misra, and R.P. Singh, *IPCBEE*, Vol. 4, p. 162, 2011.
95. T. Liu, L. Zhao, D. Sun, and X. Tan, *J. Hazard. Mater.*, Vol. 184, p. 724, 2010.
96. S. Krajangpan, J.J. Bermudez, A.N. Bezbaruah, B.J. Chisholm, and E. Khan, *Water Sci. Technol.*, Vol. 58, p. 2215, 2008.
97. H. Kim, H.J. Hong, J. Jung, S.H. Kim, and J.W. Yang, *J. Hazard. Mater.*, Vol. 176, p. 1038, 2010.
98. T. Liu, Z.L. Wang, L. Zhao, and X. Yang, *Chem. Eng. J.*, Vol. 189, p. 196, 2012.
99. W.R. Puls, C.J. Paul, and R.M. Powell, *Appl. Geochem.*, Vol. 14, p. 989, 1999.
100. R.T. Wilkin, C.M. Su, R.G. Ford, and C.J. Paul, *Environ. Sci. Technol.*, Vol. 39, p. 4599, 2005.
101. B. Flury, U. Eggenberger, and U. Mader, *Appl. Geochem.*, Vol. 24, p. 687, 2009.
102. D.W. Elliott, and W.X. Zhang, *Environ. Sci. Technol.*, Vol. 35, p. 4922, 2001.

103. R. Glazier, R. Vankatakrishnan, F. Gheorghiu, L. Walata, R. Nash, and W. Zhang, *Civil Engineering*, Vol. 73, p. 64, 2003.
104. J. Quinn, C. Geiger, C. Clausen, K. Brooks, C. Coon, S. O'Hara, T. Krug, D. Major, W.S. Yoon, A. Gavaskar, and T. Holdsworth, *Environ. Sci. Technol.*, Vol. 39, p. 1309, 2005.
105. S. O'Hara, T. Krug, J. Quinn, C. Clausen, and C. Geiger, *Remediation*, Vol. 16, p. 35, 2006.
106. F. He, D. Zhao, and C. Paul, *Water Res.*, Vol. 44, p. 2360, 2010.
107. R. Singh, V. Misra, and R.P. Singh, *Bull. Environ. Contam. Toxicol.*, Vol. 88, p. 210, 2012.
108. N.C. Muller, J. Braun, J. Bruns, M. Cernik, P. Rissing, D. Rickerby, and B. Nowack, *Environ. Sci. Pollut. Res.*, Vol. 19, p. 550, 2012.
109. W. Yan, H.L. Lien, B.E. Koel, and W.X. Zhang, *Environ. Sci.: Processes Impacts*, Vol. 15, p. 63, 2013.
110. S. Harendra, and C. Vipulanandan, *Ind. Eng. Chem. Res., Vol.* 50, p. 404, 2011.
111. B. Schrick, B.W. Hydutsky, J.L. Blough, and T.E. Mallouk, *Chem. Mater.*, Vol. 16, p. 2187, 2004.
112. Y.P. Sun, X.Q. Li, W.X. Zhang, and H.P. Wang, *Colloid. Surf. A: Physicochem. Eng. Aspects*, Vol. 208, p. 60, 2007.
113. N. Saleh, T. Phenrat, K. Sirk, B. Dufour, J. Ok, T. Sarbu, K. Matyjaszewski, R.D. Tiliton, and G.V. Lowry, *Nano Lett.*, Vol. 5, p. 2489, 2005.
114. E.D. Vechhia, M. Luna, and R. Sethi, *Environ. Sci. Technol.*, Vol. 43, p. 8942, 2009.
115. C. Cameselle, K. Reddy, K.K. Darko, and A. Khodadoust, *J. Envrion. Eng.*, Vol. 139, p. 23, 2013.
116. Y. Li, Z. Xiu, T. Li, and Z.Z. Zing, *ACS Symposium Series*, Vol. 1124, p. 307, 2013.
117. G.C.C. Yang, H.C. Tu, and C.H. Hung, *Sep. Purif. Technol.*, Vol. 58, p. 166, 2007.
118. S. Pamukcu, L. Hannum, and J.K. Wittle, *J. Environ. Sci. Health*, Vol. 43, p. 934, 2008.
119. K.R. Reddy, K. Darko-Kagya, and C. Cameselle, *Sep. Purif. Technol.*, Vol. 79, p. 230, 2011.
120. G.C. Yang, C.H. Hung, and H.C. Tu, *J. Environ. Sci. Health, A: Tox. Hazard. Subst. Environ. Eng.*, Vol. 43, p. 945, 2008.
121. Z. Li, S. Yuan, J. Wan, H. Long, and M. Tong, *J. Contam. Hydrol.*, Vol. 124, p. 99, 2011.
122. T. Suzuki, Y. Ovama, M. Moribe, and M. Niinae, *Water Res.*, Vol. 46, p. 772, 2012.
123. V. Bokare, K. Murugesan, Y.M. Kim, J.R. Jeon, E.J. Kim, and Y.S. Chang, *Bioresour. Technol.*, Vol. 101, p. 6354, 2010.
124. Y.M. Kim, K. Murugesan, Y.Y. Chang, E.J. Kima, and Y.S. Chang, *J. Chem. Technol. Biotechnol.*, Vol. 87, p. 216, 2012.
125. K. Murugesan, V. Bokare, J.R. Jeon, E.J. Kim, J.H. Kim, and Y.S. Chang, *Bioresour. Technol.*, Vol. 102, p. 6019, 2011.

126. S.M. Wang, and S.K. Tseng, *Bioresour. Technol.*, Vol. 100, p. 111, 2009.
127. R. Singh, N. Manickam, M.K.R. Mudiam, R.C. Murthy, and V. Misra, *J. Hazard. Mater.*, Vol. 258–259, p. 35, 2013.
128. M.H. Diao, and M.S. Yao, *Water Res.*, Vol. 43, p. 5243, 2009.
129. Z. Li, K. Greden, P.J.J. Alvarez, K.B. Gregory, G.V. Lowry, *Environ. Sci. Technol.*, Vol. 44, p. 3462, 2010.
130. T. Phenrat, T.C. Long, G.V. Lowry, and B. Veronesi, *Environ. Sci. Technol.*, Vol. 43, p. 195, 2009.
131. C.R. Keenan, G.R. Goth, D. Lucas, and D.L. Sedlak, *Environ. Sci. Technol.*, Vol. 43, p. 4555, 2009.
132. P.J. Chen, W.L. Wu, and K.C.W. Wu, *Water Res.*, doi.Org/10.1016/j.watres.2012.12.043.
133. Y. An, T. Li, Z. Jin, M. Dong, H. Xia, and X. Wang, *Bioresour. Technol.*, Vol. 101, p. 9825, 2010..
134. M. Yu, S. Huang, K.V. Yu, and A.M. Clyne, *Int. J. Mol. Sci.*, Vol. 13, p. 5554, 2012.
135. H. Song, and E.R. Carraway, *Environ. Sci. Technol.*, Vol. 39, p. 6237, 2005.
136. X.Q. Li, G.B. Derick, and W.X. Zhang, *J. Nanopart. Res.*, Vol. 9, p. 233, 2007.
137. D.W. Elliott, H.L. Lien, and W.X. Zhang, *J. Environ. Eng.*, Vol. 135, p. 317, 2009.
138. H.Y. Lee, C.L. Lee, and C.J.G. Jou, *Water Air Soil Pollut.*, Vol. 211, p. 17, 2010.
139. J.M. Thompson, B.J. Chisholm, and A.N. Bezbaruah, *Environ. Eng. Sci.*, Vol. 27, p. 227, 2010.
140. Y.H. Shih, and C.P. Tso, *Environ. Eng. Sci.*, Vol. 29, p. 929, 2012.
141. K.H. Shin, and D.K. Cha, *Chemosphere*, Vol. 72, p. 257, 2008.
142. X.Q. Li, J. Cao, and W.X. Zhang, *Ind. Eng. Chem. Res.*, Vol. 47, p. 2131, 2008.
143. K.S. Lin, N.B. Chang, and T.D. Chuang, *Sci. Technol. Adv. Mater.*, Vol. 9, p. 025015, 2008.
144. M. Biterna, L. Antonoglou, E. Lazou, and D. Voutsa, *Chemosphere*, Vol. 78, p. 7, 2010.
145. H.K. Boparai, M. Joseph, and D.M. O'Carroll, *J. Hazard. Mater.*, Vol. 186, p. 458, 2011.
146. T.B. Scott, I.C. Popescu, R.A. Crane, and C. Noubactep, *J. Hazard. Mater.*, Vol. 186, p. 280, 2011.
147. T. Almeelbi, and A. Bezbaruah, *J. Nanopart. Res.*, Vol. 14, p. 900, 2012.
148. J. Feng, and T. Lim, *Chemosphere*, Vol. 59, p. 1267, 2005.
149. W. Zhang, X. Quan, J. Wang, Z. Zhang, and S. Chen, *Chemosphere*, Vol. 65, p. 58, 2006.
150. Z. Zhang, Q. Shen, N. Cissoko, J. Wo, and X. Xu, *J. Hazard. Mater.*, Vol. 182, p. 252, 2010.
151. N. He, P. Li, Y. Zhou, W. Ren, S. Fan, and V.A. Verkhozina, *J. Hazard. Mater.*, Vol. 164, p. 126, 2009.
152. Y.P. Katsenovich, and F.R. Miralles-Wilhelm, *Sci. Total Environ.*, Vol. 407, p. 4968, 2009.
153. V. Nagpal, A.D. Bokare, R.C. Chikate, C.V. Rode, and K.M. Paknikar, *J. Hazard. Mater.*, Vol. 175, p. 680, 2010.

154. S. Luo, S. Yang, X. Wang, and C. Sun, *Chemosphere*, Vol. 79, p. 672, 2010.
155. Z. Fang, X. Qiu, J. Chen, and X. Qiu, *J. Hazard. Mater.*, Vol. 185, p. 958, 2011.
156. Z. Zhang, S. Hu, S.A. Baig, J. Tang, and X.J. Xu, *Colloid Interface Sci.*, Vol. 385, p. 160, 2012.
157. S.H. Joo, and D. Zhao, *Chemosphere*, Vol. 70, p. 418, 2008.
158. J. Cao, R. Xu, H. Tang, S. Tang, and M. Cao, *Sci. Tot. Environ.*, Vol. 409, p. 2336, 2011.
159. H. Zhou, J. Han, S.A. Baig, and X. Xu, *J. Hazard. Mater.*, Vol. 198, p. 7, 2011.
160. H. Kang, Z. Xiu, J. Chen, W. Cao, Y. Guo, T. Li, and Z. Jin, *Environ. Technol.*, Vol. 33, p. 2185, 2012.
161. Z. Marková, K.M. Šišková, J. Filip, J. Čuda, M. Kolář, K. Šafářová, I. Medřík, and R. Zbořil, *Environ. Sci. Technol.*, Vol. 47, p. 5285, 2013.
162. Y. Li, J. Li, and Y. Zhang, *J. Hazard. Mater.*, Vol. 227, p. 211, 2012.

15

Typical Synthesis and Environmental Application of Novel TiO_2 Nanoparticles

Tanmay Kumar Ghorai

Department of Chemistry, West Bengal State University, Kolkata, West Bengal, India

Abstract

In the 21st century, water pollution has become a major worldwide problem due in large part to the textile industry. Therefore the emphasis of this chapter will be on novel TiO_2-based nanoparticles that will continue to play an important role in the implementation of efficient, consistent and profitable techniques to remove pollutants (organic/inorganic) from wastewater. This chapter also focuses on the preparation, characterization and application of noble metal nanoparticles, transition metal ions, and nonmetal-doped titania networks. A review is presented of the effects of operating parameters on the photocatalytic degradation of textile dyes using TiO_2-based photocatalysts. The findings reveal that various parameters, such as the initial pH of the solution to be degraded, oxidizing agents, temperature at which the catalysts must be calcined, dopant(s) content and catalyst loading, exert their individual influence on the photocatalytic degradation of any dye in wastewaters. It was also found that the sol–gel or chemical methods are widely used in the production of TiO_2-based photocatalysts because of the advantage derived from their ability to synthesize nanosized crystallized powder of the photocatalysts with high purity at relatively low temperature.

Keywords: Synthesis, TiO_2 nanoparticles, photocatalysts, environmental application

15.1 Introduction

Environmental science is a multidisciplinary science involving chemistry, physics, life science, agriculture, medical science, public health, sanitary engineering, etc. In simple terms, it is the science of chemical phenomenon

**Corresponding author*: tanmayghorai66@gmail.com

Ashutosh Tiwari and Mikael Syväjärvi (eds.) Advanced Materials for Agriculture, Food, and Environmental Safety, (421–452) 2014 © Scrivener Publishing LLC

in the environment. Environmental pollution is any discharge of material or energy into air, land or water, that causes or may cause acute immediate or continuing detriment to the Earth's ecological balance or that lowers the eminence of life. Pollutants may cause primary damage with direct identifiable impact on the environment, or secondary damage in the form of minor perturbations in the delicate balance of the biological food web that are detectable only over long time periods. Factors contributing to pollution such as the industrialization of society, the introduction of motorized vehicles, and the explosion of the human population, and an exponential growth in the production of goods and services.

Although many different types of pollution have been observed (i.e., water, air, noise, thermal, land, pesticide, radiation, etc.), the focus of this chapter is water pollution. In the 21st century, water pollution is a worldwide environmental problem typically created from industrial and agricultural wastes. Most of the dyes produced by dyeing, painting, and textile industries contain different organic or inorganic contaminants, which are toxic and carcinogenic; and thus they have imposed a serious problem in the environment [1–4]. Therefore, a major focus for every scientist is the development and implementation of efficient, consistent and profitable techniques for the removal of such toxic contaminants. Among the different kinds of procedures, one of the best procedures is the photocatalytic oxidation of organic pollutants in wastewater and air, as an advanced oxidation process has been proven to be an effective technique for environmental remediation. To develop efficient photocatalytic systems, high-quality semiconductor-based materials have been actively studied in recent years. Among a wide spectrum of semiconductors, TiO_2 has attracted significant attention over the past decades due to its excellent performance as a photocatalyst under UV light irradiation. The most commonly used TiO_2 for environmental applications is the commercial Degussa P25 TiO_2 [5–6]. The photocatalyst is normally 99.5% pure TiO_2 (80% anatase and 20% rutile). The model structure of the rutile and antase TiO_2 photocatalyst [6] is illustrated in Figure 15.1a and 15.1b.

In this context, novel TiO_2 or mesoporous compounds have an important role in photochemical reactions for removal of such types of pollutants. Novel or mesoporous titanium dioxide (TiO_2) has continued to be highly active in photocatalytic applications because of its high surface area and continuous particle framework, which may be beneficial, compared to separate individual nanoparticles, in particular for catalyst recovery [7–8]. In the past decade, there have been few reports about mesoporous transition-metal oxides [9–11]. Doped TiO_2 with high surface area transfers the electrons from the dye molecules to the transfer conduction

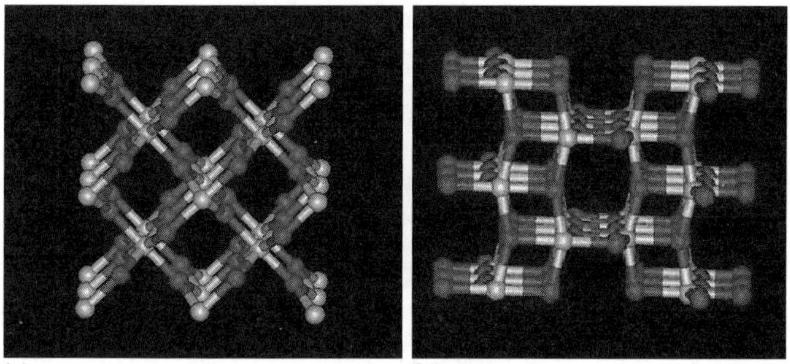

Figure 15.1 (a) Rutile structure of titanium dioxide; (b) Anatase structure of titanium dioxide.

band and generate e^-/h^+ pairs that degrade the dye molecules. Generally, nanoparticles are used for the fabrication of noble TiO_2 surface to obtain high surface areas and a mesoporous structure. However, nanocrystalline TiO_2 is a promising semiconducting material that has attracted considerable interest for its applications as electrodes [12], electronics [13], capacitors [14], optics [15, 16], sensors [17, 18], ceramics [19], solar cells [20–22], catalysis [23, 24] and photocatalysis [25, 26].

Henceforth, introduction of larger particles with a lower internal surface area on nanoparticles hinders dye adsorption. The result is an increase in the electron diffusion path and recombination rate of photogenerated carriers rather than an enhancement of the effect of light scattering. However, for specific advanced applications in photocatalysis, especially in the degradation of organic pollutants, the optical and electrical properties of TiO_2 have been extensively studied [27–32]. The possibility of carrying out selective photooxidations in nonaqueous media is interesting in the context of Green Chemistry [33, 34]. However, TiO_2 has wide band-gap energy (3.0–3.2 eV), which means that UV light (wavelength < 387 nm) can excite anatase TiO_2 to generate e^-_{cb}/h^+_{vb} pairs which cause the photocatalytic reactions.

Therefore, lots of research have been done on the photocatalytic activity of TiO_2 doped with different metal ions such as noble metals (Ag [35], Au [36], Pt [37] and Pd [38]), transition metals (Mn [39], Fe [40, 41], Co [42], Ni [43], Cu [44, 45]), and nonmetals (nitrogen, carbon, fluoride, iodine, nitrate and nitrite [46–51]), which have shown to be good photocatalyst. Ismail *et al.* found that TiO_2 microspheres with porous structures have higher photocatalytic activity and are easily recovered for reuse [52]. This breakthrough opened a novel pathway for the elevation of the photocatalysis of TiO_2 materials. Furthermore, photoactivity is highly dependent on surface area and crystallinity or crystal sizes, which are influenced by the synthetic methods of titania [53, 54].

The most recent studies show that the addition of small quantities of Nb_2O_5 into TiO_2 lattice significantly increases the photocatalytic activity and surface acidity of TiO_2 [55, 56]. Concurrently, Cr(III)- and Fe(III)-doped photocrystalline titania have also been widely studied in the field of photocatalysis [41, 57–60]. Recently, Han *et al.* studied the photocatalytic degradation of 2,4-dichlorophenol with SiO_2-TiO_2 nanomaterials [61], and Tiwari *et al.* synthesized different types of nanomaterial compounds [62]. The TiO_2-based alloy composites [26, 32] also show the photocatalytic activity of degradation of different dyes. Nonconventional synthetic procedures such as chemical vapor deposition [63], hydrothermal [64], microemulsion [65], sol-gel [66, 67], coprecipitation [68], solid state [39] and chemical methods [69] were adopted to produce such compositions.

In the present chapter, the recent developments in the syntheses and characterization of TiO_2-based noble metals, transitions metals ions and different TiO_2-based mesoporous compounds were prepared from a titanium tartarate with triethanol amine precursor by sol-gel or novel chemical method. The photocatalytic activity of doped TiO_2 materials upon visible and UV illumination will be reviewed, summarized and discussed, in particular, concerning the influence of preparation and solid-state properties of the materials. Reaction mechanisms explaining these effects will be presented and critically evaluated.

15.2 Use of Different Dyes

The chemical structures of four dyes are presented below. The degradation rate of all four dyes shows a pseudo-first-order degradation pattern.

15.2.1 Methyl Orange Degradation (MO)

Methyl orange is one of the most important classes of commercial dyes. It is stable to visible and near-UV light and provides a useful probe for photocatalytic reactions [70–72]. Because methyl orange turns yellow in an alkaline solution and red in acidic solution, it is easy to monitor and analyze by spectrophotometry. The mineralization, decolorization and decomposition of methyl orange over TiO_2 have been well studied, showing a pseudo-first-order degradation pattern [72]. The absorbance

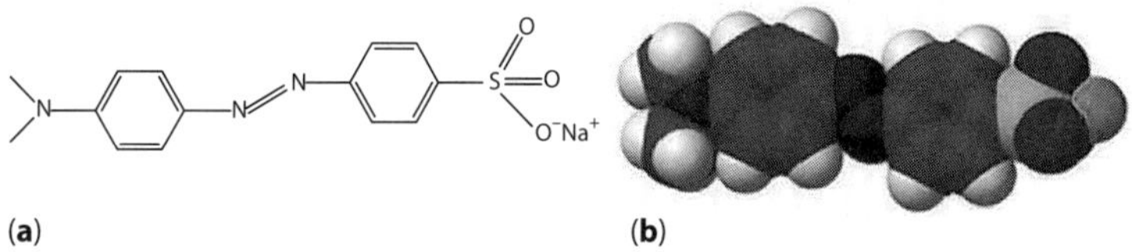

Figure 15.2 (a) The chemical structure of MO; (b) Model structure of Methyl orange.

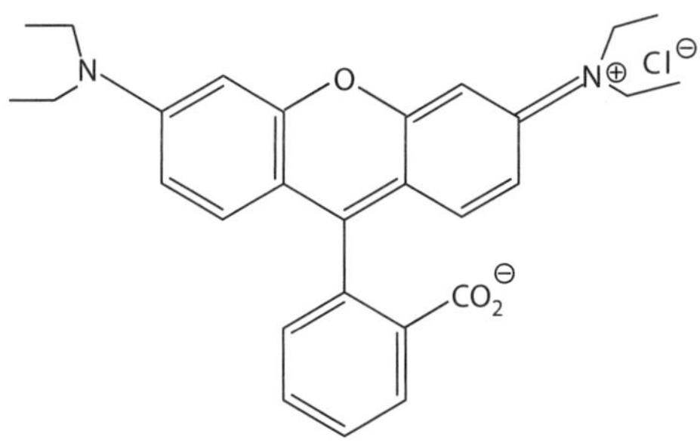

Figure 15.3 The chemical structure of Rhodamine B.

of methyl orange was detected by a UV-vis spectrophotometer at 464 nm. The chemical and model structure of methyl orange is shown in Figure 15.2a and 15.2b.

15.2.2 Rhodamine B (RB)

The chemical structure of Rhodamine B is shown in Figure 15.3. It is used as a dye or pigment and is a neutral indicator. The degradation of Rhodamine B is dependent on the catalyst and UV/visible light irradiation, and the rate of the reaction is also dependent on the loading of photocatalyst. The absorbance of Rhodamine B is 554 nm, which was detected by a UV-vis spectrophotometer.

15.2.3 Thymol Blue (TB)

The chemical structure of thymol blue ($C_{27}H_{30}O_5S$) is shown in Figure 15.4. It is used as an acid base indicator. Because thymol blue turns blue in an alkaline solution and yellow in acidic solution, its absorbance was 597 nm when analyzed by UV-vis spectrophotometer.

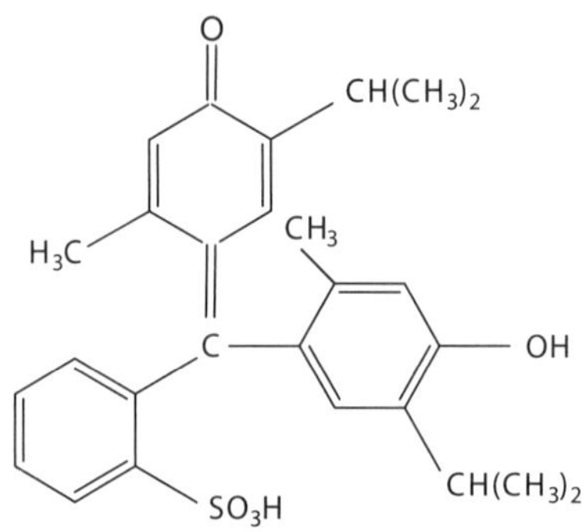

Figure 15.4 The chemical structure of Thymol blue.

15.2.4 Bromocresol Green (BG)

Bromocresol green (3,3',5,5'-tetrabromo-m-cresolsulfonphthalein) is a dye of the triphenylmethane family (triarylmethane dyes), which is used as a pH indicator and as a tracking dye for DNA agarose gel electrophoresis. In aqueous solution, both solids ionize to give the monoanionic form (yellow), that further deprotonates at higher pH to give the dianionic form (blue), which is stabilized by resonance. The absorbance of Bromocresol green is 616 nm, which was measured by a UV-vis spectrophotometer. The chemical structure and resonance form of Bromocresol green are presented in Figure 15.5.

Therefore, TiO_2 is an interesting material for photocatalytic applications and is regarded as the most efficient and environmentally benign photocatalyst, and it has been most widely used for photodegradation of various pollutants such as Rhodamine B (RhB) [73–76], Rhodamine 6G (Rh6G) [77–85], methyl orange (MO) [78, 79], methylene blue (MB) [78, 80–83], thymol blue [84], Bromocresol green [41] and phenol [75–83]. The principle of the semiconductor photocatalytic reaction is straightforward. When photons with energies > 3.2 eV, i.e., exceeding the band gap energy of TiO_2, are absorbed by the anatase particles in the mesoporous TiO_2 photocatalysts, electrons are rapidly promoted from the valence band to the conduction band leaving holes behind in the valence band [86]. The thus formed electrons and holes participate in redox processes at the semiconductor/water interface. The valance band holes migrate to the surface of the particles where they react with adsorbed hydroxide ions (or water molecules), generating adsorbed $^{\cdot}OH$ radicals. This photodecomposition process usually involves one or more radicals or intermediate species such as $HO^{\cdot}$, $O^{\cdot}$, H_2O_2 or $O_2^{\cdot-}$, which play important roles in the photocatalytic

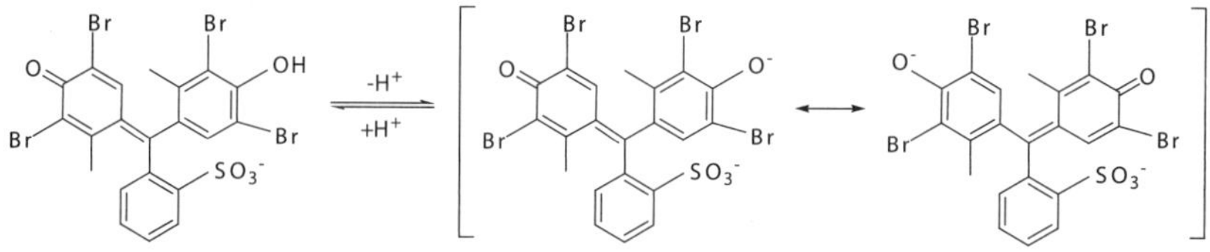

Figure 15.5 The resonance structure of Bromocresol green.

reaction mechanisms [41, 86]. The photocatalytic activity of a semiconductor is largely controlled by: (i) the light absorption properties, e.g., light absorption spectrum and coefficient, (ii) reduction and oxidation rates on the surface by the electron and hole, and (iii) the electron–hole recombination rate [86].

15.3 Synthetic Methods for Novel Titania Photocatalysts

There are different preparative methods described in the literature to obtain the preparation of nanosize transition metal-doped titanium dioxide photocatalysts. The widely known methods to prepare nanomaterials compound are: sol-gel synthesis and novel chemical method. The precursor routes play a crucial role in designing the final products and are also better and more convenient methods for the preparation of multicomponent, transition metal-doped titanium dioxide nanophotocatalysts. The precursor compounds are usually complex combinations of cations in the proper ratios, together with ionic and molecular species, containing the necessary oxygen for the formation of solid solutions or the crystalline compounds. The remainders are volatile, hence are decomposable into volatile elements. The pyrolysis of the complex combinations at temperatures between 200°C to 500°C in an appropriate inert/oxidizing/reducing atmosphere, depending on the material, gives nanoscale particles of the desired mixed-oxide system with a good cation stoichiometry. Literature reveals that various organometallic complexes (metal alkoxides, etc.), metal-hydroxides/-carbonates/-oxalates/-citrates/-nitrates, are a few of the commonly used precursor compounds in this process [87–88]. Very recently, Pramanik *et al.* prepared the nanosized transition metal-doped titanium dioxide photocatalyst by precursor decomposition method [26, 41, 89].

The precursor method involves preparation of a precursor compound. Morgan [90, 91] has suggested that the use of chemistry in the preparation of nanocrystalline materials, can avoid three major problems: diffusions, impurities and agglomeration. Nanocrystalline powders, prepared by chemical process, have better processibility with improved homogeneity, brought about by reduction in the diffusion distance between the atoms. The chemical precursors used in these processes can be developed to increase the purity, and a careful control of solvent removal could led to crushable agglomerates. However, in many cases, the chemical routes generally involve complex technique compared to the conventional ceramic methods, and an improved level of skill is required to realize their benefits [92, 93].

To date, of all the above processes, only novel chemical precursor decomposition method can precisely control the microstructure of the final products. The novel chemical precursor decomposition method:

- Generally is used to produce different types of nanosize ceramics and photocatalytic materials at ultra-low temperatures (around 250–700°C); and in conventional technique temperature varies from 800–1500°C;
- Synthesizes large quantities in a commercially viable, low-cost, chemically stable and nontoxic manner;
- Synthesizes almost any material;
- Produces extremely homogeneous alloys and composites;
- Synthesizes ultra-high purity (99.9999%) materials;
- Tailors the composition very accurately even in the early stages of the process, because the synthesis is actually performed on an atomic level;
- Precisely controls the physical, mechanical, and chemical properties of the final products.

The total synthesis was carried out in two steps by novel chemical solution decomposition method. In the first step, the stock solutions of metal nitrate [$Ni(NO_3)_2$ $9H_2O$, $Cu(NO_3)_2$ $6H_2O$, $Zn(NO_3)_2$, $Bi(NO_3)_3$], $(NH_4)_2MoO_4$ and titanium tartarate solutions were prepared. The titanium tartarate solution was prepared by the following procedure. The TiO_2 powder was dissolved in 40% HF solution in a 500 ml teflon beaker kept in a water bath for ~ 24 h. The solution was occasionally shaken during warming in the water bath. The clear fluoro complex of titanium was then precipitated with Conc. NH_4OH solution. The precipitate was filtered and thoroughly washed with 5% aq soln of NH_4OH to make the precipitate fluoride free. Then the hydroxide precipitate of titanium was dissolved in tartaric acid solution. The strength of the Ti^{4+} solution was estimated by gravimetric method.

In the second step, the equivalent amount of metal nitrate, $(NH_4)_2MoO_4$ and titanium tartarate solution were taken in a beaker as per chemical composition. The complexing agent TEA (triethanolamine) (where molecular ratio of metal ion:TEA = 1:3) was added to the homogeneous solution of constituents maintaining pH at 4–7 by nitric acid and ammonia. This mixed solution after evaporation and decompositions at 200°C, resulted in black carbonaceous light porous mass, which was followed by calcination in air at temperatures 500–750°C for 2 h at a heating rate of 5% min for different chemical compositions.

Recently, mesoporous nanoclusters of $M_xNb_xTi_{1-2x}O_{2-x/2}$ (M = Cr, Fe; x = 0.01, 0.05, 0.1, 0.2) were prepared by sol-gel method [89]. The total synthesis was carried out in two steps. In the first step, the stock solutions of ferric nitrate, titanium tartarate and niobium tartarate solutions were prepared. The solution of the titanium and niobium tartarate complexes, which are not commercially available, were prepared in the laboratory from its hydrated oxide (Nb_2O_5 $n$$H_2O$ and TiO_2 $n$$H_2O$); the details of the preparation process are discussed elsewhere [94]. In the second step, stoichiometric amounts of titanium tartarate (100mL; 0.2545 gL^{-1}), niobium tartarate (1.72mL; 0.2167 gL^{-1}), ferric nitrate (0.088 g, 0.21 mmol) with triethanolamine (3mL) and 20 mL of absolute ethanol solutions were taken in a beaker as per the predetermined chemical compositions (example shown for x = 0.01) with constant stirring for 30 min at room temperature. Then the mixture was adjusted to a pH of 9.0 with 6M NaOH solution, and stirred for 30 min, yielding a stable light yellow homogeneous emulsion. The resulting mixture was transferred into a 100 mL Teflon-lined stainless steel autoclave and heated to 150°C for 22 h under auto-generated pressure. The reaction mixture was allowed to cool to room temperature and the precipitate was filtered, washed with distilled water five times, and dried in a vacuum oven at 100°C for 15 h. For comparison, the same method was used to synthesize Fe-TiO_2, Nb-TiO_2 and pure TiO_2.

15.3.1 Photocatalytic Reactor

The photocatalytic experiment was carried out in a simple cubic photoreactor as depicted in Figure 15.6. There was a 400 W ultrahigh-pressure Hg-lamp (PHILIPS-HPL-N, G / 74 / 2, MBF-400) with a primary wavelength distribution approximately ($\lambda > 280$ nm) attached inside the photoreactor. An air circulating exhaust fan was attached to the backside of the cubic reactor to remove the hot air inside the reactor. The upper portion of

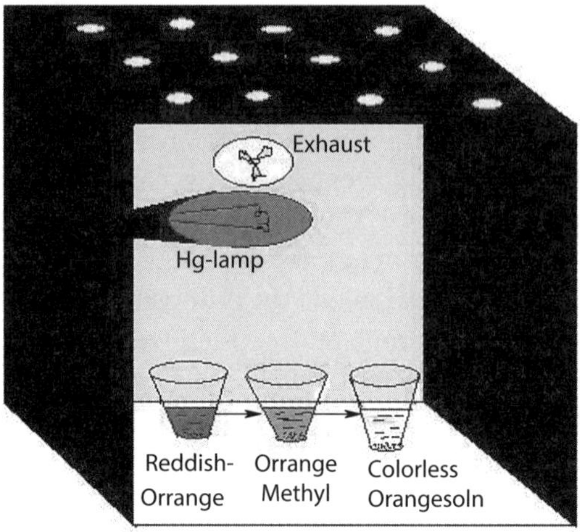

Figure 15.6 Photocatalytic reactor; color changes of methyl orange solution in presence of Hg lamp.

the reactor contained a number of small holes for cool air to enter inside the reactor. The front side of the reactor was fixed with a sliding door for collecting the reaction solutions. All four sides were made of black colored wood to protect the UV irradiation outside of the reactor. Inside the photoreactor, the distance between the 200 ml quartz glass beaker and the UV lamp was maintained at approximately 5 cm. All the experiments were carried out at 28 ± 2°C. In every case, the 5 cc solutions were collected from each of the beakers at the same time interval; after the solutions were sonicated 3–4 times they were used for UV measurement (UV-1601, SHIMADZU).

15.3.2 Sol-Gel Method

The sol-gel method is a versatile process used in making various ceramic materials [74–76]. In a typical sol-gel process, a colloidal suspension, or a sol, is formed from the hydrolysis and polymerization reactions of the precursors, which are usually inorganic metal salts or metal organic compounds such as metal alkoxides. Complete polymerization and loss of solvent leads to the transition from the liquid sol into a solid gel phase. For sol-gel processes there are three key steps [95 ,96].

Step I: The method often uses metal organic precursors where the various components are mixed together in a solvent to get a solution. This "solution," which has a multicomponent composition, is a true ionic or molecular mixture, which ensures the atomic-scale mixing (i.e., ultra homogenization) of the components in the sol-gel technique. The liquid phase for the vast majority of the oxide gels is short-chain alcohols or water, and the solutes may be inorganic nitrates, inorganic chlorides, or a wide variety of metal organic molecules.

Step II: This step is the key step in the sol-gel process. It involves formation of a sol and conversion of it to a gel so that it retains the chemical homogeneity of the samples during the desiccation. Conversion to a sol is accompanied by adjusting the activity of some species, H^+ and OH^- and other ions, which results in the formation of a dispersed solid phase. In principle, the pH, ionic strength and temperature of the precursor mixture controls the gelation of the sol. Manipulation of these parameters is an empirical procedure and it must be worked out independently for each composition.

Step III: The last step of the process involves desiccation and heat treatment of the gels to ceramic powders. The particle sizes in this process have been reported to vary from 20 to 300 nm according to the experimental conditions.

Therefore, the most convenient process is Step II in which a wet gel will form when the sol is cast into a mold, and the wet gel is converted into a dense ceramic with further drying and heat treatment. A highly porous and extremely low-density material called an aerogel is obtained if the solvent in a wet gel is removed under a supercritical condition. Ceramic fibers can be drawn from the sol when the viscosity of a sol is adjusted into a proper viscosity range. Ultrafine and uniform ceramic powders are formed by precipitation, spray pyrolysis or emulsion techniques. Under proper conditions, nanomaterials can be obtained. The controlled hydrolysis of different alkoxides, acetates, nitrates and sub-nitrates are normally used for the preparation of transition metal-doped TiO_2 or coupled oxide sol-gel process [97–102]. The alkoxides Ta/Nb/Ti-ethoxides/isopropoxide/propoxide/butoxide, etc., are used as the precursor material. Powders produced in this manner are uniform and well dispersed.

The TiO_2 nanomaterials have been synthesized with the sol-gel method from hydrolysis of a titanium precusor [103–119]. This process normally proceeds via an acid-catalyzed hydrolysis step of titanium(IV) alkoxide followed by condensation [103–119]. The development of Ti-O-Ti chains is favored with low content of water, low hydrolysis rates, and excess titanium alkoxide in the reaction mixture. Three-dimensional polymeric skeletons with close packing result from the development of Ti-O-Ti chains. The formation of $Ti(OH)_4$ is favored with high hydrolysis rates for a medium amount of water. The presence of a large quantity of Ti-OH and insufficient development of three-dimensional polymeric skeletons leads to loosely packed first-order particles. Polymeric Ti-O-Ti chains are developed in the presence of a large excess of water. Closely packed first order particles are yielded via a three-dimensionally developed gel skeleton [103–119]. From the study on the growth kinetics of TiO_2 nanoparticles in aqueous solution using titanium tartarate as precursor, it is found that the rate constant for coarsening increases with temperature due to the temperature dependence of the viscosity of the solution and the equilibrium solubility of TiO_2 [104]. Secondary particles are formed by epitaxial self-assembly of primary particles at longer times and higher temperatures, and the number of primary particles per secondary particle increases with time. The average TiO_2 nanoparticle radius increases linearly with time, in agreement with the Lifshitz-Slyozov-Wagner model for coarsening [104].

Titania adapted chromium-niobate nanocatalysts $Cr_xNb_xTi_{1-2x}O_{2-x/2}$ (x = 0.01–0.2) were synthesized by Ghorai *et al.* for the first time by the sol-gel method by using *N,N,N',N'*-tetrakis(2-hydroxyethyl)ethylenediamine ($edteH_4$) precursors in CH_3OH/H_2O medium [57]. The $Cr_xNb_xTi_{1-2x}O_{2-x/2}$ (x = 0.01) (CNT1) nanocatalysts display good photocatalytic activity for degradation of Rhodamine B in the presence of UV light because they have high surface area (S_{BET} = 162m^2 g^{-1}), small particle sizes (12 ± 1 nm) and lower band gap energy (1.85 eV) compared to TiO_2. The reaction procedure of nano $Cr_xNb_xTi_{1-2x}O_{2-x/2}$ catalysts is shown in Figure 15.7.

The finer details of the particles and their morphologies have been investigated by TEM. The bright-field (BF) electron micrograph of the CNT1 powder produced at 500°C reflects a narrow distribution of particles, with an average particle diameter of 12 ± 1 nm, which is shown in Figure 15.8. The particle sizes were obtained from multiple TEM images: Figure 15.8a (particle size distribution histogram inset in 15.8a), Figure 15.8b represents the SAED pattern of CNT1, Figure 15.8c indicates the mesoporous structures of CNT1 observed from HRTEM and Figure 15.8d represents the

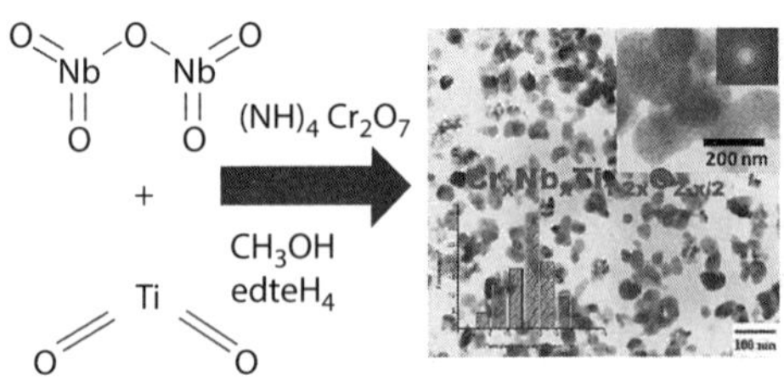

Figure 15.7 Reaction procedure of nano $Cr_xNb_xTi_{1-2x}O_{2-x/2}$ catalysts.

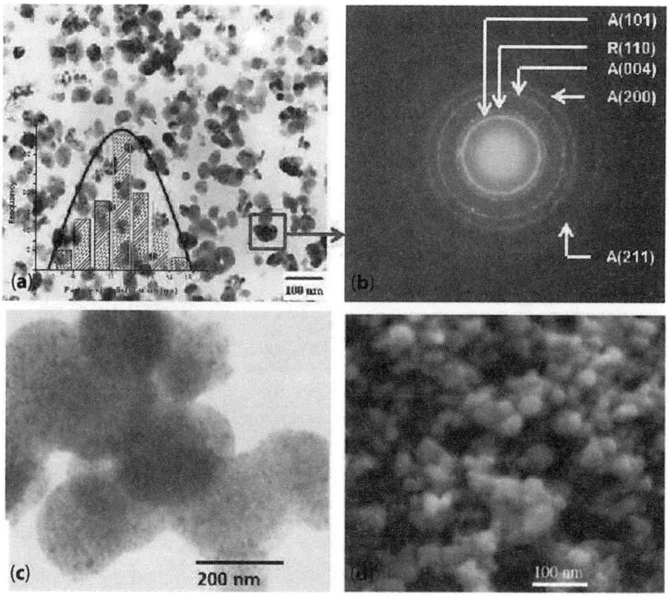

Figure 15.8 (a) TEM images of CNT1 (inset: Particle size distribution histogram); (b) SEAD pattern of CNT1; (c) Mesoporous structures of CNT1 observed from HRTEM; (d) SEM images of CNT1.

scanning electron microscopy (SEM) of CNT1. The average grain sizes of CNT1 are ~15 nm measured from SEM. Therefore, Figure 15.8 concludes that the synthesized materials are in nanometer range and developed mesoporous characters. During annealing, the decomposition of CNT powders is obtained from $edteH_4$ in methanol and its complexes precursor solutions, and finally gets mesoporous clusters of CNT. The N_2 adsorption-desorption isotherms of CNT1 nanopowders indicate the type IV nature of the curve, which is characteristic of a mesoporous material. The material possesses a high BET surface area of ~162m^2 g^{-1} with a single point pore volume of 0.159 ml g^{-1} at $p/p_o = 1$.

In order to obtain some information on the potential application of CNT1 in photocatalysis, UV-Vis spectra were done. The changes in concentration of RhB as a function of UV-light exposure time in the presence of the prepared photocatalysts are shown in Figure 15.9. The photocatalytic activity of CNT1 (27.31×10^{-3} min^{-1}) is 2.5 times higher than that of pure TiO_2 (11.68×10^{-3} min^{-1}) and other compositions of $Cr_xNb_xTi_{1-2x}O_{2-x/2}$ under UV light (Table 15.1). An increase of the different dopant (Cr and Nb) concentrations in TiO_2 solid solution, decreased the rate of photodegradation

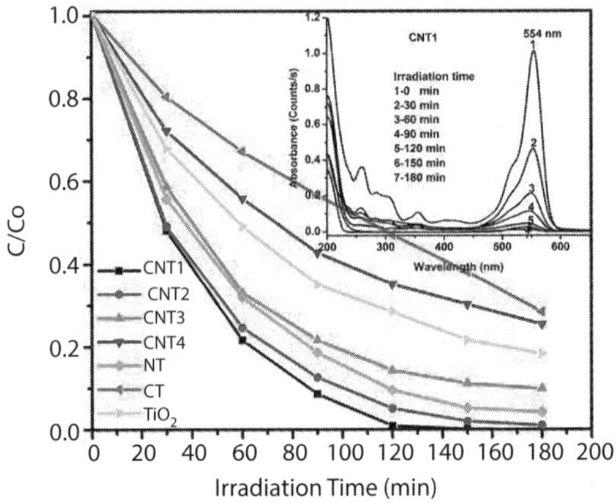

Figure 15.9 The changes in concentrations of RB solution at 554 nm in the presence of CNT1, CNT2, CNT3, CNT4, NT, CT, TiO_2 and Hg lamp.

due to the decrease of specific surface area for the different doped catalysts. The photocatalytic activity was also observed in the presence of visible light, but the catalytic activity was not so significant like UV light.

Highly crystalline anatase TiO_2 nanoparticles with different sizes and shapes could be obtained with the polycondensation of titanium alkoxide in the presence of tetramethylammonium hydroxide [120, 121]. In a typical procedure, spherical nanoclusters of $Fe_xNb_xTi_{1-2x}O_{2-x/2}$ (x = 0.01) were prepared by sol-gel method at pH 9 with triethanolamine (TEA) [122]. The pH of the solution can be adjusted by adding NaOH solution. Figure 15.10 shows representative TEM images of the novel TiO_2 nanoparticles at under pH 9 with the control shape of FNT1 [122]. Secondary amines, such as diethylamine, and tertiary amines, such as trimethylamine and triethylamine, act as complexing agents of Ti(IV) ions to promote the growth of ellipsoidal particles with lower aspect ratios. The shape of the TiO_2 nanoparticle can also be tuned from rounded-corner cubes to sharp-edged cubes with sodium oleate and sodium stearate [123]. The shape control is attributed to the tuning of the growth rate of the different crystal planes of TiO_2 nanoparticles by the specific adsorption of shape controllers to these planes under different pH conditions [123]. A prolonged heating time below 100°C for the as-prepared gel can be used to avoid the agglomeration of the TiO_2 nanoparticles during the crystallization process [124]. By heating amorphous TiO_2 in air, large quantities of single-phase anatase TiO_2 nanoparticles with average particle sizes between 7 and 50 nm can be obtained, as reported by Zhang and Banfield [125–128]. Much effort has

Table 15.1 Resultant properties of CNT1, CNT2, CNT3, CNT4, NT, CT and Pure TiO_2 clusters.

Sample ($Cr_xNb_xTi_{1-2x}O_{2-x/2}$)	**Acronym**	**Reaction Rate Constant** k**($\times 10^{-3}$ min^{-1})**	**S_{BET} (m^2/g)**	**Crystallites size (nm)**	**Band gap Energy (eV)**	**RhB decolorization, $t_{1/2}$ (min)**	**Time required for degradation of RhB(min)**
x = 0.01	CNT1	27.31	162.24	13.70	1.85	25.37	180
x = 0.05	CNT2	23.07	125.32	13.98	1.85	30.03	210
x = 0.1	CNT3	17.08	59.11	14.08	1.85	40.57	330
x = 0.5	CNT4	9.48	45.73	14.17	–	71.30	720
Nb-TiO_2	NT	18.74	52.69	12.14	2.2	36.97	270
Cr-TiO_2	CT	6.37	32.33	12.39	–	108.79	–
TiO_2	–	11.68	49	12.42	3.2	59.33	540

Reaction rate constants were measured after 50% decolorization of RhB solution; BET surface area measured by liquid nitrogen adsorption-desorption isotherm at room temperature.

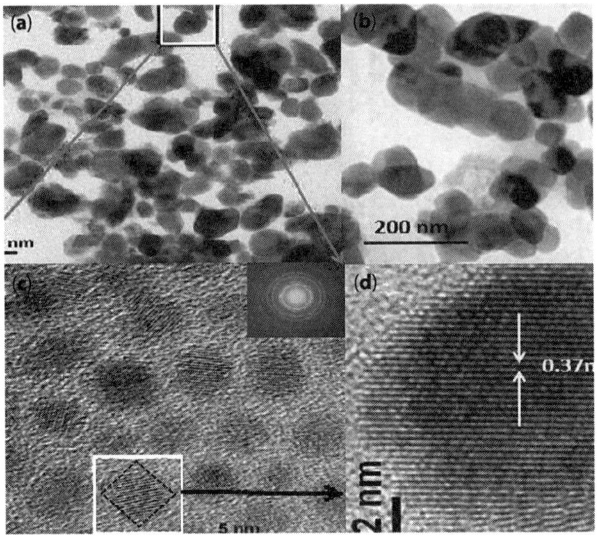

Figure 15.10 (a) TEM images of FNT1; (b) High resolution with high magnification of FNT1 TEM; (c) Enlargement of one of the spheres in Fig. 15.10a and SAED pattern of same sphere of FNT1; (d) Enlargement of one of the mesoporous spheres from (Figure 15.10c).

been exerted to achieve highly crystallized and narrowly dispersed TiO_2 nanoparticles using the sol-gel method with other modifications, such as a semicontinuous reaction method by Znaidi *et al.* [129] and a two-stage mixed method and a continuous reaction method by Kim *et al.* [130, 131].

Qiu *et al.* [132] found that a typical SEM image of the TiO_2 nanotube array with the ZnO nanorod array template was synthesized by sol-gel method. The TiO_2 nanotubes inherit the uniform hexagonal cross-sectional shape and the length of 1.5 nm and inner diameter of 100–120 nm of the ZnO nanorod template. As the concentration of the TiO_2 sol is constant, well-aligned TiO_2 nanotube arrays can only be obtained from an optimal dip-coating cycle number in the range of 2–3 cycles. A dense, porous TiO_2 thick film with holes is obtained instead if the dip-coating number further increases. The heating rate is critical to the formation of TiO_2 nanotube arrays. When the heating rate is extra rapid, e.g., above 6°C min^{-1}, the TiO_2 coat will easily crack and flake off from the ZnO nanorods due to great tensile stress between the TiO_2 coat and the ZnO template, and a TiO_2 film with loose, porous nanostructure is obtained.

In the presence of UV light, FNT1 reduces the 4-Nitrophenol (4-NP) to 4-aminophenol using a little bit of $NaBH_4$, in contrast to pure TiO_2 and other compositions of $Fe_xNb_xTi_{1-2x}O_{2-x/2}$ photocatalysts [122]. The 4-nitrophenol is

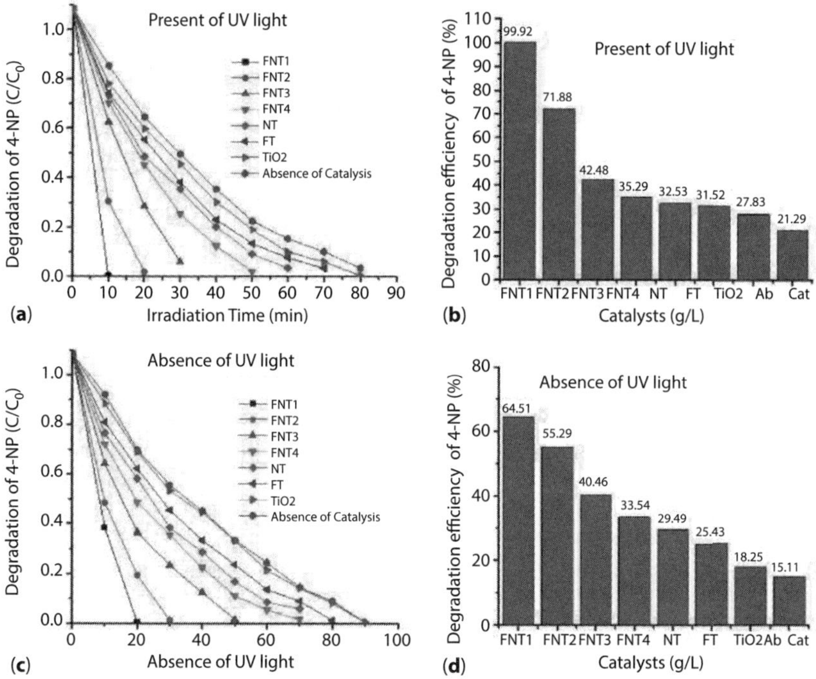

Figure 15.11 (a–d); (d) represents the reaction rate constant and degradation efficiency of 4-nitrophenol, in presence and absence of UV light and catalysts.

reduced to 4-aminophenol within 10 minutes in the presence of FNT1 and UV light, but in the absence of catalysts it takes approximate 82 minutes, as shown in Figure 15.11. The catalytic activity of FNT1 is much faster in the presence of UV light compared to the absence of UV light. Therefore, FNT1 catalysts give much faster reaction kinetics (5.89×10^{-5} min^{-1}) than that of FNT2, FNT3, FNT4 and their corresponding Fe/Nb titanates and TiO_2 due to the high porosity of the materials and small particle sizes. The FNT1 catalyst led to a significant decrease of the absorption peak at 415 nm and increase of absorbance peak at 285 nm in UV-vis spectra corresponding to 4-NP. During the reduction, the yellow color faded with the simultaneous formation of a slight shifting peak position, and a new peak arose at 294 nm assigned to 4-AmP in UV-vis spectra in Figure 15.12 [133, 134]. The complete disappearance of the UV-vis absorption peak at 415 nm of 4-NP occurred at 10 min, which meant the complete reduction of 4-NP to 4-AmP. This is probably due to efficient generation of electron and hole through mutual charge transfer from Fe(III) and Nb(V) ion in the presence of UV light, where the electron resides in the conduction band and a hole is

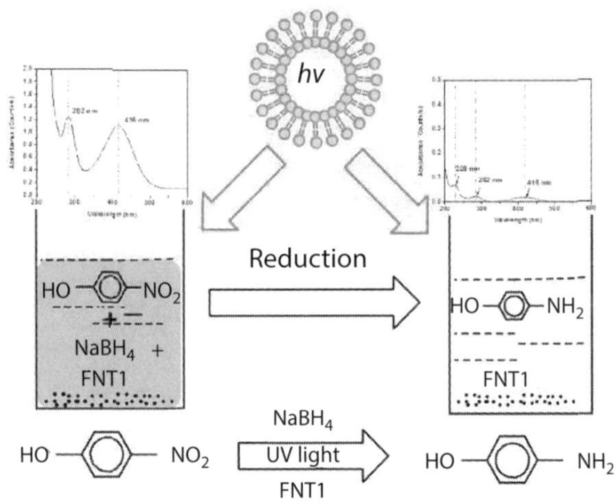

Figure 15.12 UV-Vis absorbance changes during the reaction in presence of catalyst and UV-light, and also color changes of the p-NP solution.

captured, forming Nb^{4+}. Hence, this Nb^{4+} is faster at reducing p-nitrophenol to 4-AmP. However, in the nanoparticle phase, the surface-to-volume ratio increases drastically and the surface atoms include an increasing fraction of the total particulate volume having high defect structures. Thus they are expected to show drastically improved catalytic properties. However, the reduction of 4-NP did not perform well under the condition, even with a large excess amount of $NaBH_4$, in the absence of the catalyst.

15.4 Novel Chemical Synthesis Routes

Chemical synthesis has been very useful in the synthesis of a wide range of nanostructured materials, including high-surface-area transition metals, alloys, carbides, oxides and colloids. It is well established that various factors, such as availability and cost of the required reagents; reproducibility of a particular route; necessary characteristics required in the final product; and the cost of the process, determines the choice of the preparative route. The chemical preparative routes that have been investigated so far for the preparation of nanosize metal-doped titanium dioxide photocatalyst can be categorized under two broad headings of:

a. Vapor-phase or Gas-phase reaction
b. Solution processing technique

Very few reports are available in literature on the preparation of nano-size transition metal-doped titanium dioxide photocatalyst [135–137], however, this process is important for their preparation inspite of its high cost.

a) Vapor phase technique: This process involves dissociation-vaporization of primary powders (reactants), followed by vigorous quenching of the resulting vapors onto a cold metal substrate. It results in a refined deposite of sample and/or substrate. The sample required is recovered by scrubbing the deposit from substrate. Generally, the rate and the temperature of decomposition determine the reaction kinetics and the rates at which the decomposed products can crystallize on the reaction surface [138]. There are three types of vapor phase preparation methods: (i) reactions between gas and a solid, (ii) reactions between gas and a liquid, and (iii) reactions between two or more gases.

b) Solution processing methods: These processes can also be termed as chemical synthesis methods. The solution processing method covers a broad area of preparative methods of nanocrystalline materials, which can be distinctly divided into two types: (i) precipitation from aqueous or nonaqueous solutions includes aerosol, co-precipitation, controlled hydrolysis of metal alkoxides, hydrothermal, colloid emulsion methods, etc.; (ii) evaporation or evaporative decomposition of the liquid includes spray drying, spray pyrolysis, freeze drying, sol-gel processing, the Pechini methods, combustion, etc. In general, the solution processing method offers the advantage of easy preparation of nearly any composition, maintaining compositional homogeneity and high purity. Various fine titanium dioxide-based nanophotocatalysts [139–142] have been successfully synthesized using this method.

15.4.1 Fe(III)-Doped TiO_2 Nanophotocatalyst

The most effective nanophotocatalyst, Fe(III)-doped TiO_2, was synthesized by novel chemical method. The chemical synthesis of nanosized Fe(III)-doped TiO_2 was studied on degradations of different dyes like Methyl orange (MO), Rhodamine B (RB), Thymol blue (TB) and Bromocresol green (BG) under UV light irradiation it was found that $Fe_xTi_{1-x}O_2$ ($x = 0.005$) (FT) is more photoactive compared with all other compositions of $Fe_xTi_{1-x}O_2$ and degussa P25 [41]. It was also found that among all the four dye solutions, the rate of degradation of RB is best in the presence of FT and UV light. The present synthetic procedure is a low-temperture-based method which produces nanosized, chemically homogeneous Fe(III)-doped TiO_2 with narrow distribution of particle size. Zhang *et al.* [142]

found that the optimal Fe^{3+} dopant concentration for enhancing catalytic activity strongly depends on the particle size of the TiO_2 nanocatalyst. The optimal Fe^{3+} concentration was found to decrease with increasing particle size. Yao *et al.* [143] reported that La (Fe)-doped bismuth titanate shows good photocatalytic activity. Feng [144] showed that nanosized Fe-doped TiO_2 synthesized by sol–gel pillaring technique was excellent photocatalyst for discoloration and mineralization of Orange II. For the nanosized photocatalysts, high specific surface areas and small crystal sizes are beneficial for efficient photocatalytic reactions. By acting as both hole and electron traps, Fe can enhance the photocatalytic activity of TiO_2 [145].

The general pathway of photochemical reaction is in the presence of UV light; Fe(III) in TiO_2 matrix generates Fe^{2+}, Fe^{4+} and charge-pair (Eq. 15.1), and at the same time Fe^{3+} is also regenerated through electron or hole trapping by reversible reaction. The photocatalytic reaction involves the migration (Eq. 15.2) and recombination (Eqs. 15.3, 15.4, 15.5) of the charge between the iron and titanium ions. The dye is discolored within the time of reaction through the mutual contact between the dye and the catalyst. The discoloration of the dye occurs because the dye is sensitized and the dye cation is formed. The dye cation is unstable and decomposes, injecting an electron in the conduction band of Fe^{2+}. This band-gap electron starts the production of highly oxidative radical species through the formation of super oxide radical anion $O_2^{\bullet-}$ on the surface of the catalyst as long as the dye is present. In later steps of the reaction, when long-lived colorless intermediates are present, the Fe^{2+} absorbs the light and produces the conduction-band electron and valence-band holes. The mechanism of Fe^{3+}-doped TiO_2 samples [41, 145–148] is illustrated in the following equations.

$$2Fe^{3+} \xrightleftharpoons{hv} Fe^{2+} + Fe^{4+} + e_{CB}^{-} + h_{NB}^{+} \tag{15.1}$$

$$Fe^{2+} + Ti^{4+} \rightarrow Fe^{3+} + Ti^{3+} \tag{15.2}$$

$$Fe^{4+} + OH^{-} \rightarrow Fe^{3+} + OH \tag{15.3}$$

$$Fe^{2+} + OH \rightarrow Fe^{3+} + OH^{-} \tag{15.4}$$

$$Fe^{4+}\ Ti^{3+} \rightarrow Fe^{3+} + Ti^{4+} \tag{15.5}$$

$$\text{Dye} + hv \rightarrow Dye* \tag{15.6}$$

$$Dye* + e^- (Fe^{2+}) \rightarrow Dye^{\bullet+} + e^- (Fe^{2+}) \tag{15.7}$$

$$e^-(Fe^{2+}) + O_{2(ads)} \rightarrow O_2^{\bullet-} + Fe^{2+} \tag{15.8}$$

$$Dye^{\bullet+} \rightarrow \text{Degradation products} \tag{15.9}$$

15.4.2 Metal Molybdate Incorporated Titanium Dioxide Photocatalyst

Different metal molybdates (Metal: Ni, Cu, Zn) incorporated titanium dioxide were prepared by the chemical solution decomposition (CSD) method. Ghorai *et al.* [26] found that among many transition-metal molybdates, nickel molybdate incorporated titanium dioxide (NMT) was found to be more photoactive than P25 TiO_2 and other metal molybdate doped TiO_2 ($M_xMo_xTi_{1-x}O_6$) (M= Ni, Cu, Zn; $x = 0.05$), for photocatalytic oxidation of various dye solutions (MO, RhB, TB and BG) to harmless decolorized solution at room temperature with the help of a Hg lamp (Fig. 15.13). The average particle size of NMT1 was found to be 15 nm measured from TEM (inset Fig. 15.13) and calculated band gap from adsorption edge is found to be 2.66 eV [26] (Table 15.2). The UV-Vis diffuse reflectance spectrum of $NiMoO_4$ incorporated TiO_2 and pure TiO_2 presented in inset Figure 15.13, gave distinct band gap absorption edges at 465 nm, 313 nm, 305 nm and 387 nm for doped NMT1, NMT2, NMT3 and pure TiO_2, and corresponding band gap energies are 2.66, 3.00, 3.06 and 3.20 eV respectively. By increasing the dopant concentration, the band gap increased; as a consequence the photocatalytic activities decreased. The photochemical reaction proceeds step-by-step. Firstly, the dye is sensitized by the catalyst in the presence of UV light and the dye cation is formed. Secondly, the dye cation is unstable and decomposes, injecting an electron on the conduction band of TiO_2. Electron goes to conduction band and hole is captured by Ni^{2+} producing Ni^{3+}, and helps to degrade the dye molecule faster [26].

15.4.3 Metal Molybdate Doped Bismuth Titanate (NMBT) Nanocomposites

Various compositions of nanosized metal molybdate $(MMoO_4)_x$-doped bismuth titanate ($Bi_2Ti_4O_{11}$) (M=Ni, Cu; x = 0.01, 0.05, 0.1) composites have been prepared by chemical solution decomposition (CSD) method using triethanolamine (TEA) as complexing agent [89]. Yao *et al.* have

Table 15.2 Resultant properties of FT, NMT, CMT, NMBT, CFBT, and P25 composites.

Sample	Acronym	S_{BET} (m^2/g)	Crystallites size (nm) (TEM)	Particle size (nm)	Band gap energy (eV)	Reaction rate constant (h^{-1}/min^{-1})
$Fe_xTi_{1-x}O_2$(x = 0.005)	FT	69	11.85	10	2.38	16.3(RB)
$Ni_xMo_xTi_{1-x}O_6$ (x = 0.05)	NMT1	149	11.85	15	2.66	4.4 (MO)
$Cu_xMo_xTi_{1-x}O_6$ (x = 0.05)	CMT1	101	11.89	10	3.03	2.7 (4-NP)
$(NiMoO_4)_x$ $(Bi_2Ti_4O_{11})_{1-x}$ (x =0.01)	NMBT1	71	12	30	2.82	3.2 (MO)
$(CuMoO_4)_x$ $(Bi_2Ti_4O_{11})_{1-x}$ (x =0.05)	CMBT1	63	19	32	–	10.3 (TB)
Degussa P25	P25	49	12.46	–	3.2	different

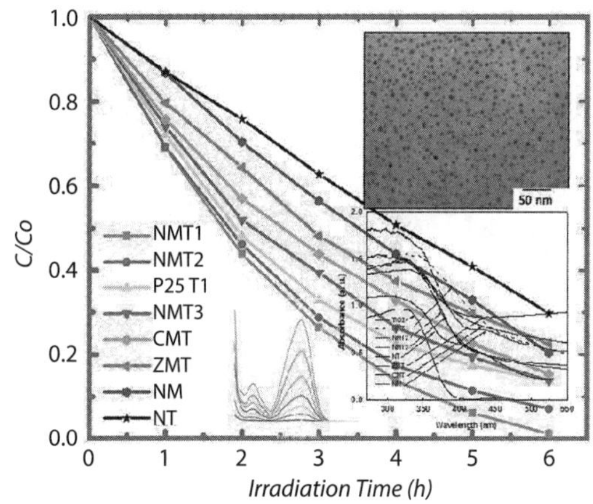

Figure 15.13 Photocatalytic activity on methyl orange and band gap energy of NMT1 and other prepared ctalysts in presence of UV light.

found that $Bi_{12}TiO_{20}$ and perovskites $Bi_4Ti_3O_{12}$ shows high photocatalytic activity against methyl orange [149–150]. It has been demonstrated that the dopant ions or oxides can increase the quantum efficiency of the heterogeneous photocatalytic property by acting as electron/hole traps or by altering the e^-/h^+ pair recombination rate, and are therefore an effective way to enhance the photocatalytic activity. Yao *et al.* have observed that 0.5 at% Ba doping in $Bi_{12}TiO_{20}$, La (Fe)-doped bismuth titanate, shows maximum photocatalytic activity compared to $Bi_{12}TiO_{20}$ [151]. Figure 15.14A shows TEM images of NM_xBT_{1-x} ($x = 0.01$) composite [89], which has maximum photocatalytic activity under the experimental condition. The average particle size of NM_xBT_{1-x} ($x = 0.01$) was found to be around 30±2 nm calculated from images of TEM using ImageJ software; the SAD (selected area diffraction) pattern as well as the high-resolution TEM of a highlighted particle from Figure 15.14B are presented in Figure 15.14C and 15.14D. The SAD pattern of NM_xBT_{1-x} ($x = 0.01$) clearly indicates that the crystal structure is close to parent compound $Bi_2Ti_4O_{11}$. The high-resolution transmission electron microscopy (HRTEM) for the sample NM_xBT_{1-x} ($x = 0.01$) shown in Figure 15.14D, clearly demonstrates the lattice fringes for the monoclinic phase of $Bi_2Ti_4O_{11}$. From the micrograph the lattice spacing in the particle is determined to be 2.969Å, which corresponds to the (113) plane of the monoclinic phase of these materials.

The UV–vis diffuse reflectance spectrum of various compositions of NM_xBT_{1-x} (x = 0.01, 0.05, 0.1) composite and of $Bi_2Ti_4O_{11}$ and $NiMoO_4$ are also presented in Figure 15.15. The band gap absorption edges of

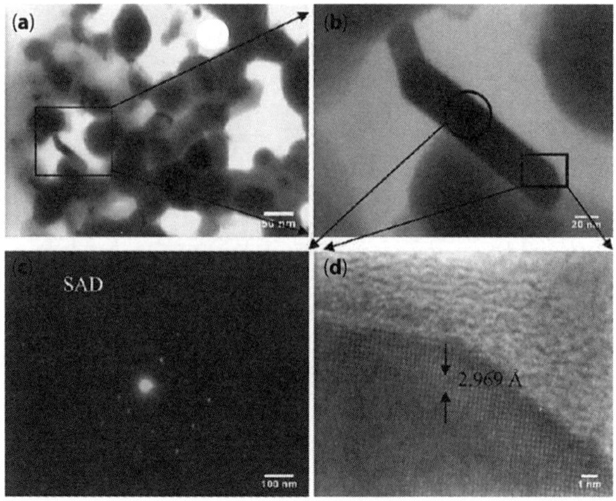

Figure 15.14 TEM images of $NiMoO_4$-doped $Bi_2Ti_4O_{11}$ nanocomposite: (A) bright field (BF) pattern, (B) highlighted particle from (A), (C) SAD pattern of the particular particle, and (D) particle distribution or arrangement by high resolution.

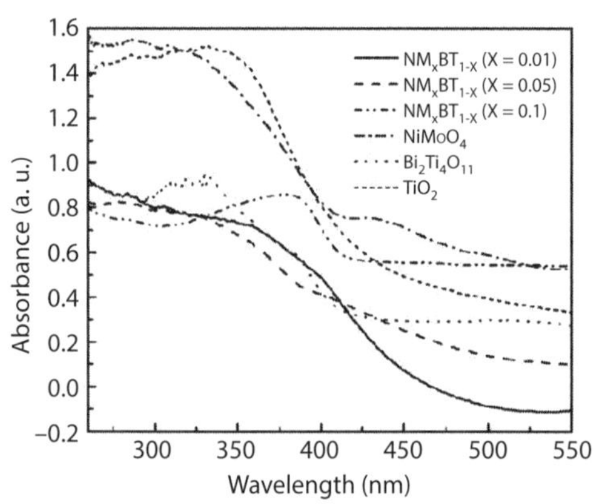

Figure 15.15 The UV-Vis diffuses reflectance spectra of NM_xBT_{1-x} (x = 0.01, 0.05, 0.1) composites, $NiMoO_{4,}$ $Bi_2Ti_4O_{11}$ and TiO_2.

NM_xBT_{1-x} (x = 0.01, 0.05, 0.1) composite are determined to be 439, 424 and 409 nm, corresponding to the band gap energy of 2.82, 2.92 and 3.03 eV, respectively [89]. Similarly, the band gap absorption edge of $NiMoO_4$ and $Bi_2Ti_4O_{11}$ is determined to be 441 and 382 nm, and the corresponding band gap energies are 2.81 and 3.24 eV. A mechanistic scheme of the charge separation and the photocatalytic activity for the photocatalysts is shown

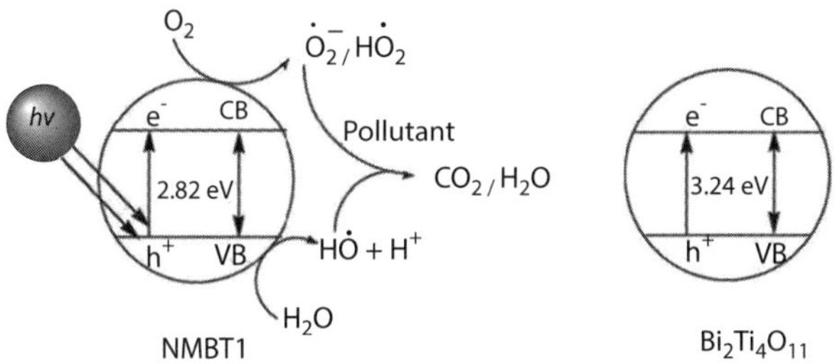

Figure 15.16 General mechanistic representation of the photocatalytic activity of NMBT on MO solution under UV-irradiation.

in Figure 15.16 [89, 152, 153]. The NMBT can be excited by the photons with the wavelengths under 409–439 nm for different compositions and produces the photogenerated electron/hole pairs, following photocatalytic activity. In the case of $Bi_2Ti_4O_{11}$ with the band gap energy of 3.24 eV, which theoretically requires the photons of wavelengths 382 nm, little photocatalytic activity is shown under UV-light compared to the NMBT photocatalyst. On the other hand, $NiMoO_4$ powder shows photocatalytic activity, which is better than $Bi_2Ti_4O_{11}$ but inferior to NMBT photocatalysts under similar experimental conditions. Of the three compositions of NM_xBT_{1-x} (x = 0.01, 0.05, 0.1) composite, the sample having x = 0.01 shows maximum photocatalytic activity. This is probably due to efficient generation of electron and hole, where electron resides in the conduction band and the hole is captured forming Ni^{3+}. With an increase of dopant concentration of Ni^{2+} (for x = 0.01–0.1) the band gap increases and the rate of reaction decreases due to inefficient formation of hole and electron. The NMBT is capable of localizing electron in the conduction band and hole in Ni^{3+} more efficiently than individual components [153]. Moreover, the high surface area of NMBT produced at lower temperature helps to produce more active centers for photocatalytic reaction.

References

1. S.G. Anju, K.P. Jyothi, J. Sindhu, Y. Suguna, and E.P. Yesodharan, *Res. J. Recent Sci.*, Vol. 1 (ISC-2011), pp. 191–201, 2012.
2. R. Zouaghi, B. David, J. Suptil, K. Djebbar, A. Boutiti, and S. Guittonneau, *Ultrason. Sonochem.*, Vol. 18, pp. 1107–1112, 2011.
3. M. Ying-Shih, S. Chi-Fanga, and L. Jih-Gaw, *J. Hazard. Mater.*, Vol. 178, pp. 320–325, 2010.

4. R.A. Torres, F. Abdelmalek, E. Combet, C. Petrier, and C. Pulgarin, *J. Hazard. Mater.*, Vol. 146, pp. 546–555, 2007.
5. A. Mills, and S. Le Hunte, *J. Photochem. Photobiol. A: Chem.*, Vol. 108, pp. 1–35, 1997.
6. A.L. Linsebigler, G. Lu, and J.T. Yates Jr., *Chem. Rev.*, Vol. 95, p. 735, 1995.
7. D. Wang, Y. Duan, Q. Luo, X. Li, J. An, L. Bao, and L. Shi, *J. Mater. Chem.*, Vol. 22, pp. 4847–4854, 2012.
8. Z. Xiong, J. Ma, W.J. Ng, T.D. Waite, and X.S. Zhao, *Water Research*, Vol. 45, pp. 2095–2103, 2011.
9. H. Luo, C. Wang, and Y. Yan, *Chem. Mater.*, Vol. 15, p. 3841, 2003.
10. H. Yun, K. Miyazawa, H. Zhou, I. Honma, and M. Kuwabara, *Adv. Mater.*, Vol. 13, p. 1377, 2001.
11. P. Kluson, P. Kacer, T. Cajthaml, and M. Kalaji, *J. Mater. Chem.*, Vol. 11, p. 644, 2001.
12. S.C. Hayden, N.K. Allam, and M.A. El-Sayed, *J. Am. Chem. Soc.*, Vol. 132, No. 41, pp. 14406–14408, 2010.
13. J. Ravez, C. Broustera, and A. Simon, *J. Mater. Chem.*, Vol. 9, pp. 1609–1613, 1999.
14. A. Arshak, K. Arshak, D. Morris, O. Korostynska, and E. Jafer, *Sensors and Actuators A*, Vol. 122, pp. 242–249, 2005.
15. S. Dutta, A.K. Patra, S. De, A. Bhaumik, and B. Saha, *ACS Appl. Mater. Interfaces*, Vol. 4, pp. 1560–1564, 2012.
16. R. Shirley, and M. Kraft, *Physical Review B*, Vol. 81, p. 075111, 2010.
17. Z. Li, H. Zhang, W. Zheng, W. Wang, H. Huang, C. Wang, A.G.M. Diarmid, and Y. Wei, *J. Am. Chem. Soc.*, Vol. 130, pp. 5036–5037, 2008.
18. X. Du, Y. Wang, Y. Mu, L. Gui, P. Wang, and Y. Tang, *Chem. Mater.*, Vol. 14, pp. 3953–3957, 2002.
19. M.T. Colomer, and J.R. Jurado, *Chem. Mater.*, Vol. 12, Iss. 4, pp. 923–930, 2000.
20. J. Yan, and F. Zhou, *J. Mater. Chem.*, Vol. 21, p. 9406, 2011.
21. H.-G. Jung, Y.S. Kang, and Y.-K. Sun, *Electrochimica Acta*, Vol. 55, pp. 4637–4641, 2010.
22. U. Bach, D. Lupo, P. Comte, J.E. Moser, F. Weissortel, J. Salbeck, H. Spreitzer, and M. Gratzel, *Nature*, Vol. 395, pp. 583–585, 1998.
23. C.T. Kresge, M.E. Leonowicz, W.J. Roth, J.C. Vartuli, and J.S. Beck, *Nature*, Vol. 359, pp. 710–712, 1992.
24. S.A. Davis, S.L. Burkett, N.H. Mendelson, and S. Mann, *Nature*, Vol. 385, pp. 420–423, 1997.
25. A. Primo, A. Corma, and H. Garcia, *Phys. Chem. Chem. Phys.*, Vol. 13, pp. 886–910, 2011.
26. T.K. Ghorai, D. Dhak, S.K. Biswas, S. Dalai, and P. Pramanik, *J. Mol. Catal. A: Gen.*, Vol. 273, pp. 224–229, 2007.
27. C. Guillard, J. Disdier, C. Monnet, J. Dussaud, S. Malato, J. Blanco, M.I. Maldonado, and J.M. Herrmann, *Appl. Catal. B*, Vol. 46, p. 319, 2003.

28. S. Malato, J. Blanco, A. Campos, J. Caceres, C. Guillard, J.M. Herrmann, and A.R. Fernandez-Alba, *Appl. Catal. B*, Vol. 42, p. 349, 2003.
29. D.D. Mulmi, T. Sekiya, N. Kamiya, S. Kurita, Y. Murakami, and T. Kodaira, *J. Phys. Chem. Solids*, Vol. 65, pp. 1181–1185, 2004.
30. M.A. Aramendıa, A. Marinas, J.M. Marinas, J.M. Moreno, and F.J. Urbano, *Catal. Today*, Vol. 101, p. 187, 2005.
31. J.C. Colmenares, M.A. Aramendıa, A. Marinas, J.M. Marinas, and F.J. Urbano, *Appl. Catal. A: Gen.*, Vol. 306, pp. 120–127, 2006.
32. T.K. Ghorai, M. Chakraborty, and P. Pramanik, *J. Alloys and Compounds*, Vol. 509, pp. 8158–8164, 2011.
33. C.E. Taylor, *Catal. Today*, Vol. 84, p. 9, 2003.
34. K.I. Shimizu, H. Akahane, T. Kodama, and Y. Kitayama, *Appl. Catal. A*, Vol. 269, p. 75, 2004.
35. H. Zhang, C. Liang, J. Liu, Z. Tian, G. Wang, and W. Cai, *Langmuir*, Vol. 28, Iss. 8, pp. 3938–3944, 2012.
36. S.T. Kochuveedu, D. Kim, and D.H. Kim, *J. Phys. Chem. C*, Vol. 116, Iss. 3, pp. 2500–2506, 2012.
37. M. Andersson, H. Birkedal, N.R. Franklin, T. Ostomel, S. Boettcher, A.C. Palmqvist, and G.D. Stucky, *Chem. Mater.*, Vol. 17, pp. 1409–1415, 2005.
38. Z. Zhang, Y. Yu, and P. Wang, *ACS Appl. Mater. Interfaces*, Vol. 4, Iss. 2, pp. 990–996, 2012.
39. T.K. Ghorai, S. Pramanik, and P. Pramanik, *Appl. Surf. Sci.*, Vol. 255, pp. 9026–9031, 2009.
40. B. Peng, X. Meng, F. Tang, X. Ren, D. Chen, and J. Ren, *J. Phys. Chem. C*, Vol. 113, pp. 20240–20245, 2009.
41. T.K. Ghorai, S.K. Biswas, and P. Pramanik, *Appl. Surf. Sci.*, Vol. 254, pp. 7498–7504, 2008.
42. J.K. Jakovljevic, M. Radoicic, T. Radetic, Z. Konstantinovic, Z.V. Saponjic, and J. Nedeljkovic, *J. Phys. Chem. C*, Vol. 113, Iss. 50, pp. 21029–21033, 2009.
43. M. Karimipour, J.M. Wikberg, V. Kapaklis, N. Shahtahmasebi, M.R.R. Abad, M. Yeganeh, M.M. Bagheri-Mohagheghi, and P. Svedlindh, *Phys. Scr.*, Vol. 84, p. 035702, 2011.
44. X. Qiu, M. Miyauchi, K. Sunada, M. Minoshima, M. Liu, Y. Lu, D. Li, Y. Shimodaira, Y. Hosogi, Y. Kuroda, and K. Hashimoto, *ACS Nano*, Vol. 6, No. 2, pp. 1609–1618, 2012.
45. T.K. Ghorai, *Open J. Phy. Chem.*, Vol. 1, pp. 28–36, 2011.
46. Y. Huang, W. Ho, S. Lee, L. Li, G. Zhang, and J.C. Yu, *Langmuir*, Vol. 24, pp. 3510–3516, 2008.
47. H. Choi, M.G. Aantoniou, M. Pelaez, A.A. Delacruz, O.A. Shoemaker, and D.D. Dionysiou, *Environ. Sci. Technol.*, Vol. 41, pp. 7530–7535, 2007.
48. J.C. Yu, J.G. Yu, W.K. Ho, Z.T. Jiang, and L.Z. Zhang, *Chem. Mater.*, Vol. 14, pp. 3808–3816, 2002.
49. S. Tojo, T. Tachikawa, M. Fujitsuka, and T. Majima, *J. Phys. Chem. C*, Vol. 112, pp. 14948–14954, 2008.

50. D.H. Wang, L. Jia, X.L. Wu, LQ. Lu, and A.W. Xu, *Nanoscale*, Vol. 4, pp. 576–584, 2012.
51. J. Virkutyte, R.S. Varma, *RSC Adv.* Vol. 2, pp. 1533–1539, 2012.
52. A.A. Ismail, and D.W. Bahnemann, *J. Mater. Chem.*, Vol. 21, p. 11686, 2011.
53. S. Senthilkumaar, K. Porkodi, and R. Vidyalakshmi, *J. Photochem. Photobiol. A*, Vol. 170, p. 225, 2005.
54. W.J. Li, E. W. Shi, and T. Fukuda, *Cryst. Res. Technol.*, Vol. 38, p. 847, 2003.
55. H. Cui, K. Dwight, and A. Weld, *J. Solid State Chem.*, Vol. 1, Iss. 15, p. 187, 1995.
56. M. Martos, B. Julian, H. Dehouli, D. Gourier, E. Cordoncillo, and P. Escribano, *J. Solid State Chem.*, Vol. 180, pp. 679–687, 2007.
57. T.K. Ghorai, and D. Dhak, *Adv. Mat. Lett.*, Vol. 4, No. 2, pp. 121–130, 2013.
58. B. Peng, X. Meng, F. Tang, X. Ren, D. Chen, and J. Ren, *J. Phys. Chem. C*, Vol. 113, pp. 20240–20245, 2009.
59. X. Chen, and S.S. Mao, *Chem. Rev.*, Vol. 107, pp. 2891–2959, 2007.
60. J. Arana, J.M.D. Rodrıguez, O. Gonzalez-Dıaz, E.T. Rendon, J.A.H. Melian, G. Colon, J.A. Navio, and J.P. Pena, *J. Mol. Catal. A*, Vol. 215, p. 153, 2004.
61. D. Han, Y. Li, and W. Jia, *Adv. Mat. Lett.*, Vol. 1, No. 3, pp. 188–192, 2010.
62. A. Tiwari, A.K. Mishra, H. Kobayashi, A.P.F. Turner, *Intelligent Nanomaterials*, Wiley-Scrivener Publishing LLC, USA, 2012.
63. M. Okumura, S. Nakamura, S. Tsubota, T. Nakamura, M. Azuma, and M. Haruta, *Catal. Lett.*, Vol. 51, pp. 53–58, 1998.
64. Y. Jing, L. Li, Q. Zhang, P. Lu, P. Liu, and X. Lu, *J. Hazard. Mater.*, Vol. 189, pp. 40–47, 2011.
65. M. Andersson, L. Osterlund, S. Ljungstrom, and A. Palmqvist, *J. Phys. Chem. B*, Vol. 106, Iss. 41, pp. 10674–10679, 2002.
66. K.B. Jaimy, S. Ghosh, S. Sankar, and K.G.K. Warrier, *Mater. Res. Bull.*, Vol. 46, pp. 914–921, 2011.
67. C. Su, B.Y. Hong, and C.M. Tseng, *Catal. Today*, Vol. 96, p. 119, 2004.
68. D.S. Lee, S.D. Han, S.M. Lee, J.S. Huh, and D.D. Lee, *Sensors and Actuators B*, Vol. 65, pp. 331–335, 2000.
69. Q.F. Zhou, H.L.W. Chan, and C.L. Choy, *J. Non-Cryst. Solids,* Vol. 254, Iss. 1–3, pp. 106–111, 1999.
70. G.T. Brown, and J.R. Darwent, *J. Phy. Chem.,* Vol. 88, pp. 4955–4959, 1984.
71. W. Nam, J. Kim, and G. Han, *Chemosphere*, Vol. 47, pp. 1019–1024, 2002.
72. N. Bao, X. Feng, Z. Yang, L. Shen, and X. Lu, *Environ. Sci. Technol.,* Vol. 38, pp. 2729–2736, 2004.
73. Y. Lu, M.R. Hoffmann, S. Yelamanchili, A. Terrenoire, M. Schrinner, M. Drechsler, M.W. Moller, J. Breu, and M. Ballauff, *Macromol. Chem. Phys.*, Vol. 210, pp. 377–386, 2009.
74. H. Xu, and L. Zhang, *J. Phys. Chem. C*, Vol. 113, pp. 1785–1790, 2009.
75. G. Tian, H. Fu, L. Jing, B. Xin, and K. Pan, *J. Phys. Chem. C*, Vol. 112, pp. 3083–3089, 2008.
76. T. Peng, D. Zhao, K. Dai, W. Shi, and K. Hirao, *J. Phys. Chem. B*, Vol. 109, pp. 4947–4952, 2005.

77. E. Beyers, P. Cool, and E.F. Vansant, *J. Phys. Chem. B*, Vol. 109, pp. 10081–10086, 2005.
78. H. Wang, J.J. Miao, J.M. Zhu, H.M. Ma, J.J. Zhu, and H.Y. Chen, *Langmuir*, Vol. 20, pp. 11738–11747, 2004.
79. J. Yu, W.H. Liu, and A. Yu, *Cryst. Growth Des.*, Vol. 8, pp. 930–934, 2008.
80. S. Zhan, D. Chen, X. Jiao, and C. Tao, *J. Phys. Chem. B*, Vol. 110, pp. 11199–11204, 2006.
81. H.W. Wang, C.H. Kuo, H.C. Lin, I.T. Kuo, and C.F. Cheng, *J. Am. Ceram. Soc.*, Vol. 89, pp. 3388–3392, 2006.
82. K. Wessels, M. Minnermann, J. Rathousky, M. Wark, and T. Oekermann, *J. Phys. Chem. C*, Vol. 112, pp. 15122–15128, 2008.
83. H. Choi, A.C. Sofranko, and D.D. Dionysiou, *Adv. Funct. Mater.*, Vol. 16, pp. 1067–1074, 2006.
84. T.K. Ghorai, D. Dhak, S. Dalai, and P. Pramanik, *Mater. Res. Bull.*, Vol. 43, pp. 1770–1780, 2008.
85. L. Chen, B. Yao, Y. Cao, and K. Fan, *J. Phys. Chem. C*, Vol. 111, pp. 11849–11853, 2007.
86. M.R. Hoffmann, S.T. Martin, W.Y. Choi, and D.W. Bahnemann, *Chem. Rev.*, Vol. 95, pp. 69, 1995.
87. K.C. Patil, S.T. Aruna, and S. Ekambaram, *Curr. Opinion in Solid State and Mater. Sci.*, Vol. 2, p. 158, 1997.
88. M. Martos, B. Julian, H. Dehouli, D. Gourier, E. Cordoncillo, and P. Escribano, *J. Solid State Chem.*, Vol. 180, pp. 679–687, 2007.
89. T.K. Ghorai, D. Dhak, S. Dalai, and P. Pramanik, *J. Alloy. Compd.*, Vol. 463, pp. 390–397, 2008.
90. P.E.D. Morgen, Better ceramics through chemistry, in: *Mater. Res. Soc. Symp. Proc., Vol. 32*, C.Y. Binker, D.K. Clark, D.R. Ulrich, eds., p. 213, 1984.
91. C.H. Lu, and W.J. Hwang, *Japanese J. Appl. Phys. Part 1: Regular Papers, Short Notes & Review Papers*, Vol. 38, No. 9B, p. 5478, 1999.
92. D.R. Uhlmann, B.J. Zelinsky, and G.E. Wrek, Better ceramics through chemistry, in: *Mater. Res. Soc. Symp. Proc., Vol. 32*, C.J. Brinker, D.E. Clerk, D.R. Ulrich, eds., p. 49, 1984.
93. M.M. Lenka, and R.E. Riman, *Chem. Mater.*, Vol. 7, p. 18, 1995.
94. A. Pathak, S. Mohapatra, S. Mohapatra, S.K. Biswas, D. Dhak, N.K. Pramanik, A. Tarafdar, and P. Pramanik, *Am. Ceram. Soc. Bull.*, Vol. 83, No. 8, pp. 9301–9306, 2004.
95. R. Roy, *Science*, Vol. 238, p. 1664, 1987.
96. C.L. Mak, B. Lai, K.H. Wong, *J. Sol-Gel Sci. Technol.*, Vol. 18, No. 3, p. 225, 2000.
97. C.M. Ling, A.R. Mohamed, and S. Bhatia, *Chemosphere*, Vol. 57, pp. 547–554, 2004.
98. N.L. Wu, M.S. Lee, Z.J. Pon, and J.Z. Hsu, *J. Photochem. Photobiol. A: Chem.*, Vol. 163, pp. 277–280, 2004.
99. M.A. Barakata, H. Schaeffer, G. Hayes, and S. Ismat-Shah, *Appl. Catal. B: Environ.*, Vol. 57, pp. 23–30, 2004.
100. S. Liao, H. Donggen, D. Yu, Y. Su, and G. Yuan, *J. Photochem. Photobiol. A: Chem.*, Vol. 168, pp. 7–13, 2004.

101. T. Sreethawong, Y. Suzuki, and S. Yoshikawa, *Int. Hydrogen Energy*, Vol. 30, pp. 1053–1062, 2005.
102. Y. Chen, K. Wang, and L. Lou, *J. Photochem. Photobiol. A: Chem.*, Vol. 163, pp. 281–287, 2004.
103. D.S. Kim, and S.-Y. Kwak, *Appl. Catal. A*, Vol. 323, pp. 110–118, 2007.
104. J. Georgieva, S. Armyanov, E. Valova, I. Poulios, and S. Sotiropoulos, *Electrochem. Commun.*, Vol. 9, p. 365, 2007.
105. J. Zhu, J. Ren, Y. Huo, Z. Bian, and H. Li, *J. Phys. Chem. C*, Vol. 111, pp. 18965–18969, 2007.
106. V. Houskova, V. Stengl, S. Bakardjieva, N. Murafa, and V. Tyrpekl, *Appl. Catal. B*, Vol. 89, pp. 613–619, 2009.
107. A.A. Ismail, L. Robben, and D.W. Bahnemann, *Chem. Phys. Chem.*, Vol. 12, pp. 982–991, 2011.
108. J.H. Pan, and W.I. Lee, *Chem. Mater.*, Vol. 18, pp. 847–853, 2006.
109. S. Liu, J. Yu, and S. Mann, *J. Phys. Chem. C*, Vol. 113, pp. 10712–10717, 2009.
110. R. Nakamura, A. Okamoto, H. Osawa, H. Irie, and K. Hashimoto, *J. Am. Chem. Soc.*, Vol. 129, pp. 9596–9597, 2007.
111. T.W. Kim, S.J. Hwang, Y. Park, W. Choi, and J.-H. Choy, *J. Phys. Chem. C*, Vol. 111, pp. 1658–1664, 2007.
112. J.C. Yu, L. Zhang, Z. Zheng, and J. Zhao, *Chem. Mater.*, Vol. 15, pp. 2280–2286, 2003.
113. K. Dai, T. Peng, H. Chen, J. Liu, and L. Zan, *Environ. Sci. Technol.*, Vol. 43, pp. 1540–1545, 2009.
114. T. Peng, D. Zhao, H.B. Song, and C.H. Yan, *J. Mol. Catal. A: Chem.*, Vol. 238, pp. 119–126, 2005.
115. X. Yang, Y. Wang, L. Xu, X. Yu, and Y. Guo, *J. Phys. Chem. C*, Vol. 112, pp. 11481–11489, 2008.
116. E. Stathatos, T. Petrova, and P. Lianos, *Langmuir*, Vol. 17, pp. 5025–5030, 2001.
117. S.Z. Chu, S. Inoue, K. Li, D. Wada, H. Haneda, and S. Awatsu, *J. Phys. Chem. B*, Vol. 107, pp. 6586–6589, 2003.
118. W. Dong, Y. Sun, C.W. Lee, W. Hua, X. Lu, Y. Shi, S. Zhang, J. Chen, and D. Zhao, *J. Am. Chem. Soc.*, Vol. 129, pp. 13894–13904, 2007.
119. Z. Wang, F. Zhang, Y. Xue, B. Cui, and J.N. Guan, *Chem. Mater.*, Vol. 19, pp. 3286–3293, 2007.
120. A. Chemseddine, and T. Moritz, *Eur. J. Inorg. Chem.*, p. 235, 1999.
121. J.H. Lee, and Y.S. Yang, *Mater. Chem. Phys.*, Vol. 93, p. 237, 2005.
122. T.K. Ghorai, A new spherical mesoporous nanoclusters of titania modified iron-niobate catalysts, communicated to: *Appl. Catal. B: Environ.*
123. T. Sugimoto, X. Zhou, and A. Muramatsu, *J. Colloid Interface Sci.*, Vol. 259, p. 53, 2003.
124. N. Uekawa, J. Kajiwara, K. Kakegawa, and Y. Sasaki, *J. Colloid Interface Sci.*, Vol. 250, p. 285, 2002.
125. H. Zhang, and J.F. Banfield, *J. Mater. Chem.*, Vol. 8, p. 2073, 1998.
126. H. Zhang, and J.F. Banfield, *J. Phys. Chem. B*, Vol. 104, p. 3481, 2000.

127. H. Zhang, M. Finnegan, and J.F. Banfield, *Nano Lett.*, Vol. 1, p. 81, 2001.
128. H. Zhang, and J.F. Banfield, *Chem. Mater.*, Vol. 17, p. 3421, 2005.
129. L. Znaidi, R. Seraphimova, J.F. Bocquet, C. Colbeau-Justin, and C. Pommier, *Mater. Res. Bull.*, Vol. 36, p. 811, 2001.
130. K.D. Kim, and H.T. Kim, *Powder Technol.*, Vol. 119, p. 164, 2001.
131. K.D. Kim, and H.T. Kim, *Colloids Surf. A*, Vol. 207, p. 263, 2002.
132. J.J. Qiu, W.D. Yu, X.D. Gao, and X.M. Li, *Nanotechnology*, Vol. 17, p. 4695, 2006.
133. S.C. Mitchel, *Aminophenols, 4th Ed, Vol. 2*, pp. 580–604, 2000.
134. H. Zhang, X. Yang, *Poly. Chem.*, Vol. 1, pp. 670–677, 2010.
135. J.A. Ayllon, A. Figueras, S. Garelik, L. Spirkova, J. Durand, and L. Cot, *J. Mater. Sci. Lett.*, Vol. 18, p. 1319, 1999.
136. S.K. Pradhan, P.J. Reucroft, F. Yang, and A. Dozier, *J. Cryst. Growth*, Vol. 256, p. 83, 2003.
137. J.J. Wu, and C.C. Yu, *J. Phys. Chem. B*, Vol. 108, p. 3377, 2004.
138. V.I. Popolitov, and Z.V. Pisma, *Tekhnicheskoi Fiziki*, Vol. 22, No. 5, p. 34, 1996.
139. L.K. Campbell, B.K. Na, and E.I. Ko, *Chem. Mater.*, Vol. 4, p. 1329, 1992.
140. J.M. Wu, H.C. Shih, and W.T. Wu, *Chem. Phys. Lett.*, Vol. 413, p. 490, 2005.
141. J.M. Wu, H.C. Shih, W.T. Wu, Y.K. Tseng, and I.C. Chen, *J. Cryst. Growth*, Vol. 281, p. 384, 2005.
142. Z. Zhang, C. Wang, R. Zakria, and J.Y. Ying, *J. Phys. Chem. B*, Vol. 102, pp. 10871–10878, 1998.
143. W.F. Yao, H. Wang, X.H. Xu, X.F. Cheng, J. Huang, S.X. Shang, X.N. Yang, and M. Wang, *Appl. Catal. A: Gen.*, Vol. 243, Iss. 1, p. 185, 2003.
144. J. Feng, R.S.K. Wong, X. Hu, and P.L. Yue, *Catal. Today*, Vol. 98, pp. 441–446, 2004.
145. J. Zhua, W. Zheng, B. Hea, J. Zhang, and M. Anpob, *J. Mol. Catal. A: Chem.*, Vol. 216, pp. 35–43, 2004.
146. M.I. Litter, and J.A. Navio, *J. Photochem. Photobiol. A: Chem.*, Vol. 98, pp. 171–181, 1996.
147. I.K. Konstantinou, T.A. Albanis, *Appl. Catal. B: Environ.*, Vol. 49, pp. 1–14, 2004.
148. W. Feng, and D. Nansheng, *Chemosphere*, Vol. 41, p. 1137, 2000.
149. W.F. Yao, H. Wang, X.H. Xu, X.F. Cheng, J. Huang, S.X. Shang, X.N. Yang, and M. Wang, *Appl. Catal. A: Gen.*, Vol. 243, Iss. 1, p. 185, 2003.
150. W.F. Yao, X.H. Xu, H. Wang, J.T. Zhou, X.N. Yang, Y. Zhang, S.X. Shang, and B.B. Huang, *Appl. Catal. B: Environ.*, Vol. 52, pp. 109–116, 2004.
151. W.F. Yao, H. Wang, X.H. Xu, Y. Zhang, X.N. Yang, S.X. Shang, Y.H. Liu, J.T. Zhou, and M. Wang, *J. Mol. Catal. A*, Vol. 202, pp. 305–311, 2003.
152. N. Daneshvar, D. Salari, and A.R. Khataee, *J. Photochem. Photobiol. A: Chem.*, Vol. 162, pp. 317–322, 2004.
153. W. Cun, Z. Jincai, W. Xinming, M. Bixian, S. Guoying. P. Ping'an, and F. Jiamo, *Appl. Catal. B: Environ.*, Vol. 39, pp. 269–279, 2002.

16

Zinc Oxide Nanowire Films: Solution Growth, Defect States and Electrical Conductivity

Ajay Kushwaha* **and M. Aslam**

Department of Physics, Indian Institute of Technology Bombay, Powai, Mumbai, India

Abstract

Zinc oxide (ZnO) nanostructures offer the most promising platform for fabrication of various optoelectronic devices. Insulating and nonmagnetic pristine ZnO becomes highly electrically conductive and magnetic due to the presence of the large concentration of impurities and defects. High surface area in nanowire films enhances the defect density and alters optical and electrical properties. In this chapter, we initially review the solution growth processes to fabricate ZnO nanowire films. Then, theoretically proposed and experimentally observed defect states and their origin are discussed. The role of defect states which unintentionally dope the ZnO is reviewed and the source of n-type and p-type conductivities is looked into. Further, we discuss the influence of defect states on the electrical conductivity and photosensitivity of zinc oxide nanowire films. The chapter concludes with a brief introduction about ZnO nanowires-based green energy conversion devices.

Keywords: Zinc oxide, nanowire, photoluminescence, defect states, surface defects, oxygen vacancies, electrical transport, green energy conversion

16.1 Introduction

Zinc oxide is a binary compound semiconductor which belongs to II-VI group and has cubic zinc-blende or hexagonal wurtzite crystal structure

**Corresponding author*: kushwaha09@gmail.com

Ashutosh Tiwari and Mikael Syväjärvi (eds.) Advanced Materials for Agriculture, Food, and Environmental Safety, (453–492) 2014 © Scrivener Publishing LLC

[1]. Zinc oxide crystal has SP^3 bonding between zinc and oxygen atoms and partial ionic crystal (iconicity ranges between covalent and ionic semiconductors), which result in wide band gap material [1]. Direct and wide band gap (3.3eV at 300 K) facilitates higher breakdown voltage, ability to sustain large electric field, lower noise generation, and high temperature and high-power operation in ZnO- based devices [1]. High excitonic binding energy (60 meV) is another important feature of ZnO, which reveals an efficient excitonic emission at room temperature [2]. Direct band gap is suitable for emission-based devices such as LEDs and lasers, while large exciton binding energy is useful for devices based on excitonic effect [2]. Zinc oxide also possess piezoelectric properties, spintronic behavior, high thermal conductivity and highly sensitive surface, which can offer a variety of device applications [2, 3]. Besides these excellent properties, low-cost and facile crystal growth makes ZnO an alternative of GaN for numerous opto-electronic devices. Nanostructures of ZnO offer various advantages and demonstrated exceptional properties. Variation in size and shape of the nanostructures can easily tune the optical, electrical and magnetic properties of ZnO, provides an opportunity for new material design as per the application need. A nonmagnetic and insulating material (ZnO) becomes magnetic and electrically conductive due to the presence of various defects states/surface states at nanoscale [4–7]. Zinc oxide nanostructured films (confinement at least in one direction) offer the most promising platform for fabrication of optoelectronics devices. Necessity of high surface area for generation and separation of large numbers of photogenerated charge carriers for efficient energy devices is also satisfied by nanowire array and opens a new pathway for more efficient device fabrication [8, 9]. Nanowire films (free-standing nanostructures on substrates) are fundamentally the most suitable structural design for optoelectronic and sensing applications. The nanowire geometry provides ultrafast charge separation as well as high surface area within the same volume of device. This justifies the potential utility of nanowire films in photovoltaic, light emitting diodes, lasers, water splitting/hydrogen generation, piezoelectric generators and thermoelectric energy conversion devices [10–17].

Zinc oxide nanowire films can be fabricated by a variety of techniques [18, 19]. The nanowire growth depends on the enhancement of the crystal growth rate in one dimension and suppression of growth in the other dimensions [20]. The free-standing 1D nanostructures typically grow outwards away from a single nucleation point, growth confined in one direction because of different growth rated along different dimensions. Numerous procedures have been utilized to grow oriented nanowires, namely physical vapor deposition, chemical vapor deposition, electrochemical deposition and solution or chemical process [18, 20–24]. Though vapor deposition process (physical methods) result in good structural control on growth and form high quality NWs films, but their inefficiency in economical and large-scale production is a main issue. On the other hand, the solution growth process offer a cost-effective, high yield production with superior control over the structural and physical properties of ZnO nanowires [18].

Defects become prominent in nanostructures due to high reactive surface area, which significantly affects the optical and electrical properties of ZnO. The presence of defect can be easily observed through photoluminescence measurement, an additional broad peak appears in the emission spectra of ZnO due to defect states. The unstructured defect band emission extends from green to red (500–750 nm), depending upon the type of defects/vacancies created during fabrication [25]. Defect emission can be categorized into three emissions as green, yellow and red emission, in which the cause of green emission is not well understood and is highly controversial. Various hypothesis such as transition of singly ionized oxygen vacancies to photoexcited hole; transition between singly charged oxygen vacancy to valence band; recombination of electrons close to conduction band with a doubly trap holes at V_o^{+2}; oxygen deficiencies, and; surface defects are counted as a cause for green emission in ZnO [25]. Defect states also play a vital role in controlling the electrical and magnetic properties of the ZnO nanowires [25]. The ZnO inherently shows n-type semiconducting nature [26], but the origin of n-type conductivity in undoped ZnO is unclear and no conclusive remarks are found in literature. Presence of native point defects as zinc interstitial/vacancies or oxygen vacancies/interstitial in ZnO lattice are possible cause for such intrinsic doping. Oxygen vacancies are the most accepted defects states for unintentional n-type doping, however, a few theoretical reports have suggested that oxygen vacancies are deep level donor and cannot contribute in conductivity [26–28]. Janotti *et al.* suggested that an impurity atom like hydrogen creates unintentional doping in ZnO during growth, which renders n-type behavior in ZnO. However, there is no clear experimental evidence which supports hydrogen doping. Most of the experimental reports still accept that oxygen vacancies are the shallow level defects abundant in ZnO crystal due to less formation energy. The role of defect states on magnetic properties has been explored quite well and it has been observed that oxygen vacancies are mainly responsible for magnetism in ZnO nanostructures at room temperature [29]. Surface defects and various interactions among defects like RKRY type, itinerant electron and small polaron models are proposed as causes for magnetism [30–32]. Analogous to magnetic properties, electrical transport properties of ZnO nanostructure can be correlated with specific defects and controlled by manipulating defect states. The surface charge states and surface defects alter the shape of band and band bending between nanowire grains/surface of the nanostructure and strongly effect the electrical transport properties [33, 25]. Solution grown ZnO nanowire consists a high density of surface defects, hence electrical properties can easily be controlled by various surface modification processes.

In this chapter, after a brief discussion of low temperature hydrothermal growth, advancements in the hydrothermal approach and other solution processes are reviewed. Further, types of defects in ZnO are illustrated and their impact on optical emission spectra is discussed. The role of defect states in electrical conductivity of ZnO is reviewed and the cause of n-type and p-type states is looked into both from a theoretical as well as experimental point of view. Influences of defect states on electrical transport properties of ZnO nanowire films are presented and the effect of surface modification is discussed. Finally, the chapter concludes with an illustration of some applications of ZnO nanowire films in energy conversion devices.

16.2 Solution Growth of ZnO Nanowire Films

Fabrication of one-dimensional nanostructure systems is a popular research topic among nanomaterial researchers. However, control of density, orientation, shape and size are the major challenges during nanowire fabrication [18]. The synthesis of highly crystalline, continuous and oriented array of nanowire is essential for studying the interesting physical properties of low-dimensional systems and integration of nanostructured devices. In the last decade, significant research has been carried out on nanowire synthesis, using VLS technique, [23, 24] vapor deposition [34], electrochemical deposition [35], molecular beam epitaxy [36] and laser ablation [37]. These different methods led to the fabrication of a broad range of one-dimensional nanostructures. Zinc oxide nanowire array growth using the solution approach was reported for the first time by Andres-Vergés *et al.* [38] and became more popular through the work of Vayssieres *et al.* [39]. The method comprises controlled synthesis of ZnO nanowires on glass and Si substrates by the thermal decomposition of methenamine and zinc nitrate in an autoclave.

Several growth mechanisms for the anisotropic growth of nanocrystals in solution have been proposed. In solution-liquid-solid (SLS) mechanism growth is initiated from seed particles. In the SLS reaction monomers are generated first by decomposition of molecular precursor at a low temperature. The seed particles/catalysts for this kind of reaction are extremely small and therefore easily activated at low temperatures. The monomers react with the nanoseeds to form supersaturated droplets [40, 41]. In the case of self-assembly, attachment growth is based on the fact that nanoparticles generated in solution have a large surface-to-volume ratio. To reduce the surface energy and total system energy, the particles may segregate together. Oriented attachment is one of the ways for this segregation process. Anisotropic growth of crystals for formation of the most elongated nanocrystals is induced due to different surface energies [40, 42, 43]. However, the difference in surface energies (the intrinsic properties of the crystal) is not large enough to cause highly anisotropic growth of long nanowires. By adding surfactants to the reaction solution, the effective surface energy of a nanocrystal can be modulated, and the surfactant molecules selectively adsorb and bound onto certain surfaces of the nanocrystal seeds [40, 44, 45]. Fabrication of nanostructured films on a large scale is the main challenge for any industrial application and solution growth provides an opportunity to grow large-scale nanowire films in a low-cost and facile process. Solution growth has been extensively explored by various researchers and excellent control on growth location or spatial growth, orientation, density and various doping has been successfully achieved [25, 46–48].

16.2.1 Low Temperature Hydrothermal Growth

Low temperature hydrothermal process has advantage that nanowire film growth can be initiated on any arbitrary substrate after deposition of ZnO seed layer. A typical synthesis is performed in an aqueous solution of zinc nitrate and hexamine (hexamethylenetetramine (HMT)), however, zinc chloride and zinc acetate is also used, and growth is carried out at low temperature (< 100°C) [29, 49]. Hexamine (HMT) releases hydroxyl ions in the solution which further react with Zn^{2+} ions to form ZnO [50]. The following equations explain the reaction of the ZnO growth as proposed by Vergés *et al.* [38].

$$(CH_2)_6N_4 + 6H_2O = 6HCHO + 4NH_3 \tag{16.1}$$

$$NH_3 + H_2O = NH_4^{\ +} + OH^- \tag{16.2}$$

$$2OH^- + Zn^{2+} = ZnO(s) + H_2O \tag{16.3}$$

It is generally supposed that HMT supplies hydroxyl ions to drive the precipitation reaction [51]. However, the role of hexamine is not fully understood, and few researchers believe that HMT acts as a buffer [52]. Ashfold *et al.* have proposed that the rate of decomposition of HMT is independent of the reaction and acts as a kinetic buffer [51]. Primarily, the pH and the concentration of Zn^{2+} ions are maintained in such a way that the ZnO growth starts with the formation of $Zn(OH)_2$. When pH is gradually increased, the concentration of the Zn ions decreases and $Zn(OH)_2$ becomes thermodynamically unstable. Sugunan *et al.* have suggested a different approach, that HMT prefers to attach on the non-polar facets rather than the polar facets, and leads to one-dimensional epitaxial growth on substrate [53]. The HMT therefore acts more like a shape-inducing polymer surfactant rather than as a buffer.

The ZnO 1D nanostructures have been successfully grown in the presence of different surfactants such as polyvinyl alcohol (PVA), polyethylene glycol (PEG), sodium dodecyl sulphate (SDS) and cetyltrimethyl ammonium bromide (CTAB) [49, 54]. The use of PVA resulted in more regular and defect free rods than PEG, SDS and CTAB. Li *et al.* have reported CTAB assisted hydrothermal growth resulted in tapered ZnO nanowires with diameter decreasing from 400 nm at the body to about 80 nm at the tip [55]. Figure 16.1 shows low temperature hydrothermal grown ZnO

Figure 16.1 Scanning electron microscopic images of low temperature hydrothermally grown ZnO nanowires. (a) Top view, (b) 20 degree tilt view, (c) cross-sectional view, and (d) high resolution tilt image.

nanowire films. The top surface of the nanowire is smooth and hexagonal in shape (Figure 16.1a). Nanowires are perpendicular to substrate surface and around 6 μm long, as shown in Figure 16.1b and 16.1c; however, variation of growth temperature leads to slightly rough surface of nanowire (Figure 16.1d) [56]. Hydrothermal methods provide various opportunities to control the structural, morphological and doping properties of ZnO nanostructures [57–59].

16.2.1.1 Shape, Size, Orientation and Density Controlled Growth

The hydrothermal growth process has shown great potential to achieve excellent control of shape, size and growth orientation on a large scale in an inexpensive and facile way. The size of ZnO nanowire ranges from a few nanometers (~20 nm) to hundreds of nanometer in diameter and several micrometers in length. Oriented growth of ZnO nanowire in a vertical as well as lateral direction of substrates is performed [60]. Precursor concentration has been shown to be a major factor to control the density of nanowire in seeded and non-seeded substrates [61–65]. Selective adsorption of citrate anions on a ZnO basal plane results in a shape selective growth and controls aspect ratio, morphology and oriented nanocolumns [60]. Pretreatment of substrates, seeding and annealing affects both diameter and orientation of the nanowire [66]. A vapor-phase hydrothermal method has been recently demonstrated to grow ZnO nanotube and nanorod array films on both sides of a zinc foil under ammonia vapor [67]. Ultra-thin ZnO nanowire growth is observed in the alkali solution of zinc prepared by zinc nitrate and NaOH mixed solution in deionized water followed by ethanol [68]. Recently, Kawai *et al.* stated that each plane of ZnO crystal has their critical concentration for nucleation, which plays an important role in the anisotropic growth of hydrothermal ZnO nanowires [69]. The Shin group showed that Au nanodots on Al-doped ZnO seed layer can also be used for density control growth of nanowire [70].

A lateral growth of ZnO nanowire on substrates was shown, and it was proposed that surface-induced variations in the local Zn^{2+}/OH^- concentration ratio within the reactive solution are responsible for this orientation [71]. The growth of highly dense and ultra-long ZnO nanowires is performed by introducing an adequate content of ammonia into the nutrient solution on seedless substrates [61]. Theoretically, predicted growth conditions/growth statistics are implemented and it was observed that the aspect ratio of nanowire can be change from 10 to 23 by using optimum growth condition [72]. Recently, growth of "nanoforest" composed of high-density, long-branched "treelike" multigeneration hierarchical ZnO nanowire has been demonstrated, which shows highly efficient PV device performance [73].

16.2.1.2 Patterned Growth of ZnO Nanowire

Spatially controlled or patterned growth has become important for many functional devices. The high-temperature-based VLS growth process has shown significant potential for patterned and highly oriented growth using special located seed particles; however, low temperature processing of devices is challenging for high temperature methods [74, 75]. To overcome this problem the hydrothermal process can be utilized to get desired control on nanowire growth and patterning by ZnO seed layer. Pattern growth can be done by using electron-beam lithography, laser interference lithography, nanoimprint lithography (NIL) or soft chemical approach by designing various self-assembled monolayers (SAM) followed by hydrothermal growth [75–79]. Electron-beam lithography (EBL) is a very promising tool, which can fabricate pattern of various shape and size with high precision in nanometer-scales [76]. The ZnO seed layer can be patterned on silicon or GaN substrates using photoresist as polymethylmethacrylate (PMMA) by spin-coating on as-prepared ZnO seed layer. A desired pattern can be written on the photoresist by electron-beam lithographic system and then developed into a desired structure and residue layer of PMMA pattern removed from the surface [79]. Figure 16.2a represents a top view SEM image of the double line of the ZnO nanorod grid with 2.5 μm line width and period of 10 μm prepared by using EBL patterning and hydrothermal growth methods. The figure also shows the parallel ZnO nanorod lines with 600 nm width and 1 μm period, and 45° tilt view of individual ZnO nanorod arrays with diameter of 250 nm and period of 1 μm with 1 μm feature size and 4 μm period. High-aspect ratio of lateral array of ZnO are fabricated by overgrowth from dot patterns created by electron beam lithography (EBL) [77].

A patterned growth of vertically aligned ZnO nanowire arrays on non-epitaxial substrate is shown at low temperature and it is observed that thin seed layer of ZnO assists the directed growth. Initially nanowires are vertically grown due to structural confinement between the resist, but the NW could grow in both a vertical and lateral fashion at a later stage [78]. Different shaped lattice patterns such as primitive rectangular, centered rectangular, oblique, and hexagonal are grown via laser interference lithography followed by hydrothermal synthesis (Figure 16.2b) [74]. Polymer templated hydrothermal growth rendered excellent control over the location and morphology of the ZnO nanostructures; nanorods, pencil-like,

Figure 16.2 (a) E-beam lightraphically patterned ZnO nanorod [76]. (b) Nanoscale two-dimensional patterns of ZnO nanorods using laser interference lithography [74]. (c) polymer-templated hydrothermal growth of patterned ZnO nanorods and various morphologies [80]. (d) Soft chemical templated (polystyrene sphere and TiO2 layer) control growth of ZnO nanowires [75].

needle-like, tubular, tree-like, and spherical shapes are obtained by controlling the growth conditions. Figure 16.2c represents various morphologies of ZnO nanostructures which utilized various growth conditions and structural polarity [74]. A direct patterning of ZnO nanoparticle (NP) seeds via microcontact printing demonstrated excellent spatial control over nanowire growth in a large area, which can also be performed on flexible substrates to be utilized in flexible electronic devices [79]. This kind of printing needs master stamps with defined structures, the concentration and viscosity of ink (seed solution) plays an important role on pattern quality [81]. Though top-down approaches to patterning have shown superb control over patterns, these processes are very time consuming and require high cost tools and excellent operating skills. Hence, the soft chemical-based self-assembled monolayer approach is becoming quite attractive for large-scale and low-cost patterning. Monolayer of PS microsphere rendered a well-defined geometry, as shown in Figure 16.2d, and resulted in a patterned growth of nanowire arrays with long-range periodicity wafer scale [75]. The SAM-based techniques again have advantages on quality of nanowire; no local destruction or chemical modification occurred, which is common in top-down approaches [82].

16.2.1.3 Growth of ZnO Nanowire on Various Substrates and Doping

In high temperature growth processes, oriented growth of ZnO nanowires are possible when substrates have minimum lattice mismatch. However, solution-based hydrothermal growth enables oriented growth on any arbitrary substrate at low temperature with/without assisting seed layer [83–87]. Various reports have shown oriented growth of ZnO nanowire on flexible substrate using the hydrothermal approach. A position-controlled growth of ZnO nanorods on transparent glass, flexible polymer and polyethylene terephthalate (PET) substrates are reported, which could be a suitable design for flexible optoelectronic devices [84]. Vertically aligned ZnO nanowire is also transferred on flexible substrate like PDMS or PMMA, after growth on flat Au(111) surface, or Si(100) wafer, or titanium, or aluminum substrate by the solution approach [83]. A large-scale wafer-size growth on plastic or silicon substrate is performed using the hydrothermal process, which reflects the industrial scale production capability [86]. GaN and sapphire have already shown highly oriented nanowire structures due to minimum lattice mismatch with ZnO. Zinc metal has also been demonstrated as substrate to grow ZnO nanostructures, and zinc plating results in an alternative to Zn substrate and enables growth on any arbitrary substrate [87].

Zinc oxide shows variable electrical properties due to intrinsic dopant as vacancies or interstitial of oxygen and zinc and various unintentional dopants like hydrogen and chlorine [88]. Detailed discussion about intrinsic doping is given in the next section. Extrinsic dopant is important to tune the optical and electrical properties of ZnO nanostructures [89]. Usually doping takes place during growth; dopant is co-dissolved in precursor solution before reaction. However, doping become a little tricky in aqueous solution growth due to formation of aqua ions which may not incorporate into the crystal lattice easily, while nonaqueous growth is relatively more facile for a good quality doping [90]. Though doping is not as easy, doped ZnO nanowire by various elements like Al, Sn, Ga, In, Sb, and Cu is successfully grown by hydrothermal process and shows excellent properties [91]. Mostly all the dopant shows n-type behavior, however, the Djurisic group has shown that p-type nature can be achieved in undoped ZnO nanorods by adjusting the properties of the seed layer, and proposed that increased concentration of zinc vacancies might be responsible for this [78].

16.2.2 Alternative Solution Growth Methods

16.2.2.1 Chemical Bath Deposition

Compound semiconductor films can be easily grown from aqueous solutions using this method. The process demonstrated good control over film thickness, deposition rate and quality of crystallites by changing the solution pH, temperature and bath concentration, however, reproducibility is a little difficult [92]. A simple growth involves zinc salts such as zinc acetate and amine group such as ethylenediamine. A hot plate is used to heat the solution with stirring and any arbitrary substrate (glass/TCO/Silicon, etc.) can be used for growth [92]. Growth of nanowire films can be performed at various temperatures. The reaction mixture was maintained constant at a desired temperature for deposition with continuous stirring. Multiple dipping is generally executed and then washed with distilled water and dried in air [92–95]. Two types of nucleation happened in solution: in homogeneous nucleation, the growth process is very fast and results in large-size particles, while heterogeneous nucleation occurs at the substrate surface and particles grow slowly [92]. By applying ZnO seeded substrate or two-step growth process, good quality nanowire films can grow. The pretreated glass substrates were also utilized to form hydrophilic surface for uniform growth over the entire substrate surface [95]. Growth of nanowire starts from small ZnO nuclei and crystalline-size nuclei mainly govern the diameter of nanowire [95]. Size, density and orientation of nanowire are easily controlled by tuning the Zn ions concentration in solution, growth time and seed layer [95]. Growth of ZnO at room temperature without using amine at neutral pH is also reported and it has been found that amine only controls the saturation index of ZnO [93]. A submillimeter channel was used to flow the solution for ZnO nanowire growth on heated glass/silicon substrate, which can improve yield and decrease waste solvent generation and result in more uniform nanowire film [96].

16.2.2.2 Spray Pyrolysis

In typical spray pyrolysis process, an aqueous solution of zinc precursor (zinc chloride or nitrate or hydride) with variable concentration is used as per desire [97–100]. The depositing substrate is kept at higher temperature in the range of 300°C to 600°C and the precursor solution is sprayed onto the substrates by a nozzle. Commercial glass slides or TCOs are used as substrates for deposition; however any arbitrary substrate can be used which can be sustained at this temperature without any damage [101]. Compressed air is used as a carrier gas and generally few ml of precursor solution is sprayed at slow rate [101]. The growth of ZnO nanowire on substrate requires slightly higher substrate surface temperature than temperature required for thin film growth. The substrate temperatures around 400°C result in ZnO films, the substrate temperature 450°C leads to hazy films, while 490°C results in oriented hexagonal rods [101]. Krunks *at al.* showed that a spray of aqueous solution of zinc chloride (with concentration of 0.05 mol/l) on ITO at 550°C results in well-shaped hexagonal ZnO nanorods with diameter of 100–150 nm and length of up to 1 μm [97]. An alcoholic precursor solution leads to high aspect ratio of nanorod at lower substrate temperature compared to the spray of aqueous solution [100]. An addition of thiocarbamide into the spray solution significantly changes the size and shape and thinner nanorods are grown. Though spray pyrolysis method provides low cost, high yield of ZnO nanorod films, this process has some limitations in control over nanowire diameter and length.

16.2.2.3 Electrochemical Growth

The electrochemical growth process is performed in a three-electrode electrochemical cell using potentiostat/galvanostat; the substrate on which film has to be grown is taken as working electrode, while two other electrodes named as counter and reference are used in this procedure. A conducting substrate (TCO/metal deposited glass/conducting polymer coated film, etc.) is required which acts as a working electrode. Electrolyte of the experimental solution contains zinc salt (zinc nitrate or zinc chloride solution) with some supporting electrolyte such as KCl [102–104]. Deposition reaction is generally carried out at a slightly higher temperature than room temperature in the range of 60–90°C. The ZnO posses intrinsic highly anisotropic crystal structures, which can facilitate templateless growth of one-dimensional nanostructure [105]. Thus template-free electrochemical growth of ZnO nanowire films is possible and does not require the complicated pre- and post-treatments as in the case of template-assisted growth and is relatively less expensive [104, 105]. Template grown NWs show poor structural and surface properties and defects of host pore leads to polycrystalline ZnO nanowires [105]. Hence, to get single crystalline ZnO nanowire, template-free electrochemical process is more suitable. Diameter of nanowire can be controlled by pore diameter in template NW, however a control of nucleation step and growth with manipulation of deposition parameter can also result in variable size nanowires [106, 107]. The nanostructured growth rate can easily be controlled by deposition potentials, current densities and salt concentrations [106]. The concentration of zinc precursor directs the size of ZnO nanowire, for example, control of the dimensions of ZnO nanowires from 25 to 80 nm is reported by controlling $ZnCl_2$ concentration [105]. The presence of hydroxide ions increase pH close to the cathode and Zn^{2+} and OH^- ions react together and form ZnO [102]. The nucleation, distribution and growth of the ZnO nanostructures are determined by the availability of Zn^{+2} ions [102]. Higher temperature deposition gives in-homogenous morphology and thicker ZnO columns. Nonaqueous solvents are also used for NW films, which may allow a large deposition potential window and good crystal quality with improved adherence on substrate [108]. A large-scale synthesis of hierarchical ZnO nanorods on a transparent conductive substrate is also shown by using DMSO-H_2O solution [109]. Zinc oxide nanorod bundle growth was observed when zinc chloride and tartaric acid precursor were used at 90°C [110]. Though template-free electrodeposition yields high crystal quality and large-scale production capability, very good control of nanowire aspect ratio and orientation is still missing.

The solution growth approach leads to large-scale and facile growth of ZnO nanowire films in any arbitrary substrate at low temperature (< 90°C). Control over nanowire orientation, shape/size, distribution, density, patterning and doping can be easily achieve by the solution approach, which shows high order of applicability in various device configuration. However, this growth process results in low crystal quality and various defects, such as point defects, voids, surface roughness, and surface species/organic residuals at surface are created [18]. In this approach, presence of intense defects states effectively alters the optical and electrical properties of nanowire and is responsible for low efficient devices. Hence, to utilize the beauty of the solution growth process for device fabrication, the control over crystal defects of ZnO nanowire is essential.

16.3 Defects and Photoluminescence Properties of ZnO

16.3.1 Defects in ZnO

Missing of any atoms of oxygen or zinc from their original position or sitting at another site creates point defect in ZnO crystal. Different types of native point defects can generate in ZnO during growth; it might be oxygen vacancies or oxygen interstitial, zinc vacancies or oxygen interstitial, or a complex of these native defects. The concentration of a defect in a crystal depends upon its formation energy E^f in the following form [111]:

$$C = N_{sites} \exp\left(\frac{-E^f}{K_B T}\right) \qquad 16.4$$

where N_{sites} is the concentration of sites in the crystal where the defect can occur, while formation energy of a point defect in a charge state q can be written as:

$$E^f(q) = E^{total}(q) - n_{Zn}\mu_{Zn} - n_O\mu_O - qE_F \qquad (16.5)$$

where E^{total} (q) is the total energy of a system containing n_{Zn} and n_O number of zinc and oxygen atoms, μ_{Zn} and μ_O are the chemical potentials for zinc and oxygen, respectively, and E_F is the Fermi energy [3]. The chemical potential is dependent upon the growth condition and decides formation energy of a defect, and low formation energy defects reveal high defect concentration in crystal. In zinc-rich growth condition, oxygen vacancies have lower formation energy, while zinc vacancies have lower formation energy in oxygen-rich condition. Oxygen vacancies can be formed in neutral, +1 or +2 charge state (V_o, V_o^{+1} and V_o^{+2}), and zinc can have 0, -1 and -2 charged defect states in ZnO crystal. Though native point defect has low formation energy and highly dominates in structure, some other defects like grain boundaries, line defect or plane defect can also be present in the ZnO system. Grain boundaries are very significant defect in nanostructured thin films; a depletion region is generally formed at each grain boundary and strongly affects optical and electrical properties [112]. Threading dislocations of 20°– 30° from the c-axis are also observed in plasma-assisted MBE grown ZnO under Zn-rich conditions [113]. These dislocation transmissions may have negative charge due to the accumulation of Zn vacancies near the dislocation core [112]. It is observed that dislocations created by plastic deformation increase the excitonic emission intensity [114]. Basal-plane stacking fault is seen in MBE grown ZnO film on sapphire, but no electronic states creation is observed [115, 112]. Some other impurity elements/unintentional dopant and intentional dopant can create defects in the ZnO crystal. Nitrogen can substitute oxygen and exists in two distinct chemical environments, attributed to N acceptors and N_2 molecules [116]. Phosphorus, arsenic, antimony, lithium, sodium and copper are studied as intentional dopant to create extra defect sites in ZnO, which introduce new states in the band gap region [112]. Lithium, chlorine, hydrogen and hydrogen related complex are generally unintentionally introduced in ZnO lattice during growth and effect electronic states. Hydrogen is considered the most prominent defect in ZnO which creates shallow donor state in electronic structure [3]. Besides these, surface-related defects become important when material reaches nanosize regime. Surface band bending, chemisorptions/photodesorption of species near surfaces, native surface defects/states, and surface roughness are mainly considered as surface-related defects [33, 116–121]. A surface accumulation of electrons due to high surface area in nanostructure materials promotes high surface conduction and contributes to optoelectical properties [120]. Thus, it can be concluded that defects introduce new states in electronic band structure and strongly affect the optical and electrical properties of ZnO nanowire films [33].

16.3.2 Photoluminescence of ZnO Nanowire

The luminescence properties are characteristics of energy band structure and lattice dynamics of the materials. The relative amount of radiative and non-radiative recombination governs the intensity of photoluminescence (PL). The PL measurement process involves excitation of semiconducting material with photons of higher energy than the band gap of that material. Absorption of photon results in creation of electron-hole pair and is followed by recombination of these excited charge carriers [122]. Excited electron recombine by interband transition and release photon with energy corresponding to the band gap. Zinc oxide has high excitonic binding energy (60 meV) which enables efficient excitonic emission in ZnO at room temperature [122, 123]. Good quality zinc oxide crystal only shows band edge transition and results in intense single peak in PL measurement [124]. The presence of defect drastically alters the emission spectra and one more broad peak in the visible region appears. In the case of nanostructured ZnO, defect-related emission becomes very prominent due to high surface effect [10]. Thus, the emission from ZnO nanostructures can be categorized as UV emission related to band edge transition and visible emission corresponds to defect states transitions.

16.3.2.1 UV Emission

Ultraviolet (UV) emission of ZnO is associated with transitions between the electrons in the conduction band and holes in the valence band, which is influenced by excitonic effects due to the coulomb interaction [125, 10,

126]. Band edge transition is related to various exciton recombinations. The associations of impurities/defects with exciton or exciton impurity complexes play an important role in the luminescence [127]. Balandin *et al.* stated that UV emission in ZnO may be attributed to the confined excitons, transverse optical phonon band of the confined excitons, donor-bound excitons, acceptor-bound excitons, or donor acceptor pairs [128]. Their group reported that recombination of the acceptor-bound excitons is mainly responsible for emission at all temperatures, while recombination of the acceptor-bound excitons is dominant in PL only at low temperature [128]. In the case of ZnO quantum dots (2 to 6 nm), excitons confined in the quantum dot and excitons bound to an ionized impurity located at the quantum-dot surface are considered as a source of emission [129]. Polarity of ZnO facets also affect the emission from free excitons and it is found that the O face is more intense than the Zn face [130]. Different shape- and size-dependent nanostructures show variable peak position of UV PL spectra and more than 10 nm shift is observed, these shifts in the UV emission peak are not due to the confinement (size is larger than Bohr radius) but due to different concentrations of native defects/surface defects [125].

16.3.2.2 Defect Emission

Figure 16.3a shows PL measurement of a CVD grown high quality ZnO nanowire film, where only one emission peak appears at room temperature. Figure 16.3b shows a PL spectra of solution grown ZnO nanowire; emission spectra shows another very intense and broad peak in the range of 450 to 750 nm and is called defect emission peak. Unstructured broad emission peak in visible range can deconvolute in many peaks.

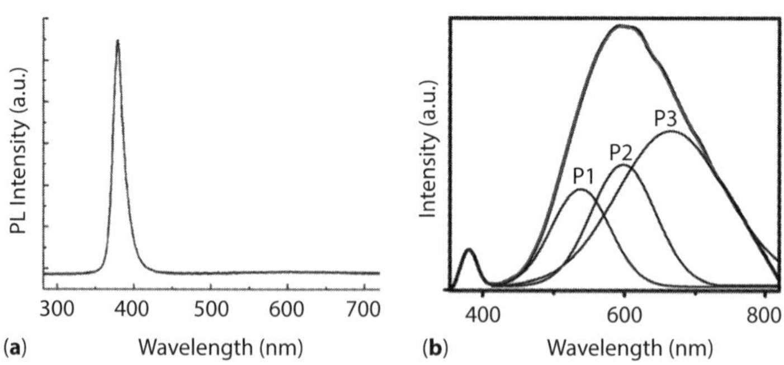

Figure 16.3 Room-temperature PL spectra of (a) high quality CVD grown ZnO nanowires [124], (b) Defect rich solution grown ZnO nanowires [29].

Figure 16.3b demonstrates that most accepted three peak fitting is related to green (P1), yellow (P2) and red emissions (P3). Green emission is the most commonly observed defect emission in ZnO nanostructures, which is often attributed to singly ionized oxygen vacancies [25]. Other opinions for green emission as donor–acceptor transitions, recombination at V_o centers (where these centers are generated by surface trapping of photogenerated holes, followed by recombination with electron in an oxygen vacancy (V_o), zinc vacancy, and surface defects are also considered [125]. Yellow emission is typically attributed to doubly charged oxygen vacancies, sometimes oxygen interstitial or impurity represents another possible candidate [131], whereas red emission is mainly attributed to oxygen interstitials.

16.4 Role of Defect States in Electrical Conductivity of ZnO

16.4.1 Defect States Responsible for n-Type Conductivity

The conductivity of a semiconductor can be given as:

$$\sigma = CT^{2/3} q_e \left(\mu_e + \mu_h\right) \exp\left({-E_g}/{2KT} \right) \tag{16.6}$$

where μ_e and μ_h is the mobility of electron and hole, E_g is the band gap of semiconductor and T is the temperature. At T = 0, there are no electrons in the conduction band, and the semiconductor does not conduct (lack of free charge carriers). At T > 0, some fraction of electrons have sufficient thermal kinetic energy to overcome the gap and jump to the conduction band [132, 133]. A semiconductor will only have free carriers for conduction if the impurity atoms/defects are ionized. The ionization of atoms leads to electrons jumping across ΔE into the conduction band from donor or accepter level and govern the electron conduction. Ionization is a process by which free charge carriers (electrons or holes) are produced in a semiconductor [134]. Shallow level states require less ionization energy, while deep level states need high ionization energy. Impurity atoms or defect states contribute to electrical conductivity when they form a shallow state in band structures. Native defect can also create shallow level states in semiconductor; in the case of ZnO, oxygen vacancies and/or zinc interstitial are potential candidates for n-type conductivity, while ZnO is inherently non-conducting. Native defects can produce n-type doping if they are creating a shallow level in band structure (defect states should be near to conduction band minima), donor defects must have low formation energy (defect states should be abundant) and electron killer center must have high formation energy [26]. Though many studies have proposed that V_O and Zn_i are responsible for n-type conductivity, some theoretical proposals have said that oxygen vacancies are deep rather than shallow and cannot contribute to conductivity [135]. It has also been proposed that zinc interstitial is thermally unstable and has high formation energy, thus it cannot help in electrical conduction, though it is a shallow donor. Janotti *et al.* have suggested that two forms of hydrogen act as electrically active impurities, interstitial hydrogen which attaches to an oxygen host atom, and substitutional hydrogen on an oxygen site, both of which are strong candidates for n-type conductivity [27].

The experimental studies have also given mixed results about n-type doping in ZnO. In 1941, Miller showed experimental result about the electrical conductivity of ZnO and proposed that the conductivity in the lower temperature is due to the ionization of interstitial zinc atom pairs, whose ionization energy is 20 mV or less. In the higher temperature range the conductivity is caused by the ionization of single interstitial zinc atoms, whose ionization energy is found to be 700 mV. He also observed free electron density of ZnO about $10^{15}cm^{-3}$ at room temperature by hall measurement [134]. In 1957, Thomas found that pairs of positive (zinc) plus negative (oxygen) ion vacancies with excess oxygen ion vacancies are capable of reproducing diversified electrical properties in ZnO [136]. Heating of ZnO crystal in zinc vapor followed by rapid quenching results in zinc interstitial and facilitates n-type conductivity [132]. A zinc self-diffusion process study indicated that oxygen vacancies are probably the intrinsic ionic defects responsible for n-type conductivity in reducing atmospheres [137]. High-energy electron irradiation in ZnO produces shallow donors at about 30 meV below the conduction band and the donor is identified as a Zn-sublattice defect, most likely the interstitial Zn_i or a Zn_i-related complex [138]. Characterization of ZnO by admittance spectroscopy has shown that residual donors such as Zn interstitials and oxygen vacancies are responsible for n-type conductivity.[139] Impurity atoms such as Al, In and Ga are shown as n-type dopant by secondary ion mass spectroscopy (SIMS) analysis [140]. Dissociated in water, O^{-2} and H^+ are also considered as sources of electrons in ZnO crystal. Some experimental reports also support hydrogen as a cause for n-type conductivity [141]. However, Kim and his group have given strong justification against the hydrogen assumption that a high concentration of electron carriers is observed, even when H contamination is minimized or when H is removed. They illustrated that annealing at high temperature at 1100°C will diffuse out H and will not make any contribution in conductivity, but experimental results have shown high electrical conduction [28]. They supported the premise that O-deficient ZnO can easily become n-type even without the introduction of any intentional dopants and a metastable shallow-donor state of V_O was suggested as an alternative source of the n-type doping. The cause of n-type intrinsic doping is very ambiguous; no clear correlation is observed among various theoretical and experimental studies and in-depth investigations are still needed.

16.4.2 Defect States Responsible for p-Type Conductivity

P-type doping of ZnO is still challenging. If we are looking at doping with respect to native defects, the following condition must be satisfied: Acceptor defects (O_i, V_{zn}) must have shallow levels (defect states should be near to valence band maxima) and formation energy for acceptor defect must be lower than donor defects. Even ZnO has p-type acceptor defects (O_i , V_{Zn}) which can supply holes at room temperature. Despite the fact that they are shallow, p-type doping is not possible because hole-killing defects have low formation energy and form readily in both Zn-rich and O-rich conditions and can easily compensate p-type doping (V_O, Zni) [26]. Thus, this asymmetry of intrinsic n-type versus p-type doping leads to n-type conductivity in ZnO and kills the possibility of p-type doping. Many other external dopants like impurities of P and As or nitrogen and formation of N_{Zn}-V_O complex or Zr, and N co-doping are theoretically suggested for p-type doping [142–145]. While in experimental works nitrogen, lithium, sodium, potassium, phosphorous, arsenic and bismuth, are shown as p-type dopants in ZnO [146–156]. Control p-type conductivity in sputter ZnO film is also demonstrated by adjusting the oxygen partial pressure in the sputtering plasma [146]. Thus, various theoretical and experimental results demonstrate that p-type doping is possible in ZnO but needs more systematic experimental work and deep investigation to achieve a good control. Stability and reproducibility are the existing challenges with p-type doping in ZnO.

16.5 Defects and Electrical Conductivity of ZnO Nanowire Films

Nanowire (NW) films have all favorable properties for various optoelectronic applications, but poor electrical conductivity of these nanowire films (glass substrate) is a major constraint. Metal doping (Al, Gd and In), sandwich of noble metal (Au, Pt and Ag) particles and surface modification by metal/organic molecules are a few important pathways to overcome this issue [50]. In the case of nanostructured films, the surface defects (grain boundary, surface adsorbed oxygen and unsaturated dangling bonds) are the most important factors for controlling the electrical conductivity of film [10, 18, 50, 157, 25, 158]. In this section of the chapter, we will discuss the atomistic engineering of the defect states which could tailor the characteristic of zinc oxide from insulating to semiconducting. We will talk about a few surface modification schemes which can alter the nanowire surface and control the defect density; and finally the impact of defect concentration on electrical properties of these nanowire films. Electrical conductivity of ZnO nanowire films can be measured using two different approaches: two-probe IV measurement or four-probe IV measurement. For two-probe measurement, a layer of metal contact can be deposited on top of ZnO nanowire array and another contact on the surface of the seed layer. This design basically helps to measure the IV response of nanowire array, where electrons moves along the nanowire axis, resulting in very fast transport [159].

However, for optoelectronic application, we also need highly electrical conducting film, thus it is important to see the electrical conductivity of nanowire films in lateral direction [160]. Similarly, if replacement of TCO with ZnO nanowire films is desired, than these nanowire films must have high electrical conductivity in lateral direction. Thus, in order to trace the nanowire film IV characteristic, a four-probe conductivity measurement can be performed, as demonstrated in Figure 16.4. The silver contacts are

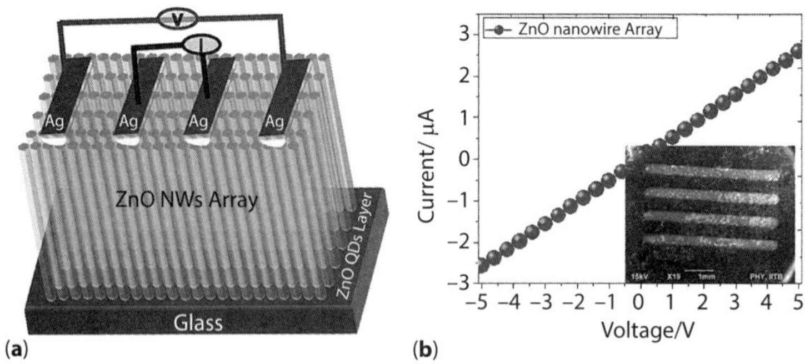

Figure 16.4 (a) Schematic diagram shows the four probe contacts for IV measurement of ZnO nanowire and (b) current-voltage plot for ZnO nanowire measured in lateral direction of films.

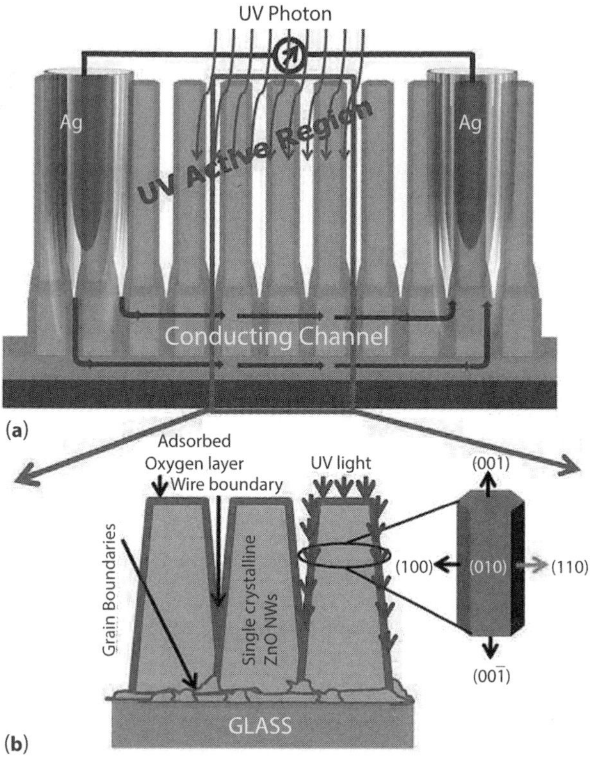

Figure 16.5 Cross-sectional schematic view of nanowire array photodetector: (a) the charge transport takes place via bottom dense part of the NWS (called the conducting channel); (b) Nanowire surfaces that have "O" species called as absorbed oxygen layer [25].

deposited using thermal coating unit which render metal semiconductor metal (M-S-M) junctions with the nanowire films. Two terminals are used to measure the current while the other two are used to apply voltage

The electron transport mechanism is quite different in nanowire array films than in planer ZnO films. A schematic cross-sectional cartoon (Figure 16.5) of a typical device depicts that nanowires are connected through Ag metal on the top and the bottom layer of NWs, acting as the conducting channel [25]. Transport properties of ZnO nanowire films depend upon the interaction among the nanowires and their orientation. The increased density of oriented (002) crystallites (nanowire) in ZnO film is associated with a reduction of large-angle grain boundaries; and it reduces scattering of charge carriers and increases the mobility of ZnO NWs film [161, 162]. The large-size grains (NW diameter) which attach to each other at the bottom and single crystalline nature of nanowire give highly efficient charge transport through the bottom of the nanowire, called the conducting channel.

16.5.1 Electrical Conductivity of Nanowire Film in Dark

We have recently shown that defects can significantly change electrical conductivity of ZnO nanowire films [25, 46, 48, 158]. The effect on electrical conductivity is probed by comparing just three post-growth treated ZnO nanowire films with variable concentrations of defects. Figure 16.6a illustrates the dark- and photocurrent measurements of as-grown, air annealed (400°C), and oxygen annealed (400°C) nanowire films. The dark conductivity of NWs array is found to increase after annealing in air and oxygen ambiance. Relatively less value of dark conductance of as-grown array is primarily observed due to the presence of surface defects/amorphous layer ($Zn(OH)_2$)

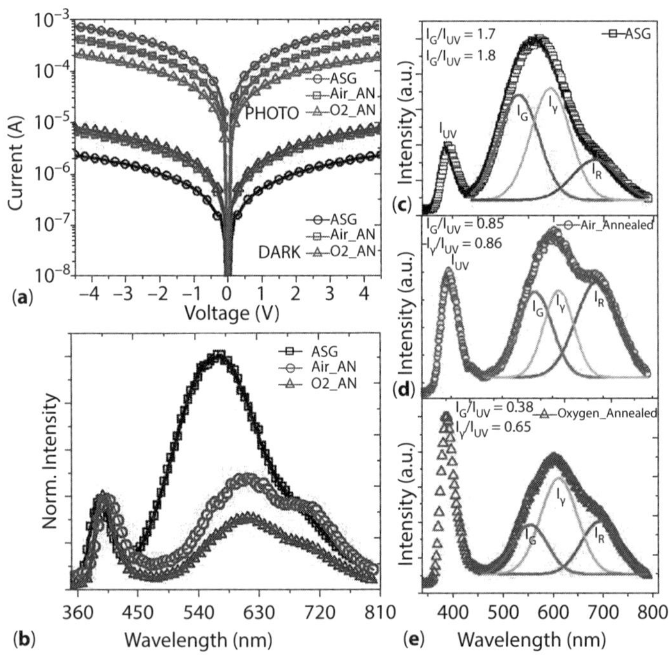

Figure 16.6 (a) Dark- and photocurrent measurement and (b) room temperature PL measurements of as-grown, air annealed (400°C) and oxygen annealed (400°C) ZnO nanowire array. Three-peak Gaussian fitting for I_G (green defect), I_Y (yellow defect), and I_R (Red defect) in PL spectrum (c) as-grown, (d) air annealed, and (e) oxygen annealed ZnO NWs [25].

accompanied by the higher concentration of oxygen vacancies [25]. These surface defects may serve as trapping/scattering centers for conduction electrons. After annealing in air and O_2 atmosphere, the amorphous layer associated with oxygen defects is partially depleted and the surface barrier shrinks, leading to an increase in conductivity. To correlate this change in electrical conductivity with defects, PL measurements of the same three samples were performed and it was found that a huge reduction in defect band occurred after annealing. Gaussian fitting of 3 peaks named as I_G (green defect), I_Y (yellow defect), and I_R (red defect) is carried out in an unstructured broad defect peak to find the type of defects. The intensity of green emission is drastically suppressed after annealing compared with other emissions, which shows that the green defects (i.e., single-charged oxygen vacancies) are mainly responsible for the poor dark conductivity [25].

The surface of ZnO nanowire film changes from hydrophobic (96°) to hydrophilic (16°) and improvement in crystal quality of the nanowire is observed under oxygen annealing. The defect band emission intensity (relative to band edge emission, I_D/I_{UV}) reduces from 1.3 to 0.2 after annealing at 600°C at 10 SCCM flow of oxygen, and results in an order enhancement in dark conductivity. Annealing in various other gas atmospheres strongly affects the defect concentration of nanowire films and resulting change in electrical conductivity. Figure 16.7 demonstrated such a result, where nitrogen and argon annealing leads to good control on defect band emission, while hydrogen annealing results in fully suppressed defect band intensity. The electrical conductivity is also increased; argon leads to a one-order increase and nitrogen leads to around a two-order increase in electrical conductance of the film. However, surprising results were obtained in hydrogen annealed nanowire films. More than a four-order increment was noted in electrical conductance of film, which is also found stable for years.

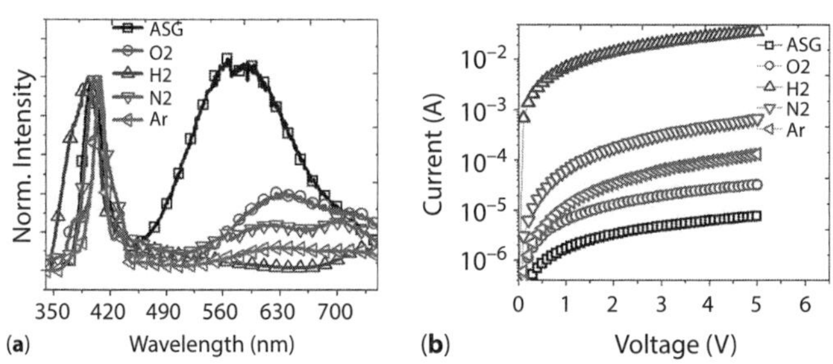

Figure 16.7 (a) Room temperature PL measurements of ZnO nanowire films annealed in oxygen, nitrogen, argon and hydrogen ambience; (b) Dark conductivity of respective samples.

16.5.2 Defect-Induced Photoconductivity in Nanowire Films

Optoelectrical properties of ZnO can be extracted by measuring photoconductivity of the sample. When ZnO nanowire films are excited with UV light, the valence band electron gets excited and contributes in electrical conductivity. The effect of defect states on photoconductivity of ZnO nanowire has been studied and it has been observed that high defect states in ZnO nanowire films enhances photocurrent [25]. As seen in Figure 16.6, that dark current was higher in annealed NWs, however photocurrent was higher in as-grown sample. These results reveal that defect induces high photocurrent in solution grown NWs films. This can be explained through the schematic of the band diagram of ZnO nanowire under dark and photoexcited conditions (Figure 16.8). Photo-excitation creates electron-hole pairs in NWs. The excited electrons fill the conduction band and contribute in NWs film conductivity. Simultaneously, surface adsorbed oxygen molecules (O^{2-}) capture the photogenerated holes available at the nearby surface and release oxygen. This leads to change in band bending by an amount with respect to the dark condition. The amount of band bending

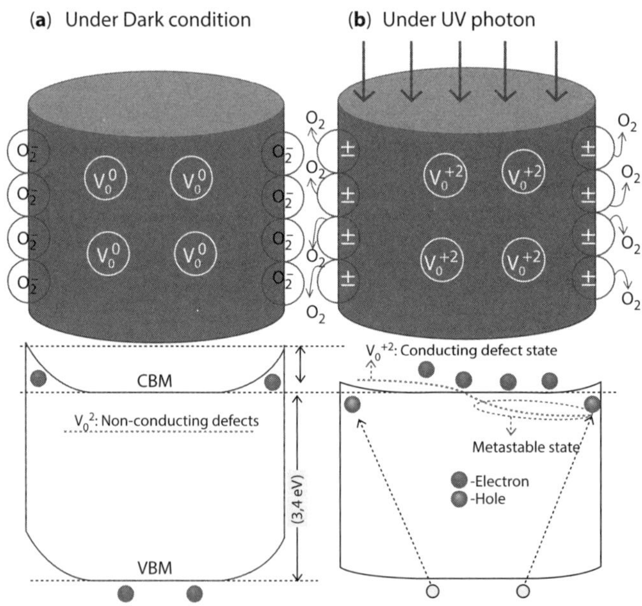

Figure 16.8 Energy band diagram of ZnO nanowire (a) under dark condition and (b) under UV light illumination condition [25].

depends upon width of depletion layer, which is associated with charged oxygen vacancies. The PL data (Figure 16.4) show that annealing in air/oxygen suppresses green defect intensity, which indicates a decrease in the concentration of oxygen vacancies and surface traps. The higher density of V_o^+ vacancies in as-grown NWs film reveals large change in band bending under UV excitation, and leads to an enhanced photoconductivity. Further, under UV excitation, neutral oxygen vacancies get converted into doubly charged oxygen vacancies (V_o^{+2}) with the release of two electrons. The light-induced V_o^{+2} vacancies are unstable shallow-level defects, and are conducting in nature. As a consequence, we consider that the higher concentration of charge oxygen vacancies (V_o^{+1}) near NW surfaces and neutral oxygen vacancies (V_o^0) in bulk (transition from V_o^0 to V_o^{+2}) of ZnO nanowire are responsible for the enhanced photocurrent in as-grown NWs. The annealed nanowire exhibited less change in band bending due to a reduced amount of surface defects, and result in relatively low photosensitivity compared with as-grown NWs. The ZnO nanowire annealed in other environments have also shown the same trend, UV light photosensitivity was reduced after post-growth annealing treatment.

16.5.3 Surface Modification and Optoelectrical Properties of ZnO Nanowires

Optical and electrical properties strongly depend on the surface of nanostructures due to high surface-to-volume ratio. Surface modification by different means can be an alternative approach to tune the optoelectrical properties of ZnO nanowire films. In the first principle study, Huang *et al.* have shown that surface properties modulate the band structure of ZnO nanowires as significantly as that of quantum confinement [163]. They stated that passivation of nanowire by H, F, Cl, and NH_2 is energetically favorable; also, electronic states can be redesigned in ZnO NWs and different band gaps can be realized after surface passivation. Surface of ZnO nanowire can be modified/decorated by coating of various chemical species, metals nanoparticles, metal oxide or semiconductor nanoparticles and various core-shell formations. Anthracene, poly(vinyl alcohol) (PVA), polystyrene sulfate (PSS), poly(styrene-co-maleic acid) (PS-co-MAc), anionically charged poly(N-isopropylacrylamide) (PNIPAM), and carboxymethylcellulose (CMC), 3-amino propyltriethoxy silane (APTES), amines and Zn-Porphyrin via Metal-Ligand are demonstrated as surface functionalizing agent [163–168]. Chemical functionalization resulting in enhancement of electrical conductivity and reduction in defect band emission. Various metal nanoparticles such as Au, Pt, Ag, Ti, Pb, Al and Zn were successfully decorated over ZnO nanowire [169–174]. A good control on photoluminescence and photoconductivity is observed by metal surface coating. The metal work function plays an important role in deciding the nature of emission spectra; enhanced/reduced UV emissions are observed for different metals. A p-type nature has also been observed in Ti-coated nanowire and it was proposed that Ti decoration on ZnO nanowires led to the formation of a TiOx layer, which reveals p-type conducting performance due to the existence of electron captured oxygen molecules [169]. Surface modifications were also performed by various metal oxide/semiconductors coatings; CdTe quantum dots, MgO nanoparticles, ZnS nanoparticles, amorphous-Si and a-SiNx:H layer have been successfully demonstrated [175–179]. The ZnO/CdSSe, ZnO/Al_2O_3, ZnO–$ZnGa_2O_4$ and ZnO/ZnSe type of core-shell structures have also been proposed to manipulate electrical properties of ZnO nanowires [180–184]. Suppressed defect emission and enhanced electrical conduction is observed in core-shell structures and they also showed relatively high stable nanowire compared with bare ZnO. Thus, we can say that, various surface treatments can lead to manipulation of the defects density, which ultimately controls the electrical conduction in nanowire films.

16.6 ZnO Nanowires for Energy Conversion Devices

Our daily activities are such that household applications, transport systems and many other industrial applications require a huge amount of energy. The conventional sources of energy such as fuel, coal and natural gases are very limited and also cause pollution in the environment [185]. The production of various greenhouse gases is the result of consumption of these conventional energy sources [185]. Sunlight, wind, rain, tides, waves and geothermal heat are renewable sources of energy and can fulfill our need for energy; however, storage and conversion of energy in the proper form is a major challenge [186]. Solar energy is very important and can be stored easily in various forms by using different energy conversion and storage devices [187]. In this section of the chapter, we will look at the application of ZnO nanowire films in various energy conversion devices. Some examples are the conversion of solar energy into electrical energy by using photovoltaic devices and conversion of solar energy to hydrogen by using water splitting process through electrochemical cell. Thermal and mechanical energy can also be utilized to generate electric power using

ZnO nanowire; piezoelectric properties-based nanoenergy generator and thermoelectric generator are possible devices.

16.6.1 Photovoltaic Applications

Zinc oxide is transparent for visible light and good electrical conductor; hence it can be used as an electrode for PV device fabrication [186]. Nanowire films provide the best geometry to enhance light trapping/scattering to improve solar light absorption, which ultimately increases the device efficiency and performance. The ZnO nanowire films are also used as anti-reflector in various conventional solar cells [188]. P-n junction-based conversional solar cells are fabricated using ZnO as n-type materials, but recently non-conventional device design has become more popular. Excitonic solar cells are the most prominent due to their facile and low-cost fabrication process, however, efficiency and instability are existing issues. Dye-sensitized solar cells and organic-based hybrid solar cells have demonstrated reasonably well in efficiency, but stability of these solar cells is a major issue [189]. Solid-state excitonic solar cells are an option to get more stable device, but poor efficiency is a problem here [190]. A typical solid-state excitonic solar cell is demonstrated in Figure 16.9; ZnO nanowire grown on ZnO seed layer works as high surface electrode to collect the photogenerated electrons and PbSe quantum dots are used as active material and absorber of sunlight, which generate electron-hole pair. Due to the difference in chemical potential of top gold contact and bottom ZnO/ITO contact, photogenerated carriers easily separated from each

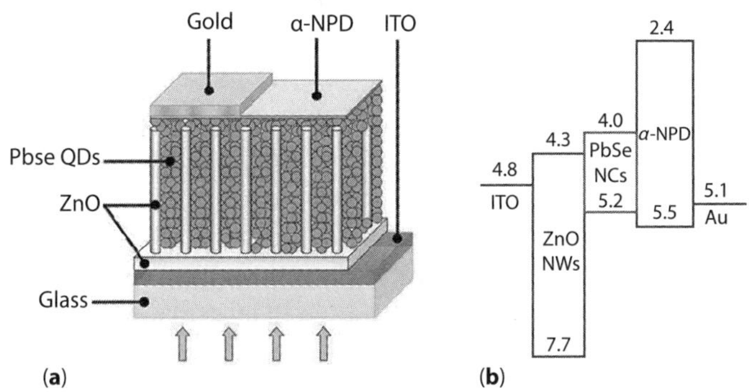

Figure 16.9 (a) Schematics of a PbSe QD / ZnO nanowire solar cell and (b) band diagram of device [190].

other. A thin ZnO film at the base of the nanowires is deposited to prevent hole recombination, while α-NPD between PbSe QDs and Au top contact is used to block PbSe-to-Au electron transfer. Aydil and coworkers have shown that significant improvement in photocurrent and power conversion can be achieved by increasing nanowire length [190].

Recently carbon nanoparticles or graphene quantum dots have also been applied to sensitize the ZnO nanowire, and demonstrated cell functioning [10]. Li-doped ZnO nanowire has shown improved short-circuit current density (Jsc) and open-circuit voltage (Voc) [191]. Reduction in radiative trap sites on the ZnO surface is attributed to more efficient charge extraction and improves the efficiency. Presence of $Zn(OH)_2$ on the surfaces of nanorods and its decomposition products also govern the photovoltaic performance [192]. Defects in ZnO strongly affect the transport properties; hence they will alter the solar cell performance. Nanowire films have all the desired properties such as high electron transport, larger open surface area and single-crystalline structure for electron transport with minimum grain boundaries for highly efficient devices. Yet, the best solar cell performance for applying nanostructures of ZnO in DSC currently reach around 7.2% efficiency and 2.5% for vertically-aligned nanowire. Therefore, further investigation is required in this direction, and defect engineering of ZnO nanowire films might be one important path to improve the quality and efficiency of PV devices [10, 192–194].

16.6.2 Water Splitting/Solar Hydrogen Generation

Hydrogen is a clean, energy efficient and abundant fuel, which can be an excellent energy source in the future; it has a very high energy value per gram mass (122 kJ/g). Among the various hydrogen generation methods, water splitting is a very economical and clean process to produce hydrogen [195]. The decomposition of water into hydrogen and oxygen is a chemical reaction with large Gibbs free energy (ΔG= 237 kJ/mol). Water splitting by photocatalysis can be achieved via two different approach, either by photochemical cell or photoelectrochemical (PEC) cell. A suspended particle solution of photocatalyst is used in the first method; one can appreciate the simplicity of the process, which uses the high reactive surface of catalyst [195]. However, the PEC method has several other advantages; for example, internal bias can be used to enhance the photocatalystic activities. Multiple material films can utilize and increase the absorption of visible light [196]. Oxidation and reduction processes occur simultaneously and can perform separately in photoelectrochemical cell; they can separately store oxygen and hydrogen, which is a relatively safer process.

The ZnO-based electrodes are utilized for hydrogen generation via photoelectrochemical process and are found to be suitable candidates for water splitting. Nanowire film-based photoanode comes with some promising properties, which may enhance the performance of electrochemical cell. High band gap (3.37 eV) and high electrochemical stability of ZnO make it an attractive material for PEC water splitting [13]. Further, the level of conduction and valence band are apart from H_2O/H_2 and OH^-/O_2 redox level and satisfy a mandatory requirement for spontaneous photo splitting of water [197]. Zhang *et al.* have studied ZnO nanostructured photoanode and observed that hydrogen generation efficiency is a function of the morphology, semiconductor electrolyte interaction and defect density [13]. Nitrogen-doped ZnO nanowire arrays have shown significant enhancement in conversion efficiency in the visible region. An order of magnitude increase in photoresponse with an overall hydrogen generation efficiency of 0.15% is reported in nitrogen doping [12]. Quantum dot monolayer sensitized ZnO nanowire-array has shown enhanced efficiency for water splitting. Metal nanoparticles-coated ZnO nanowire shows improved charge transport due to inhibition of ZnO surface charge recombination and enhanced light trapping (harvesting) due to plasmonic effects, which improved the overall efficiency [198]. A three-dimensional branched ZnO/Si heterojunction nanowire-based PEC has been recently demonstrated which has very high efficiency of solar water splitting [196].

16.6.3 Piezoelectric Nanogenerators

Due to piezoelectric nature, mechanical stress in ZnO can produce electrical voltage. Wang *et al.*, utilized this property and demonstrated a piezoelectric nanogenerator based on ZnO nanowire [15]. They proposed that the coupling of piezoelectric and semiconducting properties in zinc oxide created a strain field and resulted in charge separation across the NW after bending. A typical device design consists of a long ZnO nanowire, with one end fixed on silicon substrate using silver paste and the other end left free. When ZnO nanowire is deflected using conductive AFM tip, a change in device output voltage is observed. Thus they concluded that defection of nanowire generates electrical power in the device [14]. A more robust and simpler device design has been promoted by the Wang group to generate direct current (DC) by using ZnO nanowire array [199]. Figure 16.10 shows a schematic cartoon of a direct-current nanogenerator driven by ultrasonic waves. The device consists of vertically aligned zinc oxide nanowire array that is covered by a zigzag silicon electrode on top. Ultrasonic waves create bending and/or vibration on ZnO nanowires

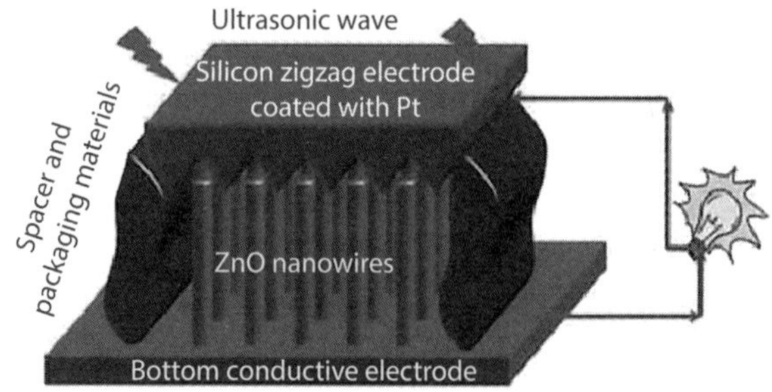

Figure 16.10 A ZnO nanowire array-based direct current generator [200].

and piezoelectric semiconducting coupling process converts mechanical energy into electricity [200].

Piezoelectric properties-based nanogenerator can be use to convert mechanical, vibrational, and/or hydraulic energy into electrical energy. Lateral nanowire film-based flexible device can open up the large-scale production and encourage practical applications of nanowire-based piezoelectric nanogenerators for self-powered nanosystems [201]. Recently, Kim and his collaborators have shown that simple muscle stretching can produce electrical current using flexible thin-film piezoelectric nanogenerator fabricated by solution process [199]. Though piezoelectric-based nanogenerators have shown a lot of promise, low output voltage, design of charge storage device, loss of free packaging and durability are a few of the existing challenges that currently need to be addressed [14].

16.6.4 Thermoelectric Energy Convertor

A Thermoelectric device can convert heat into electricity based on the Seebeck effect. The ability of a material to generate electrical power is quantified by power factor (σS^2), where S is the Seebeck coefficient, and σ is the electrical conductivity [202, 203]. However, the efficiency of a material for thermoelectric power generation is demonstrated by a dimensionless quantity, know as *figure of merit (ZT)*; $ZT = \sigma S^2 T/\lambda$, where T is temperature and λ is thermal conductivity. Thus efficient material needs high value of Seebeck coefficient and electrical conductivity, while least value of thermal conductivity [202, 203, 17]. In order to design a new material for thermoelectric application, we should have a short phonon mean free path compared with grain boundaries, which can enhance phonon-phonon scattering. Small-sized grains for enhancing phonon scattering should simultaneously have good electrical conductor. It is also observe that TEP of semiconductor material depends on the electron and hole asymmetries at the Fermi level, and is very sensitive to small changes in scattering rate and the density of states at the Fermi level [204]. Recent studies are shown that ZnO can be potential material for thermoelectric energy generators due to its excellent stability at elevated temperatures and good electrical conductivity [16, 204–207]. However, high lattice thermal conductivity is the main barrier to its practical applications. In order to increase phonon scattering, amorphous ZnO thin-film device has been tried, [205]. Doping of metallic elements enhances conductivity of ZnO, but the effect on phonon scattering is found to be mixed. Theoretical results have shown that the thermoelectric performance of the Ga-doped ZnO is strongly dependent on the Ga contents, and ZnO NW has demonstrated an increase by a factor of 2.5 at Ga content of 0.04, compared with pure ZnO NWs figure of merit [207]. It has been seen that phonon scattering is promoted by Al-induced grain refinement and it was concluded that Al-doped non-nanostructured ZnO is one of the best n-type thermoelectric oxides [17]. The thermoelectric performance of ZnO is largely dependent on oxygen vacancies, hence manipulation of defect states/vacancies in ZnO nanowire can improve figure of merit in nanowire array-based devices in order to provide a high order of efficiency. Therefore new device design needs to be address.

References

1. U. Özgür, Y.I. Alivov, C. Liu, A. Teke, M.A. Reshchikov, S. Doğan, V. Avrutin, S.-J. Cho, H. Morkoç, *Journal of Applied Physics*, Vol. 98, p. 041301, 2005.
2. A. Janotti, C.G. Van de Walle, *Reports on Progress in Physics*, Vol. 72, p. 126501, 2009.
3. C. Klingshirn, *ChemPhysChem* vol.8, p.782 , 2008.
4. V. Bhosle, J. Narayan, *Journal of Applied Physics*, Vol. 100, p. 093519, 2006.
5. C.G. Van de Walle, *Physica B: Condensed Matter*, Vol. 308–310, p. 899, 2001.
6. B.J. Jin, S.H. Bae, S.Y. Lee, S. Im, *Materials Science and Engineering: B*, Vol. 71, p. 301, 2000.
7. A. Mansingh, and V. Gupta, *Physical Review B*, Vol. 49, p. 1989, 1994.
8. I. Gonzalez-Valls, M. Lira-Cantu, *Energy & Environmental Science*, Vol. 2, p. 19, 2009.
9. M. Law, L.E. Greene, J.C. Johnson, R. Saykally, P. Yang, *Nature Materials*, Vol. 4, p. 455, 2005.
10. L. Li, T. Zhai, Y. Bando, D. Golberg, *Nano Energy*, Vol. 1, p. 91, 2012.

11. K.S. Leschkies, A.G. Jacobs, D.J. Norris, E.S. Aydil, *Applied Physics Letters*, Vol. 95, p. 193103, 2009.
12. X. Yang, A. Wolcott, G. Wang, A. Sobo, R.C. Fitzmorris, F. Qian, J.Z. Zhang, Y. Li, *Nano Letters*, Vol. 9, p. 2331, 2009.
13. A. Wolcott, W.A. Smith, T.R. Kuykendall, Y. Zhao, J.Z. Zhang, *Advanced Functional Materials*, Vol. 19, p. 1849, 2009.
14. Z.L. Wang, X. Wang, J. Song, J. Liu, Y. Gao, *IEEE Pervasive Computing Mobile and Ubiquitous* System, 2008.
15. Z.L. Wang, J. Song, *Science*, Vol. 312, p. 242, 2006.
16. M. Søndergaard, E.D. Bøjesen, K.A. Borup, S. Christensen, M. Christensen, B.B. Iversen, *Acta Materialia*, Vol. 61, p. 3314, 2013.
17. P. Jood, R.J. Mehta, Y. Zhang, G. Peleckis, X. Wang, R.W. Siegel, T. Borca-Tasciuc, S.X. Dou, G. Ramanath, *Nano Letters*, Vol. 11, p. 4337, 2011.
18. S. Xu, Z.L. Wang, *Nano Research*, Vol. 4, p. 1013, 2011.
19. A.J. Mieszawska, R. Jalilian, G.U. Sumanasekera, F.P. Zamborini, *Small*, Vol. 3, p. 722, 2007.
20. Z.L. Wang, *Journal of Physics: Condensed Matter*, Vol. 16, p. R829, 2004.
21. T. Sun, J. Qiu, *Materials Letters*, Vol. 62, p. 1528, 2008.
22. J.-J. Wu, S.-C. Liu, Advanced Materials, Vol. 14, p. 215, 2002.
23. Z. Yu, Z. Yang, S. Wang, Y. Jin, J.G. Liu, M. Gong, X. Sun, *Chemical Vapor Deposition*, Vol. 11, p. 433, 2005.
24. S.-W. Kim, S. Fujita, S. Fujita, *Applied Physics Letters*, Vol. 86, p. 153119, 2005.
25. A. Kushwaha, M. Aslam, Journal of Applied Physics, Vol. 112, p. 054316, 2012.
26. S. Zhang, S.-H. Wei, A. Zunger, *Physical Review B*, Vol. 63, p. 075205, 2001.
27. A. Janotti, C.G. Van de Walle, *Physical Review B*, Vol. 76, p. 165202, 2007.
28. Y.-S. Kim, C. Park, *Physical Review Letters*, Vol. 102, p. 086403, 2009.
29. B. Panigrahy, M. Aslam, D.S. Misra, M. Ghosh, D. Bahadur, *Advanced Functional Materials*, Vol.20, p. 1161, 2010.
30. J.M.D. Coey, M. Venkatesan, C.B. Fitzgerald, *Nature Materials*, Vol. 4, p. 173, 2005.
31. A. Durst, R. Bhatt, P. Wolff, *Physical Review B*, Vol. 65, p. 235205, 2002.
32. C. Kittel, *Physical Review*, Vol. 96, p. 72, 1954.
33. C.-Y. Chen, M.-W. Chen, J.-J. Ke, C.-A. Lin, J.R.D. Retamal, J.-H. He, *Pure and Applied Chemistry*, Vol. 82, p. 2055, 2010.
34. O. Tigli, J. Juhala, Nanotechnology (IEEE-NANO), 10.1109/NANO.2011.6144585 11th IEEE International Conference on Nanotechnology. Ieee; 2011.
35. Z. Yin, S. Wu, X. Zhou, X. Huang, Q. Zhang, F. Boey, H. Zhang, *Small*, Vol. 6, p. 307, 2010.
36. Y.W. Heo, V. Varadarajan, M. Kaufman, K. Kim, D.P. Norton, F. Ren, P.H. Fleming, *Applied Physics Letters*, Vol. 81, p. 3046, 2002.
37. A.B. Hartanto, X. Ning, Y. Nakata, T. Okada, *Applied Physics A: Materials Science & Processing*, Vol. 78, p. 299, 2004.

38. M.A. Verges, A. Mifsud, *J. Chem. Soc. Faraday Trans.*, Vol. 86, p. 959, 1990.
39. L. Vayssieres, K. Keis, S. Lindquist, A. Hagfeldt, *J. Phys. Chem. B*, Vol. 28, p. 3350, 2001.
40. N. Wang, Y. Cai, R.Q. Zhang, *Materials Science and Engineering: R: Reports*, Vol. 60, p. 1, 2008.
41. Y. Jun, J. Choi, J. Cheon, *Angewandte Chemie*, Vol. 45, p. 3414, 2006.
42. R.L. Penn, *Geochimica et Cosmochimica Acta*, Vol. 63, p. 1549, 1999.
43. C. Pacholski, A. Kornowski, H. Weller, *Angewandte Chemie*, Vol. 41, p. 1188, 2002.
44. B. Cheng, E.T. Samulski, *Chemical Communications*, 2004, p. 986.
45. L. Vayssieres, K. Keis, S. Lindquist, A. Hagfeldt, *J. Phys. Chem. B*, Vol. 28, p. 3350, 2001.
46. A. Kushwaha, M. Aslam, IEEE Explore, 10.1109/ICEmElec.2012.6636232
47. A. Kushwaha, M. Aslam, *AIP Conf. Proc.*, Vol. 1512, p. 240, 2013.
48. A. Kushwaha, H. Kalita, M. Aslam, *World Academy of Science, Engineering and Technology*, Vol. 74, p. 272, 2013.
49. B. Panigrahy, M. Aslam, D.S. Misra, D. Bahadur, *Cryst. Eng. Comm.*, Vol. 11, p. 1920, 2009.
50. L. Schmidt-Mende, J.L. MacManus-Driscoll, *Materials Today*, Vol. 10, p. 40, 2007.
51. M.N.R. Ashfold, R.P. Doherty, N.G. Ndifor-Angwafor, D.J. Riley, Y. Sun, *Thin Solid Films*, Vol. 515, p. 8679, 2007.
52. K. Govender, D.S. Boyle, P.B. Kenway, P.O. Brien, *J. Material Chemistry*, Vol. 14, p. 2575, 2004.
53. A. Sugunan, H.C. Warad, M. Boman, J. Dutta, *J. Sol-Gel Sci. Technol.*, Vol. 39, p. 49, 2006.
54. L. Tang, X.-B. Bao, H. Zhou, A.-H. Yuan, *Physica E: Low-dimensional Systems Nanostructures*, Vol. 40, p. 924, 2008.
55. D. Wang, C. Song, *J. Phys. Chem. B*, Vol. 109, p. 12697, 2005.
56. Y.F. Hsu, Y.Y. Xi, K.H. Tam, A.B. Djurišić, J. Luo, C.C. Ling, C.K. Cheung, A.M.C. Ng, W.K. Chan, X. Deng, C.D. Beling, S. Fung, K.W. Cheah, P.W.K. Fong, C.C. Surya, *Advanced Functional Materials*, Vol. 18, p. 1020, 2008.
57. Y. Gao, M. Nagai, T. Chang, J. Shyue, *Crystal Growth & Design*, Vol. 7, p. 2467, 2007.
58. L. Schlur, A. Carton, P. Lévêque, D. Guillon, G. Pourroy, *The Journal of Physical Chemistry C*, Vol. 117, p. 2993, 2013.
59. X. Wang, C. Yu, J. Wu, Y. Zhang, *Ionics*, Vol. 19, p. 355, 2012.
60. Z.R. Tian, J.A. Voigt, J.U.N. Liu, B. Mckenzie, M.J. Mcdermott, M.A. Rodriguez, H. Konishi, H. Xu, *Nature Materials*, Vol. 2, p. 821, 2003.
61. J. Tian, J. Hu, S. Li, F. Zhang, J. Liu, J. Shi, X. Li, Z. Tian, Y. Chen, *Nanotechnology*, Vol. 22, p. 245601, 2011.
62. Y. He, T. Yanagida, K. Nagashima, F. Zhuge, G. Meng, B. Xu, A. Klamchuen, S. Rahong, M. Kanai, X. Li, M. Suzuki, S. Kai, T. Kawai, *The Journal of Physical Chemistry C*, Vol. 117, p. 1197, 2013.

63. S. Xu, C. Lao, B. Weintraub, Z.L. Wang, *Journal of Materials Research*, Vol. 23, p. 2072, 2011.
64. J. Joo, B.Y. Chow, M. Prakash, E.S. Boyden, J.M. Jacobson, *Nature Materials*, Vol. 10, p. 596, 2011.
65. R. Kozhummal, Y. Yang, F. Güder, A. Hartel, X. Lu, U.M. Küçükbayrak, A. Mateo-Alonso, M. Elwenspoek, M. Zacharias, *ACS Nano*, Vol. 6, p. 7133, 2012.
66. M. Guo, P. Diao, S. Cai, *Journal of Solid State Chemistry*, Vol. 178, p. 1864, 2005.
67. H. Zhang, Y. Li, P. Liu, Y. Li, D. Yang, H. Yang, H. Zhao, *Chemistry Eur. J.*, Vol. 18, p. 5165, 2012.
68. B. Liu, H.C. Zeng, *Journal of American Chemical Society*, Vol. 125, p. 4430, 2003.
69. Y. He, T. Yanagida, K. Nagashima, F. Zhuge, G. Meng, B. Xu, A. Klamchuen, S. Rahong, M. Kanai, X. Li, M. Suzuki, S. Kai, T. Kawai, *The Journal of Physical Chemistry C*, Vol. 117, p. 1197, 2013.
70. I. Jun, N. Sunwoo, L. Sung, H. Kim, J. Woo, C.P. Shin, *Japanese Journal of Applied Physics*, Vol. 52, p. 1, 2013.
71. Y. Sun, N.A. Fox, D.J. Riley, M.N.R. Ashfold, *Journal of Physical Chemistry C*, Vol. 112, p. 9234, 2008.
72. S. Xu, N. Adiga, S. Ba, T. Dasgupta, C.F.J. Wu, Z.L. Wang, *ACS Nano*, Vol. 3, p. 1803, 2009.
73. C.P. Grigoropoulos, H.J. Sung, *Nano Letters*, Vol. 11, p. 666, 2011.
74. T.-U. Kim, J.-A. Kim, S.M. Pawar, J.-H. Moon, J.H. Kim, *Crystal Growth & Design*, Vol. 10, p. 4256, 2010.
75. J.J. Dong, X.W. Zhang, Z.G. Yin, S.G. Zhang, J.X. Wang, H.R. Tan, Y. Gao, F.T. Si, H.L. Gao, *ACS Applied Materials & Interfaces*, Vol. 3, p. 4388, 2011.
76. D. Zhang, S. Wang, K. Cheng, S. Dai, B. Hu, X. Han, Q. Shi, Z. Du, *ACS Applied Materials & Interfaces*, Vol. 4, p. 2969, 2012.
77. P. Soman, M. Darnell, M.D. Feldman, S. Chen, *Journal of Nanoscience and Nanotechnology*, Vol. 11, p. 6880, 2011.
78. S. Xu, Y. Wei, M. Kirkham, J. Liu, W. Mai, D. Davidovic, R.L. Snyder, Z.L. Wang, *Journal of the American Chemical Society*, Vol. 130, p. 14958, 2008.
79. H.W. Kang, J. Yeo, J.O. Hwang, S. Hong, P. Lee, S.Y. Han, J.H. Lee, Y.S. Rho, S.O. Kim, S.H. Ko, H.J. Sung, *The Journal of Physical Chemistry C*, Vol. 115, p. 11435, 2011.
80. K.S. Kim, H. Jeong, M.S. Jeong, G.Y. Jung, *Advanced Functional Materials*, Vol. 20, p. 3055, 2010.
81. C.H. Wang, A.S.W. Wong, G.W. Ho, *Langmuir*, Vol. 23, p. 11960, 2007.
82. A. George, A.W. Maijenburg, M.G. Maas, D.H.A. Blank, J.E. ten Elshof, *Langmuir*, Vol. 27, p. 12235, 2011.
83. S. Zhang, Y. Shen, H. Fang, S. Xu, J. Song, Z.L. Wang, Journal of Materials Chemistry, Vol. 20, p. 10606, 2010.
84. H.-H. Park, X. Zhang, K.W. Lee, K.H. Kim, S.H. Jung, D.S. Park, Y.S. Choi, H.-B. Shin, H.K. Sung, K.H. Park, H.K. Kang, H.-H. Park, C.K. Ko, *Cryst. Eng. Comm.*, Vol. 15, p. 3463, 2013.

85. C. Lu, L. Qi, J. Yang, L. Tang, D. Zhang, J. Ma, *Chemical Communications*, 2006, p. 3551.
86. L.E. Greene, M. Law, J. Goldberger, F. Kim, J.C. Johnson, Y. Zhang, R.J. Saykally, P. Yang, *Angewandte Chemie*, Vol. 42, p. 3031, 2003.
87. Y. Sheng, Y. Jiang, X. Lan, C. Wang, S. Li, X. Liu, H. Zhong, *Journal of Nanomaterials*, Vol. 2011, p. 1, 2011.
88. Y.W. Heo, M.P. Ivill, K. Ip, S.J. Pearton, M.F. Chisholm, T. Steiner, *Materials Today*, 2008, p. 34.
89. Y. Jin, Y. Ren, M. Cao, Z. Ye, *Journal of Nanomaterials*, 2012, p. 1, 2012.
90. L.E. Greene, B.D. Yuhas, M. Law, D. Zitoun, P. Yang, *Inorganic Chemistry*, Vol. 45, p. 7535, 2006.
91. S.Y. Bae, C.W. Na, J.H. Kang, J. Park, *The Journal of Physical Chemistry B*, Vol. 109, p. 2526, 2005.
92. J. Rhee, *Material Science-Poland*, Vol. 28, p. 513, 2010.
93. K.M. McPeak, T.P. Le, N.G. Britton, Z.S. Nickolov, Y.A. Elabd, J.B. Baxter, *Langmuir*, Vol. 27, p. 3672, 2011.
94. S.-H. Yi, S.-K. Choi, J.-M. Jang, J.-A. Kim, W.-G. Jung, *Journal Colloid Interface Science*, Vol. 313, p. 705, 2007.
95. G. Jia, Y. Wang, J. Yao, *Journal of Ovonic Research*, Vol. 6, p. 303, 2010.
96. K.M. McPeak, J.B. Baxter, *Crystal Growth & Design*, Vol. 9, p. 4538, 2009.
97. T. Dedova, M. Krunks, M. Grossberg, O. Volobujeva, I. Oja Acik, *Superlattices and Microstructures*, Vol. 42, p. 444, 2007.
98. D. Perednis, L.J. Gauckler, *Journal of Electroceramics*, Vol. 14, p. 103, 2005.
99. M.J. Height, L. Ma, S.E. Pratsinis, F. Krumeich, *Chem. Mater.*, Vol. 18, p. 572, 2006.
100. E. Kärber, T. Raadik, T. Dedova, J. Krustok, A. Mere, V. Mikli, M. Krunks, *Nanoscale Research Letters*, Vol. 6, p. 359, 2011.
101. M. Krunks, T. Dedova, I. Oja Açik, *Thin Solid Films*, Vol. 515, p. 1157, 2006.
102. M.D. Reyes Tolosa, J. Orozco-Messana, A.N.C. Lima, R. Camaratta, M. Pascual, M.A. Hernandez-Fenollosa, *Journal of the Electrochemical Society*, Vol. 158, p. E107, 2011.
103. B. Cao, Y. Li, G. Duan, W. Cai, *Crystal Growth & Design*, Vol. 6, p. 1091, 2010.
104. H. Kim, J.Y. Moon, H.S. Lee, *Electronic Materials Letters*, Vol. 5, p. 135, 2009.
105. J. Elias, R. Tena-Zaera, C. Lévy-Clément, *Journal of Electroanalytical Chemistry*, Vol. 621, p. 171, 2008.
106. S.K. Sharma, A. Rammohan, A. Sharma, *Journal Colloid Interface Science*, Vol. 344, p. 1, 2010.
107. T.-H. Lee, H.-J. Sue, X. Cheng, *Journal Vacuum Science & Technology B: Microelectronics Nanometer Structures*, Vol. 30, p. 06FF01, 2012.
108. D. Gal, G. Hodes, D. Lincot, H. Schock, *Thin Solid Films*, Vol. 361–362, p. 79, 2000.
109. X.-H. Lu, D. Wang, G.-R. Li, C.-Y. Su, D.-B. Kuang, Y.-X. Tong, *The Journal of Physical Chemistry C*, Vol. 113, p. 13574, 2009.

110. G.-R. Li, C.-R. Dawa, Q. Bu, F. Zhen, X.-H. Lu, Z.-H. Ke, H.-E. Hong, C.-Z. Yao, P. Liu, Y.-X. Tong, *Electrochemistry Communications*, Vol. 9, p. 863, 2007.
111. A. Janotti, C.G. Van de Walle, *Physical Review B*, Vol. 76, p. 165202, 2007.
112. M.D. McCluskey, S.J. Jokela, *Journal of Applied Physics*, Vol. 106, p. 071101, 2009.
113. A. Setiawan, Z. Vashaei, M.W. Cho, T. Yao, H. Kato, M. Sano, K. Miyamoto, I. Yonenaga, H.J. Ko, *Journal of Applied Physics*, Vol. 96, p. 3763, 2004.
114. Y. Ohno, H. Koizumi, T. Taishi, I. Yonenaga, K. Fujii, H. Goto, T. Yao, *Applied Physics Letters*, Vol. 92, p. 011922, 2008.
115. Y. Yan, G. Dalpian, M. Al-Jassim, S.-H. Wei, *Physical Review B*, Vol. 70, p. 193206, 2004.
116. C.L. Perkins, S.-H. Lee, X. Li, S.E. Asher, T.J. Coutts, *Journal of Applied Physics*, Vol. 97, p. 034907, 2005.
117. H.K. Yadav, K. Sreenivas, V. Gupta, *Applied Physics Letters* 90 (17), 172113.
118. R. Lindsay, E. Michelangeli, B.G. Daniels, T.V. Ashworth, A.J. Limb, G. Thornton, A. Gutiérrez-Sosa, A. Baraldi, R. Larciprete, S. Lizzit, *Journal of the American Chemical Society*, Vol. 124, p. 7117, 2002.
119. J.R. Sadaf, M.Q. Israr, O. Nur, M. Willander, Y. Ding, Z.L. Wang, *Nanoscale Research Letters*, Vol. 6, p. 513, 2011.
120. Y.Y. Tay, T.T. Tan, M.H. Liang, F. Boey, S. Li, *Physical Chemistry Chemical Physics*, Vol. 12, p. 6008, 2010.
121. J. Grabowska, A. Meaney, K. Nanda, J.-P. Mosnier, M. Henry, J.-R. Duclère, E. McGlynn, *Physical Review B*, Vol. 71, p. 115439, 2005.
122. V. Fonoberov, K. Alim, A. Balandin, F. Xiu, J. Liu, *Physical Review B*, Vol. 73, p. 165317, 2006.
123. W.I. Park, Y.H. Jun, S.W. Jung, G.-C. Yi, *Applied Physics Letters*, Vol. 82, p. 964, 2003.
124. W.Y. Song, J.H. Yang, D.V. Dinh, T.I. Shin, S.M. Kang, S.-W. Kim, D.H. Yoon, *Journal of Physics and Chemistry of Solids*, Vol. 69, p. 1486, 2008.
125. A.B. Djurisić, Y.H. Leung, *Small*, Vol. 2, p. 944, 2006.
126. B.K. Meyer, H. Alves, D.M. Hofmann, W. Kriegseis, D. Forster, F. Bertram, J. Christen, A. Hoffmann, M. Straßburg, M. Dworzak, U. Haboeck, A.V. Rodina, *Physica Status Solidi (B)*, Vol. 241, p. 231, 2004.
127. T. Butkhuzi, T. Chelidze, A. Georgobiani, D. Jashiashvili, T. Khulordava, B. Tsekvava, *Physical Review B*, Vol. 58, p. 10692, 1998.
128. V. Fonoberov, K. Alim, A. Balandin, F. Xiu, J. Liu, *Physical Review B*, Vol. 73, p. 165317, 2006.
129. V.A. Fonoberov, A.A. Balandin, *Applied Physics Letters*, Vol. 85, p. 5971, 2004.
130. R.E. Sherriff, D.C. Reynolds, D.C. Look, B. Jogai, J.E. Hoelscher, T.C. Collins, G. Cantwell, W.C. Harsch, *Journal of Applied Physics*, Vol. 88, p. 3454, 2000.
131. D. Li, Y.H. Leung, A.B. Djurišić, Z.T. Liu, M.H. Xie, S.L. Shi, S.J. Xu, W.K. Chan, *Applied Physics Letters*, Vol. 85, p. 1601, 2004.

132. S.E. Harrison, *Physical Review*, Vol. 93, p. 52, 1954.
133. S.L. Kakani, and K.C. Bhandari, *A Text Book of Electronics*, New Edge International, 2008.
134. P.H. Miller, *Physical Review*, Vol. 60, p. 890, 1941.
135. D.C. Look, J.W. Hemsky, *Physical Review Letter*, Vol. 82, p. 2552, 1999.
136. D.G. Thomas, *J. Phys. Chem. Solids*, Vol. 3, p. 229, 1957.
137. G.W. Tomlins, J.L. Routbort, T.O. Mason, *Journal of Applied Physics*, Vol. 87, p. 117, 2000.
138. D.C. Look, J.W. Hemsky, *Physical Review Letter*, Vol. 82, p. 2552, 1999.
139. D. Seghier, H.P. Gislason, *Journal of Materials Science: Materials in Electronics*, Vol. 19, p. 687, 2007.
140. M.D. McCluskey, S.J. Jokela, *Physica B: Condensed Matter*, Vol. 401–402, p. 355, 2007.
141. D. Bhattacharya, *J. Res. Catalysis*, Vol. 24, P. 149, 1976.
142. C.H. Park, S.B. Zhang, S.-H. Wei, *Physical Review B*, Vol. 66, p. 073202, 2002.
143. L. Liu, J. Xu, D. Wang, M. Jiang, S. Wang, B. Li, Z. Zhang, D. Zhao, C.-X. Shan, B. Yao, D.Z. Shen, *Physical Review Letters*, Vol. 108, p. 215501, 2012.
144. K. Chang, *Journal of the Korean Physical Society*, Vol. 53, p. 196, 2008.
145. X.-Y. Duan, Y.-J. Zhao, R.-H. Yao, *Solid State Communications*, Vol. 147, p. 194, 2008.
146. G. Xiong, J. Wilkinson, B. Mischuck, S. Tüzemen, K.B. Ucer, R.T. Williams, *Applied Physics Letters*, Vol. 80, p. 1195, 2002.
147. A. Krtschil, A. Dadgar, N. Oleynik, J. Bläsing, A. Diez, A. Krost, *Applied Physics Letters*, Vol. 87, p. 262105, 2005.
148. W. Liu, F. Xiu, K. Sun, Y.-H. Xie, K.L. Wang, Y. Wang, J. Zou, Z. Yang, J. Liu, *Journal of the American Chemical Society*, Vol. 132, p. 2498, 2010.
149. L. Yu, Q. Zhu, D. Fan, Z. Lan, *Journal of Wuhan University of Technology-Mater. Sci. Ed.*, Vol. 27, p. 1184, 2012.
150. Y. Yan, S. Zhang, S. Pantelides, *Physical Review Letters*, Vol. 86, p. 5723, 2001.
151. E.-C. Lee, Y.-S. Kim, Y.-G. Jin, K. Chang, *Physical Review B*, Vol. 64, p. 085120, 2001.
152. D.C. Look, D.C. Reynolds, C.W. Litton, R.L. Jones, D.B. Eason, G. Cantwell, *Applied Physics Letters*, Vol. 81, p. 1830, 2002.
153. S. Limpijumnong, L. Gordon, M. Miao, A. Janotti, C.G. Van de Walle, *Applied Physics Letters*, Vol. 97, p. 072112, 2010.
154. C.H. Park, S.B. Zhang, S.-H. Wei, *Physical Review B*, Vol. 66, P. 073202, 2002.
155. S. Dhara, P.K. Giri, *Thin Solid Films*, Vol. 520, p. 5000, 2012.
156. J.W. Lee, N.G. Subramaniam, J.G. Lee, S.S. Kumar, T.W. Kang, *EPL Europhysics Letters*, Vol. 95, p. 47002, 2011.
157. M.J.S. Spencer, K.W.J. Wong, I. Yarovsky, *Journal of Physics Condensed Matter*, Vol. 24, p. 305001, 2012.
158. A. Kushwaha, H. Tyagi, M. Aslam, *AIP Advances*, Vol. 3, p. 042110, 2013.

159. K. Mahmood, S.B. Park, *Journal of Crystal Growth*, Vol. 347, p. 104, 2012.
160. S.-H. Lee, S.-H. Han, H.S. Jung, H. Shin, J. Lee, J.-H. Noh, S. Lee, I.-S. Cho, J.-K. Lee, J. Kim, H. Shin, *The Journal of Physical Chemistry C*, Vol. 114, p. 7185, 2010.
161. T. Takenaka, K. Sakata, *Journal of Applied Physics*, Vol. 55, p. 1092, 1984.
162. S.A. Zahra, *Journal of Electron Devices*, Vol. 17, p. 1494, 2013.
163. S.-P. Huang, H. Xu, I. Bello, R.Q. Zhang, *The Journal of Physical Chemistry C*, Vol. 114, p. 8861, 2010.
164. C.S. Lao, M. Park, Q. Kuang, Y. Deng, A.K. Sood, D.L. Polla, Z.L. Wang, *Journal of American Chemical Society*, Vol. 129, p. 12096, 2007.
165. S. Dhara, P.K. Giri, *Journal of Applied Physics*, Vol. 111, p. 044320, 2012.
166. A. Bera, D. Basak, *ACS Applied Materials & Interfaces*, Vol. 1, p. 2066, 2009.
167. T. Belagodu, E.A. Azhar, H. Yu, *Nanoscale*, Vol. 4, p. 7330, 2012.
168. J.-H. Koo, J.-J. Cho, J.-H. Yang, P.-J. Yoo, K.-W. Oh, J.-H. Park, *Bulletin of the Korean Chemical Society*, Vol. 33, p. 636, 2012.
169. C.H. Kim, Y.C. Park, J. Lee, W.S. Shin, S.-J. Moon, J. Park, J. Kim, *Materials Letters*, Vol. 65, p. 1548, 2011.
170. S. Dhara, P.K. Giri, *Journal of Experimental Nanoscience*, Vol. 8, p. 332, 2013.
171. S. Dhara, P.K. Giri, *Chemical Physics Letters*, Vol. 541, p. 39, 2012.
172. A. Bera, D. Basak, *Nanotechnology*, Vol. 22, p. 265501, 2011.
173. T. Chen, G.Z. Xing, Z. Zhang, H.Y. Chen, T. Wu, *Nanotechnology*, Vol. 19, p. 435711, 2008.
174. Y.J. Fang, J. Sha, Z.L. Wang, Y.T. Wan, W.W. Xia, Y.W. Wang, *Applied Physics Letters*, Vol. 98, p. 033103, 2011.
175. R. Huang, S. Xu, X. Wang, W. Guo, C. Song, J. Song, K.M. Ho, S. Du, N. Wang, *Optics Letters*, Vol. 37, p. 211, 2012.
176. R.S. Aga, D. Jowhar, A. Ueda, Z. Pan, W.E. Collins, R. Mu, K.D. Singer, J. Shen, *Applied Physics Letters*, Vol. 91, p. 232108, 2007.
177. H.W. Kim, S.H. Shim, *Journal of the Korean Physical Society*, Vol. 49, p. 241, 2006.
178. A. Bera, D. Basak, *ACS Applied Materials & Interfaces*, Vol. 2, p. 408, 2010.
179. C. Cheng, T.-L. Wang, L. Feng, W. Li, K.M. Ho, M.M.T. Loy, K.K. Fung, N. Wang, *Nanotechnology*, Vol. 21, p. 475703, 2010.
180. Y. Myung, D.M. Jang, T.K. Sung, Y.J. Sohn, G.B. Jung, Y.J. Cho, H.S. Kim, J. Park, *ACS Nano*, Vol. 4, p. 3789, 2010.
181. J.-P. Richters, T. Voss, D.S. Kim, R. Scholz, M. Zacharias, *Nanotechnology*, Vol. 19, p. 305202, 2008.
182. D. Zhao, X. Zhang, H. Dong, L. Yang, Q. Zeng, J. Li, L. Cai, X. Zhang, P. Luan, Q. Zhang, M. Tu, S. Wang, W. Zhou, S. Xie, *Nanoscale*, Vol. 5, p. 4443, 2013.
183. M. Zhong, Y. Li, I. Yamada, J.-J. Delaunay, *Nanoscale*, Vol. 4, p. 1509, 2012.
184. Y. Zhang, Z. Wu, J. Zheng, X. Lin, H. Zhan, S. Li, J. Kang, J. Bleuse, H. Mariette, *Solar Energy Materials and Solar Cells*, Vol. 102, p. 15, 2012.
185. A.V. Herzog, T.E. Lipman, D.M. Kammen, *Renewable Energy Sources*, n.d.
186. A.I. Hochbaum, P. Yang, *Chemical Reviews*, Vol. 110, p. 527, 2010.
187. Y.-G. Guo, J.-S. Hu, L.-J. Wan, *Advanced Materials*, Vol. 20, p. 2878, 2008.

188. R. Ren, Y. Guo, R. Zhu, *Optics Letters*, Vol. 37, p. 4245, 2012.
189. A. Hinsch, European Photovoltaic Solar Energy Conference and Exhibition, Glasgow 2000 OB9.2, in: *16th European Photovoltaic Solar Energy Conference and Exhibition Glassgow*, Vol. 1702, 2000.
190. K. Driscoll, J. Fang, N. Humphry-Baker, T. Torres, W.T.S. Huck, H.J. Snaith, R.H. Friend, *Nano Letters*, Vol. 10, p. 4981, 2010.
191. P. Ruankham, T. Sagawa, H. Sakaguchi, S. Yoshikawa, *Journal of Materials Chemistry*, Vol. 21, p. 9710, 2011.
192. D.C. Iza, D. Muñoz-Rojas, Q. Jia, B. Swartzentruber, J.L. Macmanus-Driscoll, *Nanoscale Research Letters*, Vol. 7, p. 655, 2012.
193. S. Han, A.A.D.T. Adikaari, K.D.G.I. Jayawardena, N.A. Nismy, S.R.P. Silva, *Journal of Applied Physics*, Vol. 112, p. 066103, 2012.
194. Q. Zhang, C.S. Dandeneau, X. Zhou, G. Cao, *Advanced Materials*, Vol. 21, p. 4087, 2009.
195. C.-H. Liao, C.-W. Huang, J.C.S. Wu, *Catalysts*, Vol. 2, p. 490, 2012.
196. K. Sun, Y. Jing, C. Li, X. Zhang, R. Aguinaldo, A. Kargar, K. Madsen, K. Banu, Y. Zhou, Y. Bando, Z. Liu, D. Wang, *Nanoscale*, Vol. 4, p. 1515, 2012.
197. M. Gupta, V. Sharma, J. Shrivastava, A. Solanki, A.P. Singh, V.R. Satsangi, S. Dass, R. Shrivastav, *Bulletin of Materials Science*, Vol. 32, p. 23, 2009.
198. Y. Wei, L. Ke, J. Kong, H. Liu, Z. Jiao, X. Lu, H. Du, X.W. Sun, *Nanotechnology*, Vol. 23, p. 235401, 2012.
199. S.Y. Chung, S. Kim, J.-H. Lee, K. Kim, S.-W. Kim, C.-Y. Kang, S.-J. Yoon, Y.S. Kim, *Advanced Materials*, Vol. 24, p. 6022, 2012.
200. X. Wang, J. Song, J. Liu, Z.L. Wang, *Science*, Vol. 316, p. 102, 2007.
201. G. Zhu, R. Yang, S. Wang, Z.L. Wang, *Nano Letters*, Vol. 10, p. 3151, 2010.
202. D.O. Demchenko, P.D. Heinz, B. Lee, *Nanoscale Research Letters*, Vol. 6, p. 502, 2011.
203. Y. Liu, Z. Zhang, X. Wei, Q. Li, L.-M. Peng, *Advanced Functional Materials*, Vol. 21, p. 3900, 2011.
204. C.-H. Lee, G.-C. Yi, Y.M. Zuev, P. Kim, *Applied Physics Letters*, Vol. 94, p. 022106, 2009.
205. Y. Inoue, M. Okamoto, T. Kawahara, Y. Okamoto, J. Morimoto, *Materials Transactions*, Vol. 46, p. 1470, 2005.
206. Y. Yang, W. Guo, K.C. Pradel, G. Zhu, Y. Zhou, Y. Zhang, Y. Hu, L. Lin, Z.L. Wang, *Nano Letters*, Vol. 12, p. 2833, 2012.
207. L. Shi, J. Chen, G. Zhang, B. Li, *Physics Letters A*, Vol. 376, p. 978, 2012.

Index

Also of Interest

Check out these published and forthcoming related titles from Scrivener Publishing

Books marked with an * denote a title in the Advanced Materials Series

***Advanced Materials for Agriculture, Food and Environmental Safety**
Edited by Ashutosh Tiwari and Mikael Syväjärvi
Published 2014. ISBN: 978-1-118-77343-7

***Advanced Biomaterials and Biodevices**
Edited by Ashutosh Tiwari and Anis N. Nordin
Published 2014. ISBN 978-1-118-77363-5

***Biosensors Nanotechnology**
Edited by Ashutosh Tiwari and Anthony P. F. Turner
Published 2014. ISBN 978-1-118-77351-2

***Advanced Sensor and Detection Materials**
Edited by Ashutosh Tiwari and Mustafa M. Demir
Published 2014. ISBN 978-1-118-77348-2

***Advanced Healthcare Materials**
Edited by Ashutosh Tiwari
Published 2014. ISBN 978-1-118-77359-8

***Advanced Energy Materials**
Edited by Ashutosh Tiwari and Sergiy Valyukh
Published 2014. ISBN 978-1-118-68629-4

***Advanced Carbon Materials and Technology**
Edited by Ashutosh Tiwari and S.K. Shukla
Published 2014. ISBN 978-1-118-68623-2

***Responsive Materials and Methods**
State-of-the-Art Stimuli-Responsive Materials and Their Applications
Edited by Ashutosh Tiwari and Hisatoshi Kobayashi
Published 2013. ISBN 978-1-118-68622-5

Nanomaterials in Drug Delivery, Imaging, and Tissue Engineering
Edited by Ashutosh Tiwari and Atul Tiwari
Published 2013. ISBN 978-1-118-29032-3

Solar Cell Nanotechnology
Edited by Atul Tiwari, Rabah Boukherroub, and Maheshwar Sharon
Published 2013. ISBN 978-1-118-68625-6

Biomimetics
Advancing Nanobiomaterials and Tissue Engineering
Edited by Murugan Ramalingam, Xiumei Wang, Guoping Chen, Peter Ma, and Fu-Zhai Cui
Published 2013. ISBN 9781118469620

Biomedical Materials and Diagnostic Devices Devices
Edited by Ashutosh Tiwari, Murugan Ramalingam, Hisatoshi Kobayashi and Anthony P.F. Turner
Published 2012. ISBN 978-1-118-03014-1

Intelligent Nanomaterials
Processes, Properties, and Applications
Edited by Ashutosh Tiwari Ajay K. Mishra, Hisatoshi Kobayashi and Anthony P.F. Turner
Published 2012. ISBN 978-0-470-93879-9

Integrated Biomaterials for Biomedical Technology
Edited by Murugan Ramalingam, Ashutosh Tiwari, Seeram Ramakrishna and Hisatoshi Kobayashi
Published 2012. ISBN 978-1-118-42385-1

Integrated Biomaterials in Tissue Engineering
Edited by Murugan Ramalingam, Ziyad Haidar, Seeram Ramakrishna, Hisatoshi Kobayashi, and Youssef Haikel
Published 2012. ISBN 978-1-118-31198-1

Introduction to Surface Engineering and Functionally Engineered Materials
By Peter Martin
Published 2011. ISBN 978-0-470-63927-6

Encapsulation Nanotechnologies
Edited by Vikas Mittal
Published 2013. ISBN 978-1-118-34455-2

Introduction to Industrial Polypropylene: Properties, Catalysts, Processes
by Dennis P. Malpass and Elliot Band.
Published 2012. ISBN 978-1-118-06276-0

Introduction to Industrial Polyethylene: Properties, Catalysts, Processes
by Dennis P. Malpass.
Published 2010. ISBN 978-0-470-62598-9

Handbook of Bioplastics and Biocomposites Engineering Applications
Edited by Srikanth Pilla
Published 2011. ISBN 978-0-470-62607-8

Biopolymers
Biomedical and Environmental Applications
Edited by Susheel Kalia and Luc Avérous
Published 2011. ISBN 978-0-470-63923-8

Renewable Polymers: Synthesis, Processing, and Technology
Edited by Vikas Mittal
Published 2011. ISBN 978-0-470-93877-5

Polymer Nanotube Nanocomposites
Synthesis, Properties, and Applications
Edited by Vikas Mittal.
Published 2010. ISBN 978-0-470-62592-7